Valorization of Citrus Food Waste

Anamika Chauhan • Fakhar Islam
Ali Imran • Jagbir Singh Aswal
Editors

Valorization of Citrus Food Waste

Springer

Editors
Anamika Chauhan
Department of Home Sciences
Chaman Lal Mahvidhyalya Landhaura
Haridwar
Uttarakhand, India

Ali Imran
Department of Food Science
Government College University Faisalabad
Faisalabad, Pakistan

Fakhar Islam
Department of Clinical Nutrition
NUR International University
Lahore, Pakistan

Jagbir Singh Aswal
Council For Science & Technology
Jhajra Via Premnagar Dehradun
Uttarakhand, India

ISBN 978-3-031-77998-5 ISBN 978-3-031-77999-2 (eBook)
https://doi.org/10.1007/978-3-031-77999-2

This Springer imprint is published by the registered company Springer Nature Switzerland AG
The registered company address is: Gewerbestrasse 11, 6330 Cham, Switzerland

Preface

The citrus industry generates vast amounts of waste every year, predominantly in the form of peels, pulp, and seeds. This discarded biomass, often treated as a burden, is a treasure trove of bioactive compounds, essential oils, dietary fibers, and other valuable constituents. This book, *Valorization of Citrus Food Waste*, seeks to reframe citrus waste as a renewable resource with numerous applications across industries, from food and agriculture to energy and pharmaceuticals. Comprising 22 chapters, this book explores diverse techniques for transforming citrus waste into value-added products. We start with an introductory overview of citrus waste utilization and its potential, setting the stage for discussions on waste management for various citrus types, including oranges and lemons. We then delve into innovative techniques such as fermentation, torrefaction, and green extraction methods to retrieve bioactive compounds with significant commercial applications. In later chapters, we explore the synthesis of energy from citrus waste, highlighting its potential as a renewable energy source, and examine specific bioactivities of citrus components, such as their antiepileptic properties. Challenges and opportunities within the field are thoroughly discussed, offering insights into practical and economic aspects of citrus waste valorization. Topics on citrus waste in culinary applications, packaging, agriculture, and soil health reveal its versatility and utility in promoting sustainable practices. With contributions from experts across various disciplines, this book provides a comprehensive guide to harnessing the potential of citrus waste, aiming to inspire further research and innovation in food waste management. It is an invaluable resource for researchers, industry professionals, and anyone interested in advancing sustainable practices through the valorization of food waste.

This book not only highlights the technical aspects of citrus waste valorization but also emphasizes the sustainability imperative in today's food and agricultural industries. By turning waste into valuable resources, we can address critical environmental challenges, reduce landfill loads, and contribute to a circular economy. Each chapter presents unique case studies, methodologies, and emerging technologies that underscore the diversity and practicality of citrus waste applications. From nutraceuticals and bioactive extracts to bioenergy and eco-friendly packaging solutions, *Valorization of Citrus Food Waste* showcases the transformative potential of citrus waste. Readers will gain both foundational knowledge and innovative perspectives to drive sustainable practices forward.

Uttarakhand, India	Anamika Chauhan
Lahore, Pakistan	Fakhar Islam
Faisalabad, Pakistan	Ali Imran
Uttarakhand, India	Jagbir Singh Aswal

Contents

Chapter 1
Introduction of Citrus Waste Utilization

Arisha Ahmed, Ali Imran, Muhammad Afzaal, Musarrat Rasheed, Anamika Chauhan, and Fakhar Islam

1.1 Introduction

The Rutaceas family includes citrus fruits, which are among the most widely farmed and consumed fruits worldwide. Citrus is well-liked for its distinct flavor, scent, and amount of phenolics, vitamin C, and other nutrients. One of the most popular fruit crops worldwide is citrus, which includes oranges, grapefruits, lemons, pomelo, limes, tangerines, and mandarins. Eighteen percent of all citrus fruits harvested worldwide are used in industrial operations, primarily in the juice industry. However, between 50% and 70% of all citrus fruit is used in the production of industrialized citrus products, which results in significant waste (peel, pulp, and seed leftovers). Every year, between 15 and 25 thousand tons of citrus trash are produced, which could be a source for a variety of advantageous purposes [1].

Within the agro-industrial sector, the citrus processing industry is significant. About 50–60% of all citrus fruits are produced worldwide, with oranges being the most frequently farmed crop; other species also contribute. The primary byproduct of the citrus processing industry is citrus peel waste (CPW), which is distinguished

A. Ahmed
Department of Nutritional Sciences, NUR International University, Lahore, Pakistan

A. Imran · M. Afzaal
Department of Food Science, Government College University Faisalabad, Faisalabad, Pakistan

M. Rasheed
Department of Food Science and Technology, NUR International University, Lahore, Pakistan

A. Chauhan (✉)
Department of Home Sciences, Chaman Lal Mahvidhyalya Landhaura Haridwar, Uttarakhand, India

F. Islam
Department of Clinical Nutrition, NUR International University, Lahore, Pakistan

A. Chauhan et al. (eds.), *Valorization of Citrus Food Waste*,
https://doi.org/10.1007/978-3-031-77999-2_1

by its seasonal production (which frequently necessitates biomass storage), high water content, and concentration of essential oils [2].

Waste from citrus processing industry is strategically and significantly important to the agro-industrial sectors, the pharmaceutical industries, cosmetic and animal feed industries [3]. Citrus waste is highly valuable biologically and may have health benefits [4]. This is because it contains high-quality fiber, pectin, and bioactive compounds like carotenoids [5], polyphenols (flavonoids and phenolic acids), and essential oils (EOS) (d-limonene, α-terpinolene, α-pinene) that are extracted from the peel. These compounds, along with their high-quality fiber content and pectin content, make citrus waste highly valuable commercially. They can be used as an ingredient in foods, beverages, cosmetics, soaps, perfumes, and pharmaceuticals as a potential source of antibiotics like penicillin and metacillin, which are used to treat pathogenic bacteria like bacillus subtilis, salmonella typhi, staphylococcus aureus, Escherichia coli and klebsiella pneumonia [6]. Before disposing of peel trash in a landfill, it is crucial to extract the oils from it since they have a negative impact on bacterial decomposition and fermentation. According to Murdock and Allen, yeasts get contaminated by the oils found in orange peels. The demand for natural food additives has sharply increased as people's awareness of health and a healthy diet has grown in recent years. Customers anticipate a safe and health-promoting diet. Fresh citrus fruits are consumed, as well as other processed fruit items (Fig. 1.1).

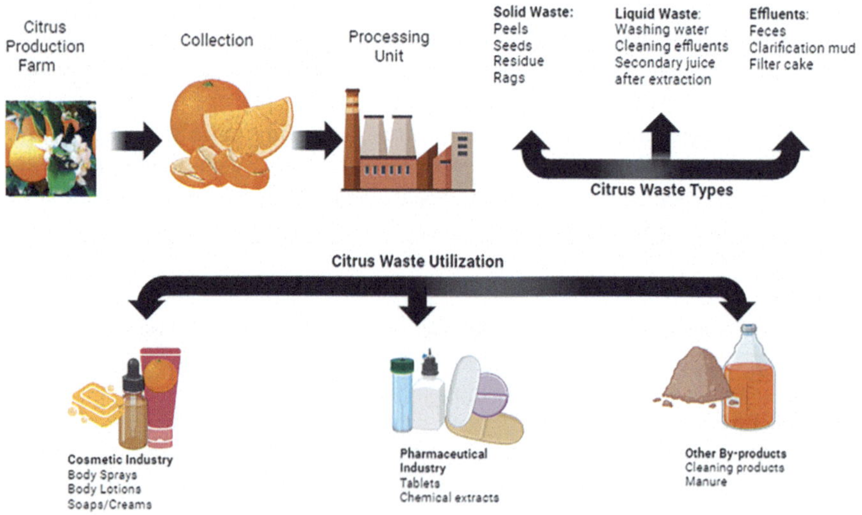

Fig. 1.1 Citrus waste utilization

1.2 Historical Perspectives of Citrus Waste

The usage of citrus trash from historical views sheds light on how technologies and methods have changed throughout time. It is easier to grasp the background and developments that have influenced present-day and future use techniques when one is aware of this evolution.

1.3 Ancient and Traditional Uses

1.3.1 Agricultural Practices

Citrus peels and pulp were frequently utilized straight as compost or animal feed in the prehistoric era. Farmers realized that citrus trash might improve crop output and soil fertility by providing valuable nutrients.

1.3.2 Medicinal Uses

Citrus peels have been used in traditional medicine in many cultures for a variety of diseases, including skin and digestive disorders. For instance, dried citrus peels, or chenpi, were used in traditional Chinese medicine to promote better digestion and decrease inflammation.

1.4 Early Industrialization (Nineteenth to Early Twentieth Century)

- **Citrus Byproducts in Animal Feed:** Citrus waste started to be processed into animal feed on a wider scale with the expansion of the citrus industry in the nineteenth century, and advancements in drying and pelletizing technology made it easier to store and transport citrus pulp for animals.
- **Essential Oils and Flavorings:** Citrus peel distillation began to acquire industrial importance as a means of obtaining essential oils. These oils were utilized in cleaning supplies, fragrances, and food flavorings.

1.4.1 *Mid-Twentieth Century Developments*

- **Pectin Production:** Innovations in technology made it possible to extract pectin from citrus peels in the middle of the twentieth century. In the food business, pectin has become a useful gelling agent, especially for making jams and jellies.
- **Biogas Production:** Anaerobic digestion of citrus waste was investigated experimentally in the 1960s and 1970s to see if it could produce biogas. Despite their initial lack of widespread adoption, these research set the foundation for upcoming bioenergy applications.

1.4.2 *Late Twentieth Century to Early Twenty-First Century*

- **Environmental Awareness:** The need for more sustainably managed agricultural waste was brought to light by the growth of environmental movements in the late twentieth century. More people now see citrus waste as a resource than an issue that needs to be disposed of.
- **Comprehensive Utilization Strategies**: With an emphasis on making the most of citrus waste components, integrated waste management strategies started to take shape. This includes utilizing leftovers for compost or bioenergy, peels for essential oils, and pulp for animal feed [7].
- **Developments in Nutraceuticals:** Studies on the advantages citrus components, such flavonoids and limonoids, have for health have resulted in the creation of nutritional supplements and nutraceuticals made from citrus waste.

1.4.3 *Recent Innovations and Current Trends*

- **Biorefineries:** The concept of biorefineries, which convert biomass into a spectrum of marketable products and energy, gained traction. Citrus waste biorefineries focus on producing biofuels, chemicals, and materials from citrus byproducts.
- **Sustainable Packaging:** Advances in materials science have led to the development of biodegradable plastics and films from citrus-derived pectin and other polysaccharides. These innovations address the growing demand for sustainable packaging solutions.
- **Circular Economy Models:** Modern approaches emphasize integrating citrus waste utilization into circular economy models, where waste is minimized, and resources are continuously reused and recycled. This shift is driven by both economic incentives and environmental regulations [8].

1.5 Challenges and Lessons Learned

- **Economic Viability:** Large-scale adoption has frequently been hampered by high processing costs and little markets for byproducts, making the economic sustainability of citrus waste reuse initiatives historically a major obstacle.
- **Technological Barriers:** The effective extraction and conversion of valuable chemicals from citrus trash was limited by early technological restrictions. Overcoming these obstacles has required persistent creativity and research.
- **Regulatory and Market Dynamics:** Utilization options for citrus trash have been impacted by changes in market needs and regulatory changes. For instance, the nutraceutical and cosmetic businesses based on citrus have grown as a result of growing customer desire for sustainable and natural products [9].

1.6 Valorization of Citrus Waste

Knowing the historical vantage points on the use of citrus waste emphasizes the gains made over centuries and emphasizes the role that innovation, sustainability and economic viability will have in advancing the field going forward.

The current chapter primarily focuses on the advancements in the industrial residual waste valorization of citrus juice. According to Chavan et al. (2018), waste valorization is the process of turning waste and biomass into energy, fuel, and other more usable goods including chemicals, materials, and fuels with an emphasis on sustainability objectives and environmental indicators [10]. Wastes from the processing of citrus fruits include distillery effluents, liquid wastes, and solid wastes. According to Sharma et al. (2017), solid waste is made up of residue, rags, peels, and seeds. Because citrus trash has a high vitamin C content and can be used to make juice and other nutrient-dense beverages, it is important to handle and manage citrus waste with greater care [11].

When seeds, pulp, and peel—which are regarded as leftovers or residues—are burned, disposed of in water, or both, they can contaminate the environment and cause oxygen depletion and water contamination [12]. Because citrus processing wastes may also contain larger levels of essential oils and water, disposing of them may have an adverse effect on the environment and the economy. When these items are valued scientifically, both financial gain and environmental projections are possible [3].

According to Okino et al. (2016), citrus waste utilized as a biofertilizer is a rich source of carotenoids and phenols that can be used to extend the shelf life of food and beverages. Citrus essential oil has a significant amount of terpenes, which have potent antifungal properties [13]. According to Chavan et al. (2018), citrus seeds are typically discarded as waste since they are deemed useless. According to Ammerman and Arrington (1961), the average percentage of seed in dried citrus pulp was 4.8% [10]. These wasted seeds have a high protein content, making them a cost-effective protein supplement for animals. Citrus seeds also contain about 30% oil (by weight) and have a lot of potential for use as biodiesel [14].

Utilizing citrus processing waste to create high value added products like biogas, anaerobic digestion, fermentation, pectic oligosaccharides, enzymes, bio-ethanol, bio-polymers and thermochemical processes like syngas, gastification, incineration, and physico-chemical processes like biodiesel and trans-esterification has seen significant trends in recent years [15].

As a byproduct of the citrus industry, citrus trash offers a lot of potential for value-added goods and sustainable use. Citrus trash has a bright future because of the continuous research and technical advancements that are fostering the creation of environmentally friendly and commercially viable uses. These developments support environmental sustainability and generate new sources of income in addition to helping to reduce trash.

1.7 Proximate Composition of Citrus by-Products

Composition (g/100 g)	Citrus peel	Citrus seed	Citrus pulp
Crude protein	10.2	3.1	8.6
Total ash	3.33	2.5	6.5
Crude fiber	57.0	5.50	4.9
Crude fat	2.22	52.0	4.9
Moisture content	75.3	–	85.7
References	[16]	[17]	[16]

1.8 Research and Technological Advancements

- **Innovative Extraction Techniques:** The development of effective extraction techniques, including supercritical fluid extraction, is crucial to optimize the quantity and purity of important chemicals from citrus waste.
- **Biotechnological methods**: Using microbial fermentation and genetic engineering, citrus waste can be transformed into high-value goods including enzymes, bioactive peptides, and biofuels.
- **Circular Economy Models**: To improve sustainability and lessen environmental impact, citrus waste usage should be incorporated into circular economy frameworks [18].

1.9 Health Benefits

Citrus fruits and their byproducts are known to be abundant in vitamins, minerals, and dietary fibers, or nonstarch polysaccharides, which are vital for human nourishment, growth, and development as a whole. Other physiologically active, nonnutritive substances found in citrus fruits, such as flavonoids, carotenoids, vitamins, and minerals, have recently come under investigation. These chemicals may help lower the risk of several chronic illnesses, including age related macular degeneration and cardiovascular disorders. Citrus fruit bioactive chemicals are rich in medicinal qualities and function as anti-inflammatory, anti-tumor, anti-cancer and antioxidant agents. Their antiviral, antiplatelet aggregation, and anti-inflammatory properties are also well-known. It has been documented that certain bioactive substances create hepatic protective enzymes and prevent the material present in the human cells.

A few "not peer reviewed" studies published recently linked citrus hesperidin and COVID-19 illness. The performance of hesperidin is based on its interactions with receptors that may have an inhibitory effect on viral infection and replication, including SARS-CoV-2 main protease (PDB:6Y84) [19], crystal structure main protease (PDB:6LU7), Spike glycoprotein-RBD (PDB:6LXT), and PD-ACE2 (PDB:6VW1) [20]. Citrus waste has been suggested by various research as a viable natural source of hesperidin due to its abundance in citrus peel. Clinical trials should be conducted with this medicine to determine its preventive or therapeutic activity against COVID-19, given its interactions with SARS-CoV-2 receptors [21].

Citrus press cake extracts have been shown to dramatically lower the amount of melanin in cells by inhibiting tyrosinase activity and the transcription factors TRP-1 and TRP-2 in a dose-dependent way [22] (Fig. 1.2).

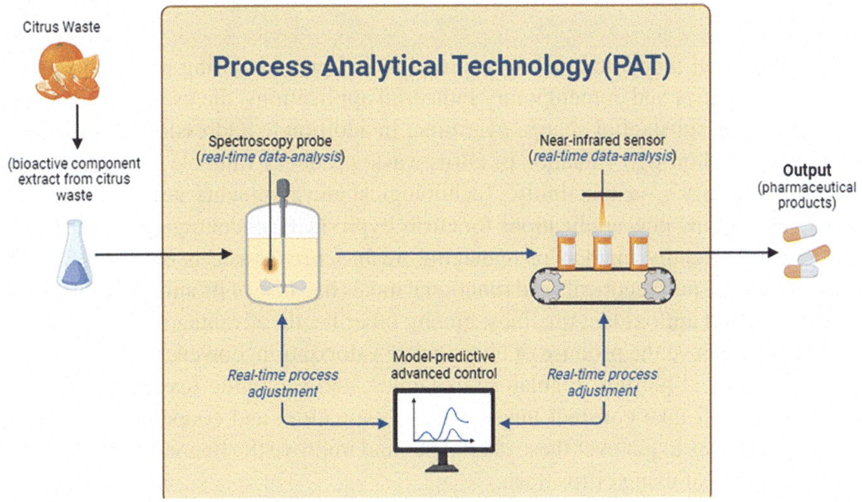

Fig. 1.2 Manufacturing of Pharmaceutical Products from Citrus Waste

The other vital parts of citrus species that can resist damage from photooxidation are called carotenoids. There have been reports of the biological actions of vitamin C and flavonoids working in concert. Lastly, the amount and bioavailability of flavonoids and vitamin C that are taken directly affects their health benefits. Chronic poisoning can be caused by synthetic antioxidants. As a result, natural antioxidants derived from plants become more significant. Natural, potent, harmless antioxidants called polyphenols scavenge free radicals without causing any negative side effects [23].

1.10 Ideas for Valorizing Citrus Waste

1. Research and Development Expenditure
2. Putting Circular Economy Models into Practice
3. Improving Technological Abilities
4. Consumer awareness and market development
5. Support for regulations and policy development
6. Assessment of the Environmental Impact
7. Financial Viability and Business Plans

Using these recommendations can help generate high-value products and save the environment by utilizing the potential of citrus trash to build a profitable and sustainable enterprise.

1.11 Conclusion

From traditional medicinal and agricultural techniques to cutting-edge biotechnological techniques and contemporary industrial applications, the use of citrus waste has undergone substantial change over time. In addition to addressing environmental issues, the thorough valuation of citrus waste produces valuable goods that support the industry's sustainability. Technological developments and research are constantly finding new applications for citrus byproducts, encouraging a more environmentally friendly method of managing trash. Citrus waste is essential for the development of nutraceuticals and functional meals because of its anti-inflammatory, anti-tumor, and antioxidant qualities, among other health advantages.

To fully achieve the promise of citrus waste valorization, however, issues including technological hurdles, regulatory dynamics, and economic feasibility must be resolved. It will take constant innovation, investigation, and cooperation between interested parties to get over these challenges and improve the financial and ecological advantages of using citrus trash.

References

1. Anticona, M., et al. (2020). High biological value compounds extraction from citrus waste with non-conventional methods. *Food, 9*(6), 811.
2. Zema, D., et al. (2018). Valorisation of citrus processing waste: A review. *Waste Management, 80*, 252–273.
3. Varmie, E., & Thakur, M. (2021). Utilization of citrus processing waste: A review. *Pharma Innovation Journal, 10*, 682–697.
4. Singh, B., et al. (2020). Phenolic composition, antioxidant potential and health benefits of citrus peel. *Food Research International, 132*, 109114.
5. Mahato, N., et al. (2018). Citrus waste derived nutra–/pharmaceuticals for health benefits: Current trends and future perspectives. *Journal of Functional Foods, 40*, 307–316.
6. Kumar, K. A., et al. (2011). Antimicrobial activity and phytochemical analysis of citrus fruit peels-utilization of fruit waste. *International Journal of Engineering Science and Technology, 3*(6), 5414–5421.
7. Suri, S., Singh, A., & Nema, P. K. (2022). Current applications of citrus fruit processing waste: A scientific outlook. *Applied Food Research, 2*(1), 100050.
8. Panwar, D., et al. (2021). Unraveling the scientific perspectives of citrus by-products utilization: Progress towards circular economy. *Trends in Food Science & Technology, 111*, 549–562.
9. Lucia, C., et al. (2022). Challenges and opportunities for citrus wastewater management and valorisation: A review. *Journal of Environmental Management, 321*, 115924.
10. Chavan, P., Singh, A. K., & Kaur, G. (2018). Recent progress in the utilization of industrial waste and by-products of citrus fruits: A review. *Journal of Food Process Engineering, 41*(8), e12895.
11. Sharma, K., et al. (2017). Converting citrus wastes into value-added products: Economic and environmently friendly approaches. *Nutrition, 34*, 29–46.
12. Wadhwa, M., & Bakshi, M. (2013). Utilization of fruit and vegetable wastes as livestock feed and as substrates for generation of other value-added products. *RAP Publication, 4*(2013), 67.
13. Okino Delgado, C. H., & Fleuri, L. F. (2016). Orange and mango by-products: Agro-industrial waste as source of bioactive compounds and botanical versus commercial description—A review. *Food Reviews International, 32*(1), 1–14.
14. Rashid, U., et al. (2013). Biodiesel from Citrus reticulata (mandarin orange) seed oil, a potential non-food feedstock. *Industrial Crops and Products, 45*, 355–359.
15. Nizami, A.-S., et al. (2017). Waste biorefineries: Enabling circular economies in developing countries. *Bioresource Technology, 241*, 1101–1117.
16. Matsuo, Y., et al. (2019). Proximate composition and profiles of free amino acids, fatty acids, minerals and aroma compounds in Citrus natsudaidai peel. *Food Chemistry, 279*, 356–363.
17. Akpata, M., & Akubor, P. (1999). Chemical composition and selected functional properties of sweet orange (Citrus sinensis) seed flour. *Plant Foods for Human Nutrition, 54*, 353–362.
18. Panwar, D., Panesar, P. S., & Chopra, H. K. (2021). Recent trends on the valorization strategies for the management of citrus by-products. *Food Reviews International, 37*(1), 91–120.
19. Das, S., et al. (2021). An investigation into the identification of potential inhibitors of SARS-CoV-2 main protease using molecular docking study. *Journal of Biomolecular Structure and Dynamics, 39*(9), 3347–3357.
20. Utomo, R. Y., Ikawati, M., & Meiyanto, E. (2020). Revealing the potency of citrus and galangal constituents to halt SARS-CoV-2 infection.
21. Meneguzzo, F., et al. (2020). Accelerated production of hesperidin-rich citrus pectin from waste citrus peel for prevention and therapy of COVID-19.
22. Kim, S. S., et al. (2013). Down-regulation of tyrosinase, TRP-1, TRP-2 and MITF expressions by citrus press-cakes in murine B16 F10 melanoma. *Asian Pacific Journal of Tropical Biomedicine, 3*(8), 617–622.
23. Khan, U. M., et al. (2021). [Retracted] citrus genus and its waste utilization: A review on health-promoting activities and industrial application. *Evidence-based Complementary and Alternative Medicine, 2021*(1), 2488804.

Chapter 2
Volarization of Citrus Fruit Waste and by Products Through Fermentation Process

Zahra Maqbool, Mahum, Maryam Maqsood, Anam Matloob, Hamda, Shakeela Khalid, and Waseem Khalid

2.1 Introduction

The rising mandate for ready-to-eat and processed food puts pressure on releasing massive quantities of waste, and practice of artificial food additives during processing puts customers' health at risk. Due to reviving taste and beneficial aspects citrus is considered as the best consumed fruits worldwide [17]. *Poncirus* and *Fortunella,* are two common species of Citrus genera, make up the group of plants that produce citrus fruits. The primary species include *Citrus sinensis and Aurantium* (sweet and bitter oranges), *Citrus maxima* (pummelos), *Poncirus trifoliate* (trifoliate orange), *Fortunella spp* (kumquats), *Citrus paradise* (grapefruits), *Citrus latifolia, Citrus*

Z. Maqbool
Department of Food Science, Government College University Faisalabad, Faisalabad, Pakistan

Mahum · M. Maqsood
Department of Food Science and Technology, Faculty of Food and Home Sciences, Muhammad Nawaz Shareef University of Agriculture, Multan, Pakistan

A. Matloob
Zhejiang University, College of Biosystems Engineering and Food Science, Hangzhou, China

Hamda
Department of Plant Breeding and Genetics, Faculty of Agriculture Science and Technology, Bahauddin Zakariya University, Multan, Pakistan

S. Khalid
Department of Zoology, University of Sargodha, Sargodha, Pakistan

W. Khalid (✉)
Department of Organic Chemistry, Faculty of Chemical Sciences and Technologies, University of Castilla La Mancha, Ciudad Real, Spain

University Institute of Food Science and Technology, The University of Lahore, Lahore, Pakistan

A. Chauhan et al. (eds.), *Valorization of Citrus Food Waste*,
https://doi.org/10.1007/978-3-031-77999-2_2

aurantifolia, and *Citrus limettioides* (limes), and *Citrus Limon* (lemons). Owed to current processing and harvesting approaches, the business of citrus fruit with an annual production of more than 146 million tons, has spread in more than 140 countries worldwide [2].

Citrus fruit processing is mostly done for food manufacturing (such as marmalades and candied peels) and juice extraction. The chemical and culinary industries also employ citrus fruits to make pectin, dietary fiber, biofuels, flavorings, and essential oils (EOs). Citrus fruits are made up of peel, pomace, and seeds, which are thrown into the environment after processing constitutes almost half of the weight of the fresh fruit. Citrus fruits produce a large amount of processing waste. A multitude of potent phytochemicals, vitamins, folic acid, potassium, and other elements found in citrus waste can help shield people from serious illnesses [1].

Pectin, which has the ability to decrease cholesterol and blood sugar, is prevalent in citrus pomace and peels. Valorizing citrus waste in order to boost retrieval of pectin would therefore be an enduring tactic that might aid in waste reduction. Crumbling citric waste yields an enormous amount of improved, expensive compounds, including antioxidants, flavonoids, organic acids, phenolic, pectin, and essential oil (limonene). Pectin's gelling, thickening, and stabilizing qualities make it a popular ingredient in meals, medications, textiles, cosmetics, and personal care items because of its vast array of nutritional and functional qualities [17].

The large amount of citrus waste when burned and disposed of in landfills becomes a source of environmental pollution. The mishandling of a potential bio resource cause financial losses and also affects human health by producing airborne pathogenic microbes. Citrus fruits are the most commonly processed fruits, with roughly 50% of the fruit remaining wasted. By valuing this waste through advanced processes, there are numerous functional and financial benefits as well as a potential reduction in environmental pollution [18]. Citrus fruits are preferred due to their high nutritional content and health benefits. In industrialized countries fruit processing industries utilizes about 20% of the energy and imposes a heavy burden on the environment. Inadequate waste disposal practices also contribute to greenhouse gas emissions [2]. Citrus peel waste can be valorized using both biological and physicochemical processes. The earlier methods of processing are quite energy-intensive and laborious, which is undesirable [18]. In the last 10 years, new, so-called "green," technologies have arisen to limit essential bioactive and pectin waste from citrus fruits while causing less environmental damage and producing higher yields in a shorter amount of time.

Fruit waste valorization is a practical and successful waste management technique. Wastes from citrus fruits are abundant in bioactive substances. In order to address environmental and human health issues emerges due to improper handling of citrus waste; it will take time of using innovative environmentally friendly methods to recover the useful bioactive compounds. Citrus peel pectin has a higher capacity for gelling, thickening, absorbing water, making it useful in a variety of industrial, food [22].

2.2 Citrus Waste Composition

Increased consumption of citrus fruits leads to environmental pollution by adding their waste. More over many fruit segments are unattractive and unclean, discarding them improperly poses a risk to the environment. Manufacturing of citrus juice add approximately half of the fresh fruit mass waste. This includes 50–55% of the entire fruit bulk (peels), wastewater, pomace, and 20–40% of the overall fruit mass (seeds). Peels, pulp, seeds, and bits of ruined fruit are all covered by citrus wastewater [16].

2.3 Types of Citrus Waste

2.3.1 Citrus Fruit Peel

Citrus fruits are mostly preferred as fresh produce or in the form of juice and their peel off are wasted. However, the peel is enriched with antioxidant compounds. The waste is then sold as cattle feed after drying and mixing with other dried pulps. The two parts of peels are soft, white center layer called albedo or mesocarp and a colored periphery known as flavedo or epicarp [11].

Citrus fruits have variable amounts of significant phytonutrients in different sections of the fruit. Compared to other fruit sections, the peel, a good source of limonene, pectin and molasses is often overlooked, has a higher concentration of secondary components that exhibit strong antioxidant properties. Regardless of great nutritious, the rind which makes up 50% of the fruit's original weight is typically dried, combined with dried pulp, and fed to calves or thrown out and treated as garbage [3].

It is important to pay particular attention to their potential application in the pharmaceutical business or as a component of functional meals, as the amount of these substances in the peel of the majority of the studied fruits is significantly higher than that of the pulp. Citrus fruit peels can be used to make extruded morning cornflakes with a high concentration of calcium, potassium and phosphorus, or mineral preparations with a range of composition and qualities [20].

2.3.2 Citrus Fruit Pulp

Auraptene, bergamottin, imperatorin, and heraclenin are only a few of the naturally occurring prenyloxycoumarin compounds found in citrus fruit peels, along with other macro- and microminerals that add to the fruit's nutritional and medicinal value. Peels are a waste product of citrus fruit, therefore their availability and low cost make them a potential source of nutraceuticals. Due to their high bioactive chemical content, citrus peels can be utilized to make functional foods or organic "dietary supplements" that include minerals in addition to dietary fiber and antioxidants [3].

There are other sections of the fruit that also contain these elements, even though citrus pulp is known to be a source of certain mineral elements for human nutrition, such as potassium, calcium, phosphorus, or magnesium. Because they are typically the inedible sections of the fruit, such as peels, these other fruit parts are not recognized in nutrition [22].

2.3.3 Citrus Fruit Seeds

During processing, about 20–25% of citrus fruits constitute seeds that may be a valuable commodity due to their essential oil content and protein content. Research showed that seeds are enriched with strong substances, including essential oils, tocopherols, carotenoids, polyphenols, and limonoids. Additionally, citrus seeds are nutritionally valuable and highly demanded by cosmetic and pharmaceutical industries [4].

Numerous researchers worked with the utilization of waste citrus seeds, either to extract the essential oil limonene or to incorporate the oil from the seeds into culinary products. Citrus seed byproducts such as seed oil and seed meal are the most common. Essential oils are said to be effectively obtained from citrus seeds. The citrus variety ranges from 20 to 40% (by weight) has the biggest influence on the oil content of citrus seeds [12].

2.3.4 Citrus Fruit Juice Processing Residue

The primary output of the citrus processing sector is juice, which is a common element in many dishes and beverages (such as soft drinks). Additional items include oils and essences that are used to make food-grade goods, jams, candied peel, potpourris, marmalades, and flavoring compounds for beverages and health drinks [22].

The waste's methanol extract of processed citrus juice is potentially used in medicine and nutraceutical industry. Citrus juice manufacturing waste contains more flavonoids and phenols than pulp. The flavonoid composition of processed waste of citrus juice differed from that of citrus fruit pulp, according to the results of UPLC tests, and it was mostly rich in neo-hesperedin and hesperidin. It was also discovered that the processing waste from citrus juice has remarkable antioxidants. The findings suggested that citrus wastes served as a cheap and potential dietary and medicinal complement as well as a practical source of free antioxidants [6].

2.4 Chemical Composition

2.4.1 Nutritional Composition of Citrus Fruit

Due to their delicious and aromatic qualities, high nutritional content, and numerous health-promoting properties resulting from their rich supply of phytonutrients, citrus is expended in great amounts throughout the world. Citrus fruits have excellent nutritional qualities because they are rich in water, carbohydrates, vitamins, minerals, and dietary fibers, among other essential components. A variety of essential oils, synephrines, flavonoids, limonoids, and carotenoids provide protection for a range of illnesses, such as cancer, cardiovascular, digestive as well as inflammatory conditions. There are considerable differences in the photochemistry of citrus kinds, growth phases and fruit parts [7].

2.4.2 Bioactive Compounds in Citrus Fruit

Phytonutrients principally PMFs (polymethoxylated flavones) such as 5-demethyl nobiletin, tangeritin and nobiletin are copious and richly available in citrus. These fruits are also enriched with provitamin, violaxanthin esters (xanthophylls), β-cryptoxanthin (carotenoids) as well as β-citraurin (apocarotenoids). These bioactive substances decrease the chances of metabolic syndrome by lowering the body's levels of reactive oxygen species (ROS) and inflammatory mediators, which includes cancer, cardiovascular disease (CVD), diabetes and neurological illnesses. Citrus essential oil is also a valuable commodity in the economy because of its flavorful, antioxidant, and antibacterial qualities. It is also rich in limonoids, such as limonin, and terpenes, such as D-limonene [15].

2.4.3 Fermentation Processing

Along with environmental degradation and health risks, the development and improper handling of processed fruit waste offer a serious challenge to waste controlling techniques worldwide. One such concerning fruit waste is citrus waste, which is abundant in numerous compounds with added value, such as pectin [14].

Four primary processes are typically included in the citrus processing industries. Citrus fruits are shipped to production factories after being harvested. The fruits are then graded in a first stage, and then, once the dirt has been removed, they are stored so that the second step water-washing cleaning can begin. The third process involves pressing and centrifuging the fruit juice to extract the juice. After heating the juice to activate the pectin enzymes, it is concentrated in syrup tanks. To ensure the highest quality vitamin C, enzymes and citric acid are added. After that, the juice is sealed in bags, sterilized by boiling, and labeled [21] Fig. 2.1.

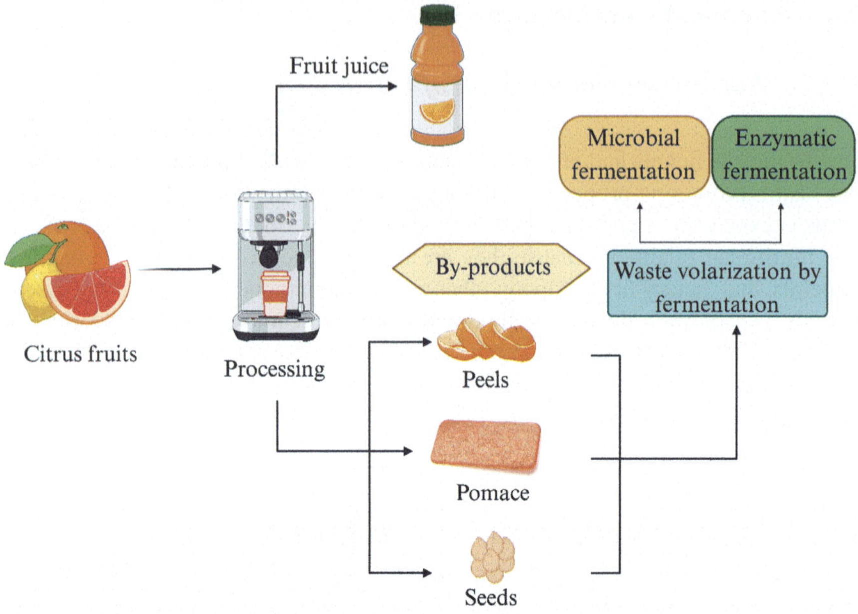

Fig. 2.1 Citrus fruit juice processing and waste volarization by fermentation

2.4.4 Microbial Fermentation

A significant content of citrus waste (CW), consisting of inedible pulp and peel, is produced during juice production. Dried corn gluten pellets have a low protein content (<10%), making them nutritionally negligible even if they can be utilized in cattle feed. Moreover, an uneven mineral ratio (a high calcium to phosphorus proportion) increases the risk of milk fever. As a result, landfills are typically used to dispose of CW. However, groundwater pollution happens as a result of its organic matter composition, which contains substances such as free sugars and polyphenols with elevated biological oxygen demands. Moreover, CW's acidity causes soil contamination, while polyphenols' antimicrobial properties stop organic compounds from biodegrading [10].

2.4.5 Enzymatic Process

Fermentation, enzymatic treatments comprising substrate modification and extraction are the steps in the bio refinery technique to valorizing CW. Currently, the most widely used commercially relevant technology is recovery of essential components, such as pectin, limonene and flavonoids. Thanks to progresses in microbial fermentation and metabolic engineering of corn starch has drawn consideration recently for its potential to produce industrially valuable compounds and fuels from sugars, including ethanol, methane, and mucic acid. A lot of effort has gone into creating yeast strains that can metabolize non-fermentable sugars, especially those generated from galacturonic acid found in pectin, in order to boost practicality. Getting accessible sugars at a reasonable price requires effective enzyme scarification [8].

The value of CW as a biomass that can be converted into biochemical and fuels has drawn a lot of interest. In order to produce bio-based fuels and chemicals that do not conflict with human use, biomass must be chosen in a sustainable and environmentally friendly manner. Citrus peel waste (CPW), which is growing yearly, has the potential to be a desirable biomass due to its low lignin concentration, which can help bioprocesses by lowering their costs significantly when compared to lignocellulose biomass [5].

2.5 Factors Influencing Fermentation

2.5.1 Temperature

The dried peels of citrus fruits (CPs), food items with additional value, were evaluated for their quality, energy consumption and drying behavior in relation to the drying temperature and residual moisture content. The CPs sections were dehydrated in a hot-air dryer between 40 and 70 °C by providing balanced air velocity of 1 m/s up to the final dampness level for the particular drying circumstances was obtained or the requisite humidity for storage. Cakes made from mixtures with varying percentages of dried CPs (0%, 10%, 15%, and 20%) were also assessed for their chemical makeup and sensory qualities. The ideal moisture content and drying temperature ranges were 10 ± 0.2% w.b. and 50–60 °C, respectively. These parameters minimized drying time and energy usage while optimizing product quality. Conversely, the antioxidants like carotenoids, vitamin C decreases and essential oils were significantly increased when CPs was dehydrated to a final moisture level of 5.4 ± 0.2%. Cake formulations including 15% dried mandarin and orange peels had increases in dietary fiber (33.5% and 29.6%), crude fat (2.9% and 4.6%), and ash (30.6% and 29.0%), but small decreases in protein and total carbohydrate content. Cakes with a moisture level of 10% weight-based and 15% dried mandarin and orange peels dried could be highly palatable and healthful [9].

2.5.2 pH and Time

Dehydrated CPs at greater temperatures and to a final humidity content of 5.4 ± 0.2% significantly increased the removal of essential oils, antioxidants, vitamin C and carotenoids. Cake recipes including 15% dried mandarin and orange peels showed slight drops in protein and total carbohydrate content but increases in dietary fiber (33.5% and 29.6%), crude fat (2.9% and 4.6%), and ash (30.6% and 29.0%). Cakes that have a moisture content of 10% weight-based and 15% dried orange or mandarin peel may be very tasty and nutritious. Furthermore, the pH solution was modified by using sulphuric acid, and it was found that the electrical conductivity value varied between 1.46 and 2.06 S/m throughout warming. As a result, it was resolute that as the process stock period grows so does the overall amount of energy utilized. By turning the system on and off, it was discovered that the average power value had decreased [13].

- **Substrate preprocessing**

In this work, we investigated the pretreating and up gradation procedures in a particular sort of cultivated feedstock substrate called citrus waste (CW), which is a significant source of pollution that leads to excessive acidification of surface water. We examined the impacts of biodegradation preprocessing (BP) for the first time and created a multi-stage optimization technique to gradually boost the output of methane and biomass. The findings demonstrated that using a biodegradation preprocessing, CW's pH can be considerably increased. With combined types of Aspergillus niger CCTCC 206113 and Phanerochaete chrysosporium ATCC 20696, the most notable results of lowering CW hard waste while attaining the maximum biogas production were reported. These findings imply that CW solid waste can be substantially catalyzed by biodegradation pretreatment, which can also produce high-efficiency bioenergy while relieving the environmental pollution issues caused by [19].

2.5.3 Valorization of Citrus Waste

A. **Biofuel production**

1. **Bioethanol**

   Bioethanol is a viable form of bio-fuel. The combustion of bioethanol produces a significant amount of thermal efficiency and power. Initially, Grohmann and Baldwin extract sugars by performing hydrolysis of citrus waste in the presence of enzymes, which were subsequently fermented to produce bioethanol [23]. Since then, CW, specifically lemon, mandarin, orange, and grapefruit peels, has been widely planned for its prospective in producing bioethanol [24].

   Production of bioethanol primarily involves three key steps: preheated, hydrolysis, and fermentation. The Preheated aims to raise the release of fermentable polysaccharides. However, during second event known as the saccharification process or hydrolysis, the polysaccharides are broken down into simpler monosaccharides. The final stage of producing bioethanol is the fermentation process, which converts simple sugars into ethanol [21].

Continuous consideration is given to the consumption of citrus waste for the synthesis of bioethanol. Recent studies have utilized various preheated methods, comprising auto hydrolysis, grinding, steam explosion and dilute-acid before the conduction of acid/enzyme based hydrolysis and fermentation. An instance of this is the experiment performed by Boluda Aguilar et al. [25]. In their study, they produce ethanol by investigating the utilization of waste of mandarin peel. This was done by exposing the waste material to preheated step by using steam explosion method, then hydrolysis in the presence of particular enzyme and finally uses *Saccharomyces cerevisiae* for fermentation.

The orange peel waste was reported to be 50–60 L/1000 kg after undergoing ethanol concentration. The study documented an ethanol production of 0.495 grams/grams, a productivity of 4.85 grams/L/ hour. Joshi et al. [26] consumed waste of orange peel as a substrate for producing ethanol. They used steam explosion preheated method and then hydrolysis and fermentation with *Saccharomyces cerevisiae* NCIM 3495 2877 and they claimed the ethanol content as 0.25 g/g. In a recent study, John [27] employed *Citrus limetta* (musambi) peels as a substrate to produce bioethanol. They utilized acid-catalyzed steam preheated method, then hydrolysis and fermentation. The researchers claimed ethanol yield (64%) attained at pH 4 after treatment of 48 hours.

2. **Biogas**

Biogas is produced after anaerobic decomposition of organic substances and is composed of carbon dioxide and methane [21]. Agro-waste, such as citrus trash, can be used to make biogas. Citrus waste is rich in minerals that can enhance the production of methane, as demonstrated by Bozym et al. [28] in 2015. The methane output from citrus waste is influenced by specific parameters, including pH, substrate availability and temperature. However, the essential oil present in citrus waste poses a significant obstacle during anaerobic break down. Ruiz and Flotats [29] documented the suppressive effect of limonene, a constituent of citrus essential oil, at concentrations exceeding 200 mg/kg. Wikandari et al. [30] employed a bioreactor using a membrane of hydrophilic poly-vinylidine difluoride (PVDF) to prevent the presence of D-limonene throughout the production of methane gas from CW under high temperature settings. The study shown that the membrane bioreactor had a better methane gas production rate of 0.33 Nm^3 /kg volatile solids, whereas the classic free cell reactor produce less methane gas at a rate of 0.05 Nm^3 /kg. Consequently, it has been noted that removing limonene prior to the digestion process results in increased methane production from citrus waste.

B. **Organic acids**

1. **Citrate**

Citrate, also known as 1, 2, 3-tricarboxylic acid or 2-hydroxy propane is highly valued for its non-toxic properties, Generally Recognized as Safe

(GRAS) status, and enjoyable sour flavor. It is widely regarded as significant regular organic acids. The reference is is linked to the publication of Sawant et al. [31] in 2018. Similar to malic acid, it is present in various fruits including grapes, berries, oranges, limes, tangerines and lemons. The widespread utilization of this substance as an antioxidant, emulsifier, buffer, acidulant, and preservative in the food industry, along with its use in the medical, pharmaceutical sectors, chemical and contributes to its substantial global worth in the market. The production of this acid occurs either by chemical or physical means, both of which are intricate, costly, and ecologically unfriendly [32, 33, 34]. To satisfy the substantial requirement for this acid, which greatly surpasses its existing availability; it is imperative to employ efficient and uninterrupted production techniques. Microorganisms capable of producing citric acid, including bacteria such as yeasts like *Hansenula anomala, Candida citroformans, C. lipolytica C. guilliermondii, C. tropicalis, C. oleophila, C. intermedia, Y. lipolytica*, and *Saccharomycopsis lipolytica. Bacteria such as Mucor piriformis, Bacillus subtilis, B. licheniformis, Corynebacterium* species and *Arthrobacter paraffinens*; as well as fungi such as *Aspergillus carbonarius, A. awamori, A. flavus, A. foetidus, A. aculeatus, A. fonsecaeus, A. phoenicis, A. niger, Penicillium janthinellum, P. restrictum, Trichoderma viride, Ustulina vulgaris* and *Talaromyces* species [35], are currently being utilized in various industries.

2. **Lactic acid**

Lactate commonly called lactic acid, also known as 2-hydroxypropanoic acid, is an organic acid that is naturally found in various sources and has been utilized by humans for many years [36]. It has obtained Generally Recognised as Safe (GRAS) status and has been utilized in the cosmetic, chemical, pharmaceutical sectors and food industry [37]. Furthermore, it plays a significant role in the manufacturing of specifically polylactic acid, PLA (polylactide polymers) and other bioplastics [38].

They exhibit more susceptibility to degradation compared to other polymers. Lactate exists in two enantiomeric forms and possesses a chiral carbon atom. It can be generated through chemically, resulting through fermentative method in a racemic mixture of DL-lactate. The reference is [37]. *S. cerevisiae*, a type of yeast, is seen as a desirable substitute for LAB since it has Generally Recognized as Safe (GRAS) status, can tolerate different pH levels, is strong and resilient, has uncomplicated feeding needs, and is frequently employed in industry. Nevertheless, as stated before, it does not generate substantial amounts of organic acids [39]. Consequently, scientists have also examined unconventional yeasts as potential hosts for the synthesis of lactate from diverse substrates, yielding varying degrees of success (as discussed by Sauer et al. [38]). The conversion of L-malic acid into microbial L-lactate can be achieved by many enzymatic routes [40].

C. **Functional foods and nutraceuticals**

1. **CW extracts**

   When vegetable and fruit substances are disposed of in the environment, they disrupt the ecosystem because of their great moisture content, which makes them more susceptible to microbial destruction. These by-products contain abundant total phenolic components (TPC) and dietary fibers that possess cardio protective, antimicrobial, and anticancer effects. Citrus fruits, including oranges, limes, lemons, and grapefruit, are commonly consumed as juice or fresh pulps. However, during the handling of these fruits, constituting around 50% of the weight of the fresh fruit a major extent of by-products such as pulp, seeds and peels are generated [41]. By-products can provide constituents such as pectins, flavonoids, limonene and fibres, by extraction. The primary flavonoids present in citrus seeds and peels are hesperidin, naringin, eriocitrin, and narirutin, each exhibiting varying degrees of antioxidative properties [42]. Nimolin, nomilinic acid and Limonin, are prominent limonoids found in citrus fruit peels, and they have been proven to possess antimicrobial, antiviral, and antibacterial properties [43].

2. **Probiotics production**

   Microencapsulation has emerged as a promising approach for safeguarding probiotic cultures against unfavorable conditions present in derivatives of citrus fruits. Regardless of the considerable phytonutrients of citrus fruits, their low pH and inadequate levels of certain amino acids and peptides pose limitations in the development of citrus fruit probiotic products [44]. The process of microencapsulation has been extensively studied for the manufacture of microcapsules composed of probiotic substances. These capsules comprise a central region that contains active components and a polymeric outer layer that serves as a safeguarding obstruction that controls the transfer of chemicals into and out of the microcapsules [45].

   The research on microencapsulation practices is the subject of substantial research with the goal of improving the procedure and maintaining the vitality of the probiotic culture. The selection of an appropriate process for making microcapsules is contingent upon various aspects, including the physicochemical qualities of the chosen stuffs, the preferred particle size, and the intended release [46]. The experiment performed by Da Silva et al. [44] demonstrated that *Lactobacillus acidophilus* LA-02 containing microcapsules were successfully produced using the complex coacervation method and subsequently crosslinked with transglutaminase. These microcapsules were then added to various fruit juices. The results revealed that the microcapsules effectively protected *L. acidophilus*, confirming the survival and sustainability of the probiotic. This work demonstrates that the utilization of the complicated coacervation approach presents a feasible solution for addressing the difficulties associated with probiotic development in citrus fruits.

D. **Animal feed**

1. **Citrus waste and animal feed**

The global harvest of citrus trash occurs on a big scale, covering a significant amount of land. This covers about 50–60% of the entire invention level and industrial importance of citrus waste [47]. The economic value of citrus waste is significant due to the presence of diverse biologically active combinations, such as essential oils, carotenoids, dietary fibre, flavonoids, ascorbic acid and polyphenols. Additionally, citrus waste contains substantial quantities of minerals, vitamins and trace elements, that may have remedial properties for animals. Additionally, it possesses elevated concentrations of sugars that are well-suited for the process of fermentation, specifically for the manufacture of bioethanol [48].

The solid waste produced during citrus processing can be utilized as components in animal feed. Many enterprises globally are opting to dispose of their trash in either dried or fresh form as a cost-effective and convenient method, mostly for the purpose of producing animal food stuff [49]. In the process of producing dry silage, the pulp undergoes two rounds of pressing to decrease the moisture content to a range of 65–70% (w.b.). It is then dried until it reaches a weight of 90 grammes per 100 grammes and subsequently pelletized for the purpose of marketing. Citrus residue has a significant quantity of pectin, a substance that has the ability to bind water. Lime and either calcium oxide or calcium hydroxide are introduced throughout the drying process. Lime and either calcium oxide or calcium hydroxide are used to achieve a pH level between 5.5 and 6.5. Additionally, they disrupt the water affinity of pectin when mixed in the pulp. This aids in the purification of water and carbohydrates found in the pulp [50].

2. **Nutritional benefits**

Flavonoids are phenolic chemicals that are abundantly found in citrus peel and has nutraceutical benefits mostly due to their potent antioxidant properties. Flavanones, the primary category of flavonoids, are found in citrus fruits. They can be categorized on the basis of presence or absence of glycoside group as aglycone flavanones or glycoside [51]. The composition of flavonoid varies in orange peel significantly depending on the geographical location where the plant is grown. The orange peel extract was found to have a total flavonoid concentration ranging 14.0–31.9 mg/g [52]. The best prevalent glycoside flavanones are naringin, hesperidin, and neohesperidin found in citrus peel, whereas aglycone flavanones was found less when compared to glycosides [53]. The results reported by Chen et al. [52] revealed that the most prevalent flavonoid is hesperidin discovered in orange peel extracts, among the other four flavonoids namely narirutin, hesperidin, nobiletin, and tangeritin.

2.6 Challenges and Opportunities

A. **Technical challenges**

1. **Fermentation optimization**

The preheated and hydrolysis methods are not directly applicable to pectin-rich CPW biomass that used in the present biorefinery of lignocellulosic biomass. First and primarily, the employment of pre-engineered processes for the treatment of lignocellulosic biomass has been demonstrated to be effective across several physical properties [61]. This biomass has a low moisture content of around 10–15% by weight, whereas CPW biomass is a high-moisture substance containing 75–85% water by weight. Furthermore, CPW exhibits a lower abundance of lignin like chemicals and a higher abundance of pectin, in comparison to lignocellulosic biomass [62].

These characteristics suggest that a technique designed for the biomass may not be appropriate for the refining of CPW. The total engineering procedure for CPW requires the adaptation of development settings, reactor architecture, kinetic parameters, and hydrolysis practices to suit the material's precise chemical and physical features. Evidently, there is a requirement for novel technologies and progress. Insufficient data has been published regarding the engineering and scientific parameters necessary for designing the preheated and catalytic hydrolysis stages of a planned refinery that utilizes CPW. The citation is from Grohmann and Bothast [63]. Significant data deficiencies have been familiar and specifically addressed in order to finalize the advancement of a viable CPW refinery expertise. Following steps are followed;

1. The impact of size and surface area of the element on the rates of catalytic hydrolysis and the products of sugar.
2. Researching chemical preheated parameters to enhance the availability of cellulases and pectinases on structural polysaccharides.
3. Assessment of the impact of primary resolvable free-sugars on the rate of catalytic hydrolysis.
4. Implementation of enhanced catalyst complexes and supplementary catalysts to enhance catalytic hydrolysis.
5. Key manufacturing considerations for expanding and improving the efficiency of the methods, including the study of reaction rates, the design of reactors, and the analysis of mass and energy flows in all stages of the refinery.

2. **Waste pre-treatments**

Any refinery process used to form monosaccharides from cellulosic biomass will include a chemical and physical preheated step. This step is necessary to enhance the ability of the cellulose fraction to absorb acid and to make it more available for catalysis. Preheated facilitates the efficient breakdown of the structural polysaccharides in order to generate a mixture consisting of 5- and 6-carbon sugars (including reducing, neutral, and acidic sugars) during the subsequent hydrolysis process. Efforts are currently underway to actively search for enhanced preheated methods in order to optimize these procedures. Various studies have been led in the past 10 years to examine the impact of particle size on cellulose hydrolysis in biomass containing lignocellulose. These investigations have utilized both pure cellulose and, lignocellulose constituents like soft wood, cotton dregs, and newspaper [59]. The outcomes have shown that reducing the size of element led to a significant decrease in the time required to achieve maximum hydrolysis conversion, by several hours. A study conducted by Dasari and Berson [60] has shown that lowering the size of particles in sawdust slurries can potentially decrease the amount of time needed for the catalytic hydrolysis phase in converting lignocellulose stuff into ethyl alcohol.

B. **Economics and industrial perspectives**

The economic feasibility of converting CPW into bioethanol in most reported research and patents relies heavily on the retrieval of co-products, which will be concisely argued. Pourbafrani et al. [54] documented the simultaneous generation of pectin, methane and D-limonene as a means to support the economic viability of ethyl alcohol production. This was achieved by employing enzymatic and acid hydrolysis methods. This finding indicates that there is an extra economic limitation when expanding the use of these technologies. This limitation is related to the impact of the production costs affectivity of ethyl alcohol on the recovery of co-products. According to the given assumptions, with a bulk of plant is around 152,000 cubic metres annual and an ethanol price of US$475.51 per cubic metre, Grohmann [55] estimated gross margin incomes of approximately US$90.0 per cubic metre excluding D-limonene and when D-limonene was included as a co-product is US$169.0 per cubic metre. Related findings were informed by Zhou et al. [56].

C. **Environmental considerations**

The drying process gives rise to environmental concerns, specifically air emissions. These emissions including D-limonene are estimated to be around 9000–18,000 tonnes annually, which is classified as a volatile organic compound (VOC) [55]. The Environmental Protection Agency [58] mandates in accordance with the Federal Clean Air Act Amendments of 1990, that citrus juice manufacturers must submit a Title V permit if their limonene productions surpass annual 91 tonnes [57].

The employment of organizations to limit and decrease regulated discharges in the drying process due to the consumption of different fuels and natural gas (such as particulate matter, carbon monoxide, nitrogen oxides, sulphur dioxide and other harmful air pollutants like formaldehyde, methanol, and CO2) negatively affects process economics due to the additional environmental burdens it creates [57]. From an environmental engineering perspective, using CPW as a representative biomass has the latent to assist as a nearby, renewable source of raw materials for producing bioethanol and specialty chemicals. This can help decrease greenhouse gas emissions associated with transporting biomass to other areas and also replace some of the consumption of nonrenewable fuels when CPW is used as biomass for bioethanol production. Implementing a commercial technique might eradicate the secretions produced by the drying process of wet CPW to transform it into calf fodder, while also creating space for other dense litters in local municipal landfills.

2.7 Future Directions

The citrus processing industries produce huge bulks of by-products, which are seen as a menace to the surroundings due to their unique features and carbon-based composition. Therefore, there is an immediate need for their effective management [64]. Nevertheless, these byproducts, which contain valuable composites, can be employed for regaining or reprocessing into useful yields using various valorization strategies. This allows for their utilization in the pharmaceutical industries, food, and cosmetics, resulting in both added value and environmental management benefits [65].

Various environmentally-friendly methods have been created to efficiently utilize citrus byproducts for extracting pectin, dietary fibers, polyphenols, and essential oils utilizing sustainable technology. These byproducts can then be transformed into valuable products using biotechnological processes. Nevertheless, research has shown that the existence of limonene negatively affects the effective utilization of citrus by-products. Consequently, several pre-treatment techniques are typically necessary to eliminate it [66].

In addition, ionic solvents and deep-eutectic solvents are presently becoming recognized as sustainable diluters for extracting valuable compounds. This is because they have very little volatility, are highly miscible in water, and possess excellent solvent properties. Furthermore, researchers have devised diverse hybrid methodologies by combining distinct extraction strategies to optimize product retrieval and achieve substantial yields. Furthermore, biotechnological pathways such as anaerobic digestion and fermentation have arisen as efficient and cost-effective means of converting citrus byproducts into valuable bio-products [67].

The formation of biofuels from citrus by-products by biotransformation is a highly favorable method of valorization. However, its implementation is now limited to laboratory settings. Therefore, it is necessary to undertake a thorough investigation of the process and make efforts to step up the transformation methods on commercial scale. In addition, the efficiency of these bio products can be improved by employing appropriate pretreatments to reduce the harmful impacts of essential oils.

Advanced methods such as utilizing genetically engineered microorganisms, employing continuous cell recycle reactors, and implementing co-culture fermentation processes can be employed to achieve optimal input of bio-products derived from citrus by-products [68]. In order to enable a cost-effective expansion of production and application of value-added ingredients, it is necessary to address and resolve certain technical, scientific, and economic obstacles [65]. An example of an important consideration is the utilization of innovative technology for the transportation, collection and processing of by-products from fruits and vegetables [66]. Furthermore, additional study is needed to analyze the economic and technological features of pretreating by-products, such as densification and drying, prior to transporting them to treating plants. Indeed, the elevated production expenses associated with usual important components derived from vegetable and fruit by-products may impede their utilization in the food industries. Consequently, future research efforts should prioritise the decrease of costs associated with these ingredients. The source cited is Wu [69].

2.8 Conclusion

In conclusion, the process of citrus fruit waste and by-products' volarization by means of fermentation offers a promising solution for the management of agricultural waste and the production of desired products. Through the utilization of fermentation techniques, such as anaerobic digestion or aerobic fermentation, citrus leftover is efficiently transformed into valuable resources such as bioethanol, organic acids, biogas and various other biochemicals. This not only reduces the environmental pollution caused by waste disposal but also presents opportunities for the advance of viable biomass industries. However, the success of volarization processes relies on various factors including substrate composition, fermentation conditions, and microbial activities, which necessitate further research and optimization. Additionally, economic feasibility and scalability are crucial aspects that need to be addressed to ensure the widespread adoption of volarization technologies in citrus waste management practices. Overall, the volarization of citrus fruit waste through fermentation holds great potential for mitigating waste-related environmental concerns and promoting the transition towards a more circular and sustainable agricultural industry.

References

1. Albrigo, L. G., Stelinski, L. L., & Timmer, L. W. (2019). *Citrus*. CABI.
2. Chavan, P., Singh, A. K., & Kaur, G. (2018). Recent progress in the utilization of industrial waste and by-products of citrus fruits: A review. *Journal of Food Process Engineering, 41*, e12895.
3. Czech, A., Zarycka, E., Yanovych, D., Zasadna, Z., Grzegorczyk, I., & Kłys, S. (2020). Mineral content of the pulp and peel of various citrus fruit cultivars. *Biological Trace Element Research, 193*, 555–563.
4. Hernández-Montoya, V., Montes-Morán, M., & Elizalde-González, M. P. (2009). Study of the thermal degradation of citrus seeds. *Biomass and Bioenergy, 33*, 1295–1299.

5. Kim, I. J., Jeong, D., & Kim, S. R. (2022). Upstream processes of citrus fruit waste biorefinery for complete valorization. *Bioresource Technology, 362*, 127776.
6. Kim, J. H., & Kim, M. Y. (2016). The potential use of citrus juice waste as sources of natural phenolic antioxidants. *Journal of Applied Pharmaceutical Science, 6*, 202–205.
7. Lu, X., Zhao, C., Shi, H., Liao, Y., Xu, F., Du, H., Xiao, H., & Zheng, J. (2023). Nutrients and bioactives in citrus fruits: Different citrus varieties, fruit parts, and growth stages. *Critical Reviews in Food Science and Nutrition, 63*, 2018–2041.
8. Mamma, D., & Christakopoulos, P. (2014). Biotransformation of citrus by-products into value added products. *Waste and Biomass Valorization, 5*, 529–549.
9. Marey, S., & Shoughy, M. (2016). Effect of temperature on the drying behavior and quality of citrus peels. *International Journal of Food Engineering, 12*, 661–671.
10. Patsalou, M., Chrysargyris, A., Tzortzakis, N., & Koutinas, M. (2020). A biorefinery for conversion of citrus peel waste into essential oils, pectin, fertilizer and succinic acid via different fermentation strategies. *Waste Management, 113*, 469–477.
11. Rafiq, S., Kaul, R., Sofi, S., Bashir, N., Nazir, F., & Nayik, G. A. (2018). Citrus peel as a source of functional ingredient: A review. *Journal of the Saudi Society of Agricultural Sciences, 17*, 351–358.
12. Rosa, A., Era, B., Masala, C., Nieddu, M., Scano, P., Fais, A., Porcedda, S., & Piras, A. (2019). Supercritical CO2 extraction of waste citrus seeds: Chemical composition, nutritional and biological properties of edible fixed oils. *European Journal of Lipid Science and Technology, 121*, 1800502.
13. Sabanci, S., Çevik, M., & Göksu, A. (2021). Investigation of time effect on pectin production from citrus wastes with ohmic heating assisted extraction process. *Journal of Food Process Engineering, 44*, e13689.
14. Sabater, C., Ruiz, L., Delgado, S., Ruas-Madiedo, P., & Margolles, A. (2020). Valorization of vegetable food waste and by-products through fermentation processes. *Frontiers in Microbiology, 11*, 581997.
15. Saini, R. K., Ranjit, A., Sharma, K., Prasad, P., Shang, X., Gowda, K. G. M., & Keum, Y.-S. (2022). Bioactive compounds of citrus fruits: A review of composition and health benefits of carotenoids, flavonoids, limonoids, and terpenes. *Antioxidants, 11*, 239.
16. Salazar Díaz, D. A. (2020). Citrus waste composition and bioactive compounds: a review.
17. Sharma, P., Vishvakarma, R., Gautam, K., Vimal, A., Gaur, V. K., Farooqui, A., Varjani, S., & Younis, K. (2022). Valorization of citrus peel waste for the sustainable production of value-added products. *Bioresource Technology, 351*, 127064.
18. Spreen, T. H., Gao, Z., Fernandes, W., Jr., & Zansler, M. L. (2020). Global economics and marketing of citrus products. In *The genus citrus*. Elsevier.
19. Su, H., Tan, F., & Xu, Y. (2016). Enhancement of biogas and methanization of citrus waste via biodegradation pretreatment and subsequent optimized fermentation. *Fuel, 181*, 843–851.
20. Suri, S., Singh, A., & Nema, P. K. (2022). Current applications of citrus fruit processing waste: A scientific outlook. *Applied Food Research, 2*, 100050.
21. Zema, D., Calabrò, P., Folino, A., Tamburino, V., Zappia, G., & Zimbone, S. (2018). Valorisation of citrus processing waste: A review. *Waste Management, 80*, 252–273.
22. Zema, D. A., Calabro, P. S., Folino, A., Tamburino, V., Zappia, G., & Zimbone, S. M. (2019). Wastewater management in citrus processing industries: An overview of advantages and limits. *Water, 11*, 2481.
23. Grohmann, K., & Baldwin, E. A. (2021). Hydrolysis of orange peel with pectinase and cellulase enzymes. *Biotechnology Letters, 14*, 1169–1174.
24. Boluda-Aguilar, M., & Lopez-Gomez, A. (2019). Production of bioethanol by fermentation of lemon (Citrus limon L.) peel wastes pretreated with steam explosion. *Industrial Crops and Products, 41*, 188–197.
25. Boluda-Aguilar, M., Garcia-Vidal, L., del Pilar, G.-C. F., & Lopez-Gomez, A. (2010). Mandarin peel wastes pretreatment with steam explosion for bioethanol production. *Bioresource Technology, 101*, 3506–3513.

26. Joshi, S. M., Waghmare, J. S., Sonawane, K. D., & Waghmare, S. R. (2015). Bio-ethanol and bio-butanol production from orange peel waste. *Biofuels, 6*, 55–61.
27. John, I., Pola, J., Thanabalan, M., & Appusamy, A. (2020). Bioethanol production from musambi peel by acid catalyzed steam pretreatment and enzymatic saccharification: Optimization of delignification using taguchi design. *Waste and Biomass Valorization, 11*, 2631–2643.
28. Bozym, M., Florczak, I., Zdanowska, P., Wojdalski, J., & Klimkiewicz, M. (2015). An analysis of metal concentrations in food wastes for biogas production. *Renewable Energy, 77*, 467–472.
29. Ruiz, B., & Flotats, X. (2016). Effect of limonene on batch anaerobic digestion of citrus peel waste. *Biochemical Engineering Journal, 109*, 9–18.
30. Wikandari, R., Millati, R., Cahyanto, M. N., & Taherzadeh, M. J. (2014). Biogas production from citrus waste by membrane bioreactor. *Membranes, 4*, 596–607.
31. Sawant, O., Mahale, S., Ramchandran, V., Nagaraj, G., & Bankar, A. (2018). Fungal citric acid production using waste materials: A minireview. *Journal of Microbiology, Biotechnology and Food Sciences, 8*, 821–828. [CrossRef].
32. Gupta, G. K., De, S., Franco, A., Balu, A., & Luque, R. (2015). Sustainable biomaterials: Current trends, challenges and applications. *Molecules, 21*, 48. [CrossRef].
33. Yin, X., Shin, H.-D., Li, J., Du, G., Liu, L., & Chen, J. (2017). Comparative genomics and transcriptome analysis of Aspergillus Niger and metabolic engineering for citrate production. *Scientific Reports, 7*, 41040. [CrossRef].
34. Yu, B., Zhang, X., Sun, W., Xi, X., Zhao, N., Huang, Z., Ying, Z., Liu, L., Liu, D., Niu, H., et al. (2018). Continuous citric acid production in repeated-fed batch fermentation by Aspergillus Niger immobilized on a new porous foam. *Journal of Biotechnology, 276–277*, 1–9. [CrossRef].
35. Papagianni, M. (2017). Advances in citric acid fermentation by Aspergillus Niger: Biochemical aspects, membrane transport and modeling. *Biotechnology Advances, 25*, 244–263. [CrossRef.
36. Castillo Martinez, F. A., Balciunas, E. M., Salgado, J. M., Domínguez González, J. M., Converti, A., & Oliveira, R. P. D. S. (2021). Lactic acid properties, applications and production: A review. *Trends in Food Science and Technology, 30*, 70–83. [CrossRef].
37. Abdel-Rahman, M. A., Tashiro, Y., & Sonomoto, K. (2019). Recent advances in lactic acid production by microbial fermentation processes. *Biotechnology Advances, 31*, 877–902. [CrossRef].
38. Sauer, M., Porro, D., Mattanovich, D., & Branduardi, P. (2020). 16 years research on lactic acid production with yeast—Ready for the market? *Biotechnology & Genetic Engineering Reviews, 27*, 229–256. [CrossRef].
39. Abbott, D. A., Zelle, R. M., Pronk, J. T., & van Maris, A. J. A. (2019). Metabolic engineering of Saccharomyces cerevisiae for production of carboxylic acids: Current status and challenges. *FEMS Yeast Research, 9*, 1123–1136. [CrossRef] [PubMed].
40. Su, J., Wang, T., Wang, Y., Li, Y. Y., & Li, H. (2014). The use of lactic acid-producing, malic acid-producing, or malic acid-degrading yeast strains for acidity adjustment in the wine industry. *Applied Microbiology and Biotechnology, 98*, 2395–2413. [CrossRef].
41. Khao, T. H., & Chen, B. H. (2013). Fruits and vegetables, Chapter 18. In M. Chandrasekaran (Ed.), *Valorization of food processing by-products* (p. 517557). Taylor and Francis Group.
42. Manthey, J. A., Grohmann, K., & Guthrie, N. (2021). Biological properties of citrus flavonoids pertaining to cancer and inflammation. *Current Medicinal Chemistry, 8*, 135–153.
43. Dijlas, S., Canadanovic, B. J., & Cetkovic, G. (2019). By-products of fruit processing as a source of phytochemicals. *Chemical Industry & Chemical Engineering, 15*, 191–202.
44. Da Silva, T. M., Sonza Pinto, V., Soares, V. R. F., Marotz, D., Cichoski, A. J., Queiroz Zepka, L., Jacob Lopes, E., de Bona da Silva, C., & de Menezes, C. R. (2021). Viability of microencapsulated Lactobacillus Acidophilus by complex Coacervation associated with enzymatic crosslinking under application in different fruit juices. *Food Research International, 141*, 110190. [CrossRef].

45. Craparo, E. F., Drago, S. E., Giammona, G., & Cavallaro, G. (2020). Production of polymeric micro- and nanostructures with Tunable properties as pharmaceutical delivery systems. *Polymer, 200*, 122596. [CrossRef].
46. Bonaccorso, A., Russo, N., Romeo, A., Carbone, C., Grimaudo, M. A., Alvarez-Lorenzo, C., Randazzo, C., Musumeci, T., & Caggia, C. (2021). Coating Lacticaseibacillus Rhamnosus GG in alginate systems: An emerging strategy towards improved viability in Orange juice. *AAPS PharmSciTech, 22*, 123. [CrossRef] [PubMed].
47. Satari, B., & Karimi, K. (2018). Citrus processing waste environmental impacts, recent advances, and future perspectives in total valorization. *Resources, Conservation & Recycling, 129*, 153–167.
48. Varmie, E. B., & Thakur, M. (2021). Utilization of citrus processing waste: A review. *The Pharma Innovation Journal, 10*, 682–697.
49. Moreira, O. S., Reis, R. B., & Lana, A. M. Q. (2018). Produc¸ão e Composic¸ão do Leite de Vacas Alimentadas com Polpa Cítrica em Substituic¸ão ao Milho. Anais da Reunião anual da Sociedade Brasileira de Zootecnia 41, CD-ROM.
50. Wing, J. M. (2019). *Citrus feedstuffs for dairy cattle*. University of Florida, IFAS, 829.
51. Roowi, S., & Crozier, A. (2011). Flavonoids in tropical citrus species. *Journal of Agricultural and Food Chemistry, 59*, 12217–12225.
52. Chen, X.-M., Tait, A. R., & Kitts, D. D. (2017). Flavonoid composition of orange peel and its association with antioxidant and anti-inflammatory activities. *Food Chemistry, 218*, 15–21.
53. Molina-Calle, M., Priego-Capote, F., & Luque De Castro, M. D. (2015). Development and application of a quantitative method for determination of flavonoids in orange peel: Influence of sample pretreatment on composition. *Talanta, 144*, 349–355.
54. Pourbafrani, M., Forgács, G., Horváth, I. S., Niklasson, C., & Taherzadeh, M. J. (2020). Production of biofuels, limonene and pectin from citrus wastes. *Bioresource Technology, 101*, 4246–4250.
55. Grohmann, K. (2017). *Final technical report, project title: Citrus waste biomass program*. USDA Agricultural Research Service CRADA 58– 3K95–4–1053. Available at: www.osti.gov/bridge/servlets/purl/898345– 9P4Slw/898345.pdf. Accessed 12 January 2012.
56. Zhou, W., Widmer, W. W., & Grohmann, K. (2019). Economic analysis of ethanol production from citrus peel waste. *Proceedings of the Florida State Horticultural Society, 120*, 310–315.
57. Braddock, R. J., & Miller, W. M. (2021). Volatile organic compounds from citrus feed mill emissions. *Journal of Food Process Engineering, 24*, 1–15.
58. USEPA (eds). (2020). 40 CFR part 52, approval and promulgation of implementation plans: Florida: Citrus juice processing. *Federal Register, 69*, 4459–4463.
59. Neilson, M. J., Kelsey, R. G., & Shafizadhe, F. (2018). Enhancement of enzymatic hydrolysis by simultaneous attrition of cellulosic substrates. *Biotechnology and Bioengineering Symposium, 24*, 293–304.
60. Dasari, R. K., & Berson, R. E. (2017). The effect of particle size on hydrolysis reaction rates and rheological properties in cellulosic slurries. *Applied Biochemistry and Biotechnology, 136–140*, 289–299.
61. Lohrasbi, M., Pourbafrani, M., Niklasson, C., & Taherzadeh, M. J. (2010). Process design and economic analysis of a citrus waste biorefinery with biofuels and limonene as products. *Bioresource Technology, 101*, 7382–7388.
62. Stewart, D. A., Widmer, W. W., Grohmann, K., & Wilkins, M. R. (2006). *Ethanol production from citrus processing wastes*. US Patent Application Publication No.US/2006/0177916-A1.
63. Grohmann, K., & Bothast, R. J. (1994). Pectin rich residues generated by processing of citrus fruits, apples and sugar beets. In M. E. Himmel, J. O. Baker, & R. P. Overend (Eds.), *Enzymatic conversion of biomass for fuel production, ACS Symp. Ser. No. 566* (pp. 372–390). American Chemical Society.
64. Chemat, F., Rombaut, N., Sicaire, A.-G., Meullemiestre, A., Fabiano-Tixier, A.-S., & Abert-Vian, M. (2017). Ultrasound assisted extraction of food and natural products. Mechanisms,

techniques, combinations, protocols and applications. A review. *Ultrasonics Sonochemistry, 34*, 540–560. https://doi.org/10.1016/j.ultsonch.2016.06.035
65. Gullón, P., Gullón, B., Astray, G., Carpena, M., Fraga-Corral, M., Prieto, M. A., & Simal-Gandara, J. (2020). Valorization of by-products from olive oil industry and added-value applications for innovative functional foods. *Food Research International (Ottawa, ON), 137*(June), 109683. https://doi.org/10.1016/j.foodres.2020.109683
66. Anal, A. K. (2018). *Food processing by-products and their utilization*. Wiley.
67. Rao, M., Bast, A., & de Boer, A. (2021). Valorized food processing by-products in the EU: Finding the balance between safety, nutrition, and sustainability. *Sustainability, 13*(8), 4428–4418. https://doi.org/10.3390/su13084428
68. Socas-Rodríguez, B., Alvarez-Rivera, G., Valdes, A., Ibanez, E., & Cifuentes, A. (2021). Food by-products and food wastes: Are they safe enough for their valorization? *Trends in Food Science & Technology, 114*, 133–147. https://doi.org/10.1016/j.tifs.2021.05.002
69. Wu, D. (2016). Recycle technology for potato peel waste processing: A review. *Procedia Environmental Sciences, 31*, 103–107. https://doi.org/10.1016/j.proenv.2016.02.014

Chapter 3
Citrus Peel Waste Management of Oranges

Ali Mohammad Yousef Jebreen, Oshin Sahni, and Areesha

3.1 Introduction

Citrus fruits, particularly oranges, have long been celebrated for their vibrant taste and nutritional benefits, delightful palates and nourishing bodies worldwide. However, the journey of these juicy fruits doesn't end when they're squeezed into refreshing beverages or enjoyed as snacks. Rather, it extends beyond consumption, leaving behind a significant byproduct: citrus peel waste [1].

The disposal of citrus peel waste presents a considerable challenge due to its sheer volume and potential environmental impact. In recent years, there has been a growing recognition of the need to address this issue effectively. Not only does citrus peel waste contribute to landfill overflow, but its decomposition also releases methane, a potent greenhouse gas. Furthermore, improper disposal can lead to soil and water contamination, affecting ecosystems and human health [2].

Yet, within this challenge lies a remarkable opportunity. Citrus peel waste holds untapped potential for various value-added applications, ranging from biofuel production and animal feed supplementation to the extraction of high-value compounds for pharmaceutical and cosmetic industries [3]. By reimagining citrus peel waste as a valuable resource rather than a burdensome byproduct, we can not only mitigate environmental harm but also create economic opportunities and foster sustainable practices [5].

A. M. Y. Jebreen (✉)
Department of Medical Therapeutic Nutrition, Faculty of Allied Medical Sciences, Palestine Ahliya University, Bethlehem, Palestine

O. Sahni
Department of Food Science, University of Tennessee, Knoxville, TN, USA

Areesha
Department of Clinical Nutrition, NUR International University, Lahore, Pakistan

A. Chauhan et al. (eds.), *Valorization of Citrus Food Waste*,
https://doi.org/10.1007/978-3-031-77999-2_3

This book chapter aims to explore the multifaceted landscape of citrus peel waste management, with a particular focus on oranges. Through an interdisciplinary lens, we will delve into innovative approaches, emerging technologies, and best practices for harnessing the full potential of citrus peel waste. By examining case studies, research findings, and industry insights, we seek to provide readers with a comprehensive understanding of the challenges and opportunities in this vital field.

3.2 Orange Peel Waste Generation

The production of discarded orange peels is an important aspect of orange processing, which poses environmental hurdles alongside avenues for waste management and resource utilization, as outlined below:

1. Orange peel waste generation extent: Orange peel waste is primarily generated during the industrial processing of oranges to extract juice which accounts for about 50–60% of the processed fruits. The amount of waste generated varies depending on factors such as the type of fruit, processing methods, and seasonal changes in fruit availability.
2. Environmental impact: Improper disposal of orange peel waste can lead to environmental pollution, including soil and water pollution. The high organic content of orange peels can contribute to methane emissions if disposed of in landfills, further exacerbating greenhouse gas emissions [2].
3. Potential solutions and use: There are various methods of managing orange peel waste, including composting, anaerobic digestion to produce biogas, and extracting valuable compounds for use in food, medicine, and cosmetics [9]. It also contains valuable bioactive compounds such as flavonoids, phenols and essential oils, which have antioxidant and antimicrobial properties.
4. Value-added products from orange peel: Research has explored the possibility of using orange peel waste to produce value-added products such as pectin, dietary fiber supplements, animal feed additives and natural dyes [7].

These applications not only reduce waste, but also create economic opportunities and contribute to the circular economy.

3.3 Extraction Techniques for Maximized Juice Yield

Maximizing juice yield from orange is essential for efficient fruit processing and optimal utilization of resources. Several extraction techniques have been developed to achieve high juice yields while maintaining product quality.

1. **Mechanical Extraction**: Mechanical extraction methods involve the use of equipment such as orange juicers or presses to physically crush the fruit and

extract the juice. These methods are commonly used in small to medium-scale citrus processing operations and are relatively simple and cost-effective [11].

2. **Enzymatic Treatment**: Enzymatic treatment involves the use of enzymes such as pectinases to break down cell walls and release juice from the fruit pulp. Enzymatic treatment can increase juice yield, especially in fruits with high pectin content, and improve the clarity and quality of the juice.
3. Hydrothermal Treatment: Hydrothermal treatment, including processes such as steam extraction or hot water immersion, can help soften the fruit tissues and facilitate juice extraction. This method is particularly effective for processing whole fruits or orange segments and can result in higher juice yields compared to mechanical methods alone [16].
4. **Ultrasound-Assisted Extraction**: Ultrasound-assisted extraction involves the application of high-frequency ultrasound waves to disrupt cell structures and enhance juice release. This method can increase juice yield and reduce extraction time, leading to higher efficiency and lower energy consumption.
5. **High-Pressure Processing (HPP):** High-pressure processing applies high pressures to the fruit pulp to rupture cell membranes and release juice. HPP can improve juice yield and preserve the sensory and nutritional properties of the juice by minimizing heat exposure [15] (Table 3.1).

These techniques vary in their efficiency, equipment requirements, and suitability for different scales of operation. Choosing the appropriate extraction method depends on factors such as desired juice quality, processing capacity, and available resources [20].

Table 3.1 Summarizes some common extraction techniques used in processing orange fruits to increase juice production

Extraction Technique	Description
Mechanical Pressing	Involves applying pressure to orange fruits to rupture juice sacs and release juice, can be performed using hydraulic or screw presses.
Reamer or Hand Squeezing	Manual method where orange fruits are squeezed by hand or pressed against a reamer to extract juice, often used for small-scale operations or home use.
Enzymatic Treatment	Enzymes such as pectinase or cellulase are used to break down cell walls and release juice from orange fruit tissues, can enhance juice yield and quality.
Steam Distillation	Utilizes steam to heat orange peel, releasing essential oils which can be collected separately from juice. Often used for extraction of essential oils rather than juice.
Pneumatic Pressing	Pneumatic or air-driven presses exert pressure on oranges to extract juice efficiently. Can be gentle on fruit tissues, resulting in higher juice quality.
Ultrasonic Extraction	Ultrasonic waves are used to disrupt cell structures in oranges facilitating juice extraction. Can enhance juice yield and reduce processing time.
Cold-Press Extraction	Involves pressing oranges at low temperatures to minimize heat damage and preserve juice quality. Can result in higher nutrient content in juice.
Centrifugal Extraction	Oranges are crushed and spun at high speeds in a centrifuge, separating juice from pulp and solids. Often used in large-scale juice processing plants.

3.4 Utilization of Orange Peels

The utilization of orange peels is a multifaceted topic, with various industries finding innovative ways to repurpose this waste product [12]. Here are some common and emerging applications as shown in Table 3.2:

(a) **Composting**

Utilizing orange peels in composting is a sustainable approach to manage organic waste while enriching soil with valuable nutrients.

1. Nutrient Content of Orange Peels:

   Orange peels are rich in organic matter, nitrogen, phosphorus, potassium, and various micronutrients. Their high carbon-to-nitrogen ratio makes them ideal components for composting [18, 8].
2. Composting Process:

   Orange peels can be composted using traditional aerobic composting methods, where organic materials are decomposed by microorganisms in the presence of oxygen. Proper aeration, moisture, and carbon/nitrogen balance are essential for successful composting [10].

Table 3.2 Utilization of orange peels [6]

Utilization of Orange Peels	Description
Food Industry	Pectin Extraction: Oranges are rich in pectin, a polysaccharide used as a gelling agent in food products like jams, jellies, and desserts. Flavorings and Essences: Essential oils extracted from oranges are used as flavorings in beverages, confectionery, and baked goods. Dietary Fiber: Dried and powdered orange peels are used as a source of dietary fiber in functional food products.
Animal Feed	Orange peels can be dried, ground, and incorporated into animal feed as a source of nutrients and dietary fiber for livestock and poultry.
Biofuel Production	Orange peels contain fermentable sugars that can be converted into biofuels such as ethanol through processes like enzymatic hydrolysis and fermentation.
Pharmaceuticals and Cosmetics	Orange peel extracts are utilized in the pharmaceutical and cosmetic industries for their antioxidant, antimicrobial, and skin-brightening properties. Compounds like limonene extracted from oranges are used in the production of fragrances and cleaning products.
Biodegradable Packaging	Orange peel extracts can be used to produce biodegradable packaging materials as an eco-friendly alternative to conventional plastics.
Soil Amendment	Composted orange peels can be used as organic soil amendments to improve soil structure, moisture retention, and nutrient content in agricultural and horticultural applications.
Wastewater Treatment	Orange peel extracts have been studied for their potential use in wastewater treatment processes due to their adsorption properties for heavy metals and organic pollutants.
Art and Crafts	Dried orange peels can be used in art and craft projects for their decorative and aromatic qualities, such as potpourri or homemade candles.

3. Benefits of Orange Peel Compost:

   Composting orange peels yields a nutrient-rich soil amendment that improves soil structure, fertility, and water retention. It enhances microbial activity, suppresses plant diseases, and reduces the need for chemical fertilizers [13].
4. Considerations and Challenges:

   Orange peels are acidic, which may affect the pH of compost. To mitigate this, they can be balanced with alkaline materials such as agricultural lime or supplemented with other organic wastes [4].

(b) **Anaerobic Digestion**

Anaerobic digestion of orange peels provides a sustainable solution to waste management while producing valuable products such as biogas and organic fertilizers. Biogas production: Orange peels are rich in organic matter, consisting mainly of carbohydrates such as pectin and cellulose. Anaerobic digestion breaks down these complex compounds through microbial activity, producing biogas as a byproduct. Biogas is a renewable energy source consisting mainly of methane and carbon dioxide, and can be used to generate electricity, heating, or as a fuel for cars.

1. Organic Fertilizer: The digested material remaining after anaerobic digestion is a nutrient-rich organic fertilizer. It contains essential macro and micronutrients, improving soil structure and fertility. Application of orange peel digestate enhances the organic matter content of the soil and promotes plant growth, reducing dependence on chemical fertilizers.
2. Process optimization: Various factors affect the efficiency of anaerobic digestion of orange peels, including substrate composition, operating conditions (temperature, pH, retention time), and microbial consortium. Optimization studies help enhance biogas production and digestate quality, ensuring economic viability and environmental sustainability [19].
3. Technological Innovations: Advances in anaerobic digestion technologies, such as high-speed reactors, continuous stirred tank reactors (CSTR), and anaerobic membrane bioreactors (AnMBR), provide efficient processing of citrus peels while improving biogas production and digestate quality [3].
4. Environmental Benefits: Anaerobic digestion of orange peels reduces greenhouse gas emissions by diverting organic waste from landfills and reducing methane generation during decomposition. In addition, they contribute to the circular economy by converting waste into valuable resources and promoting sustainable practices in the food industry [25].

(c) **Extraction of Valuable Compounds**

Orange peels are rich sources of bioactive compounds such as flavonoids, limonoids, phenolic acids, and essential oils, which have various applications in food, pharmaceutical, cosmetic, and nutraceutical industries. Here's an overview of the utilization of orange peels and the extraction of valuable compounds: [14]. Essential Oils Extraction: Orange peels contain essential oils that are extracted mainly through techniques such as steam distillation, cold pressing, and solvent extraction. These essential oils are used in flavoring, fragrance, and pharmaceutical industries.

1. Flavonoids Extraction: Flavonoids are polyphenolic compounds found in orange peels with antioxidant, anti-inflammatory, and anticancer properties. Techniques such as solvent extraction, ultrasound-assisted extraction, and supercritical fluid extraction are employed for their extraction.
2. Limonoids Extraction: Limonoids are unique phytochemicals present in orange peels with potential health benefits including anticancer and cholesterol-lowering effects. Extraction methods include solvent extraction, solid-phase extraction, and chromatographic techniques.
3. Pectin Extraction: Orange peels are a rich source of pectin, a polysaccharide widely used in the food industry as a gelling agent, thickener, and stabilizer. Pectin extraction is usually carried out using acid or enzyme-assisted extraction methods.
4. Phenolic Compounds Extraction: Orange peels contain various phenolic compounds with antioxidant and anti-inflammatory properties. Extraction techniques for phenolic compounds include solvent extraction, microwave-assisted extraction, and enzyme-assisted extraction.
5. Biofuel Production: Orange peels can also be utilized for biofuel production due to their high content of sugars and cellulose. Various pretreatment methods are employed to enhance enzymatic hydrolysis and fermentation processes for biofuel production.

3.5 Environmental Impacts and Considerations

Managing orange peel waste involves several environmental considerations due to its large-scale generation and potential impact on ecosystems, soil, water, and air quality. Here are some key environmental impacts and considerations [26].

1. Greenhouse Gas Emissions: Orange peel waste decomposition releases methane, a potent greenhouse gas. Landfilling or improper disposal of citrus peels can contribute to greenhouse gas emissions.
2. Soil and Water Pollution: Dumping orange peels in landfills can lead to leachate production, which may contaminate soil and groundwater with organic acids and other compounds. Implementing composting or anaerobic digestion can reduce the potential for pollution and provide nutrient-rich soil amendments
3. Loss of Biodiversity: Improper disposal of orange peel waste in natural habitats can disrupt ecosystems, affecting native flora and fauna. Developing effective waste management strategies, such as composting or utilizing orange peels for animal feed, can mitigate these impacts.
4. Resource Consumption: Landfilling orange peels requires significant land and resources for disposal. Implementing circular economy principles, such as

recycling orange peels for value-added products like pectin, essential oils, or biofuels, can reduce resource consumption and promote sustainability

5. Air Quality: Open-air disposal or burning of orange peel waste can emit air pollutants such as particulate matter and volatile organic compounds. Implementing controlled combustion or utilizing advanced conversion technologies like pyrolysis or gasification can minimize air pollution.

By addressing these environmental considerations through sustainable waste management practices, the negative impacts of orange peel waste on the environment can be minimized while unlocking its potential as a valuable resource.

3.6 Economic Feasibility Analysis

Conducting an economic feasibility analysis of orange peel waste management involves assessing various factors such as the cost of waste disposal, potential revenue streams from waste utilization, and the overall economic viability of different waste management strategies [17]. Here's a structured approach to conducting such an analysis:

1. Cost of Waste Disposal:
   - Determine the current cost incurred in disposing of orange peel waste. This may include transportation, landfill fees, and any other associated costs.
   - Consider any potential future increases in disposal costs due to environmental regulations or changes in waste management policies.
2. Revenue Streams from Waste Utilization:
   - Investigate potential revenue streams from utilizing orange peel waste. This could include:
   - Production of animal feed: Orange peel can be dried and used as feed for livestock.
   - Extraction of essential oils: Orange peel contains valuable essential oils that can be extracted and sold to industries such as cosmetics, food flavoring, and pharmaceuticals.
   - Biofuel production: Orange peel waste can be converted into biofuels through processes such as anaerobic digestion or fermentation.
   - Dietary fiber production: Orange peel contains dietary fiber which can be extracted and used in food products.
   - Estimate the potential revenue from each of these streams based on market prices and demand.
3. Investment and Operational Costs:
   - Analyze the investment required to implement different waste management strategies. This includes infrastructure costs, equipment costs, and operational expenses.

- Estimate ongoing operational costs such as labor, maintenance, and raw material procurement.

4. Market Analysis:
    - Conduct a market analysis to assess the demand for products derived from orange peel waste. Identify potential customers and market trends that could impact demand and pricing.
    - Consider the competitiveness of the market and the potential for market growth or saturation.
5. Risk Assessment:
    - Identify and assess the risks associated with each waste management strategy. This may include market risks, regulatory risks, and technical risks.
    - Develop mitigation strategies to address these risks and minimize their impact on the project's economic feasibility.
6. Financial Projections:
    - Develop financial projections for each waste management strategy, including revenue forecasts, expense forecasts, and cash flow projections.
    - Calculate key financial metrics such as net present value (NPV), internal rate of return (IRR), and payback period.
7. Risk Analysis:
    - Identify potential risks and uncertainties associated with citrus peel waste management. This could include market volatility, regulatory changes, technological limitations, and environmental concerns.
    - Develop strategies to mitigate these risks and assess their impact on the project's feasibility.
8. Sensitivity Analysis:
    - Conduct sensitivity analysis to assess how changes in key variables, such as feedstock prices, product demand, and operating costs, would affect the project's financial performance.

3.7 Technological Considerations

Managing orange peel waste involves several technological considerations to minimize environmental impact and potentially create value from what would otherwise be discarded material. Here are some technological aspects to consider.

1. Biological Treatment: Utilizing microorganisms such as bacteria, fungi, or enzymes to break down orange peels through processes like composting, anaerobic digestion, or fermentation [21].

2. Extraction of Essential Oils: Employing techniques such as steam distillation or solvent extraction to extract essential oils from orange peels, which have various industrial applications [23].
3. Production of Biofuels and Biochemicals: Converting orange peel waste into biofuels (e.g., ethanol) or biochemicals (e.g., citric acid) through processes like enzymatic hydrolysis or fermentation.
4. Utilization in Functional Foods and Nutraceuticals: Incorporating orange peel extracts or powders rich in bioactive compounds (e.g., flavonoids, limonoids) into functional foods or nutraceuticals [24].
5. Application in Biodegradable Materials: Utilizing orange peel extracts or cellulose derived from citrus peels in the production of biodegradable materials such as bioplastics or packaging films.
6. Adsorption and Water Treatment: Using orange peel waste as an adsorbent material for removing pollutants from water and wastewater through processes like adsorption or biosorption [22].

These technological considerations demonstrate the diverse ways in which orange peel waste can be managed sustainably while potentially generating economic value.

3.8 Conclusion and Future Directions

The management of orange peel waste presents a multifaceted challenge with significant environmental, economic, and social implications. Over the years, various approaches have been explored to mitigate the adverse effects of orange peel waste, ranging from traditional disposal methods to innovative valorization techniques. Through this review, it becomes evident that orange peel waste possesses substantial potential as a valuable resource rather than a mere byproduct.

Several studies have demonstrated the feasibility of utilizing orange peel waste for the extraction of valuable compounds such as essential oils, pectin, flavonoids, and dietary fibers. Additionally, the utilization of orange peel waste as a feedstock for biofuel production, composting, and animal feed supplementation showcases its versatility in different sectors. Furthermore, the application of orange peel waste in wastewater treatment, soil improvement, and as a source of functional food ingredients underscores its significance in addressing environmental challenges and promoting sustainable development.

However, despite the progress made in orange peel waste management, several challenges and opportunities remain. The scalability and economic viability of certain valorization techniques warrant further investigation, particularly in the context of small-scale industries. Moreover, the development of efficient collection, transportation, and processing systems is crucial to streamline orange peel waste utilization practices and maximize resource recovery. Additionally, comprehensive life cycle assessments are essential to evaluate the environmental impact and sustainability of different orange peel waste management strategies.

In conclusion, the effective management of orange peel waste requires a holistic approach encompassing technological innovation, policy support, and stakeholder engagement. By harnessing the potential of citrus peel waste as a valuable resource, it is possible to create a more sustainable and circular citrus industry while simultaneously mitigating environmental pollution and fostering socio-economic development.

Below are some future directions which can be taken to mitigate waste production of orange peels:

1. Technological Innovation: Continued research and development efforts are needed to explore novel technologies for the efficient extraction of valuable compounds from orange peel waste. Advanced extraction techniques, such as supercritical fluid extraction and enzymatic hydrolysis, hold promise for improving extraction yields and product quality.
2. Waste Valorization: Further investigation into the potential applications of orange peel waste, such as its utilization in bioplastics, nutraceuticals, and pharmaceuticals, can open up new avenues for waste valorization. Collaborative research initiatives between academia, industry, and government agencies can facilitate the development and commercialization of innovative citrus peel waste-derived products.
3. Circular Economy Initiatives: Implementation of circular economy principles, including closed-loop systems and resource recovery strategies, can enhance the sustainability of orange peel waste management practices. Initiatives aimed at promoting waste-to-energy conversion, nutrient recycling, and material reuse can contribute to the development of a circular orange industry.
4. Policy Support: Policy interventions, such as tax incentives, subsidies, and regulatory frameworks, can incentivize investments in orange peel waste management infrastructure and technology. Additionally, government support for research and development initiatives focused on waste valorization and sustainable agriculture can foster innovation in the orange sector.
5. Public Awareness and Education: Increasing public awareness about the environmental and economic benefits of orange peel waste valorization is essential for garnering support and participation from stakeholders. Educational campaigns, community outreach programs, and consumer awareness initiatives can help promote responsible waste management practices and stimulate demand for citrus peel waste-derived products.

By addressing these future directions, stakeholders can work towards creating a more sustainable and resource-efficient citrus industry while simultaneously mitigating environmental pollution and promoting socio-economic development.

References

1. Chaemsanit, S., Chiewchan, N., & Devahastin, S. (2016). Production of dietary fiber powders from orange peel and lime pomace: Effects of drying methods on physicochemical and functional properties. *LWT-Food Science and Technology, 66*, 484–490.
2. Chhaya, U., Godhani, N., & Pandya, A. (2019). Extraction of Pectin from citrus fruit peels and utilization of Pectin as pharmaceutical excipients: A review. *Pharmacognosy Reviews, 13*(25), 114–121. https://doi.org/10.4103/phrev.phrev_25_19
3. Cuetos, M. J., et al. (2016). Anaerobic co-digestion of organic wastes: Performance and kinetic evaluation. *Waste Management, 49*, 536–544.
4. Ehsani, A., & Jasour, M. S. (2019). Valorization of citrus peel waste as a source of high added value compounds: A review. *Journal of Cleaner Production, 234*, 963–972.
5. Capanoglu, E., Nemli, E., & Tomas-Barberan, F. (2022). Novel approaches in the valorization of agricultural wastes and their applications. *Journal of Agricultural and Food Chemistry, 70*(23), 6787–6804.
6. Giwa, S. O., Muhammad, M., & Giwa, A. (2018). Utilizing orange peels for essential oil production. *Journal of Engineering and Applied Sciences, 13*(1), 17–27.
7. El-Banna, H. A., El-Adawy, M. M., Shehata, A. M., & El-Shenawy, S. M. (2020). Utilization of orange peel waste powder in beef burger: Impact on quality attributes, nutritional value, and sensory properties. *Food, 9*(10), 1488. https://doi.org/10.3390/foods9101488
8. Jebreen, A., & Anzam, N. (2024). Enriched of Orange Jam (Marmalade) with vitamin C. Journal of Agri-Food Science and Technology, *(5*(2), 94–101.
9. Garcia-Castello, E. M., Rodriguez-Lopez, A. D., Mayor, L., & Martinez-Navarrete, N. (2012). Influence of grapefruit albedo dietary fibre on the quality of low-fat meatballs. *Food Chemistry, 134*(1), 1–7. https://doi.org/10.1016/j.foodchem.2012.01.089
10. García-Gómez, A., Báez-Benítez, G., Martín-Cabello, G., & Moreno Casco, J. (2016). Agricultural use of citrus peels: Present and future perspectives. *Waste and Biomass Valorization, 7*(4), 847–856. https://doi.org/10.1007/s12649-016-9691-3
11. Gómez, P. A., & Lorenzo, J. M. (2018). Effect of the addition of citrus fibre on the quality characteristics of pork burger. *LWT - Food Science and Technology, 91*, 552–558. https://doi.org/10.1016/j.lwt.2018.01.026
12. Lashermes, G., Houot, S., Barriuso, E., Steyer, J. P., & Mosnier, A. (2009). Influence of organic waste characteristics on their suitability for composting. *Journal of Environmental Quality, 38*(4), 1393–1404. https://doi.org/10.2134/jeq2008.0300
13. Ogunwande, G. A., Osunsami, O. O., & Oni, P. I. (2003). Kinetics of biodegradation of some citrus fruit wastes during composting. *Bioresource Technology, 88*(2), 143–149. https://doi.org/10.1016/S0960-8524(03)00046-2
14. Li, Q., Putra, N. R., Rizkiyah, D. N., Abdul Aziz, A. H., Irianto, I., & Qomariyah, L. (2023). Orange pomace and peel extraction processes towards sustainable utilization: A short review. *Molecules, 28*(8), 3550.
15. Puértolas, E., Cárcel, J. A., & Barba, F. J. (2016). How pulsed electric fields affect the quality of fresh and processed fruit and vegetables: A review. *Critical Reviews in Food Science and Nutrition, 56*(1), 98–115. https://doi.org/10.1080/10408398.2012.701197
16. Ribeiro, J. S., Fernandes, F. A. N., Rodrigues, S., & Fernandes, A. N. (2017). Effects of ultrasound and enzymatic treatments on juice extraction and quality of strawberry and cashew apple juices. *Food and Bioprocess Technology, 10*(3), 536–547. https://doi.org/10.1007/s11947-016-1821-4
17. Negro, V., Ruggeri, B., Fino, D., & Tonini, D. (2017). Life cycle assessment of orange peel waste management. *Resources, Conservation and Recycling, 127*, 148–158.
18. Shiralipour, A., & McConnell, D. B. (2002). Composting citrus processing wastes. *Biocycle, 43*(10), 58–59.
19. Ismail, A. M. S. (1996). Utilization of orange peels for the production of multienzyme complexes by some fungal strains. *Process Biochemistry, 31*(7), 645–650.

20. Ángel Siles López, J., Li, Q., & Thompson, I. P. (2010). Biorefinery of waste orange peel. *Critical Reviews in Biotechnology, 30*(1), 63–69.
21. Zhang, Y., He, Y., & Zhang, Y. (2020). Microbial transformation of citrus peel waste: A review. *Journal of Bioscience and Bioengineering, 130*(1), 1–9.
22. Zhou, Y., & Zhu, H. (2017). Study on adsorption and decolorization properties of citrus peel waste for dye wastewater treatment. *Water, Air, & Soil Pollution, 228*(11), 428.
23. Gavahian, M., Chu, Y. H., & Mousavi Khaneghah, A. (2019). Recent advances in orange oil extraction: An opportunity for the valorisation of orange peel waste a review. *International Journal of Food Science & Technology, 54*(4), 925–932.
24. Mahato, N., Sharma, K., Sinha, M., & Cho, M. H. (2018). Citrus waste derived nutra–/pharmaceuticals for health benefits: Current trends and future perspectives. *Journal of Functional Foods, 40*, 307–316.
25. Mohsin, A., Hussain, M. H., Zaman, W. Q., Mohsin, M. Z., Zhang, J., Liu, Z., et al. (2022). Advances in sustainable approaches utilizing orange peel waste to produce highly value-added bioproducts. *Critical Reviews in Biotechnology, 42*(8), 1284–1303.
26. Joglekar, S. N., Pathak, P. D., Mandavgane, S. A., & Kulkarni, B. D. (2019). Process of fruit peel waste biorefinery: A case study of citrus waste biorefinery, its environmental impacts and recommendations. *Environmental Science and Pollution Research, 26*, 34713–34722.

Chapter 4
Valorizing Citrus Waste: Innovative Strategies for Food Safety, Economic Viability, and Environmental Sustainability

Osman Tuncay Agar, Ali Imran, and Hafiz Suleria

4.1 Introduction

Citrus Processing Waste (CPW), a major byproduct of the citrus industry, is predominantly high in water content and essential oils. The global production of CPW is estimated to be in the millions of tons annually, posing significant economic and environmental challenges related to its disposal. Traditionally, CPW has been utilized mainly as animal feed, capitalizing on its nutritional content, and less frequently as an organic soil conditioner or for composting. However, the concept of valorization transforming waste into valuable products offers an expanded range of uses that provide substantial environmental and economic benefits [1, 2].

The global citrus market is significant, with over 120 million tons of citrus fruits produced annually, resulting in up to 20 million tons of residues each year. This by-product of the citrus industry poses both environmental and economic challenges due to its high organic waste content. However, these residues are also a valuable

O. T. Agar (✉)
School of Agriculture, Food and Ecosystem Sciences, Faculty of Science,
The University of Melbourne, Parkville, VIC, Australia

Department of Pharmacognosy, Faculty of Pharmacy, Suleyman Demirel University,
Isparta, Turkiye
e-mail: tuncayagar@sdu.edu.tr

A. Imran
Department of Food Science, Government College University Faisalabad,
Faisalabad, Pakistan

H. Suleria
School of Agriculture, Food and Ecosystem Sciences, Faculty of Science,
The University of Melbourne, Parkville, VIC, Australia

A. Chauhan et al. (eds.), *Valorization of Citrus Food Waste*,
https://doi.org/10.1007/978-3-031-77999-2_4

source of bioactive compounds such as essential oils, phenolic acids, flavonoids, carotenoids, and pectin, which can be valorized for various applications, including bioenergy and biochemical production [3, 4].

Advancements in technology have expanded the potential applications of CPW. Techniques such as microwave, ultrasound, and pulsed electric fields have facilitated the extraction of valuable compounds like pectin, dietary fibers, flavonoids, and citric acid, which are in growing demand across the food, cosmetics, and pharmaceutical sectors. Additionally, innovative uses of CPW in the production of biomethane and bioethanol have been developed, addressing the challenges posed by the toxic effects of essential oils on the microbial communities vital for these processes [3, 5].

Recent advances have focused on environmentally friendly extraction methods like microwave and ultrasound-assisted extractions to improve the valorization process. These methods not only enhance the efficiency of extracting valuable compounds but also align with green chemistry principles, reducing energy consumption and minimizing solvent use [1]. Valorization techniques can convert citrus waste into products such as biofuels, nanoparticles, biosorbents, and biofertilizers, contributing to a bio-economy and mitigating environmental impacts [6].

Despite these technological advances, several challenges persist. Economic sustainability and the inhibitory effects of components such as limonene on energy valorization processes remain significant obstacles. Moreover, the adoption of green technologies for extraction not only highlights the effectiveness and environmental benefits of these methods but also underscores the profitability of extracting high-quality bioactive compounds [2, 6].

The valorization of citrus waste not only provides an opportunity to tackle the significant environmental issues associated with citrus residue disposal but also represents a sustainable step towards circular economic models, where waste materials are converted into valuable resources. Moreover, the valorization of CPW encompasses not only environmental and technological benefits but also significant socio-economic impacts. The comprehensive utilization of CPW can spur rural development by creating job opportunities and enhancing local economies. This approach aligns with broader sustainable development goals, aiming to integrate environmental stewardship with economic and social growth [1].

This chapter will delve into various innovative and sustainable valorization strategies for CPW, emphasizing their role in enhancing food safety and contributing to environmental sustainability and economic growth. It will particularly focus on the comprehensive application of green technologies in the citrus waste sector and include case studies such as the anaerobic digestion of orange peel. These strategies not only mitigate environmental pollution but also facilitate the recovery of valuable resources like biomethane, thereby promoting a bioeconomy.

4.2 Characteristics of Citrus Processing Waste (CPW)

Citrus Processing Waste (CPW), the by-product of the citrus processing industry, is primarily composed of peels, seeds, and pulp. Characterized by high water content and substantial concentrations of essential oils, CPW presents both economic and environmental management challenges. The composition of CPW can vary seasonally, which necessitates adaptive storage solutions to effectively manage this variability. CPW is rich in biologically valuable compounds such as carotenoids, polyphenols, essential oils, and pectins, holding for applications in the food, pharmaceutical, and cosmetic industries. However, the disposal of these bioactive compounds presents significant economic and environmental challenges. The process of anaerobic digestion, particularly of orange peel, demonstrates the potential of CPW for energy resource recovery, transforming waste into biomethane and exemplifying its role in sustainable waste management practices [5, 7].

Improper disposal of CPW can lead to significant environmental pollution and pose health hazards. Yet, its valorization, aligned with circular economy principles, aims for zero waste and the recovery of valuable components. Despite its economic potential, challenges remain in the efficient and sustainable extraction and utilization of these compounds. Consequently, there is a shift towards exploring greener and more effective extraction techniques. Continuous research and development underscore the necessity for innovative approaches to manage and valorize CPW, transforming it from a by-product into a resource for bio-based products and energy [8–10].

The seasonal nature of citrus fruit harvesting and processing directly affects the production and availability of CPW, leading to fluctuations in waste quantities. This variability often necessitates the implementation of strategic storage solutions to manage the waste effectively. Efficient storage and processing methods are crucial to preserve the quality and utility of CPW for various applications [5, 11].

The high water content and essential oil content significantly influence the handling and potential valorization of CPW. High water content complicates waste processing and storage, requiring methods to reduce moisture levels for efficient utilization or disposal. Conversely, the essential oils in CPW, particularly in the peels, while posing challenges, also offer valuable opportunities. Although these oils can inhibit certain biochemical processes, such as biomethane production, they hold potential for various industrial applications. Extracting these oils through environmentally friendly methods can significantly enhance the economic value of CPW, converting a waste product into a valuable resource. It is critical to understand and manage the water content and essential oils in CPW to maximize its utility and minimize environmental impacts [12, 13].

Recent advances in CPW valorization focus on green extraction techniques such as microwave and ultrasound-assisted extractions, which improve process efficiency and environmental sustainability. Integrating these techniques into a biorefinery model addresses CPW's economic and environmental challenges, promoting sustainable waste management [1].

González-Miquel and Díaz (2020) highlight the use of citrus waste for extracting bioactive compounds using sustainable methods. They emphasize the potential of citrus waste as a source of valuable compounds, including essential oils, phenolic acids, flavonoids, and carotenoids, and note the necessity for innovative extraction processes to replace traditional methods that frequently employ hazardous solvents [3].

Santiago et al. (2020) evaluate the environmental aspects of valorizing citrus waste into D-limonene, an essential oil with wide-ranging industrial uses. They assess various extraction and purification methods to optimize the environmental profile of the processes, highlighting the importance of environmental considerations in CPW valorization strategies, particularly for high-value compounds such as D-limonene [14].

For a succinct summary of CPW valorization methods and their environmental impacts, Table 4.1 provides a consolidated overview, showcasing the transition of CPW from by-product to a valuable resource in sustainable production systems.

Table 4.1 Overview of CPW Valorization Techniques and Outputs

Technique	Output/Product	Applications	Environmental Impact
Anaerobic digestion	Biomethane	Energy production	Reduces organic waste, produces renewable energy, lowers greenhouse gas emissions
Microwave-assisted extraction	Pectin, Flavonoids	Food preservatives, Pharmaceuticals	Less energy-intensive, minimizes solvent use, lowers carbon footprint
Ultrasound-assisted extraction	Essential oils	Cosmetics, Food flavoring	Eco-friendly, enhances yield and purity, reduces energy consumption
Supercritical fluid extraction	High-purity extracts	Food additives, Pharmaceuticals	Eliminates toxic solvent use, high efficiency, environmentally benign
Precomposting and Vermicomposting	Vermicompost	Soil conditioner, Organic fertilizer	Enhances soil quality, supports sustainable agriculture
Bioconversion to Biofuels	Ethanol, other biofuels	Renewable energy sources	Reduces dependency on fossil fuels, supports energy sustainability
Extraction of Bioactive Compounds	Limonene, other compounds	Nutraceuticals, Cosmetics	Reduces chemical use, employs green extraction technologies
Drying for Biomass Energy	Dried biomass	Bioenergy production	Provides a renewable source of energy, reduces waste volume
Activated Carbon Production	Activated carbon	Water treatment, adsorbent	High sorption capacity, useful in pollution control

4.3 Current Uses of Citrus Waste

Citrus Waste, particularly Citrus Peel Waste (CPW), plays a significant role in sustainable waste management practices. As the primary residue of citrus processing, CPW is a nutrient-rich by-product, providing an environmentally friendly and cost-effective option for animal feed. This application not only alleviates the environmental burden associated with waste disposal but also adds considerable value to the agricultural sector by offering a nutritious feed source [5].

Beyond its use in animal feed, CPW is employed as an effective organic soil conditioner. Its application, especially in regions proximate to citrus processing facilities, enhances the organic matter content in the soil. Utilizing CPW for compost production is a viable method for improving soil quality, infusing the soil with essential nutrients, and enhancing its physical properties. These improvements lead to healthier crop growth and advance sustainable agricultural practices. The positive impact of citrus waste on soil conditioning is supported by studies demonstrating its beneficial effects on soil parameters and crop productivity, thus highlighting CPW's agricultural value [15, 16].

Transitioning from these traditional uses, CPW is increasingly explored for its potential in bioenergy production, particularly for biomethane and bioethanol. Anaerobic digestion of CPW, especially orange peel, presents a promising approach for biomethane production. This process, involving the breakdown of organic matter in an oxygen-free environment, results in methane generation. However, overcoming the inhibitory effects of essential oils, such as limonene, on the microbes involved in methane production is crucial for optimizing biomethane yield from CPW [17, 18]. Additionally, the potential for bioethanol production from CPW is being examined, leveraging its high sugar content. Yet, fermentation inhibitors like D-limonene require removal for efficient bioethanol production. Advances in pretreatment and fermentation technologies are critical for enhancing the conversion efficiency of CPW into bioethanol, providing a sustainable alternative to fossil fuels [10].

Vermicomposting represents an innovative approach to citrus waste valorization. By introducing precomposting and vermicomposting processes, it's possible to degrade orange peel waste efficiently using earthworms and microorganisms, further enhancing its value as a bioresource. The addition of a balanced mix of vegetable waste and eggshells to the orange peel waste optimizes the nutrient content for earthworm activity, ensuring the production of high-quality vermicompost. This method provides significant environmental benefits, including substantial reductions in waste and improvements in soil health [19].

Recent studies highlight the industrial potential of biological valorization of CPW into fuels and chemicals, emphasizing its role as an alternative biomass due to its low lignin content. Technical challenges, such as the high pectin content that consists of non-fermentable sugars mainly galacturonic acid, are noted. Metabolic engineering of yeast and other microbial strains to ferment CPW-derived sugars into value-added products such as ethanol and mucic acid is being actively explored, underscoring the need for further research to improve fermentation efficiency and diversify product profiles [2].

Overall, the utilization of CPW spans a range of sustainable practices, from traditional uses like animal feed and soil conditioning to innovative applications in bioenergy production and vermicomposting. By leveraging CPW for biomethane and bioethanol production, as well as for enhancing soil health through composting and vermicomposting, these initiatives contribute significantly to sustainable agricultural practices, waste reduction, and renewable energy solutions. Such diverse applications underscore CPW's role in promoting environmental sustainability and exemplify the principles of the circular economy. The continued exploration of CPW's potential in biological valorization into fuels and chemicals further highlights the need for ongoing research to optimize these processes and expand the product portfolio, reinforcing its value as an alternative biomass resource.

4.4 Valorization for Food Safety

In the realm of food safety, the innovative valorization of citrus waste, specifically peels, seeds, and pulp, is increasingly vital. These by-products are rich in bioactive compounds such as pectins, antioxidants, and essential oils, which are highly valued in food applications for their antimicrobial and antioxidant properties [3].

Recent advancements in green extraction technologies have highlighted environmentally friendly and efficient methods for extracting these bioactive compounds. Techniques such as microwave-assisted extraction, ultrasound-assisted extraction, and pulsed electric fields have proven effective. These methods are preferred over traditional ones due to their lower energy consumption and minimal use of toxic solvents, making them sustainable and economically viable [6].

Additionally, techniques like supercritical fluid extraction and pressurized water extraction have been explored for their efficiency in extracting high-value compounds from citrus wastes. These techniques are notable for the yield and purity of the compounds they extract, which are critical for food safety applications [8].

In the food industry, these extracted bioactive compounds find significant applications as natural preservatives, flavor enhancers, and ingredients in functional foods. Their antimicrobial and antioxidant properties are instrumental in extending the shelf life of food products and enhancing their nutritional profile [20].

The valorization of citrus waste through these innovative extraction techniques not only supports environmental sustainability by reducing waste but also bolsters food safety by providing high-quality bioactive compounds for the food industry. This approach aligns with the increasing demand for natural and sustainable ingredients in food production. Furthermore, the utilization of citrus seeds and peels in food applications is an evolving field, offering new avenues to enhance food safety and sustainability. Citrus seeds, particularly from mandarin and grapefruit, are rich in fatty acids, phytosterols, tocopherols, and other bioactive compounds. These components are used for producing food enhancers and antioxidants. Optimizing extraction conditions is crucial for improving yield and minimizing environmental impacts [21].

Citrus peels, containing flavonoids, limonoids, alkaloids, essential oils, and pectin, hold great potential in food product development. Their health-promoting effects make them suitable for use in functional foods, medicines, and even biofuels, demonstrating their versatility [22].

The recent research highlights the importance of fruit waste valorization, where tremendous amounts of by-products from fruit production and processing sectors are utilized to reduce economic losses and environmental impact. These by-products contain a variety of valuable bioactive compounds, such as dietary fiber, flavonoids, and antioxidants, which are effectively used as value-added products in various industrial applications. The review discusses advancements in extraction techniques and the significant role of these bioactive compounds in developing nutraceutical products, functional foods, and food additives, enhancing food safety and promoting sustainable food production practices [23].

In summary, the valorization of citrus seeds and peels for food safety applications marks a notable progression in sustainable food production. Harnessing their bioactive compounds, these by-products offer health benefits and aid in the development of functional foods, thereby promoting a more sustainable and health-conscious approach in the food industry.

4.5 Safety and Regulatory Considerations

In the pursuit of valorizing food by-products, particularly those derived from citrus processing, ensuring the safety of the resultant materials for human and animal consumption is of utmost importance. This process involves the transformation of waste materials into consumable products—an economically and environmentally commendable goal. However, this transformation raises significant concerns about the potential introduction of chemical contaminants or pathogens, which could pose serious health risks [1, 24].

4.5.1 Safety Evaluation of Valorized By-Products

The comprehensive safety evaluations of these valorized by-products are critical. Such evaluations should include detailed physicochemical and microbiological assessments, alongside rigorous screenings for toxic contaminants. These are indispensable for confirming the suitability of valorized by-products for consumption. Moreover, developing effective decontamination methods, adopting processing techniques that inhibit the generation of harmful substances, and implementing robust analytical methods constitute essential elements of these evaluations. However, the complexities associated with these assessments, coupled with a lack of specific legislation addressing the safety and suitability of such new products, present substantial challenges in safeguarding consumer health [24].

4.5.2 Regulatory Framework and Compliance

Adherence to existing regulatory frameworks and guidelines is critical for the safe use of animal by-products in feed materials. The presence of contaminants, such as heavy metals or residual chemicals, must be meticulously monitored. Legislation typically imposes stringent requirements for handling by-products, striving to balance environmental protection with the promotion of agricultural and industrial advancement. These regulations are fundamental in ensuring the biological safety of both the environment and end consumers [25].

4.5.3 Challenges in Quality Control

The quality control of animal products, including those derived from by-products, is a complex and highly regulated endeavor. This control process spans the assessment of chemical residues for safety and, in the case of antimicrobial products, the microbiological safety—including monitoring changes in bacterial pathogen loads and resistance patterns. Developing analytical methods that can distinguish between authorized and unauthorized feed materials remains a significant challenge in ensuring the safety of food and feed [26].

4.5.4 International and National Regulatory Frameworks

The regulatory landscape for food safety is shaped significantly by international standards, such as those set by the FAO/WHO Food Standards Programme, which includes the Codex Alimentarius Commission (Codex). These guidelines influence global trade in agricultural products and the food safety policies of World Trade Organization (WTO) member countries. Nonetheless, the implementation and enforcement of these guidelines are subject to national legislation and regulations. The approaches to food safety in countries like France and the U.S. exemplify how international standards are adapted and enforced at the national level [27].

4.5.5 *Private Food Safety Standards in the EU*

Within the European Union (EU), food safety regulation comprises a complex amalgam of public laws and private food standards. Notable private standards such as BRC, IFS, and FSSC 22000, initiated by retailers and manufacturers, serve to complement public regulations. While not legally binding, these standards often become essential requirements in specific markets, enhancing public standards and verified by accredited third-party certification bodies. The EU legal framework fosters the proliferation of these private standards, which are pivotal in modernizing food regulation and enhancing safety [28].

Ensuring the safety of valorized by-products for consumption involves navigating a complex landscape of international and national regulations, as well as private standards. This requires a concerted effort from various stakeholders, including regulatory bodies, industry players, and certification agencies, to ensure compliance, quality control, and consumer safety in the food supply chain.

4.6 Sustainability and Environmental Impact of Citrus Waste Valorization

The valorization of Citrus Processing Waste (CPW), including peels, seeds, and pulp, represents a significant advancement toward enhancing environmental sustainability and reducing the ecological footprint of the citrus processing industries (Fig. 4.1). The seasonal production, high water content, and presence of essential oils in CPW pose substantial economic and environmental challenges. Yet, the valorization of this waste material offers extensive environmental benefits, marking it as a crucial element of a sustainable bioeconomy [1, 2].

4.6.1 *Reduction in Environmental Pollution*

Currently, CPW is primarily repurposed as animal feed and an organic soil conditioner, which enhances the organic matter content of soil. Innovations in its valorization for biomethane or bioethanol production aim to address the inhibition caused by essential oils on microbial communities involved in these processes. By transforming CPW into renewable energy sources or valuable compounds for industries like food, cosmetics, and pharmaceuticals, we can significantly mitigate the environmental impacts associated with its disposal [5].

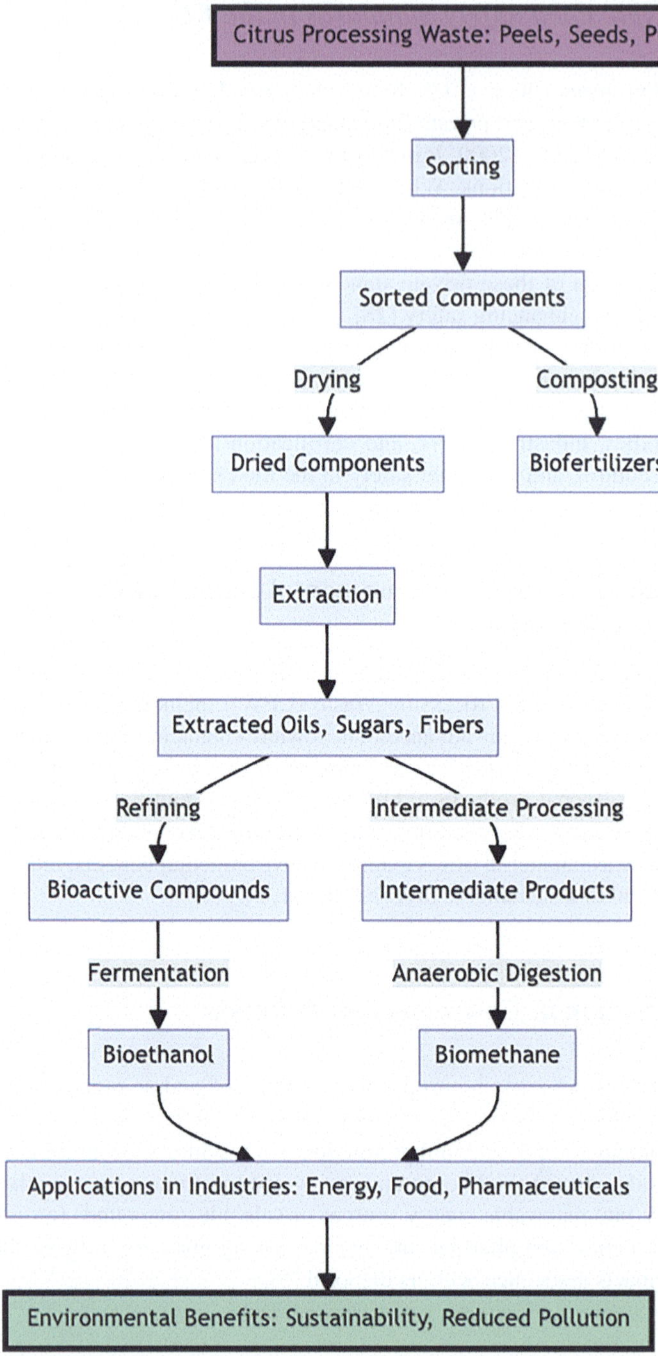

Fig. 4.1 Flow diagram of CPW valorization process

4.6.2 Contribution to Bioeconomy and Circular Economy

The comprehensive valorization of CPW not only mitigates negative environmental impacts but also promotes the development of a bioeconomy, embracing the principles of a circular economy. Through the conversion of citrus waste into valuable products such as nanoparticles, bio-sorbents, and biofertilizers, and the adoption of green extraction techniques like microwave- and ultrasound-assisted extractions, citrus waste is transformed into a resource. These practices are integral to the evolving biorefinery models, which are essential for minimizing environmental footprints and enhancing the economic viability of producing bio-based products from what was previously regarded as waste [1].

4.6.3 Advancement in Green Extraction Technologies

The implementation of novel, environmentally friendly extraction methods for obtaining bioactive compounds from citrus wastes illustrates significant progress towards environmental sustainability. These green technologies reduce energy consumption and solvent use, offering a sustainable alternative to traditional extraction methods. Such advancements facilitate the efficient utilization of citrus waste, enabling the production of high-quality bioactives, which further contribute to environmental sustainability and economic gains [6].

4.6.4 Future Perspectives on Sustainable Practices

Looking ahead, the valorization of CPW holds a promising future with ongoing research aimed at enhancing the efficiency and sustainability of extraction and conversion processes. The focus is on innovative technologies and methods that minimize energy use, reduce toxic solvent reliance, and enhance the recovery of valuable bioactive compounds. These developments are vital for scaling up sustainable practices, positioning CPW valorization as a central component in achieving broader environmental sustainability and economic growth within the citrus processing sector and beyond [1, 6, 7].

The manifold environmental benefits of citrus waste valorization encompass pollution reduction, contributions to a sustainable bioeconomy, and support for circular economy principles. Through forward-thinking valorization strategies, citrus waste is reimagined from a disposal challenge to a valuable resource, highlighting the crucial role of sustainable practices in the food processing industry and beyond.

4.7 Case Studies and Practical Applications

The process of valorizing citrus waste, as depicted in Fig. 4.2, is a multidimensional approach that not only fosters environmental sustainability but also embodies the principles of a successful circular economy. This section will explore various case studies and practical applications that illustrate the real-world implementation of the strategies represented in the figure. Each case study underscores the viability and effectiveness of these innovative practices, highlighting their contributions to sustainable development and environmental stewardship.

4.7.1 *Environmental Benefits*

The comprehensive review by Zema, Calabrò [5] emphasizes the environmental advantages of using citrus processing waste (CPW) for biomethane and bioethanol production, as well as its traditional uses as animal feed and organic soil conditioner. These practices effectively reduce the environmental impacts associated with waste disposal, transforming citrus waste into valuable resources that decrease pollution and enhance environmental health.

Fig. 4.2 Schematic overview of citrus waste valorization in a sustainable agricultural system

4.7.2 Sustainable Practices

Satari and Karimi [1] detail recent strides in the complete valorization of CPW, underscoring the shift towards a bioeconomy. Their research delves into the use of citrus waste in producing nanoparticles, bio-sorbents, and biofertilizers through green extraction techniques. These methods not only bolster environmental sustainability but also demonstrate the economic feasibility of transforming citrus waste into diverse bio-based products, showcasing the practicality of integrated biorefinery models.

4.7.3 Practical Applications

A notable case study in China, explored by Chen, Zhou [29], conducts a life cycle analysis of citrus residue management systems. The findings indicate that valorization strategies, such as those employed in feed systems and biorefineries, can markedly reduce environmental impacts compared to traditional waste treatment methods. This case study highlights the potential of valorization to enhance performance in reducing global warming, eutrophication, and human toxicity, emphasizing the critical role of energy efficiency and renewable energy utilization.

4.7.4 Future Perspectives

Suri, Singh [30] review recent developments in the valorization of citrus fruit waste for recovering value-added compounds, aiming for a zero-waste scenario. This approach aligns with the zero-waste philosophy, advocating for the use of citrus waste as a resource to extract significant valuable components, which encourages a transition towards a more sustainable and circular economy.

These case studies and practical applications of citrus waste valorization offer a promising outlook for environmental sustainability and the advancement of sustainable practices. Through effective management of citrus waste, the citrus processing industry can significantly mitigate environmental pollution, foster a thriving bioeconomy, and support circular economy principles, thereby contributing to a more sustainable future.

4.8 Technological and Economic Challenges in Citrus Waste Valorization

In addressing the valorization of Citrus Processing Waste (CPW), there are several technological and economic barriers that must be considered, along with identifying areas that warrant further research to enhance the efficiency and applicability of these valorization processes.

4.8.1 Technological and Economic Barriers

One of the primary technological challenges in the valorization of CPW is the development of cost-effective and scalable technologies that can efficiently extract valuable compounds from citrus waste. Current extraction techniques, such as microwave-assisted extraction and pulsed electric fields, while effective, often come with high capital and operational costs, which can limit their scalability and accessibility for smaller producers [6].

Furthermore, the presence of inhibitors like limonene in citrus waste can significantly impact the efficiency of bioconversion processes, such as anaerobic digestion. These compounds can inhibit microbial activity, thereby reducing the yield of desired products like biomethane and bioethanol [5]. Developing robust microbial strains or engineering existing ones to withstand such inhibitors is necessary but also adds to the complexity and cost of the process.

Economically, the fluctuating availability of CPW, due to the seasonal nature of citrus production, poses a challenge for the continuous operation of valorization facilities. This variability can lead to periods of underutilization of processing capabilities, affecting the overall profitability of valorization projects. Moreover, the market demand for bio-based products derived from CPW is still developing, and without strong market drivers, the economic incentive to invest in advanced valorization technologies may be limited.

4.8.2 Research Gaps and Potential Areas for Future Development

Research into improving the efficiency of existing technologies and developing new methods for extracting valuable compounds from CPW is essential. This includes enhancing the yield and purity of extracted compounds while minimizing energy use and solvent reliance. There is also a need for research into integrated biorefinery approaches that can process multiple feedstocks or convert CPW into a variety of products, thus maximizing resource efficiency and economic returns. Another critical area for research is the environmental impact assessment of CPW valorization. Life cycle analyses are needed to fully understand the environmental benefits and potential drawbacks of various valorization strategies. Such assessments will help in making informed decisions about which technologies are truly sustainable and contribute positively to the circular economy. Lastly, there is a considerable gap in the regulatory framework concerning the use of CPW-derived products in food and pharmaceutical industries. Research aimed at ensuring the safety and compliance of these products can help in establishing clearer regulations, which in turn would facilitate market acceptance and consumer trust. By addressing these technological and economic barriers and filling the research gaps, the valorization of CPW can be significantly advanced, contributing to environmental sustainability, economic growth, and the development of a robust bioeconomy.

4.9 Conclusion

This chapter has highlighted the various strategies for valorizing Citrus Processing Waste (CPW), focusing on their environmental, economic, and food safety benefits. Innovative extraction techniques such as microwave-assisted extraction, ultrasound-assisted extraction, and pulsed electric fields have proven effective in isolating valuable bioactive compounds from CPW—pectins, flavonoids, and essential oils. These compounds enhance food safety by serving as natural preservatives and antioxidants, thus extending the shelf life of food products and enriching their nutritional content. Furthermore, the use of citrus peels in developing functional foods aligns with the growing global preference for natural and organic products and underscores the shift toward sustainable food industry practices. Looking ahead, the potential for CPW valorization is substantial and expanding. Continued advancements in technology and methodology are expected to improve the efficiency of these processes and reduce operational costs, making them accessible to a wider range of producers. As awareness of sustainability increases, demand for bio-based products derived from CPW is likely to rise, fostering further innovation and investment in the sector. Moreover, exploring CPW for energy production like biomethane and bioethanol offers a path to diminish reliance on fossil fuels, significantly contributing to environmental sustainability. To realize the full potential of CPW valorization, it is imperative to overcome current technological and economic challenges, fill existing research gaps, and establish a supportive regulatory framework that ensures product safety and efficacy. Enhanced collaboration among researchers, industry stakeholders, and policymakers will be essential in promoting the adoption of CPW valorization technologies. Ultimately, the citrus processing industry is poised to play a crucial role in advancing sustainable practices that benefit the environment, economy, and society at large.

References

1. Satari, B., & Karimi, K. (2018). Citrus processing wastes: Environmental impacts, recent advances, and future perspectives in total valorization. *Resources Conservation and Recycling, 129*, 153–167.
2. Jeong, D., et al. (2021). Recent advances in the biological valorization of citrus peel waste into fuels and chemicals. *Bioresource Technology, 323*, 124603.
3. González-Miquel, M., & Díaz, I. (2020). Chapter 6 - Valorization of citrus waste through sustainable extraction processes. In M. R. Kosseva & C. Webb (Eds.), *Food industry wastes* (2nd ed., pp. 113–133). Academic.

4. Maqbool, Z., et al. (2023). Citrus waste as source of bioactive compounds: Extraction and utilization in health and food industry. *Molecules, 28*(4), 1636.
5. Zema, D. A., et al. (2018). Valorisation of citrus processing waste: A review. *Waste Management, 80*, 252–273.
6. Putnik, P., et al. (2017). Innovative "Green" and novel strategies for the extraction of bioactive added value compounds from citrus wastes-a review. *Molecules, 22*(5), 680.
7. Negro, V., Ruggeri, B., & Fino, D. (2018). Recovery of energy from orange peels through anaerobic digestion and pyrolysis processes after D-limonene extraction. *Waste and Biomass Valorization, 9*(8), 1331–1337.
8. Anticona, M., et al. (2020). High biological value compounds extraction from citrus waste with non-conventional methods. *Food, 9*(6), 811.
9. Russo, C., et al. (2021). The second life of citrus fruit waste: A valuable source of bioactive compounds. *Molecules, 26*(19), 5991.
10. Sharma, K., et al. (2017). Converting citrus wastes into value-added products: Economic and environmently friendly approaches. *Nutrition, 34*, 29–46.
11. Dorado, C., Cameron, R. G., & Manthey, J. A. (2019). Study of static steam explosion of Citrus sinensis juice processing waste for the isolation of sugars, Pectic hydrocolloids, flavonoids, and Peel oil. *Food and Bioprocess Technology, 12*(8), 1293–1303.
12. Patsalou, M., et al. (2020). A biorefinery for conversion of citrus peel waste into essential oils, pectin, fertilizer and succinic acid via different fermentation strategies. *Waste Management, 113*, 469–477.
13. Teigiserova, D. A., et al. (2021). A step closer to circular bioeconomy for citrus peel waste: A review of yields and technologies for sustainable management of essential oils. *Journal of Environmental Management, 280*, 111832.
14. Santiago, B., et al. (2020). Identification of environmental aspects of citrus waste valorization into D-limonene from a biorefinery approach. *Biomass & Bioenergy, 143*, 105844.
15. Consoli, S., et al. (2023). Sustainable use of citrus waste as organic amendment in Orange Orchards. *Sustainability, 15*(3), 2482.
16. Belligno, A., et al. (2005). Effects of industrial orange wastes on soil characteristics and on growth and production of durum wheat. *Agronomy for Sustainable Development, 25*(1), 129–135.
17. Ruiz, B., & Flotats, X. (2016). Effect of limonene on batch anaerobic digestion of citrus peel waste. *Biochemical Engineering Journal, 109*, 9–18.
18. Lotito, A. M., et al. (2018). Biomethanization of citrus waste: Effect of waste characteristics and of storage on treatability and evaluation of limonene degradation. *Journal of Environmental Management, 215*, 366–376.
19. De Medina-Salas, L., et al. (2020). Valorization of Orange peel waste using precomposting and vermicomposting processes. *Sustainability, 12*(18), 7626.
20. Mahato, N., et al. (2019). Modern extraction and purification techniques for obtaining high purity food-grade bioactive compounds and value-added co-products from citrus wastes. *Food, 8*(11), 523.
21. Zayed, A., Badawy, M. T., & Farag, M. A. (2021). Valorization and extraction optimization of seeds for food and functional food applications. *Food Chemistry, 355*, 129609.
22. Liu, N., et al. (2021). A review of chemical constituents and health-promoting effects of citrus peels. *Food Chemistry, 365*, 130585.
23. Nirmal, N. P., et al. (2023). Valorization of fruit waste for bioactive compounds and their applications in the food industry. *Food, 12*(3), 556.
24. Socas-Rodríguez, B., et al. (2021). Food by-products and food wastes: Are they safe enough for their valorization? *Trends in Food Science & Technology, 114*, 133–147.
25. Jedrejek, D., et al. (2016). Animal by-products for feed: Characteristics, European regulatory framework, and potential impacts on human and animal health and the environment. *Journal of Animal and Feed Sciences, 25*(3), 189–202.

26. Miller, M. A. (1999). Quality control and safety of animal products. *Canadian Journal of Animal Science, 79*(4), 533–538.
27. Hill, V. (2014). Government regulation of food quality: International and in France and the U.S. In V. Hill (Ed.), *A Kaizen approach to food safety: Quality management in the value chain from wheat to bread* (pp. 53–82). Springer International Publishing.
28. Havinga, T. (2018). Private food safety standards in the EU. In H. Bremmers & K. Purnhagen (Eds.), *Regulating and managing food safety in the EU: A legal-economic perspective* (pp. 11–37). Springer International Publishing.
29. Chen, W. H., Zhou, Y. M., & Chen, Y. P. (2021). The environmental impacts of citrus residue management in China: A case study in the three gorges reservoir region. *Waste Management, 133*, 80–88.
30. Suri, S., Singh, A., & Nema, P. K. (2021). Recent advances in valorization of citrus fruits processing waste: A way forward towards environmental sustainability. *Food Science and Biotechnology, 30*(13), 1601–1626.

Chapter 5
Extraction of Bioactive Compounds from Citrus Through Novel Innovative Techniques

Megh R. Goyal and Anamika Chauhan

Abbreviations

UAE	Ultrasonic-Assisted Extraction
SFE	Supercritical Fluid Extraction
MAE	Microwave-Assisted Extraction
PEF	Pulsed Electric Fields
EAE	Enzyme-Assisted Extraction
CO_2	Carbon dioxide

5.1 Introduction

Citrus fruits, encompassing oranges, grapefruits, lemons, limes, and tangerines, are a staple in many diets worldwide. Beyond their refreshing taste and juicy flesh, citrus fruits are a treasure trove of bioactive compounds, including flavonoids, vitamin C, and carotenoids [1]. These compounds have garnered significant scientific interest due to their association with a multitude of health benefits [2]. Studies have linked citrus bioactive compounds to antioxidant activity, which helps combat cellular damage caused by free radicals [3]. They also exhibit anti-inflammatory properties, potentially aiding in the management of chronic inflammatory conditions [4].

M. R. Goyal
Retired Faculty in Agricultural and Biomedical Engineering from College of Engineering, University of Puerto Rico, Mayaguez, Puerto Rico

Apple Academic Press Inc., Palm Bay, FL, USA

A. Chauhan (✉)
Department of Home Sciences, Chaman Lal Mahvidhyalya Landhaura Haridwar, Uttarakhand, India

A. Chauhan et al. (eds.), *Valorization of Citrus Food Waste*,
https://doi.org/10.1007/978-3-031-77999-2_5

Moreover, research suggests a role for citrus bioactive compounds in preventing the development of certain cancers [5].

However, unlocking the full potential of these health-promoting compounds relies on efficient and sustainable techniques for their extraction from the complex citrus matrix. Traditional extraction methods, such as solvent extraction, steam distillation, and maceration, have long been employed. However, these methods come with drawbacks. They often require harsh chemicals or high temperatures, which can degrade the bioactives themselves [6]. Additionally, solvent extraction can be environmentally unfriendly due to the use and disposal of organic solvents [7].

In response to these limitations, a new wave of innovative extraction techniques has emerged, offering a more efficient, sustainable, and targeted approach. This review delves into the exciting world of these novel techniques, exploring their principles, advantages, and effectiveness in extracting high-quality citrus bioactive compounds.

5.1.1 Importance of Citrus Fruits in Nutrition and Medicine

Category	Description	Citations
Nutritional value	Rich in vitamin C (immune function, collagen production)	[1]
Nutritional value	Good source of folate (cell division, DNA synthesis), potassium (blood pressure, muscle function), and fiber (digestion, gut health)	[8, 9]
Bioactive compounds	Flavonoids and carotenoids with potent antioxidant properties (neutralize free radicals, reduce chronic disease risk)	[3]
Bioactive compounds	Possible anti-inflammatory effects (arthritis, heart disease)	[10]
Bioactive compounds	Potential role in preventing certain cancers (further research needed)	[11]

5.1.2 Overview of Bioactive Compounds in Citrus (E.G., Flavonoids, Vitamin C, Carotenoids) and Their Health Benefits

5.1.2.1 Flavonoids

Flavonoids are the most abundant class of bioactive compounds in citrus fruits, concentrated primarily in the peels and membranes [1]. These diverse molecules boast a wide range of health benefits:

- **Antioxidant Activity:** Flavonoids are potent antioxidants, neutralizing free radicals that damage cells and contribute to chronic diseases like cancer and heart disease [3]. Studies have shown citrus flavonoids, such as hesperidin and naringenin, to possess significant free radical scavenging properties [12].

- **Anti-inflammatory Effects:** Chronic inflammation is a hallmark of various health problems. Citrus flavonoids exhibit anti-inflammatory properties by suppressing the production of inflammatory mediators [10]. Research suggests that naringenin may be effective in managing inflammatory conditions like arthritis [13].
- **Improved Cardiovascular Health:** Flavonoids in citrus fruits can contribute to a healthy heart. Studies indicate that citrus flavonoid intake may help lower blood pressure and improve cholesterol levels, reducing the risk of cardiovascular disease [14].

5.1.2.2 Vitamin C

Citrus fruits are renowned for their abundance of vitamin C, a crucial nutrient for optimal health:

- **Immune Function:** Vitamin C plays a vital role in supporting the immune system. It enhances the production and function of white blood cells, the body's defense against infection [8].
- **Collagen Production:** Vitamin C is essential for collagen synthesis, a protein that provides structure and support to skin, bones, and connective tissues.
- **Antioxidant Activity:** Vitamin C acts as an antioxidant, protecting cells from free radical damage and potentially reducing the risk of chronic diseases [8].

5.1.2.3 Carotenoids

Citrus fruits are a good source of carotenoids, a group of pigments with various health benefits:

- **Improved Vision:** Beta-carotene, a type of carotenoid found in citrus fruits, can be converted to vitamin A, which is essential for maintaining healthy vision.
- **Reduced Risk of Chronic Diseases:** Carotenoids possess antioxidant properties, potentially lowering the risk of chronic diseases like cancer and heart disease. Studies suggest that citrus carotenoids, such as beta-carotene and beta-cryptoxanthin, may be protective against certain cancers.

5.1.3 Challenges with Traditional Extraction Methods (E.G., Low Efficiency, Solvent Use, Thermal Degradation)

- **Low Efficiency:** Conventional techniques, such as solvent extraction and maceration, often require large volumes of solvents and extensive processing times to achieve moderate yields of bioactive compounds [7]. This inefficiency translates to higher costs and increased processing time.

- **Solvent Use:** Many traditional methods rely on harsh organic solvents like hexane, acetone, and dichloromethane for extraction [6]. These solvents can pose health risks to workers handling them and contribute to environmental pollution if not disposed of properly [15].
- **Thermal Degradation:** Traditional methods often involve high temperatures, such as during steam distillation or prolonged boiling during maceration [16]. These high temperatures can degrade heat-sensitive bioactive compounds like vitamin C and certain flavonoids, compromising the quality and functionality of the final extract [1].
- **Selectivity Issues:** Traditional methods may not be very selective, extracting a broad range of compounds from the citrus matrix alongside the desired bioactives [17]. This can necessitate further purification steps, adding complexity and cost to the process.

5.2 Traditional Extraction Methods

5.2.1 Brief Description of Conventional Extraction Techniques (e.g., Solvent Extraction, Steam Distillation, Maceration)

- **Solvent Extraction:** This method involves submerging plant material in a solvent, typically organic solvents like ethanol, methanol, or acetone, for an extended period. The solvent dissolves the desired compounds along with other soluble materials from the plant matrix. The extract is then concentrated by evaporating the solvent, leaving behind a residue rich in bioactive compounds. While effective for a broad range of compounds, solvent extraction can be inefficient, requiring large solvent volumes and lengthy processing times [1]. Additionally, the use of harsh organic solvents raises health and environmental concerns if not handled and disposed of properly [15].
- **Steam Distillation:** This technique utilizes steam to extract volatile compounds, such as essential oils, from citrus rinds. The plant material is placed in a chamber, and steam is passed through it. The steam carries the volatile compounds, which are then condensed and collected as the essential oil [16]. Steam distillation is a well-established method for citrus essential oils but may not be suitable for extracting heat-sensitive, non-volatile bioactive compounds like certain flavonoids or vitamin C.
- **Maceration:** This simple method involves soaking plant material in a solvent, often water or a diluted alcohol solution, at room temperature for an extended period [18]. Over time, the desired compounds passively diffuse from the plant material into the solvent. Maceration is a low-cost and low-tech approach but suffers from limitations in efficiency. It often requires long extraction times and may not achieve high yields of target compounds [7].

5.2.2 Limitations and Impact on Bioactive Compounds.

Traditional technique	Limitations	Impact on bioactive compounds	Citations
Solvent extraction	Requires large volumes of solvents	*Lower yields:* Large solvent volumes are needed to achieve adequate extraction, but this can lead to dilution of the target bioactive compounds, resulting in lower final concentrations in the extract.	[1, 7]
		Degradation: Heat-sensitive bioactive compounds, such as certain flavonoids and vitamin C, can be degraded by the harsh organic solvents commonly used in this method. This degradation reduces the quality and functionality of the final extract.	[7, 17]
		Health risks: Exposure to harsh organic solvents during extraction and handling poses health risks for workers.	[15]
		Environmental concerns: Improper disposal of used organic solvents can lead to environmental pollution, harming ecosystems and human health.	[15]
Steam distillation	Effective for volatile compounds only	*Ignores non-volatile bioactives*: This method is primarily suited for extracting volatile compounds like essential oils. Non-volatile bioactive compounds, such as many flavonoids and phenolic acids, are not effectively extracted by steam distillation and remain in the plant material.	[16, 17]
	High temperatures	*Heat-sensitive compound degradation*: The high temperatures involved in steam distillation can degrade heat-sensitive bioactive compounds similar to solvent extraction.	[17]
Maceration	Low efficiency	*Lower yields:* The passive diffusion process in maceration is slow and inefficient, leading to lower yields of target bioactive compounds compared to other techniques.	[1, 18]
	Long extraction times	*Degradation over time:* Long extraction times needed for maceration can increase the risk of degradation of some bioactive compounds due to factors like enzyme activity or oxidation.	[18]

5.3 Novel Innovative Techniques

5.3.1 Ultrasonic-Assisted Extraction (UAE)

In response to the limitations of traditional extraction methods, innovative techniques are emerging that offer a more efficient, sustainable, and targeted approach to extracting bioactive compounds from citrus fruits. One such technique is Ultrasonic-Assisted Extraction (UAE). UAE harnesses the power of ultrasound;

sound waves at frequencies beyond human hearing. During UAE, the plant material is submerged in a solvent, and ultrasound waves are introduced into the mixture. The cavitation phenomenon, the rapid formation and collapse of microscopic bubbles in the solvent, is a key player in UAE. These collapsing bubbles create turbulence and shear forces within the solvent, promoting enhanced mass transfer and disrupting the cell walls of the plant material. This disruption facilitates the release and diffusion of entrapped bioactive compounds into the solvent, leading to increased extraction efficiency [19, 20].

Compared to traditional techniques, UAE boasts several advantages:

- **Enhanced Efficiency:** UAE significantly reduces extraction times compared to methods like maceration. The cavitation phenomenon enhances mass transfer and facilitates faster release of target compounds [21].
- **Reduced Solvent Use:** UAE can achieve comparable extraction yields with lower solvent volumes compared to traditional methods. This not only minimizes costs but also reduces environmental impact and health risks associated with solvent usage [22].
- **Improved Selectivity:** By adjusting the ultrasound parameters (frequency, power, and time), UAE can be tailored to selectively target specific bioactive compounds present in the citrus matrix [23].
- **Preservation of Bioactive Compounds:** The shorter extraction times associated with UAE minimize the risk of degradation due to heat or prolonged exposure to solvents, potentially leading to higher quality extracts with preserved bioactivity [24].

Studies have demonstrated the effectiveness of UAE in extracting citrus bioactive compounds. For example, one study compared UAE with solvent extraction for obtaining flavonoids from orange peels. UAE achieved a higher yield of flavonoids in a shorter time compared to the traditional method. Another study reported that UAE successfully extracted limonoids, a class of bioactive compounds with potential anti-cancer properties, from grapefruit seeds with greater efficiency than conventional techniques [24].

5.3.2 Supercritical Fluid Extraction (SFE)

Supercritical Fluid Extraction (SFE) emerges as another innovative technique for extracting valuable compounds from citrus fruits. This method utilizes a supercritical fluid, a substance heated and pressurized beyond its critical point, where the distinction between liquid and gas phases blurs. Carbon dioxide (CO_2) is the most commonly used supercritical fluid due to its abundance, safety, and low critical temperature (31 °C) [25].

In SFE applied to citrus, the finely ground citrus material is packed into an extraction chamber. Supercritical CO_2 is then pumped through the chamber, dissolving the desired bioactive compounds. By manipulating pressure and temperature, the selectivity of SFE can be tuned. This allows for the targeted extraction of specific compounds of interest, minimizing the extraction of unwanted materials present in the citrus matrix [1]. Compared to traditional methods, SFE offers distinct advantages:

- **Solvent-Free Operation:** Unlike conventional techniques that rely on harsh organic solvents, SFE primarily uses CO_2, a natural and environmentally friendly solvent. After extraction, the pressure is released, allowing the CO_2 to return to its gaseous state, leaving behind some solvent-free extract rich in bioactives [26]. This eliminates concerns associated with solvent residues and simplifies the purification process.
- **Enhanced Selectivity:** As mentioned earlier, SFE allows for fine-tuning of extraction parameters to target specific bioactive compounds. This selectivity ensures high-quality extracts enriched in the desired components [27].
- **Preservation of Bioactives:** The low operating temperatures associated with SFE minimize the risk of thermal degradation often encountered in traditional methods. This helps preserve the heat-sensitive nature of many citrus bioactive compounds, such as vitamin C and certain flavonoids [28].

Studies have demonstrated the effectiveness of SFE in extracting various citrus bioactives. For instance, one study reported successful extraction of limonoids from grapefruit seeds using SFE, achieving higher yields compared to solvent extraction [27]. Another example showed the isolation of valuable essential oils from orange peels with SFE, demonstrating its suitability for extracting volatile compounds [28].

5.3.3 Microwave-Assisted Extraction (MAE)

Microwave-Assisted Extraction (MAE) stands out as another innovative technique gaining traction in the field of citrus bioactive compound extraction. This method utilizes microwave energy to heat the solvent-sample mixture, accelerating the extraction process. Unlike conventional heating methods that transfer heat from the outside in, microwaves interact directly with polar molecules within the sample and solvent. This volumetric heating results in rapid and uniform heating throughout the mixture [19]. In the context of citrus extraction, the polar nature of water in the solvent and certain functional groups within the bioactive compounds allows for efficient microwave absorption [1].

Studies have demonstrated the effectiveness of MAE compared to traditional techniques. For instance, one study compared MAE with maceration for extracting flavonoids from orange peels. MAE achieved significantly higher yields of flavonoids in a much shorter extraction time compared to the traditional method [29]. Another study reported that MAE successfully extracted essential oils from lemon peels with comparable yields to hydrodistillation, a traditional technique, but in a fraction of the time [30]. These comparisons highlight the efficiency and time-saving benefits of MAE. MAE offers advantages in terms of both the quantity and quality of the extracted bioactives:

- **Increased Extraction Yields:** The rapid and uniform heating achieved by microwaves facilitates faster diffusion and release of target compounds from the plant matrix, leading to higher extraction yields compared to traditional methods [21].
- **Preserved Bioactive Quality:** The shorter extraction times associated with MAE minimize the risk of degradation due to prolonged heating or exposure to solvents. This can lead to extracts with a higher content of heat-sensitive bioactive compounds like vitamin C and certain flavonoids [24].

5.3.4 Pulsed Electric Fields (PEF)

Pulsed Electric Fields (PEF) is an emerging and non-thermal extraction technique gaining recognition for its potential in isolating valuable compounds from citrus fruits. This section delves into the principles, effects, and recent findings of PEF in citrus extraction. PEF utilizes short, high-voltage electrical pulses applied to the plant material immersed in a solvent. These pulses create a temporary disruption in the cell membranes of the plant tissue. The specific parameters, such as pulse duration, frequency, and electric field strength, can be controlled to target specific effects [31].

The primary effect of PEF is the permeabilization of cell membranes. When exposed to the electric field, the phospholipid bilayer of the cell membrane experiences a temporary rearrangement, creating pores that allow for the leakage of intracellular components, including the desired bioactive compounds, into the surrounding solvent [32]. This enhanced permeability facilitates a more efficient release of target compounds compared to traditional methods. Studies have demonstrated the positive impact of PEF on citrus bioactive compound extraction. Recent findings highlight these benefits:

- **Increased Extraction Yields:** Research suggests that PEF treatment can significantly enhance the extraction yield of various citrus bioactives, including flavonoids, polyphenols, and essential oils [33, 34]. This increase is attributed to the improved cell membrane permeability induced by the electric pulses.
- **Preserved Bioactive Quality:** The non-thermal nature of PEF minimizes the risk of thermal degradation often encountered in conventional techniques. This can lead to extracts with a higher content of heat-sensitive bioactive compounds [11].

Some specific examples of PEF's effectiveness:

- A study reported that PEF pretreatment of orange peel before pressing significantly increased the extraction of naringin and hesperidin, two valuable flavonoids [33].
- Another study demonstrated that PEF treatment of lemon peel residues enhanced the extraction yield of total polyphenols compared to the control group.

These findings suggest that PEF holds promise as a sustainable and efficient approach for extracting valuable bioactive compounds from citrus fruits.

5.3.5 *Enzyme-Assisted Extraction (EAE)*

Enzyme-Assisted Extraction (EAE) stands out as another innovative technique gaining traction in the field of citrus bioactive compound extraction. This method harnesses the power of specific enzymes to gently break down the cell walls of plant material, facilitating the release of targeted compounds.

The effectiveness of EAE lies in the use of enzymes that can specifically target and degrade the components of the cell wall that hinder the release of desired bioactives. For citrus fruits, enzymes like pectinases, cellulases, and hemicellulases are often employed [1]. These enzymes break down complex carbohydrates like pectin, cellulose, and hemicellulose, respectively, which form a major part of the citrus cell wall. This enzymatic breakdown creates pores and weakens the cell wall structure, allowing for the easier diffusion and extraction of valuable bioactives trapped within the cells [4]. EAE offers several advantages, particularly for selective extraction of specific citrus bioactives:

- **Enhanced Selectivity:** By choosing enzymes with specific substrate preferences, EAE allows for targeted degradation of cell wall components hindering the release of the desired compounds. This can lead to extracts enriched in specific bioactives of interest, such as flavonoids or limonoids [7].
- **Gentler Processing:** Unlike harsh solvent extraction methods, EAE operates under mild conditions (typically room temperature and neutral pH). This gentle processing minimizes the risk of degradation of heat-sensitive bioactive compounds like vitamin C and certain flavonoids [17].
- **Environmentally Friendly:** EAE often utilizes enzymes derived from natural sources like microorganisms. These enzymes are biodegradable and pose minimal environmental concerns compared to harsh organic solvents used in traditional methods [15].

Compared to conventional techniques like solvent extraction or maceration, EAE offers distinct benefits:

- **Higher Yields:** By facilitating the release of entrapped bioactives, EAE can achieve higher extraction yields compared to traditional methods that rely on passive diffusion [35].
- **Reduced Solvent Use:** EAE can achieve comparable extraction efficiencies with lower solvent volumes compared to traditional methods. This not only minimizes costs but also reduces environmental impact and health risks associated with solvent usage [20].
- **Improved Quality:** The milder processing conditions associated with EAE minimize the risk of degradation, leading to extracts with a higher content of bioactive compounds with preserved functionality.

5.4 Comparative Analysis

Aspect	Novel techniques	Traditional techniques
Efficiency (extraction time)	Shorter extraction times due to enhanced mass transfer mechanisms (e.g., ultrasound, microwave)	Longer extraction times due to reliance on passive diffusion
Yield of bioactive compounds	Comparable or higher yields achievable for targeted compounds	Variable yields, may not extract all available bioactives
Quality of bioactive compounds	Milder conditions help preserve bioactivity and functionality	Harsh solvents or high temperatures can degrade heat-sensitive bioactives
Environmental impact	Lower solvent usage, some techniques use water or eco-friendly solvents	High solvent usage, potential for solvent disposal issues and worker exposure
Sustainability	Lower energy requirements, some techniques utilize renewable resources (enzymes)	Higher energy consumption for heating and solvent evaporation

5.5 Challenges and Future Perspectives

While novel techniques offer significant advantages, they also face some challenges and limitations. Optimizing the various parameters (e.g., power levels, pulse duration, enzyme selection) for each novel technique and target compound is crucial for maximizing efficiency and yield. This requires specialized knowledge and experimentation. Integrating novel techniques into existing industrial processing lines can be challenging, requiring modifications to equipment and infrastructure. While some techniques offer targeted extraction, further research is needed to refine selectivity for specific bioactive compounds within the complex citrus matrix. Novel equipment (e.g., ultrasonic generators, microwave reactors) can be expensive compared to traditional setups, impacting initial investment costs. While generally lower than traditional methods, the energy consumption of some novel techniques (e.g., high-power ultrasound) needs further optimization for cost-effectiveness.

Emerging Technologies
High-Voltage Pulsed Electric Discharge (HVPED): This technique utilizes short, high-voltage electrical pulses to disrupt cell walls, potentially offering a rapid and efficient extraction method. Supercritical Fluid Extraction (SFE): This method utilizes pressurized CO2 as a solvent, offering a clean and environmentally friendly approach for extracting heat-sensitive bioactives. Combinations of Techniques: Combining novel techniques (e.g., UAE-EAE) may offer synergistic effects, further enhancing extraction efficiency and selectivity.

Future Research Directions
Optimization and Scale-up: Continued research is necessary to optimize process parameters for each technique and scale them up for industrial applications. Integration and Automation: Developing cost-effective methods to integrate novel

techniques into existing processing lines and automating processes is crucial for industrial adoption. Exploration of Emerging Technologies: Further research on emerging techniques like HVPED and SFE for citrus bioactive extraction holds promise. Bioactive Characterization: Developing advanced analytical techniques to characterize and identify novel bioactive compounds extracted using these methods is essential.

5.6 Conclusion

Citrus fruits brim with valuable bioactive compounds, but traditional extraction methods often fall short in efficiency, yield, and quality. This chapter explored novel techniques like UAE, MAE, PEF, and EAE as game-changers. These methods boast significantly faster extraction times, potentially higher yields of targeted bioactives, and milder conditions that preserve their health benefits. Additionally, they offer a more sustainable approach with lower solvent usage and potentially lower energy requirements. While challenges like optimization and cost-effectiveness remain, future research focused on scaling up these techniques and exploring emerging options like HVPED and SFE holds immense promise. By embracing these advancements, the citrus industry can unlock a new era of efficient and sustainable extraction, paving the way for the development of novel health-promoting products.

References

1. Li, S., Luo, C., & Liu, H. (2019). Chapter 1 – Bioactive compounds in citrus fruits: Extraction and application. In H. Liu (Ed.), *Citrus Fruits* (pp. 1–22). Academic.
2. Wang, Y., Zheng, Q., & Luo, Y. (2016). Health benefits of citrus fruits. *International Journal of Molecular Sciences, 17*(3), 3102.
3. Du, G., Li, M., Zheng, J., & Li, Z. (2017). Characterization of phenolic profiles and antioxidant activity of grapefruit (Citrus paradisi) peel extracts from different cultivars. *Food Chemistry, 221*, 174–181.
4. Enzymes-assisted extraction of plants for sustainable and functional applications. https://www.mdpi.com/1422-0067/23/4/2359
5. Tanaka, T., Miyoshi, N., Yoshida, T., Matsumoto, S., & Kuwata, T. (2018). Citrus limonoids suppress the proliferation of prostate cancer PC-3 cells in vitro and in vivo. *International Journal of Oncology, 53*(4), 1323–1332.
6. Yilmaz, Y., & Toledo, R. T. (2004). Chemical composition of orange peel oil. *Developments in Food Science, 43*, 127–141.
7. Routray, W., & Orsat, V. (2012). Novel food processing technologies to improve food safety and quality. *Critical Reviews in Food Science and Nutrition, 52*(6), 455–472.
8. National Institutes of Health. (2023, September 26). Vitamin C. Office of dietary supplements. https://ods.od.nih.gov/factsheets/VitaminC%20-HealthProfessional/
9. U.S. Department of Agriculture. (2020). Full Report (All Nutrients): Oranges, raw, florida (typical). https://fdc.nal.usda.gov/fdc-app.html#/food-details/578523/nutrients

10. Hosseinzadeh, H., Nassiri-Asl, M., & Fallahzadeh, H. (2016). Anti-inflammatory effects of citrus peel flavonoids: An overview of recent findings. *Current Pharmaceutical Design, 22*(9), 1301–1310.
11. Tovar, L., Martinez, A., Garcia-Diaz, J., & Rodriguez-Amaya, D. B. (2006). Application of innovative technologies to enhance anthocyanin extraction from purple corncobs. *Journal of Agricultural and Food Chemistry, 54*(22), 8589–8593. https://doi.org/10.1021/jf061820i
12. Smeriglio, S., Denaro, M., & Russo, M. (2017). Citrus flavonoids: Intake, health effects and food sources. *Journal of the Science of Food and Agriculture, 97*(8), 2309–2326. https://doi.org/10.1002/jsfa.8200
13. Liu, X., Zhao, M., Ji, H., Wang, Z., & Zhang, Z. (2016). Potential beneficial effects of Naringenin on osteoarthritis. *Mediators of Inflammation, 2016*, 1–10. https://doi.org/10.1155/2016/8478152
14. Li, S., Pan, M. H., Li, J., Ye, J., Xu, H., Lin, S., et al. (2017). Citrus flavonoids and cardiovascular diseases. *Nutrients, 9*(12), 1277. https://doi.org/10.3390/nu9121277
15. Commission, E. (2012). Guidance document on the implementation of the legislation on the classification, labelling and packaging of substances and mixtures (CLP regulation) (OJ L 149, 11.6.2015, p. 1).
16. Darwis, Y., & Nawawi, Y. A. (2017). Techniques for extraction of essential oils from citrus fruits: A review. *International Journal of Food Engineering, 13*(2), 117–126.
17. Wang, L., & Weller, C. L. (2006). Recent advances in extraction of nutraceuticals from plants. *Trends in Food Science & Technology, 17*(6), 300–312.
18. Elik, A., Baysal, T., & Şahin, S. (2011). Extraction of phenolic compounds from black cumin seeds. *Journal of Agricultural and Food Chemistry, 59*(13), 7302–7308.
19. Chemat, F., Zafaroudis, N., Allaf, T., & Hemissi, S. (2011). Applications of ultrasound in green chemistry. *Green Chemistry, 13*(4), 82–108. https://doi.org/10.1039/C0GC01188G
20. Li, H., Qiao, K., & Xu, X. (2014). Applications of innovative technologies for enhancing mass transfer during natural product extraction. *Food & Bioproducts Processing, 92*(4), 484–491. https://doi.org/10.1016/j.fbp.2014.01.004
21. Rostagno, M. A., Palma, M., & Baranauskas, R. A. (2017). Ultrasound-assisted extraction of polyphenols from basil. *Food Research International, 99*, 1017–1023. https://doi.org/10.1016/j.foodres.2017.07.007
22. Jafari, S. M., Azarpira, H., Hejazi, M. A., & Cserháti, T. (2019). A comprehensive review on the application of ultrasound in assisted extraction of bioactive compounds from plants. *Ultrasonics Sonochemistry, 100*, 149–169. https://doi.org/10.1016/j.ultsonch.2019.03.024
23. Rostagno, M. A., Prado, J. M., & Martínes, M. L. (2015). Ultrasound-assisted extraction of polyphenols from Pequi (Caryocar brasiliense) bagasse. *Journal of Food Engineering, 167*, 381–387. https://doi.org/10.1016/j.jfoodeng.2015.07.018
24. Wang, J., Sun, B., Cao, Y., Tian, Y., & Liu, Y. (2008). Comparison of conventional and ultrasound-assisted extraction of schizandrin from Schisandra chinensis (Turcz.) Baill. *Food Chemistry, 108*(2), 877–885. https://doi.org/10.1016/j.foodchem.2007.10
25. Mendiola, M. A., Herrero, M., Ibañez, E., & Castro-Puyana, F. (2001). Extraction of essential oils from mandarin (Citrus unshiu Marc.) peel by supercritical CO_2. *Food Chemistry, 73*(1), 23–30. https://doi.org/10.1016/S0308-8146(00)00300-2
26. Reverchon, E., & De Rosa, M. P. (2017). Supercritical fluid extraction and fractionation of natural materials: Fundamentals, applications and perspectives. *The Royal Society of Chemistry.* https://doi.org/10.1039/9781788011251
27. Herrero, M., Ibañez, E., & Mendiola, M. A. (2006). Supercritical fluid extraction: Citric acid as a polar modifier for enhancing the extraction of limonoids from grapefruit seeds. *Journal of Chromatography A, 1103*(1-2), 142–148. https://doi.org/10.1016/j.chroma.2005.11.084
28. Zhang, H., Chen, S., Wang, J., & Xie, Z. (2016). Supercritical fluid extraction of antioxidants from citrus peels. *Journal of Food Processing and Preservation, 40*(2), 307–314. https://doi.org/10.1080/01458885.2014.982917

29. Wang, W., & Xu, S. Y. (2007). Microwave-assisted extraction of flavonoids from Chinese wolfberry leaves. *Food Chemistry, 106*(2), 986–992. https://doi.org/10.1016/j.foodchem.2007.05.079
30. Jiao, B., Zhang, H., Sun, Y., & Liu, Z. (2016). Microwave-assisted extraction of lemon essential oil: Influence of extraction time and microwave power on yield and quality. *Separation and Purification Technology, 166*, 272–278. https://doi.org/10.1016/j.seppur.2016.04.024
31. Donsì, F., & Ferrari, G. (2001). Applications of pulsed electric fields in the food industry. *Trends in Food Science & Technology, 12*(6–7), 185–190. https://doi.org/10.1016/S0924-2244(01)00094-2
32. Vorobiev, E., & Lebovka, N. (2008). Pulsed electric field (PEF) effects on milk: Opportunities for dairy processing. *International Dairy Journal, 18*(6), 701–706. https://doi.org/10.1016/j.idairyj.2008.01.004
33. Luengo, E., Liquete-Fernández, E., Garcia-Linares, F., Muñoz-Serrano, A., & Martín-Cabanas, M. J. (2014). Pulsed electric field-assisted extraction of polyphenols from orange peel. *Journal of Food Engineering, 143*, 8–14. https://doi.org/10.1016/j.jfoodeng.2014.04.019
34. Wang, L., Li, S., Chen, S., & Liao, X. (2017). Pulsed electric field-assisted essential oil extraction from orange peels: Chemical composition and antimicrobial activity. *LWT - Food Science and Technology, 84*, 327–334. https://doi.org/10.1016/j.lwt.2017.06.014
35. Enhancing efficiency of enzymatic-assisted extraction method for evaluating bioactive compound analysis in mulberry: An optimization approach. https://www.mdpi.com/2073-4395/13/10/2548

Chapter 6
Synthesis of Energy from Citrus Waste Through Torrefaction Method

Njideka Ebere Njoku, Godswill Kodili Elemuo, Serah Ogechi Alagbaoso, and Friday Pascal Okezie

6.1 Introduction

Fruit processing industries are integral to the global food supply chain, transforming fresh fruits into a variety of products to suit what the consumer wants for easy usage and taste [1]. It plays an essential part in converting fresh fruits into various valuable products like juices, concentrates, jams, and canned fruits. These industries contribute majorly to economic growth, employment, and food availability worldwide. They also generate waste products such as seeds, peels, pomace, stones, and meal of oil seed, rags and kernels [2].

Among the various fruits processed, citrus fruits hold a prominent position owing to their widespread cultivation and versatile applications in food and beverage industries. Food and Agriculture Organization (FAO) gave insight that consumer's demand have shifted from processed foods to natural foods of higher quality that meet their nutritional needs and also support good health [3]. Watermelon, Citrus, mango, grape, apple and banana are well known and highly produced fruits globally [4]. Wastes in both solid and liquid forms are largely generated by industries that process fruits which brings about pollution when not adequately disposed or utilized [2].

Extremely diversified waste is generated from the fruits and vegetable processing industries, which resulted from the wide processes, different variety and the

N. E. Njoku (✉) · G. K. Elemuo
Department of Food Science and Technology, Federal University of Technology, Owerri, Imo, Nigeria

Federal University of Technology Owerri, Multidisciplinary Research Team (FUTO MRT), Owerri, Nigeria
e-mail: njideka.njoku@futo.edu.ng

S. O. Alagbaoso · F. P. Okezie
Department of Food Science and Technology, Federal University of Technology, Owerri, Imo, Nigeria

A. Chauhan et al. (eds.), *Valorization of Citrus Food Waste*,
https://doi.org/10.1007/978-3-031-77999-2_6

increase of the product [5]. These industries contribute essentially to the economy by providing employment and meeting consumer demand for processed fruit products. However, major concerns faced by fruit processing industries is the output of substantial level of waste during the manufacturing process [3].

Disposing citrus waste improperly is not eco-friendly and can have negative effect in our environment. The improper disposition of this citrus waste can lead to water and soil pollution, habitat degradation and greenhouse gas emissions. Organic compounds contained in Citrus waste has capacity to leach into the soil and water, changing their chemical composition and affecting ecosystem health. Suitable management of citrus waste is vital to retard greenhouse gas emissions, mitigate environmental pollution, and maximize the extraction of essential compounds for industrial applications [6].

In a world grappling with the twin challenges of environmental sustainability and energy security, the need for hands—on solution has led the unconventional sources of energy to be major focus for innovative researchers. Among these, citrus waste—a byproduct of the flourishing citrus industry—stands out as a promising candidate for energy synthesis. Citrus production globally is estimated at annual generation of millions of tons [7], the disposal of citrus waste presents a pressing environmental concern. However, amidst this challenge lies an untapped reservoir of energy waiting to be harnessed.

Torrefaction, a thermal treatment process under an inert atmosphere, has emerged as a transformative technique for interconverting green biomass to a high-energy fuel source [8]. By subjecting citrus waste to controlled heating devoid of oxygen, torrefaction enhances the density of the energy, increasing reduction in the moisture composition, and improving its grindability and storability. Moreover, the resultant torrefied biomass exhibits properties akin to traditional fossil fuels, making it more suitable for generation of power and thermal applications [8].

This chapter embarks on a journey through the synthesis of energy from citrus waste via the torrefaction method. Drawing upon recent advancements and real-world applications, we delve into the scientific principles, process optimization strategies, and potential socio-economic and environmental implications of this innovative approach. From laboratory experiments to industrial-scale implementations, we uncover the promise and challenges of harnessing citrus waste as a sustainable energy resource. Through this exploration, we aim to not only shed light on the transformative potential of torrefaction but also inspire an innovation for more robust and resource-efficient energy solutions.

6.2 Citrus Waste Management

Food waste in the past was defined in different ways by different researchers but it is defined by [9] as "food of acceptable quality suitable for consumption but not consumed as it is wasted on or before deterioration. It can as well be defined as leftovers after processing at the industry or locally. Generally, food waste fall into four categories. Viz; food processing and its storage, production of food and its harvesting, retail counters and domestic food waste [10].

Citrus waste management poses a critical issue for citrus processing industries worldwide which is due to increased demand processed and ready to eat foods [11]. Suitable management of citrus waste is crucial to mitigate environmental pollution, reduce greenhouse gas emissions, and maximize the extraction of essential compounds for industrial applications. Several methods have been adopted to address this challenge, including waste reduction, recycling, and utilization for value-added products.

One of the primary approaches to citrus waste management is the conversion of waste into by-products such as animal feed, pectin, essential oils, and biofuels [11]. Additionally, citrus waste can be used as a material for the manufacture of enzymes, organic acids, and biodegradable plastics, further enhancing its value [12].

6.3 Characteristics of Citrus Waste

Citrus waste exhibits diverse characteristics depending on the kind of fruit processed and the processing methods employed. Different citrus fruits grown around the world which are known sources of citrus wastes includes lemons, oranges, grapefruits, limes and mandarins [13].

Citrus peels, for example, contain high levels of pectin, flavonoids, and essential oils, making them valuable for various industrial applications. It is produced widely and has the capability of being changed into fuels and chemicals biologically [14].

On the other hand, citrus seeds can serve as biodiesel as a result of their high oil content (up to 30% by weight). It is also rich in proteins that can be picked and utilized in food and pharmaceutical industries. This can also be utilized as supplement for live stocks. According to [15] citrus fruits seeds contain fat: 52%, crude fibre: 5.5%, crude protein: 3.1%, carbohydrate: 28.5% and ash: 25%.

Pomace of citrus is that solid residue left after the juice and other products from citrus fruits must have been extracted. The pomace of citrus fruits consists of pulp, seeds and rags fruits. It is known that pomace is a dependable source of essential oil, pectin, limonene as well as sugar [13].

The composition of waste from citrus also varies in the aspect of moisture content, pH, and organic matter content, influencing its suitability for different utilization pathways. Understanding the properties of waste from citrus fruit is crucial for building effective waste control strategies and maximizing its potential value [15].

6.4 Environmental Impact of Citrus Waste

The improper disposal of citrus waste is not eco-friendly and can have negative effect in our environment [6]. These includes water and soil pollution, habitat degradation and greenhouse gas emissions. Citrus waste is made up of organic compounds that has the ability to leach into the soil and water, altering their chemical composition and affecting ecosystem health.

Furthermore, the decomposition of citrus waste in landfills releases methane, which is a vital greenhouse gas that contributes to climate change. By using sustainable waste control systems, such as composting, anaerobic digestion, and bioconversion, the environmental impact of waste from citrus can be mitigated. Reduction of the environmental footprint of waste from citrus fruits are known to be the driving force for using biomass as a sustainable renewable resource for bio refineries [16].

6.5 Biomass Conversion

Citrus waste literally is waste from citrus fruits. It could consist of: (a) unwanted or unutilized parts discarded by the juice manufacturing industry, such as peel and pressed pulp; and (b) unwholesome or damaged fruits rejected for production or commercial purposes. Since they do not enter the food chain, they are deemed to be waste. Citrus waste biomass can accumulate to large quantities and thus, can present challenges with its handling, storage and even transportation. This necessitates an effective form by which it can be utilized easily. Hence, its conversion from fossil to liquid and gaseous fuels. Biomass is the lone natural source that offers fixed carbon, which is the major ingredient for the fuel production. Hence, the possibility to convert it to liquid, solid and/or gaseous fuels distinct from other energy sources. Biomass is gotten from the plant material via photosynthesis, that is, by the reaction between the carbon oxide, air, and water, to obtain carbohydrates, the major building block of biomass feedstock.

$$6CO_2 + 6\mathrm{H_2OC_6H_{12}} \rightarrow \mathrm{O_6} + 6\mathrm{O_2}$$

Citrus is an herbaceous biomass. Herbaceous biomasses are derived from plants consisting of non-woody stem. The die at the expiration of the growing season. Waste from citrus comprises seeds, peels, pulp and pith residue. Fermentable sugars are found primarily in the residue from pith and pulp while the peels on the other hand, contain appreciable quantities of limonene and other bioactive entities.

Biomass that are herbaceous in nature are classified into two: agricultural residues and energy crops. Energy crops include the waste originating from citrus fruit. Citrus waste can possess many applications, most of which are still undiscovered. Consequently, it has immense potential for the future. The unceasing exigence for fossil-sourced fuels encourages usage of citrus waster as a renewable feedstock because it presents a green and high-value alternative from both economic and environmental points of view.

Biomass not only provides fuel and energy, but it also is sourced to produce countless biomaterials and biochemicals that are applied in paints, detergents, industrial adhesives, bioplastics, and composite materials. Aside from energy and fuels, biomass may be used to generate carbon-based compounds known as bioproducts. Glycerin, sugars and sugar alcohols, furfurals, cellulose fiber and derivatives, carbonaceous materials, resins, bioplastics, and so on are examples of these products.

Biomass can be converted to energy and bio products. This can be achieved through thermal, biological, and physical process. Thus, the term 'Biomass conversion" connotes the production of heat, fuels, or electric currents by the controlled combustion of, or non-combustion thermal techniques, on specific materials. Heat is generally produced through direct combustion, while liquid fuels are produced through chemical conversion. Biological conversion is used to generate both liquid and gaseous fuels.

Through gasification, hydrogen fuel is manufactured from biomass. This gas offers varied potency possibilities for both domestic and industrial applications.

Citrus-based bioethanol manufacture has been studied as a valorization alternative. Although an encouraging bioethanol yield was recorded [16] the resources required to set up a bioethanol plant is huge in contrast to a biogas plant [17]. Additionally, valorization choice is not as effectual as biogas (methane) produced by digesting the waste from citrus anaerobically. The biogas production prospective from citrus is about twice that of its bioethanol production [17]. Thus, anaerobic digestion is a better alternative for citrus valorization although inherent (essential) oil content of citrus (CEO) inhibits the bioprocess. These essential oils are liquids that contain, among other constituents, the volatile aroma compounds of citrus plants. They are found in small vesicles located in the flavedo or exocarp of citrus fruits. They have some antimicrobial properties.

Although the composition of CEO differs depending on the fruit variety, fruit seasonality, degree of ripeness and geological origin, generally CEO contains a wide variety of compounds (85–99% volatile compounds, 1–15% non-volatile compounds). The volatile compounds consist of monoterpenes (limonene primarily), sesquiterpenes and sesquiterpenoids such as aldehydes (citral), alcohols (linalool), acids, ketones, and esters [18]. The major components are monoterpenes, especially limonene. The properties of CEO are essentially influenced by the physical, chemical, and biological properties of limonene [19].

Limonene cannot be hydrolysed because of the inherent chemical resistance of its ethylene group and cyclohexane ring to hydrolysis. However, some microorganisms (*Diplodia gossypina*, *Corynespora cassiicola*, *Penicillium digitatum*) have experienced biological degradation. Limonene is aerobically biodegradable up to 94% [20]. Methane production is achievable using limonene as the only carbon source and acetate, formate and hydrogen produced as intermediates for methane production. Although the identification of the microorganisms that participate in limonene degradation in methanogenic cultures have recorded some progress, the exact degradation pathway remains unknown.

Biomass conversion, which is allied to agriculture, food, energy, biofuel, additive, dyes, etc., is realized by two core ways, (i) biochemical and (ii) thermochemical routes.

6.5.1 *Biochemical Method*

This involves using bacteria, microorganisms, chemicals and enzymes (from bacteria or other microorganisms) to break down the bio-waste to gaseous and/or liquefied fuels (example biogas and bioethanol) and energy (methane and ethanol). Vital strategic technologies in biochemical biomass conversion are modalities necessary for efficient polysaccharides break down to monosaccharides and the subsequent translation of choice monosaccharides to desired bio-products. This requires two fundamental routes:

(i) Digestion (anaerobic and aerobic)
(ii) Fermentation.

6.5.1.1 Anaerobic Digestion of Citrus Waste

In anaerobic digestion, the biomass is converted to biogas, and through fermentation to ethanol using the by-product and organic wastes from citrus plant. The main products of anaerobic digestion are methane and carbon dioxide as well as a solid residue. The oxygen used by the microorganisms is accessed from the biomass itself instead of from ambient air.

The anaerobic digestion of wastes from different citrus fruits have been studied: waste from whole orange peel, comminuted with phosphorus and nitrogen as additives, (b) orange albedo in which the oil-containing flavedo has been removed before its juice extraction; (c) hand-peeled albedo; and (d) orange peel pellets. The digestion process occurring at 37 °C halted after 2–3 weeks. Subsequently, some aromatic acids resulted. The acids (benzoic, phenylacetic and phenylpropionic acid) themselves were not poisonous rather perceived as lethal subsequent to their relations with other compounds. This suggests that limonene and other components of CEO may be lethal for the anaerobic digestion process. The use of citrus peel press liquors and peel oil concentration differ in their conduct during anaerobic digestion, although this also depends on the reactor configuration. An up-flow anaerobic sludge blanket (UASB) reactor was more effective and showed higher resistance to toxicity than a continuously stirred tank reactor (CSTR) configuration.

Within mesophilic temperature (35–39 °C) of batch processes, CEO starts to inhibit anaerobic digestion at a concentration of 10 mg·L-1 [20]. The extent of this inhibition depends on factors such as the type of substrate used, and the type of inoculum applied. The variation is more significant in semi-continuous operations. Far-reaching divergences are observed in anaerobic digestion operated in a semi-continuous mode.

Most studies used citrus waste as the only substrate for the anaerobic digestion; adding nutrients and buffering solutions to make up for the lack of nutrients and the low pH of the citrus waste. For instance, cold-pressed orange oil, distilled orange oil, cold-pressed lemon oil, non-volatile distilled lemon oil residue and limonene have all been used as additives in the anaerobic digestion of substrates that paused after 2–3 weeks in all cases. In semi-continuous anaerobic digestion, the concentration of the inhibitor in the feed, and the rate of loading the digester are important considerations, the duo being critical in determining the dosage rate of inhibitor into the digester. Very few studies have been able to determine the maximum CEO dosage in terms of CEO per unit volume of the digester and day [20]. Mandarin waste and industrial orange waste are applied in anaerobic digestion.

6.5.1.2 Fermentation

In fermentation, the biomass is partially transformed to sugars using acid or enzymes. The sugar is then converted into ethanol or other chemicals with the help of yeasts. Pulp, peels, seeds and residues from pith constitute the wastes from citrus. The residues from the pith and the pulp consists of significant quantity of sugars that can be fermented. The peels consist of appreciable amounts of limonene in addition to bioactive molecules that do not favor fermentation. After the separation of limonene and bioactive molecules, the waste is then processed further to give ethanol, fuels, biogas and biosorbents through biotransformation. Biotransformation processes include microbial fermentation [21]. Due to its global scale of production, citrus biomass provides a promising source of natural, workable and cost-effective feedstock for generating environmentally friendly fuel. Wastes from citrus comprises approximately 70% carbohydrates, which in addition to constituent fermentable sugars, can be treated to provide ethanol and biogas. Biofuels minimize discharge of carbon dioxide by 80%, unlike fuel from petroleum or gasoline.

Wastes from citrus comprise high moisture (75–85%), thus drying them is a challenge. Asides the economic implication of drying them, the process also releases huge amounts of carbon dioxide, volatile organic compounds, nitrogen oxides, formaldehydes, sulfur dioxide and other suspended particulate matter.

Different fuels obtained from fermentation include:

6.5.1.2.1 Ethanol

Ethanol is produced by hydrolyzing citrus wastes enzymatically [21]. This release sugars that were fermented by yeast and bacteria (*Saccharomyces cerevisiae* and *Escherichia coli* respectively) to yield ethanol. Enzymes such as pectinase, cellulase, and β-glucosidase [22], have been used for the hydrolysis. Appreciable ethanol can only be produced if the hydrolysis happens post-removal of limonene. Otherwise, limonene, with its antimicrobial nature will inhibit the progression of fermentation. Subsequently, the hydroxylate (comprising carbohydrates and sugars) are hydrolyzed and fermented to ethanol.

Fermentation process is possible with or without enzymatic hydrolysis. Fermenting to synthesize ethanol is achieved by the following ways:

(i) Batch Fermentation: Here, the vessel used for the process is initially provided with rationed nutrients and microbial inoculation (at time zero). The fermentation factors (temperature, pH, pressure) are regulated to maximize products. The batch system is a closed one that prohibits more nutrient addition throughout the whole activity, by the end of which the entire nutrients are consumed. The mechanism is facilitated by selecting choice microbial strains and adjusting the nutrients utilized. The synthesized ethanol quantity, despite these measures, is restricted because the microorganism's growth is time constrained since the oxygen and carbon supply (from the nutrients and biomass) is exhaustible. Intensifying the flow of oxygen in the medium expediates the process. On completion, the medium is garnered for further processing to required product quality.

Merits of batch fermentation:

(i) The process requires a short duration.
(ii) It is easy to split-up the batch material for easy traceability.
(iii) The chance of contamination is slim because of its closed system characteristic.
(iv) Uncomplicated handling and operations.

It has some disadvantages too. They include:

(i) The product is mingled with unwanted bio-substrate, toxins from cell and nutrient metabolism.
(ii) It gives low yield consequent of short productive duration.

(ii) Fed-batch fermentation**:** otherwise called the semi-continuous process. Here, a measured proportion of nutrients and substrate are incorporated via a duct at pre-determined intervals for the duration of the activity. Fed-batch system is partially open permitting identifiable growth of microorganisms for a prolonged time. This sustains a growth curve for a moderately longer period. This also necessitates an enhanced feed rate. While this method grants certain control measures and longer operational time, adequate culture life span, increases the quantity of products stock and the population of functional microbial colonies [23], its fermentation progression could be hindered through by-products that are toxic.
(iii) Continuous-culture fermentation (CCF): Here, the batch reactor receives a regular supply of nutrients and substrate till a steady state growth phase is attained and products are collected. The expended medium (containing both the expected products and the by-products) is used to collect the synthesized ethanol alongside the spent residues and microbial cells. As a result, this procedure is often conducted in a reactor with constant in and out stream mechanism, programed with short intervals. CCF is more advantageous than fed-batch technique, although, the constant provision of replacement substrate coupled with protracted development phase heightens its contamination risk. Furthermore, controlling genetic alterations in the microorganisms, alongside

batch-wise isolation of synthesized products is demanding. Also, sustaining a steady microbial denseness within the reactor for a protracted period is challenging [23].

(iv) Separate hydrolysis and fermentation (SSF): this method allows the conduct of the hydrolysis distinct from the fermentation. Here, after the hydrolysis is achieved enzymatically, the medium is conveyed to the fermentation tank. Post-fermentation, the unreactive sugars and microbes are left behind as the ethanol is distilled off. The isolation of these all-important interactions ensures that each occur optimally. Conversely, the sugars resulting from the hydrolysis repress enzymatic activities (of cellulase and β-glucosidase enzymes), retarding the effectual conclusion of the hydrolysis. Thus, more enzymes may be supplied to the reactor for encouraging generation of fermentable sugars. Utilizing the SHF technique, 90.6% ethanol yield from pretreated mandarin peels was observed [24].

(v) Simultaneous saccharification and fermentation (SSF): Application of enzymatic hydrolysis and microbial fermentation is utilized for ethanol synthesis, both occurring in a single fermenter. This strategy advances ethanol synthesis with concentrated enzyme and substrate presence and allowing lengthened activity duration by the duo of enzymes and microbes. The products from hydrolysis (glucose, xylose), prevent more saccharification and hydrolysis process. As such, they need to be expended through the fermentation powered by microorganisms. Consequent to the concurrent occurrence of these processes, there prevails brief operation time, the system requires a relatively shorter period of operation, minimal requirement of enzyme, enhanced product yield, effectual ethanol synthesis from fermentable sugars and regulation of the sugars preventive reaction to enzymatic action. Furthermore, maintaining sterile conditions are not of high priority since the hydrolysis-resultant sugars from enzymatic activities for ethanol synthesis are speedily expended by microbes. SSF with citrus peels has been shown to give higher yields than SHF of same quantity of same material [21]. The remnant residue from distilling ethanol was also in smaller supply in SSF. This residue can be transformed to pulp pellet. Producing ethanol through these processes depends basically upon pretreatment. Also, syngas, a process by-product, can re-convert to ethanol [25;17]. Ethanol synthesis through citrus biomass hydrolysis, under SSF involves less economic costs and it entails pre-treatment of citrus bagasse and other waste (example enzymatic hydrolysis) to breakdown organic materials pre-microbial fermentation. Conversion of biomass feedstocks through fermenting is a vital issue because it allows to produce varied substances under mild conditions. Citrus waste is an interesting biomass source in biofuel and energy production, excluded in competition of demand with any food item. The extent of fermentation on the citrus waste largely depends on composition and biomass feedstock configuration. SSF is constrained by temperature, pH, duration of the fermentation activities, concentration of enzymes, etc.

(vi) Continuous immobilized fermentation**:** immobilized microbes are utilized in a progressive fermentation here. The microbes are recovered at the expiration of the fermentation and re-used. The microbes are packed within a column. A 12-fold higher yield of ethanol was observed after introducing a fermentation medium to a reactor for immobilized cells in which sufficient cells of Saccharomyces cerevisiae strains were packed to immobilize them [26]. A column for removing limonene was used to introduce the medium and the fermentation progressed for 10 days in an incubator regulated at 30 °C.

6.5.1.2.2 Biodiesel

This type of fuel is acquired from organic sources, which include animal fats and vegetable oils. They are non-lethal, organically degradable and when incinerated, discharge minimal quantities of harmless gases asides CO_2. Biodiesel (and glycerol) are sourced from waste peel oil extracted from citrus and subjected to trans-esterification, filtration, and distillation [27].

6.5.1.2.3 Biogas

This gas is essentially a blend of methane and carbon dioxide intermixed with minimal measures of oxygen, hydrogen, nitrogen, and hydrogen sulphide. Also, it is a natural gas, synthetically produced in a tank through fermentation after being digested anaerobically. This digestion is of two basic steps: acidogenesis, and methanogenesis. The citrus biomass is hydrolyzed via four major pathways. They are:

(i) hydrolysis by acids where the biomass is allowed to interact with acids.
(ii) saccharification by enzymes, where the biomass is first separated from the limonene before being allowed to interact with (fungi-originated) enzymes.
(iii) Fermentation, where the biomass post hydrolysis, is allowed to interact with microorganisms.
(iv) The enzymatic hydrolysis by enzymes or saccharification is allowed by bringing some enzymes on board, for example cellulase, lipase, protease, etc.

The hydrolysis progresses to some weeks. The conclusion depends on the type of substrate used. Non-complex polysaccharides, for instance carbohydrates, need shorter duration (maximum of a day) to complete their hydrolysis while complex polysaccharides such as cellulose and proteins require weeks to conclude theirs. The former comprise of weak bonds and low energy is required to break their bonds. Following action of acidogenic bacteria on them, they disintegrate to give off breakdown products for example carbon dioxide alongside other intermediate products like alcohol. They are acted upon by acetogenic bacteria to also produce carbon dioxide. Acetogenic bacteria enjoys a symbiotic relationship with methanogenic bacteria which absorb hydrogen, synthesizing methane and maintaining an environment of low hydrogen partial pressure. At high hydrogen partial pressure, hydrogenotrophic bacteria available in the biomass substrate synthesize methane from hydrogen and carbon dioxide. This symbiotic interaction between these microorganisms in a bioreactor allow for methane synthesize as the major biogas [28]. However, these microorganisms are impacted by temperature prevailing within the reactor (20–50 °C), pH of reactor, substrate concentration, etc. Steady pH ensures improved microbial activity. The optimum quantity of methane is produced within 5–25 days. Limonene can reduce biogas yield because of its antimicrobial nature. However, pretreatment (such as steam explosion) puts this in check [29].

Biogas reactors digest biomass either by batch digestion method, or by semi-continuous digestion method. With the former, reactor-maintained temperature of 25–50 °C supports maximum growth and activity of the microbes. The digestion occurs anaerobically. In contrast, the latter makes use of the residue from the biomass gotten post the pre-treatment. This process entails that the material being digested should be stirred continuously with intermittent supervision.

Both methods yield methane and ethanol as products. Wastes from citrus consists of micronutrients, in sufficient concentration to support activities of the methanogenic bacteria [29, 30].

Producing energy via microbial digestion of waste has high requirements of costs because it requires pre-treatments (power-driven stirring, limonene removal, drying, etc.) and purchasing of costly enzymes. As a newer alternative, hydrolysis of biomass with dilute acid was proposed [31]. Although still a pilot scale study, it yields smaller quantities of simple sugars and carbohydrates on digestion. On the other hand, (SSF) yielded ethanol of 76–94% content. Its biogas product contains about 65–70% of methane. SSF method also handled the challenges posed by contamination and operational costs. Removal of d-limonene before digesting the biomass increases the worth of methane produced.

Waste biomass from citrus peel is a maintainable and organic resource to produce HMF and dimethyl furan. Dimethyl furan contains 40% more energy concentration than ethanol. Producing HMF entails depolymerization of hemicellulose, pectins, and cellulose- insoluble carbohydrates, summing up to 60% of citrus peel's total carbohydrates [32]. The insoluble carbohydrates are decomposed by treating them with catalysts and ionic solvents [33].

HMF is synthesized from citrus biomass by hydrolyzing glucan to glucose, then isomerizing glucose to fructose, and later dehydrating fructose to HMF. This process has the advantage of offering low processing cost, less technical complexity, available raw material, and its appreciable market potential.

The natural character of the select ionic solvents bear on the disintegration of C-O-C bonds available in cellulose (insoluble carbohydrate) and facilitates the production rate of HMF. Ionic solvents have the advantage of being less toxic and non-inflammable than other conventional chemicals. Gas discharge from green-houses can possibly be reduced to about 134% from usage of ethanol fuel synthesized from citrus biomass as a vehicle fuel. Methane synthesized from same source is pin-pointed as a possible replacement for electricity producing natural gas. This should lower the gas discharge from greenhouses by about 77%. D-limonene as a by-product, is utilized as green organic solvent to replace acetone. The residual reactor digestate is used to produce natural biological manure for crop cultivation rather than artificial fertilizers. The discharge of gases from greenhouses released while producing inorganic fertilizers from non-renewable resources will reduce.

6.5.1.2.4 Biorefinery

This refinery merges the conversion of biomass with system equipment to manufacture biofuel amongst other chemicals. This biofuel is for power and energy utilization. In biorefinery biofuel synthesis, waste from citrus biomass needs to be pretreated via size reduction operations, filtration, centrifugation, extraction (using solvents), supercritical carbon dioxide extraction, steam explosion etc. it is then treated with chemicals (chemical treatment) to guarantee the appropriate pH. Enzymes are utilized to hydrolyze the complex polysaccharides into monosaccharides, organic acids and alcohol. These monosaccharides are converted to ethanol, energy and lactic acid asides other chemical byproducts such as levulinic acid. The residue is utilized in livestock feed or farm manure manufacture. The wastewater is useful for controlling air discharges and solid waste valorization. It is treated and reused for domestic or agrarian intents.

It is advisable to site biorefineries close to citrus processing plants to achieve important viable fuel production for energy needs. It is noteworthy that the biorefinery requirements for treating citrus biomass rich in pectin is quite different from that for treating other agricultural biomass rich in lignin, viz-a-viz hydrolysis operations for citrus waste, design of the biomass reactor, kinetic parameters and the general engineering operations.

6.6 Energy Conversion Technologies – Combustion, Gasification, and Pyrolysis

6.6.1 Thermochemical Method

Thermal treatment alternatives (combustion, gasification or pyrolysis) cannot be economically applied to citrus waste due to its high-water content (especially for fresh citrus waste). Although they would be technically feasible, they would not be efficient from an energy or an economic point of view because they would require a prior dehydration step.

6.6.1.1 Combustion

The process of combustion is a widely applied biomass conversion technology that is in usage since the origin of human civilization. Thus, it is amongst foremost methods used for biomass conversion. Combustion provides for a reaction between the hydrocarbons in the citrus waste and oxygen reacting together. Combustion converts biomass into heat, water, and carbon dioxide through an exothermic reaction in open air. However, improper oxygen quality cause to incomplete combustion, releasing pollutants such as CO and SO_2, and particulate matter into the atmosphere [34]. In the process of combustion, a solid fuel particle is introduced into a high-temperature environment, in which it undergoes drying and devolatilization, ultimately forming a residual char. Subsequently, this residual char is oxidized by substances like oxygen, carbon dioxide, and water, leaving behind an ash residue [34]. The process could be applied in two broad ways, that is either by direct conversion of the whole biomass feedstock or by biochemical conversion in which some portions of the biomass remained. Typically, direct combustion takes place in a furnace, steam turbine, or boiler, within a temperature range spanning from 800 to 1000 °C.

During the process, depending on citrus waste configuration and the process parameters, the biomass feedstock is converted to CO_2 and water including smaller quantities of other species. In fact, the nature of the citrus waste biomass has substantial influence on its combustion performance.

The amount of heat energy released during the process depends on the energy content of the feedstock as well as the conversion efficiency of the reaction. The major share of energy in the biomass is formed by the assembly of organic matter during photosynthesis and respiration in plants. The presence of high volatile matter in the citrus waste greatly influences the thermal decomposition of the biomass feedstock and as well as the combustion performance of the solid fuels. This is because, large portion of the biomass feedstock must be vapourized before the homogeneous combustion reaction takes place and the remaining char will then undergo heterogeneous combustion reaction. The main elements that constitute the biomass feedstock are carbon, hydrogen, and oxygen. Citrus waste being an agricultural waste contains higher amounts of ash forming minerals. The biomass carbohydrate structure of the citrus waste oxygenates the biomass. During the combustion process, part of the oxygen required is supplied by the organically bonded oxygen from the biomass, while the rest can be supplied through air injection into the system. The low gross calorific value of wastes from citrus indicates that the carbon existing in biomass feedstock is in partly oxidized form. Combustion process, unlike biochemical and other thermochemical conversion technologies, endeavors to disintegrate the entire fuel to simple products.

6.6.1.2 Gasification

Hydrogen is produced through pyrolysis of dried citrus biomass with the use of an electrically powered furnace [21]. Torrefaction and slowly conducted pyrolysis of biomass were carried out at about 200–650 °C to procure bio-oil and organic char using orange and lemon peel wastes after degrading them thermally. High temperature (325 °C) stable fuel using orange peels and lemon peels with energy densities of 1.58 and 1.56 respectively have been reportedly produced via the torrefaction process. Subjecting these peel residues to pyrolysis at 400–650 °C also generated high energy tars and biological chars [21]. Pyrolysis converts residual biomass thermo-chemically into solid biofuels, which comprise of chars (solid), tar (vapors) and gases. Some o.4 g of chars was synthesized from the pyrolysis of 2 grams of dried lemon waste [21].

6.6.1.3 Pyrolysis

Pyrolysis is the thermal decomposition of waste from citrus. It is similar to combustion, but it occurs either in total absence of oxygen, in a limited supply of oxygen or in a closed atmosphere with very high temperature especially in situations where partial combustion is needed to provide thermal energy. Pyrolysis involves reactions, such as dehydration, dehydrogenation, isomerization, charring, aromatization, and oxidation. These reactions yield various products, such as steam, hydrogen, carbon oxides, coal, aliphatic and aromatic hydrocarbons, pitch (tar), polymers, hydrogen, and coal [35]. These products are classified as biochar, bio-oil, and biogas/syngas. Pyrolysis has to its advantage, its low energy needs, high energy recovery and different resulting products [36]. The characteristics of the products, however, depends on process parameters such as temperature, nature of the citrus waste molecules, its particle size, heating rate and duration of residence. Various citric wastes produce different gas mixture properties, especially in terms of their energy value. Pyrolysis can directly decompose or convert waste thermally to products exhibiting any of the three nature states- gaseous, liquid and solid. Thus, citrus waste is transformed to solid (biochar/charcoal), gas (syngas), and liquid (bio-oil). The bio-oil is a liquid fuel with possibility to replace fuel oil in electricity generation and application in static heating. Bio-oil is easily transported and stored. This makes it better than the fuel gas produced by gasification [37].

6.7 Importance of Pre-treatment of Citrus Waste for Bioenergy Conversions

Citrus waste biomass in its natural form is not easily utilized in energy production hence pretreatment is often required to overcome its recalcitrant nature and make the biomass pliable for conversion for maximum product recovery. Pretreatment is an important step for both biochemical and thermochemical cellulose conversion processes. Used for disrupting and changing the configuration of cellulosic citric waste biomass, thereby making its cellulose readily accessible to the enzymes that convert the carbohydrate polymers into fermentable sugars. Pretreatment process increases the porosity of biomass, improves hydrolysis of the cellulose and assists in achieving the desired fractionation. Thus, pretreatment exposes the surface of the cellulose to enzyme attack, improving enzymatic digestibility, and facilitates subsequent processes. It also prepares the biomass for microbial attack. It improves biomass characteristics to enhance its energy utilization efficiency. A critical step that impacts the overall process [38]. The pre-treatment essentially decreases the fermentation time and time needed for production of ethanol.

The physical pre-treatments include milling which is done for biomass size reduction and surface area increase, drying, steam explosion and distillation. Size reduction of particles improves the product yield of hydrolysis by 5–25%. The milling process involves the use of different mills, such as hammer mill, roll mill, etc.,

In the process, the biomass lignocellulose is made more pliable to biodegradation, and the polymerization degree is decreased. The factors that affect the results obtained are biomass configuration, mill used and duration of operation. Treating the biomass hydrothermally at elevated pressure and temperature is also a physical pre-treatment process. Although these processes can all be combined for better results.

Conversely, chemical pre-treatment consists of treating the citrus biomass with acids or alkalis. These decompose the biomass discriminately. Strong acids are used to remove lignin and hemicellulose while alkalis counteract the pH. Chemicals used to treat the biomass are removed post-process because they produce inhibitor products that counter fermentation activities.

Explosion using steam is an example of physical pre-treatment. Here, steam of 160–260 °C interacts with the waste within a pressurized reactor for 5–20 min. The steam completely disrupts the wastes cell structure, rupturing the undigestible polysaccharide cell walls into smaller digestible compounds to encourage enzymes to act. The pressure is released at the end of the stipulated reaction time. The reactor used is one that allows for a fast reduction in pressure [29]. Although this technique is energy intensive, it helps to expediate assimilation of acid and enzymes that will work on the undigestible molecules, permits speedy biomass disintegration, ensures effectual discharge of sugars and improved cellulose hydrolysis [23]. Reduction of the waste's particle size decreases the required hydrolysis time. *D*-limonene concentration reduced after a popping treatment of citrus peels, which enabled a smooth fermentation process [24]. Popping treatment ruptures the wastes cell walls and reduces their particle size. Improved ethanol production from raw peel waste was also recorded. In addition, explosion treatment with steam assisted in removing and/or recovering limonene. The biomass separated out after the steam treatment. This helps in isolating limonene.

Explosion method using steam is made more efficacious by adding acids as catalyst. However, its demerits are:

(a) Non-uniform disruption of cell walls, which causes partial digestion.
(b) Development of materials that inhibit fermentation.
(c) The waste's xylan fraction is lost.

6.7.1 Limonene

Removing or recovering limonene is a very important step done before fermentation of biomass while manufacturing biogas/biofuel. Aeration and biological pretreatments are used to remove limonene while steam distillation, cold pressing, sorption, steam explosion, centrifugation, liquid extraction (using organic solvents) are for recovering limonene. The outcome is usually similar to watery emulsion and is centrifuged to separate out the oils. With the exception of the cold pressing method, all other techniques, when operated at increased temperatures, elicit chemical alterations in the composition of the materials, especially in volatile biological molecules. They may disappear entirely if exposed for a long time to the increased temperature [39]. The recovery process is more cost effective than the removal process [40].

Biological treatment: this pretreatment is by either fungi (such as some species of *Sporotrichum*, *Aspergillus*, *Penicillium* and *Fusarium*) activity or via enzymes procured from these fungal strains. The use of such enzymes sourced from fungi decreased the amount of limonene. These enzymes are used to facilitate anaerobic digestion [17]. After the decrease, with the application of micro-organisms, fermentation progresses to ethanol synthesis.

The quantity of limonene is minimized or eliminated by treating the peel waste from citrus with *P. digitatum*, which causes the biological decomposition of limonene, resulting in production of α-terpeneol [17, 41]. Carvone and carveol produced with α-terpeneol, possess anti-microbial properties and thus, they prevent fermentation from going on [42]. There are other compounds with stronger antimicrobial characteristics than limonene, found within the essential oils. For instance, α-terpeniol, when acting on some bacterial strains (for example *E. coli* and *S. aureus)* displays more antimicrobial properties than limonene [43]. Another disadvantage against limonene is the accumulation of fatty acids it produces, which cause system failure [44].

In fact, for recovery of limonene, ethanol extraction method was discovered to be the best approach, with steam distillation following and lastly, biological treatment. The former is achieved by adding the mixture of ethanol (70%) and water (30%) to the biomass. The solvent phase comprising of limonene separates out after stirring the mixture for about 1–2 hours. Steam distillation method, in addition to extracting limonene, also isolates appreciable quantities of bioactive compounds.

6.8 Potential Benefits of Utilizing Citrus Waste for Energy Synthesis

Fruit residues should not be wasted, rather they should be transformed to useful products. Energy synthesis commencing from citrus waste provides alternative use as eco-friendly fuel. Waste from citrus is a biological resource used via anaerobic digestion and fermentation for synthesizing biofuels that include as bioethanol, biogas and biodiesel [45]. Thus, these wastes are utilized as per renewable source of energy. Another use of citrus seed oil is as a source for biofuel. Recently, seed oil utilized in biofuel production has gained importance. Its use in production of biodiesel is currently being investigated. The seed oil from *Citrus reticulata* (mandarin) and *Citrus reticulate* seeds investigated for this cause [15, 46] successfully produced biodiesel with 76.93% from the *Citrus sinensis* seeds.

Bioethanol as a renewable fuel has found use as a substitute for gasoline. Some countries have substituted gasoline with bioethanol and diesel with biodiesel [47]. Bioethanol can also be an additive in gasoline and a fuel in cars that operate with a gasoline engine [48]. Bioethanol mixed with gasoline in various proportions is utilized for its properties are comparable to other fuels burned to obtain other energy forms. Consequently, bioethanol can also be utilized for electricity (power devices), transportation (in aviation, land transport, engine lubricants) and heat generation (heating systems, solvent for industrial use, cooking systems) [47]. Although it has more octane number, which enables it to release more engine power, it has a lower calorific value and so requires more fuel to cover same distance with gasoline [47].

Lemon, grapefruit and orange contain citric acid. The production of bioelectricity using a galvanic cell and microbial-fuel cell technique is examined at a laboratory scale. They can therefore be used as batteries to illuminate LEDs and power clocks, calculators etc. [49]. Under certain conditions, the citric acid contained in citrus fruit can also be subjected to certain conditions by which they are made to perform as electrolytes, thereby generating electricity the same way as a galvanic battery.

Asides exploring bioelectricity generation from (microbial fuel cell) *Citrus* peel waste as a sustainable energy producer, photoluminescent carbon-based dots sourced orange juice which exhibited chemical inertness, low toxicity and easy operationality, can also have potential use as a search material in cellular imaging. Citrus waste (mesoporous cellulose from residues sourced from orange peels, nanocellulose from *Citrus* waste and cellulose nanofibrils from mandarin peels) can also have application in nanomaterials development. Pectins from citrus wastes are utilized in microsphere and nanobacterial cellulose development [50].

6.8.1 Relationship with Other Biomass

Currently, citrus waste appears promising for producing fuel and energy. Orange peel can possibly be converted into hydrogen. The biomass's organic conformation is an important parameter for evaluating the quality of syngas. Studies show that the distribution of gas products from citrus peel gasification is analogous to pine sawdust gasification. Minimal discrepancies were observed in concentrations of major gaseous products. For instance, H_2 makes up the largest proportion of syngas while the H_2 concentration of 57.70% and 58.53% are obtained respectively for pine sawdust.

However, biomass is a very complex material, because various types of biomasses have different characteristics in terms of ash, fixed carbon and volatile matters. In comparison with pinewood biomass, the syngas yield and gasification efficiencies obtained from citrus peel were more appreciable than those obtained from pinewood. The start and end temperatures of citrus peel steam gasification were lesser than that of pinewood. Citrus peel is easier to degrade thermally than woody biomass.

Development of biogas from citrus organic waste is promising because its constituent mineral ions enhance methane yield, increasing its methane potential [51].

The heterogeneity of biomass accounts for differences in the properties of biochar produced by different biomass in the pyrolysis step, which have an important influence on biomass gasification.

At present, some scholars have compared the gasification product properties (such as gas yield, and composition) between citrus peel and lignocellulosic biomass [51]. In addition to ash, the pore structure and the degree of graphitization also affect the reactivity of biochar, which in turn influence the gasification performance of any biomass.

6.9 Overview of Pyrolysis Processes of Conversion & Products

Based on the temperature and heating rate applied, pyrolysis is grouped into three categories:

- Torrefaction, or mild pyrolysis
- Slow pyrolysis
- Fast pyrolysis

The intended outcome of the process influences the choice of pyrolysis type to be undertaken.

6.9.1 Torrefaction

Torrefaction is a thermochemical pyrolysis process in which the citrus waste is excited to 200 to 350 °C in oxygen exclusion, at atmospheric pressure, low particle rates of heating and a one-hour reactor duration. This causes the natural configuration of the citrus waste to partly decompose and alter. The biopolymers—cellulose, hemicellulose and lignin partially decompose to release organic volatiles, creating a hydrophobic, soft, dry and black residual solid regarded as torrefied biomass or char (bio-coal), are easily pulverized [52]. Ultimately, producing water, carbon dioxide, methanol, carbon monoxide and acetic acid. Torrefaction improves the energy density and fuel properties while reducing the weight of the biomass [52] achieved by removing moisture and other unneeded volatile substances.

6.9.2 Slow Pyrolysis

Pyrolysis carried out slowly is best for biochar production. The process yields some solid charcoal and gas. The decomposition process used in slow pyrolysis results in char at decreased temperatures, unhurried heating rates, and lengthy vapor residence times.

6.9.3 Fast Pyrolysis

Fast pyrolysis provides bio-oil which is a liquid fuel [34] Flash method of pyrolysis entails heating speedily (1–2 seconds) the waste materials to 500 °C at a rate of more than 200 °C per second. Bio-oil, also called pyrolysis oil is a dark brown, viscous substance with various chemical compositions, including acids, aldehydes, alcohols, aldehydes, phenols, and oligomers derived from lignin. The content obtained from citrus may not be appreciable as citrus waste is generally low in lignin. Bio-oil with low calorific value [53], is likely to replace crude oil. Improving pyrolysis oil is attained physically, chemically and through catalytic procedures. Pyrolysis oil has about 75 weight percent yield. It is also cost-effective, energy efficient, and the technology is environmentally friendly [54, 55].

6.10 Application of Citrus biofuels in Industries

citrus waste has application in animal feed and organic compost production, as well as in bio-fuels and bioethanol production. The solid and concentrated liquid citrus waste could be converted into 'citrus waste-based molasses', industrial alcohol and/or ethanol, lactic acid, feed yeast, and vinegar

Citrus peel has been studied as a green source of bio-adsorbent. Bio-adsorbents are compounds that eradicate toxic substances and heavy metals. Nowadays, agro-waste is widely utilized as a low-cost bio-adsorbent for the removal of heavy metals from the solution. A low-cost, environmentally friendly activated carbon was prepared from Citrus limetta peels through carbonization and alkali chemical activation to remove ranitidine in pharmaceutical wastewater. Citrus limetta peels, orange peel and lemon peels have all been used in developing economic, eco-friendly bio-adsorbents [45].

Citrus waste is a natural resource utilized as a renewable source of energy for biofuel production such as bioethanol, biodiesel, and biogas; utilizing fermentation and anaerobic digestion for the bio-conversion of citrus peels into biofuel and through trans-esterification of essential oil with alcohol for biodiesel production. For example, bioethanol was prepared from mandarin peel utilizing pretreatment with steam explosion and microbial fermentation; bioethanol was generated using orange peel waste by simultaneous pretreatment with acid-catalyzed steam explosion and separate enzymatic hydrolysis, with fermentation with *Saccharomyces cerevisiae*. Citrus limetta peels was applied in producing bioethanol via the acid-catalyzed steam pretreatment, followed by enzymatic hydrolysis and fermentation.

Citrus wastes have enormous industrial potential, from plastic-like food coatings polymers and packaging to effective composites with antioxidant and antimicrobial activities. For example, pectin can be the base polymer for a new form of food packaging or coating, being a possible substitute for plastic. The pectins from citrus (such as lemons and lime peels) are utilized in producing film coating that enhances food properties. Pomelo peels produce a biodegradable flexible and transparent films suitable for oil packaging [56].

Citrus by-products can also be used for nanoparticle stabilization. In this function, they can add directly or indirectly to the food. Nanoparticles, such as commercial ZnO nanoparticles, are used in food packaging to reinforce the polymeric matrix. Zinc oxide nanoparticles can be strengthened with *Citrus sinensis* peel extract to achieve a film with better antimicrobial activity, better crystallinity and cytotoxicity.

Plastics used for food packaging are of enormous environmental concern by their non-biodegradable nature. Citrus waste and its extracts have successfully leveraged on the high permeability to moisture and gas that some conventional packaging materials such as kraft paper offer. Mandarin peel in synergy with hydrophobic leaf extracts was used to enhance this gas and water barrier challenge. Polyhydroxybutyrate (PHB) is a biodegradable thermoplastic that gives an amber color with little transparency. Addition of *D*-limonene extracted from citrus waste to PHB improves its properties. Addition of *D*-limonene to PLA and poly-hydroxybutyrate (PHB) (75:25 ratio) film, gives a transparent and flexible film with added advantages of enhanced oxygen barrier and water- resistant properties, that can be applied in biodegradable food use.

Extracts from *Citrus* wastes can be used as functional additives in the food packaging polymeric matrix to inhibit or suspend food spoilage, and to produce bacterial cellulose to substitute conventional polymeric matrix of the food package.

Citrus waste found application in production of bacterial cellulose because of its high carbohydrate content. Bacterial cellulose is a polymer of β-1,4-linked glucose produced by aerobic bacteria. With its high characteristic cellulose purity, crystallinity, water holding capacity and tensile strength, it can be applied to production of fruit jellies and cocktails in the food industry. Bacterial Cellulose has currently been produced from mandarin, lemon, orange and grapefruit peels [56].

Currently, application of citrus waste to the synthesis of iron and silver nanoparticles permitting cytotoxicity and enzymes like pectinase and larvicides have been suggested [45].

6.11 Torrefaction Process as a Pyrolytic Treatment for Citrus Waste

6.11.1 Forms of Torrefaction Processes

Torrefaction processes represent tailored methods employing thermochemical techniques to diminish the moisture and volatile constituents within biomass. This results in enhanced fuel characteristics, including improved grindability, a more uniform composition, heightened energy density, hydrophobic qualities, and the cessation of biological activity. Torrefaction stands as a promising preliminary step in transforming biomass into solid fuel with a greater energy density, suitable for subsequent thermal conversion systems [57].

6.11.1.1 Oxidative Torrefaction

Oxidative torrefaction has become increasingly important for preserving mass and maintaining a balanced energy density. Through oxidative torrefaction, elemental carbon buildup can be controlled, and surface functional groups can be removed, thereby improving fuel properties. This technology is environmentally friendly and facilitates the conversion of biomass into biochar. Torrefaction processes have demonstrated efficiency within a temperature range of 200 °C to 300 °C and residence times of 30 to 60 minutes, employing various oxygen contents.

6.11.1.2 Wet Torrefaction

Wet torrefaction serves as a pre-treatment method for biomass, typically conducted in hot compressed water or hydrothermal media, operating within a temperature range of 180 °C to 260 °C. To mitigate energy dissipation resulting from latent vaporization heat, it's essential to maintain the water in its liquid phase throughout the process. This technological approach to wet torrefaction offers a means to enhance the quality of moist biomass, transforming it into hydro-char solid fuel while minimizing high power consumption.

6.11.1.2.1 Application of Wet Torrefaction

They yield the same mass of torrefied product as inert torrefaction but necessitate lower temperatures due to the heightened reactivity of water compared to inert gas. Wet torrefaction stands as a non-toxic process, demanding only a simple chemical reactor setup.

6.11.1.3 Steam Torrefaction

In steam torrefaction, steam is generated through fuel combustion in a boiler, and then condensed steam undergoes circulation in regenerative stages. Biomass undergoes steam treatment at temperatures exceeding 260 °C for approximately 10 minutes. The challenge in applying steam torrefaction lies in preventing steam condensation throughout the process, which necessitates a substantial mass flow of steam. Additionally, complex processing is required before reintroducing steam into the steam cycle. Various heating streams can be derived from the steam cycle to indirectly provide torrefaction heat, sourced from boiler hot flue gases and live streams.

6.12 Impact of Torrefaction on Biomass Structure and Properties

Torrefaction process parameters, such as temperature and duration, exert a significant influence on biomass structure, inducing structural alterations in key constituents like cellulose, hemicellulose, and lignin. Common transformations involve deacetylation and decomposition of hemicellulose, cellulose degradation, demethyloxylation of lignin, and cleavage of aryl ether linkages. Through torrefaction, enhancements in the chemical composition, physical attributes, storage capacity, and energy properties of biomass are realized. Achieving a temperature of 300 °C can lead to an energy content increase of approximately 30%.

6.13 Characterization of Torrefied Citrus Waste

The ash (% wt), carbon (% wt) hydrogen (% wt), Nitrogen (% wt), Oxygen (% wt), gross calorific (MJkg-1) and Net calorific value (MJkg-1) for orange peel torrefied at 275 °C is 9.73, 67.05, 4.46,2.04, 16.64, 25.60 and 24.63 respectively [58].

6.14 Benefits of Torrefied Citrus Waste Compared to Green Biomass

Torrefied citrus waste improves its fuel properties, higher energy density, hydrophobic behavior, homogenized composition, hydrophobic behavior and elimination of biological activities. The green biomass of citrus has significant relevance as natural ingredients for packaging materials, medicine, cosmetics, and biofuels. In addition, they can be utilized as natural antioxidants, enzymes, pectin and essential oil.

6.15 Effect of Thermal Treatment on the Energy Yield and the Dry Mass of the Torrefied Biomass (Citrus Waste)

Citrus waste contains inherent physicochemical properties that limits its direct conversion to energy in its raw state. Some of these properties includes the strong absorption of water from the environment, high moisture and ash compositions, low energy and bulk density, increased alkalinity and acidity forming residues and high carbon—oxygen ratios [59, 60]. Also, citrus waste contains networks of fibrous structures which poses difficulty during the comminution process of the biomass

Torrefaction processes underscores, quality pretreatment techniques employed to improve the green properties of the biomass prior to conversions into energy and energy applications [61].

Torrefaction is a thermal treatment which employs direct combustion and coal activated combustion on the biomass usually carried out in a reactor [62]. The treatment disrupts the fibrous structure and the cohesiveness of the biomass, it raises the calorific value and lowers the carbon—oxygen ratios and its moisture retentiveness [60]. It is expected that 80–90% energy and 70–90% dry mass of the biomass be retained after torrefaction processes, the other 10–30% of the dry mass is usually converted into torrefaction gases containing about 10% energy yield of the biomass [63].

Torrefaction methods is an isothermal combustion process that leads to heat transfers and reaction rates of depolymerization, incomplete devolatilization and carbonization. Combustion processes employ high temperature variations in the reactor which are cooled after the process. This process increases the bulk density of the biomass, easing transport and storage [64]. When green biomass is torrefied, clean combustion producing less smokes is achieved with a better energy yield with increased dry mass retention than green biomass fed to the gasifiers [65].

There is a strong relationship between temperature of combustion and the residence time during torrefaction, studies investigated by Strandberg et al. [66] found that increased combustion of biomass resulted to decreased energy consumption during milling, energy and dry mass yields, volatile components, cellulose and hemicellulose structure, hydrogen content and angle of repose.

6.16 Thermochemical Properties of the Biomass Influencing Heat Transfer and Reaction Rates

During thermochemical treatment processes, some thermochemical properties of the biomass can significantly influence both the heat transfer and reaction rates, some which includes the following:

6.16.1 Vapour Residence Period

In the processing of liquids products obtained from pyrolysis, short—time residence periods are considered to be useful [67]. Depolymerization otherwise known as cracking of vapours is achievable when pyrolysis pressures are applied [68]. The problems with the cracking of vapours are always associated with the heat transfer challenges at the particle surface, which makes it impossible to achieve optimum biomass conversion in this period. In other words, this signifies that reducing the residence time would not proffer workable solution to achieving high product yield.

Hence, the residence time needs to be optimized considering a wide array of factors in order to achieve high quality product yield [69].

6.16.2 Bulk Density of the Biomass

The study investigated by Haykiri-Acma [70] found that high bulk density of biomass results to the recovery of large amount of solid char in a slow pyrolysis process of hazelnut shell with linear heating rate of 20 °K/min ambient temperature to 1173 °K under dynamic atmospheric nitrogen, yet the combustion process will still require more activation energy. Hence, as bulk density decreases, solid char yields, activation energy and values of decomposition rates are also reduced. Several researchers including Şensoz et al. [71] and Natarajan and Ganapathy Sundaram [72] have examined the effect of bulk density on the product yield in a pyrolysis process. They obtained similar findings that bio—oil products obtained are dependent on the bulk density of the biomass, although the particle size of the feed used depend largely on the type of pyrolizer and the nature of the biomass [69]. According to Akhtar and Saidina Amin [69], particle size less than 2 mm are desirable for industrial applications to improve the rate of conversions [73].

6.16.3 Moisture Content of the Green Biomass

Green biomass needs to be dried because the pyrolysis process is efficient when the moisture content of the green biomass is less than 10% [74]. The hydroscopic nature of the green biomass affects the physicochemical properties of the biofuel products and the behaviour of the biomass during pyrolysis process [75]. The optimum lower moisture content of this green biomass is necessary as fully dried biomass may pose heat transfer challenges due to its poor heat conductivity. The optimum lower water activity is needed since water serves as a medium for heat transfers assisting pyrolysis, and also as a reactant to biomass species [69].

6.16.4 Rate of Combustion

The combustion rates categorize pyrolysis into slow, fast and flash. When the combustion rate is rapid, the biomass is fragmented, yielding less solid char and increased syngas products [75]. The studies carried out by Uzun et al. [73] indicated that for rapid combustion rate employed on the pyrolysis of soybean cake, there was corresponding increased yields of bio—oil product and the general pyrolysis conversions. Also, the water content of the bio—oils decreased with increasing combustion. The combustion enthalpy of citrus green biomass obtained with an isoperibolic calorimeter reported by Pulidori et al. [76] is presented in Table 6.1.

6.16.5 Mineral Content of Biomass

Green biomass contains inherent mineral matter of about 4 to 10% and inorganic metals in trace amount like calcium, potassium, sodium, phosphorus and magnesium values obtained from the pyrolysis ash [77]. When the metals are alkaline, the speed up thermal degradation of biomass during pyrolysis, allowing them to reduce temperature degradation thereby affecting product yields significantly [78]. Generally, the characteristic of the mineral matter composition of biomass affects the reactivity of the solid char during secondary pyrolysis reactions. It also aids the extraction of char and syngas by the substantial reduction of the volume of bio—oils extracted [69].

6.16.6 Sweeping Gas

Researchers has established that inert gasses like atmospheric nitrogen and water vapours are used to purge vapors from hot pyrolysis rapidly. Akhtar and Saidina Amin [69] found that the nitrogen flow rates affect the residence time for the produced vapour phase of the pyrolysis process. He further reported that the product yields of solid char, bio—oils and syngas were not significantly impacted when the sweeping gas flow is decreased, but the volume of water increased significantly. When rapid conversions are not achieved by the sweeping process, organic compounds that are unstable undergo secondary dehydration [79].

Table 6.1 The combustion enthalpy of citrus green biomass

Biomass	Biomass residue	Combustion enthalpy $-\Delta_c H$ $(kJ/g)^a$
Citrus waste	Orange peels green biomass	15.1
	Tangerine peels green biomass	16.5
	Grapefruit peels green biomass	17.2
	Pomelo peels green biomass	15.5
	Lemon peels green biomass	15.9

[a]: RSD < 2%

6.17 Optimal Temperature Range for Biomass Stability

The optimal operating conditions are highly variable, therefore, the conditions that ensures biomass stability are to be established. Depending on the temperature employed, the thermochemical processes are categorized into drying, torrefaction, pyrolysis and gasification [80]. Each of the categories has different yields as products and intermediates. When the temperature of the process is increased, energy values may also increase. Figure 6.1 showed the optimum temperature range for biomass stability during combustion and their intermediate products.

Drying of the biomass, which is the first phase of the biomass conversion is achieved within the temperature range of 25 to 150 °C. The next combustion process is the torrefaction process which deals with the heating of the biomass in an inert/anaerobic medium. This varies with temperatures ranging from 200 to 300 °C and combustion rate that is <50 °C/min applied with ambient pressure [81]. The several factors affecting torrefaction process are the temperature of the system, concentration of oxygen present, the residence time, bulk density of the biomass and the type of reactor used [82]. The major product obtained from the torrefaction of the biomass is the solid char (biochar). When higher temperature above 300 °C are employed in the torrefaction process, it may result to much devolatilization of the biomass due to the sudden pyrolysis process being initiated. The study investigated by Chin et al. [83] employed the use of response surface methodology to optimizing the torrefaction conditions. They observed an increased material degradation related to the interaction between temperature and incidence time.

Pyrolysis precedes torrefaction with increasing temperature ranging from 300 to 600 °C. The major products of pyrolysis are activated carbons, bio—oils, bio—crude or pyrolysis oil (liquid products). The final phase is the gasification which produces syngas within the temperature above 600 °C. Studies carried out by Fu et al. [84] found that gas yield increased significantly when the temperature was increased for biomass obtained from rice and cotton straw, maize stalk and rice husk while the biochar and the bio—oil yields decreased significantly. The studies reported further that as the temperature of pyrolysis increased, syn-gasses like methane and carbon monoxides expanded incrementally.

6.18 Conclusion

Management of citrus waste is a multifaceted challenge that requires innovative solutions to decrease waste generation, increase resource utilization, and mitigate environmental impact. By adopting sustainable practices and leveraging the value of citrus waste, fruit processing industries can contribute to a more circular economy and a healthier planet. Inadequate citrus waste and other waste control pose adverse effects to our health by spreading of diseases [10]. Thermal working of citrus wastes through the torrefaction process has been established as important in

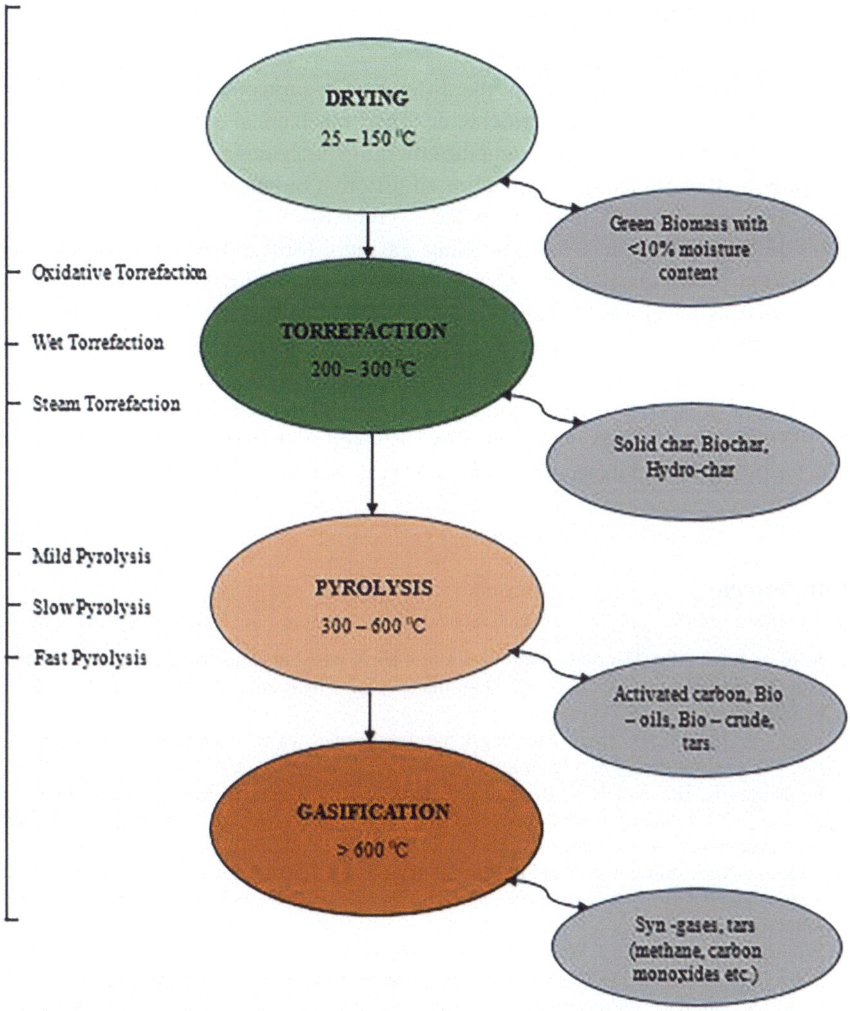

Fig. 6.1 The optimum temperature range for biomass stability during combustion and their intermediate products

the biomass conversion with strong relationship to increasing the energy and dry mass yields. This process is invariably affected by the degree of temperature employed for every phase of the thermochemical processes including the drying, torrefaction, pyrolysis and gasification. The product yields of every thermochemical phase like the solid char, liquids (bio—crude, bio—oils and activated carbons) and the syngas produced increases or decreases based on the temperature dependency of the combustion process when all the physicochemical factors affecting the heat transfers and reaction rates are considered.

6.19 Emerging Prospects

The many techniques employed in the different categories of biomass conversions including the thermochemical, biochemical and mechanical or physical processes has been improved over the years and the emerging processes are particularly geared at improving the efficiency of utilization of different biomasses for biofuel conversions. Such emerging techniques includes combustion for thermochemical processes [85], gasification processes using gasifiers [86] and devolatilization and carbonation processes [87]. Some researchers have employed biochemical processes such as anaerobic digestion systems [88] and fermentation of biomass to produce ethanol [89]. Comminution processes has been investigated for mechanical and physical processes [58] and studies on densification processes has been reported [90]. Recent prospects of calculating kinetic data of biomass pyrolysis using TG/DTG/DTA techniques was investigated by El-Sayed et al. [91] where typical parameters for pyrolysis and thermodynamics were estimated.

References

1. Joshi, V. K., & Attri, D. (2006). Optimization of apple based medium and fermentation conditions for pigment production by Rhodotorula species. *Proceedings of the National Academy of Sciences, India, 76B*, 171.
2. Joshi, V. K. (2020). Fruit and vegetable processing waste management-an overview. *International Journal of Food Fermentation and Technology, 10*(2), 67–94.
3. Ejigayehu, T., Tilahum, A. T., Ruchira, N., Jyoti, R. R., Difo, V. H., Tessema, A., & Markos, M. U. (2023). Fruits by-products and their industrial applications for nutritional benefits and health promotion. A comprehensive review. *Sustainability, 15*, 7840.
4. Food and Agriculture organization of the United Nations. (2019). *The state of food and agriculture, moving forward on food loss and waste reduction*. Food and Agriculture organization of the United Nations.
5. William, P. T. (2005). In J. Wally (Ed.), *Water treatment and disposal* (p. 9). Great Britain.
6. Behzod, S., & Keikhosro, K. (2018). Citrus processing wastes; environmental impacts, recent advances, and future perspectives in total valorization. *Resourses, Conservation and Recyling, 129*, 153–167.
7. Caterina, L., Vito, A. D., Luigi, B., Antonino, G., Eristanna, P., Michelle, T., Gaspare, V., & Santo, F. C. (2022). Challenges and opportunities for citrus wastewater management and valorization; A review. *Journal of Environmental Management., 312*, 1–3.
8. Tamelova, B., Malafak, J., & Velebil, J. (2018). Energy valorization of citrus peel waste by torrefaction treatment. *Agronomy Research, 16*(1), 276–285.
9. Brian, L., Craig, H., James, L., Lisa, K., Richard, W., & Tim, S. (2013). *Reducing food loss and waste-installment 2 of creating a sustainable food future*. World Resource Institute.
10. Ansuman, S., Akanksha, D., Partvati, M., Umesh, K., Rajesh, K. S., & Supriya, T. (2024). Insights into the management of food waste in developing countries; with special reference to India. Food waste generation and management strategies and policies. *Environmental Science and Pollution Research., 31*, 17887–17913.
11. Mini, S., Manbir, S., & Rajan, S. (2023). The green potential of citrus waste; value addition and sustainability. *International Clinical and Medical Case Reports, 2*(17), 1–10.

12. Sicari, V., Pellicano, T. M., Lagana, V., & Poiana, M. (2018). Use of orange by-products (dry peel) as an alternative gelling agent for marmalade production; evaluation of antioxidant activity and inhibition of HMF formation during different storage temperature. *Journal of Food Processing and Preservation, 42*, e13429.
13. Zahra, M., Waseem, K., Hafiz, T. A., Hyrije, K., Zaryab, J., Sadeq, K. A., Laila, A. A., Bader, D. M. D., Mohammed, A., Mohamed, A., & Ammar, A. (2023). Citrus waste as source of bioactive compounds. *Extraction and Utilization in Health and Food Industry., 28*(4), 1636.
14. Jeong, D., Park, H., Jang, B. K., Ju, Y., Shin, M. H., Oh, E. J., & Kim, S. R. (2021). Recent advances in the biological valorization of citrus peel waste into fuels and chemicals. *Bioresource Technology., 323*, 124603.
15. Rashid, U., Ibrahimc, M., Yasin, S., Yumus, R., Taufiq-Yap, Y. H., & Gerhard, K. (2013). Biodiesel from citrus reticulate (mandarin orange) seed oil, a potential non—Food feedstiock. *Industrial Crops and Products, 45*, 355–359.
16. Boluda-Aguilar, M., & López-Gómez, A. (2013). Production of bioethanol by fermentation of lemon (Citrus Limon L.) peel wastes pretreated with steam explosion. *Industrial Crops and Products, 41*, 188–197.
17. Ruiz, B., & Flotats, X. (2014). Citrus essential oils and their influence on the digestion process: An overview. *Waste Management, 34*(11), 2063–2079. ISSN 0956-053x, https://www.sciencedirect.com/science/article/pii/S0956053X14002827
18. Smith, D. C., Forland, S., Bachanos, E., Matejka, M., & Barrett, V. (2001). Qualitative analysis of citrus fruit extracts by GC/MS: An undergraduate experiment. *The Chemical Educator, 6*(1), 28–31.
19. Bakkali, F., Averbeck, S., Averbeck, D., & Idaomar, M. (2008). Biological effects of essential oils- A review. *Food and Chemical Toxicology, 46*(2), 446–475. https://doi.org/10.1016/j.fct.2007.09.106
20. Fuertes, B. R. (2015). *Effect of limonene on anaerobic digestion of citrus waste and pretreatments for its improvement*. A PhD Thesis. Universitat Politecnica De Valencia.
21. Mahato, N., Sharma, K., Sinha, M., Dhyani, A., Pathak, B., Jang, H., Park, S., Pashikanti, S., & Cho, S. (2021). Biotransformation of citrus waste-I: Production of biofuel and valuable compounds by fermentation. *PRO, 9*, 220. https://doi.org/10.3390/pr9020
22. Stewart, D. S., Widmer, W. W., Grohmann, K., & Wilkins, M. R. (2006). Ethanol production from citrus processing waste. U.S. patent US20060177916A1.
23. John, I., Muthukumar, K., & Arunagiri, A. (2017). A review on the potential of citrus waste for d-limonene, pectin, and bioethanol production. *International Journal of Green Energy, 14*, 599–612.
24. Choi, I. S., Kim, J.-H., Wi, S. G., Kim, K. H., & Bae, H.-J. (2013). Bioethanol production from mandarin (Citrus unshiu) peel waste using popping pretreatment. *Applied Energy, 102*, 204–210.
25. Boluda-Aguilar, M., García-Vidal, L., del Pilar González-Castañeda, F., & López-Gómez, A. (2010). Mandarin peel wastes pretreatment with steam explosion for bioethanol production. *Bioresource Technology, 2010*(101), 3506–3513.
26. Choi, I. S., Lee, Y. G., Khanal, S. K., Park, B. J., & Bae, H.-J. (2015). A low-energy, cost-effective approach to fruit and citrus peel waste processing for bioethanol production. *Applied Energy, 140*, 65–74.
27. Bull, O., & Obunwo, C. (2014). Bio-diesel production from oil of orange (Citrus sinensis) peels as feedstock. *Journal of Applied Sciences and Environmental Management, 18*, 371–374.
28. Deublein, D., & Steinhauser, A. (2008). *Biogas from waste and renewable resources: Formation of biogas* (pp. 87–147). Wiley-VCH.
29. Forgács, G. (2012). *Biogas production from citrus wastes and chicken feather: Pretreatment and co-digestion*. Chalmers University of Technology.
30. Martín, M. A., Siles, J. A., Chica, A. F., & Martín, A. (2010). Biomethanization of orange peel waste. *Bioresource Technology, 101*, 8993–8999.

31. Rezzadori, K., Benedetti, S., & Amante, E. R. (2012). Proposals for the residues recovery: Orange waste as raw material for new products. *Food and Bioproducts Processing, 90*, 606–614.
32. Marín, F. R., Soler-Rivas, C., Benavente-García, O., Castillo, J., & Pérez-Alvarez, J. A. (2007). By-products from different citrus processes as a source of customized functional fibres. *Food Chemistry, 100*, 736–741.
33. Chun, J. A., Lee, J. W., Yi, Y. B., Hong, S. S., & Chung, C. H. (2010). Catalytic production of hydroxymethylfurfural from sucrose using 1-methyl-3-octylimidazolium chloride ionic liquid. *Korean Journal of Chemical Engineering, 27*, 930–935.
34. Kataya, G., Cornu, D., Bechelany, M., Hijazi, A., & Issa, M. (2023). Biomass waste conversion technologies and its application for sustainable environmental development—A review. *Agronomy, 13*, 2833. https://doi.org/10.3390/agronomy13112833
35. Saravanan, A., Senthil Kumar, P., Khoo, K. S., Show, P., Femina Carolin, C., Fetcia Jackulin, C., Jeevanantham, S., Karishma, S., Show, K., & Lee, D. (2021). Biohydrogen from organic wastes as a clean and environment-friendly energy source: Production pathways, feedstock types, and future prospects. *Bioresource Technology, 342*, 126021.
36. Angin, D., & Şensöz, S. (2014). Effect of pyrolysis temperature on chemical and surface properties of biochar of rapeseed (Brassica napus L.). *The International Journal of Phytoremediation, 16*, 684–693.
37. Dhyani, V., & Bhaskar, T. (2018). A comprehensive review on the pyrolysis of lignocellulosic biomass. *Renewable Energy, 129*, 695–716. https://doi.org/10.1016/j.renene.2017.04.035
38. Al Afif, R., Pfeifer, C., & Pröll, T. (2020). Bioenergy recovery from cotton stalk. In M.-R. Ansari (Ed.), *Advances in cotton research*. IntechOpen. IntechOpen. https://doi.org/10.5772/intechopen.88005
39. Mahato, N., Sharma, K., Koteswararao, R., Sinha, M., Baral, E., & Cho, M. H. (2019a). Citrus essential oils: Extraction, authentication and application in food preservation. *Critical Reviews in Food Science and Nutrition, 59*, 611–625.
40. Mahato, N., Sinha, M., Sharma, K., Koteswararao, R., & Cho, M. H. (2019b). Modern extraction and purification techniques for obtaining high purity food-grade bioactive compounds and value-added co-products from citrus wastes. *Food, 8*, 523.
41. Bausbia, N., Vian, M. A., Ferhat, M. A., Meklati, B. Y., & Chemat, F. A. (2009). New process for extraction of essential oil from citrus peels: Microwave hydrodiffusion and gravity. *Journal of Food Engineering, 90*, 409–413.
42. Badee, A., Helmy, S., & Morsy, N. (2011). Utilisation of orange peel in the production of α-terpineol by Penicillium digitatum (NRRL 1202). *Food Chemistry, 126*, 849–854.
43. Pasqua, R. D., Hoskins, N., & Betts, G. (2006). Changes in membrane fatty acids composition of microbial cells induced by addition of thymol, carvacrol, limonene, cinnamaldehyde, and eugenol in the growing media. *Journal of Agricultural and Food Chemistry, 54*, 2745–2749.
44. Kaparaju, P., & Rintala, J. (2006). Thermophilic anaerobic digestion of industrial orange waste. *Environmental Technology, 27*, 623–633.
45. Suri, S., Singh, A., & Nema, P. K. (2022). Current applications of citrus fruit processing waste: A scientific outlook. *Applied Food Research, 2*(1), 100050. https://doi.org/10.1016/j.afres.2022.100050
46. Ezekoye, V., Adinde, R., Ezekoye, D., & Ofomatah, A. (2019). Syntheses and characteriszation of biodiesel from citrus sinensis seed oil. *Scientific African, 6*, e00217. https://doi.org/10.1016/j.sciaf.2019.e00217
47. Pilco, M. E. B., Masabanda, P. G. B., & Pilco, C. J. (2022). Potential use of citrus (Peel) waste to obtain bioethanol. *International Journal of Current Microbiology and Applied Sciences*. ISSN: 2319-7706, *11*(04), 207–214.
48. Aguilar, R. N. (2010). Azúcar, coproductos y subproductos en la diversificación de la agroindustria de la caña de azúcar. *Revista VIRTUALPRO, 106*.

49. Khan, A. M., & Obaid, M. (2015). Comparative bioelectricity generation from waste citrus fruit using a galvanic cell, fuel cell and microbial fuel cell. *Journal of Energy in Southern Africa, 26*(3), 90–99.
50. Ellouze, I. (2022). *Citrus* bio-wastes: A source of bioactive, functional products and non-food uses. In M. F. Ramadan & M. A. Farag (Eds.), *Mediterranean fruits bio-wastes*. Springer. https://doi.org/10.1007/978-3-030-84436-3_9
51. He, J., Yang, Z., Guo, M., Gu, L., Zhang, L., Yan, Y., & Ran, J. (2022). Experimental study on the key factors affecting the gasification performance between different biomass: Compare citrus peel with pine sawdust, *International Journal of Hydrogen Energy*, 47, 71, pp: 30428–30439, ISSN 0360-3199, https://doi.org/10.1016/j.ijhydene.2022.07.004.
52. Garba, A. (2020). Biomass conversion technologies for bioenergy generation: an introduction. In *Biotechnological Applications of Biomass* (pp. 1–16). https://doi.org/10.5772/intechopen.93669
53. Rahman, M. M., Liu, R., & Cai, J. (2018). Catalytic fast pyrolysis of biomass over zeolites for high quality bio-oil—A review. *Fuel Processing Technology, 180*, 32–46. https://doi.org/10.1016/j.fuproc.2018.08.002
54. Bridgwater, A. V. (2012). Review of fast pyrolysis of biomass and product upgrading. *Biomass and Bioenergy, 38*, 68–94. https://doi.org/10.1016/j.biombioe.2011.01.048
55. Jahirul, M. I., Rasul, M. G., Chowdhury, A. A., & Ashwath, N. (2012). Biofuels production through biomass pyrolysis—A Technological Review. *Energies, 5*(12), 4952–5001. https://doi.org/10.3390/en5124952
56. Andrade, M. A., Barbosa, C. H., Shah, M. A., Ahmad, N., Vilarinho, F., Khwaldia, K., Silva, A. S., Ramos, F. (2022, December 25). *Citrus* by-products: Valuable source of bioactive compounds for food applications. Antioxidants (Basel).12(1):38. https://doi.org/10.3390/antiox12010038. PMID: 36670900; PMCID: PMC9855225.
57. Wei, Y., Shengji, W., Hui, W., Pengyu, M., Toshiniri, S., Yukitaka, K., & Jie, Z. (2017). Effect of wet and dry torrefaction process on fuel properties of solid fuels derived from bamboo and Japanese cedar. *BioResources, 12*(4), 8629–8640.
58. Tamelova, B., Malatak, J., & Velebil, J. (2018). Energy valorisation of citrus peel waste by torrefaction treatment. *Agronomy Research, 16*(1), 276–285.
59. Kelz, J., Zemann, C., Muschick, D., Krenn, O., Hofmeister, G., Weissinger, A., Gölles, M., & Hochenauer, C. (2017). Evaluation of the combustion behaviour of straw, poplar and maize in a small-scale biomass boiler. In *Proceeding of the 25th European biomass conference and exhibition*.
60. Tamelova, B., Malaťák, J., & Velebil, J. (2018). Energy valorisation of citrus peel waste by torrefaction treatment. *Agronomy Research, 16*(1), 276–285.
61. Orisaleye, J. I., Jekayinfa, S. O., Pecenka, R., Ogundare, A. A., Akinseloyin, M. O., & Fadipe, O. L. (2022). Investigation of the effects of Torrefaction temperature and residence time on the fuel quality of corncobs in a fixed-bed reactor. *Energies, 15*, 5284. https://doi.org/10.3390/en15145284
62. Gent, S., Twedt, M., Gerometta, C., & Almberg, E. (2017). Fundamental theories of Torrefaction by thermochemical conversion. In *Theoretical and applied aspects of biomass torrefaction* (pp. 41–75). Elsevier.
63. Bates, R. B., & Ghoniem, A. F. (2012). Biomass torrefaction: Modeling of volatile and solid product evolution kinetics. *Bioresource Technology, 124*(460), 469–469.
64. Phanphanich, M., & Mani, S. (2011). Impact of torrefaction on the grindability and fuel characteristics of forest biomass. *Bioresource Technology, 102*, 1246–1253.
65. Chen, D., Zheng, Z., Fu, K., Zeng, Z., Wang, J., & Lu, M. (2015). Torrefaction of biomass stalk and its effect on the yield and quality of pyrolysis products. *Fuel, 159*, 27–32.
66. Strandberg, M., Olofsson, I., Pommer, L., Wiklund-Lindström, S., Åberg, K., & Nordin, A. (2015). Effects of temperature and residence time on continuous torrefaction of spruce wood. *Fuel Processing Technology, 134*, 387–398.

67. Mabrouki, J., Guedri, K., Abbassi, M. A., Omri, A., & Jeguirim, M. (2016). Simulation of the fast pyrolysis of Tunisian biomass feedstocks for bio-fuel production. *Comptes Rendus Chimie, 19*(4), 466–474. https://doi.org/10.1016/j.crci.2015.09.020
68. Patwardhan, P.R., Shanks, B.H., Brown, R. C., Fox, R. O., Koziel, J. A., & Trahanovsky, W. S. (2010). Understanding the product distribution from biomass fast pyrolysis by major: Chemical engineering program of study committee: 1–162.
69. Akhtar, J., & Saidina Amin, N. (2012). A review on operating parameters for optimum liquid oil yield in biomass pyrolysis. *Renewable and Sustainable Energy Reviews, 16*(7), 5101–5109. https://doi.org/10.1016/j.rser.2012.05.033
70. Haykiri-Acma, H. (2006). The role of particle size in the non-isothermal pyrolysis of hazelnut shell. *The Journal of Analytical and Applied Pyrolysis, 75*(2), 211–216.
71. Şensoz, S., Angin, D., & Yorgun, S. (2000). Influence of particle size on the pyrolysis of rape-seed (*Brassica napus* L.): Fuel properties of bio-oil. *Biomass and Bioenergy, 19*(4), 271–279.
72. Natarajan, E., & Ganapathy Sundaram, E. (2009). Pyrolysis of rice husk in a fixed bed reactor. *World Academy of Science, Engineering and Technology, 56*(8), 504–508.
73. Uzun, B. B., Putun, A. E., & Putun, E. (2006). Fast pyrolysis of soybean cake: Product yields and compositions. *Bioresource Technology, 97*(4), 569–576.
74. Wang, X., Chen, H., Luo, K., Shao, J., & Yang, H. (2008). The influence of microwave drying on biomass pyrolysis. *Energy & Fuels, 22*(1), 67–74.
75. Kan, T., Strezov, V., & Evans, T. J. (2016). Lignocellulosic biomass pyrolysis: A review of product properties and effects of pyrolysis parameters. *Renewable and Sustainable Energy Reviews, 57*(2016), 1126–1140. https://doi.org/10.1016/j.rser.2015.12.185
76. Pulidori, E., Gonzalez-Rivera, J., Pelosi, C., Ferrari, C., Bernazzani, L., Bramanti, E., Tine, M. R., & Duce, C. (2023). Thermochemical evaluation of different waste biomasses (citrus peels, aromatic herbs, and poultry feathers) towards their use for energy production. *Thermo., 3*, 66–75. https://doi.org/10.3390/thermo3010004
77. Rahimi, Z., Anand, A., & Gautam, S. (2022). An overview on thermochemical conversion and potential evaluation of biofuels derived from agricultural wastes. *Energy Nexus, 7*, 100125. https://doi.org/10.1016/j.nexus.2022.100125
78. Fahmi, R., Bridgwater, A. V., Donnison, I., Yates, N., & Jones, J. M. (2008). The effect of lignin and inorganic species in biomass on pyrolysis oil yields, quality and stability. *Fuel, 87*(7), 1230–1240.
79. Beaumont, O., & Schwob, Y. (1984). Influence of physical and chemical parameters on wood pyrolysis. *Industrial and Engineering Chemistry Process Design and Development, 23*(4), 637–641.
80. Knapczyk, A., Francik, S., Jewiarz, M., Zawiślak, A., & Francik, R. (2021). Thermal treatment of biomass: A bibliometric analysis—The Torrefaction case. *Energies, 14*, 162. https://doi.org/10.3390/en14010162
81. Basu, P. (2013). *Biomass gasification, pyrolysis and Torrefaction: Practical design and theory* (2nd ed., pp. 87–145). Academic.
82. Nhuchhen, D. R., Basu, P., & Acharya, B. (2014). A comprehensive review on biomass torrefaction. *International Journal of Renewable Energy and Biofuels, 2014*, 506376.
83. Chin, K. L., H'ng, P. S., Wong, W. Z., Lim, T. W., Maminski, M., Paridah, M. T., & Luqman, A. C. (2013). Optimization of torrefaction conditions for high energy density solid biofuel from oil palm biomass and fast-growing species available in Malaysia. *Industrial Crops and Products, 49*, 768–774.
84. Fu, P., Yi, W., Bai, X., Li, Z., Hu, S., & Xiang, J. (2011). Effect of temperature on gas composition and char structural features of pyrolyzed agricultural residues. *Bioresource Technology, 102*(17), 8211–8219. https://doi.org/10.1016/j.biortech.2011.05.083
85. Ibitoye, S. E., Jen, T. C., Mahamood, R. M., & Akinlabi, E. T. (2021). Improving the combustion properties of corncob biomass via torrefaction for solid fuel applications. *Journal Of Composites Science, 5*, 260.

86. Martínez, L. V., Rubiano, J. E., Figueredo, M., & Gómez, M. F. (2020). Experimental study on the performance of gasification of corncobs in a downdraft fixed bed gasifier at various conditions. *Renewable Energy, 148*, 1216–1226.
87. Wakudkar, H., & Jain, S. (2022). A holistic overview on corn cob biochar: A mini-review. *Waste Management & Research, 40*, 1143–1155.
88. Zou, H., Jiang, Q., Zhu, R., Chen, Y., Sun, T., Li, M., Zhai, J., Shi, D., Ai, H., Gu, L., et al. (2020). Enhanced hydrolysis of lignocellulose in corn cob by using food waste pretreatment to improve anaerobic digestion performance. *Journal of Environmental Management, 254*, 109830.
89. Lou, H., He, X., Cai, C., Lan, T., Pang, Y., Zhou, H., & Qiu, X. (2019). Enhancement and mechanism of a lignin amphoteric surfactant on the production of cellulosic ethanol from a high-solid corncob residue. *Journal of Agricultural and Food Chemistry, 67*, 6248–6256.
90. Jekayinfa, S. O., Pecenka, R., & Orisaleye, J. I. (2019). Empirical model for prediction of density and water resistance of corn cob briquettes. *International Journal of Renewable Energy Technology, 10*, 212–228.
91. El-Sayed, S. A., Khass, T. M., & Mostafa, M. E. (2023). Thermal degradation behaviour and chemical kinetic characteristics of biomass pyrolysis using TG/DTG/DTA techniques. In *Biomass conversion and biorefinery*. Springer. https://doi.org/10.1007/s13399-023-03926-2

Chapter 7
Biowaste from Citrus Reticulata Peels as an Effective Antiepileptic

Kenganora Mruthunjaya, S. N. Manjula, Neeraj Kumar Fuloria, Himanshu Sharma, Amit Anand, and Rashmi Pathak

7.1 Introduction

Recurrent seizures that result from abnormal electrical activity in the brain are the hallmarks of the neurological disorder known as epilepsy. A significant portion of the epileptic population still faces severe side effects from current antiepileptic drugs (AEDs) or inadequate seizure control despite advancements in medical research. Consequently, there is a continuous search for novel therapeutic molecules derived from plants [1]. Citrus fruits, particularly Citrus reticulata, which is frequently called mandarin or tangerine, are known to contain a variety of bioactive compounds. Peels from Citrus reticulata are among the numerous citrus fruits that have gained attention because of their high concentration of bioactive compounds,

K. Mruthunjaya · A. Anand
Department of Pharmacognosy, JSS College of Pharmacy, JSS Academy of Higher Education and Research, Mysuru, Karnataka, India
e-mail: Kmruthunjaya@jssuni.edu.in

S. N. Manjula
Department of Pharmacology, JSS College of Pharmacy, JSS Academy of Higher Education and Research, Mysuru, Karnataka, India
e-mail: snmanjula@jssuni.edu.in

N. K. Fuloria
Department of Pharmaceutical Chemistry, Faculty of Pharmacy, AIMST University, Bedong, Kedah, Malaysia
e-mail: nfuloria@aimst.edu.my

H. Sharma (✉)
Teerthanker Mahaveer College of Pharmacy, Teerthanker Mahaveer University, Moradabad, UP, India

R. Pathak
Department of Pharmacy, Invertis University, Bareilly, UP, India

A. Chauhan et al. (eds.), *Valorization of Citrus Food Waste*,
https://doi.org/10.1007/978-3-031-77999-2_7

such as flavonoids, terpenoids, and essential oils [2]. These compounds exhibit many pharmacological properties, including neuroprotective, anti-inflammatory, and antioxidant properties. Worldwide, citrus reticulata, often known as mandarin or tangerine, is a fruit that is farmed. Peels from citrus reticulata are often thrown away in large quantities as waste by the food processing industry [3]. Conversely, terpenoids, flavonoids, and essential oils are just a few of the bioactive compounds found in these peels that have anti-inflammatory, neuroprotective, and antioxidant properties. For this reason, researchers are now examining whether biowaste from citrus reticulata peels might help develop novel antiepileptic drugs [4]. Citrus reticulata peels provide biowaste, a unique opportunity for sustainable usage, throughout the fruit processing process. Researching alternative uses for citrus peels is a sustainable undertaking, since their disposal as waste contributes to environmental problems. The bioactive components of citrus reticulata peels appear like a promising avenue for medical use, particularly in the treatment of epilepsy [5].

7.1.1 Objectives

The primary objective is to provide a detailed report on the potential of citrus reticulata peel biowaste as a novel and effective source of antiepileptic medicines [6]. This chapter diligently examines the chemical composition, extraction methods, and pharmacological properties of the bioactive compounds found in citrus reticulata peels to establish the scientific foundation for employing these biowastes as therapeutic agents in the treatment of epilepsy [7]. The chapter addresses the sustainability of employing citrus peels as a major resource, which further encourages the adoption of ecologically friendly approaches in pharmaceutical research. This chapter's goal is to deepen awareness of the possible advantages of natural sources for neurological illnesses by offering a comprehensive examination of citrus reticulata peels' antiepileptic qualities [8]. This will contribute to the advancement of waste utilization and pharmacological understanding. This study aims to investigate the potential of Citrus reticulata peel biowaste as an effective antiepileptic drug [9]. By extracting and examining the chemicals present in the peels, we want to shed light on the mechanisms of action of the bioactive molecules and examine their efficacy in preclinical models of epilepsy. This work not only advances the field of epilepsy therapies but also addresses the growing demand for sustainable approaches to biowaste management in the food industry [10].

7.1.2 Significance of Utilizing Citrus Reticulata Peels

Around the world, tangerines and mandarin oranges, also known as Citrus reticulata, are popular citrus fruits that are consumed in enormous amounts. Citrus reticulata peels are a tasty addition to many recipes. Many dishes, beverages, and desserts get a distinct citrus flavor from the essential oils present in the zest, or colored part of the peel [11]. Citrus reticulata peels are used to extract essential oil, which is used in aromatherapy. It's a popular choice for aromatherapy applications like massage oils and diffusers because of its regarded uplifting and calming effects [12]. Because of its potential health benefits, citrus reticulata peels have been used in traditional medicine. They are used to alleviate nausea, improve overall health, and treat gastrointestinal issues since they are thought to possess antioxidant properties [13]. Vitamin C and dietary fiber are two of the many nutrients found in citrus reticulata peels. Fruit peels contribute extra nutrients to meals, even if the meat of the fruit is consumed more often [14]. Citrus reticulata is used to make candied citrus peels, a popular dessert. Boiling the peels in sugar syrup and then coating them in sugar results in a sweet and tart dessert. Orange peels contain natural oils used as a basis for do-it-yourself cleaning solutions [15]. Their efficiency in eliminating oil and filth makes them a safer alternative when compared to certain industrial cleaning products. Citrus reticulata peels are used to reduce food waste. Instead of throwing the peels away, recycled for a variety of purposes, which encourages more ecologically responsible eating practices [16].

7.2 Epilepsy and Anti-epileptic Agents: An Overview

Abrupt, abnormal electrical outbursts in the brain that occur often are the hallmarks of the neurological disorder known as epilepsy [17]. Short attention spans and full-body spasms are only two ways that these seizures manifest. It is essential to treat epilepsy using anti-epileptic medications, commonly known as anticonvulsants or anti-seizure medications. These medicines attempt to lessen or completely eradicate seizures by altering brain electrical activity [18]. There are many different kinds of anti-epileptic drugs, and each has a unique mechanism of action and potential side effects. Carbamazepine and phenytoin belong to a prominent family of anti-epileptic drugs that block sodium channels [19]. By stabilizing the membranes of neuronal cells and inhibiting abnormal electrical activity, these medications work. Another class of medications that increases the inhibitory effects of GABA, a neurotransmitter that helps to regulate brain activity, includes benzodiazepines and valproic acid. Contemporary anti-epileptic drugs, such as levetiracetam and lamotrigine, target different pathways and have better-to-better side effects [20]. They are crucial in situations when traditional medications fail or have intolerable adverse effects. A single medication or a combination of medications is used in highly personalized treatment regimens for epilepsy. Frequent adjustments and monitoring are necessary to maintain optimal seizure control while minimizing adverse effects [21].

7.2.1 *Effect of Citrus to Control the Epileptic*

The use of citrus fruits or their components in controlling or treating epilepsy is not widely supported by clinical evidence in mainstream medical literature. However, some aspects related to citrus and their compounds have been of interest in research concerning their potential health benefits, including neurological effects. Citrus fruits are rich in antioxidants such as vitamin C and flavonoids. Antioxidants are important for overall brain health and have been studied for their neuroprotective effects. They may potentially help in reducing oxidative stress which is believed to be a factor in the development and progression of epilepsy. Some studies have explored the role of flavonoids, compounds found abundantly in citrus fruits, in neuroprotection. These compounds might influence the central nervous system by interacting with neurotransmitters and modulating brain function, which could hypothetically impact conditions like epilepsy. However, direct effects specifically targeting epilepsy are not conclusively proven and need more clinical backing. Some anecdotal evidence and preliminary studies suggest that aromatherapy, including the use of citrus oils (like lemon or orange), might have beneficial effects on stress and anxiety, which are often triggers for epileptic seizures in susceptible individuals. Nonetheless, this is not a direct treatment for epilepsy but rather a complementary approach to manage some associated conditions. While not specific to citrus, the ketogenic diet, a high-fat, low-carbohydrate diet, has been recognized as an effective treatment for epilepsy in some cases. Citrus fruits, being low in fat and high in carbohydrates, are not a major component of this diet, but could be included in moderation depending on individual dietary allowances. Epilepsy is a neurological illness characterized by recurrent, spontaneous seizures that affects millions of people globally. Seizures, which result from abnormal electrical activity in the brain, produce convulsions and momentary unconsciousness in addition to other symptoms. Siang et al. (2022) investigated various fruits exhibit anti-convulsant properties through neurotransmitter modulation and anti-oxidant effects, though further human trials are needed to confirm efficacy [22]. Several factors, including heredity, brain damage, infections, and anatomical abnormalities, result in epilepsy. The use of AEDs to lessen or halt seizures is the cornerstone of contemporary epilepsy treatment approaches. People with epilepsy can get a break from their abnormal brain electrical activity with the use of these medications [23]. AED is best depending on the type of seizure, the patient's age, and if any underlying medical conditions are present. Conventional AEDs, such as phenytoin, carbamazepine, and valproic acid, have long been essential in the management of seizures. AEDs from more recent generations, such as topiramate, lamotrigine, and levetiracetam, provide substitutes with different modes of action and better adverse effect profiles [24]. To provide the best possible seizure control, a combination treatment of several AEDs required in some circumstances. Apart from drug regimes, examples of treatment approaches include the ketogenic diet, surgery for those who qualify, and neuromodulation procedures like vagus nerve stimulation [25].

7.2.2 Challenges and Limitations of Existing Antiepileptic Drugs

Inadequate seizure control, side effects, drug interactions, individual variability, tolerance development, cognitive and behavioral effects, restricted options for particular seizure types, pregnancy considerations, and financial/accessibility issues are some of the challenges faced by currently available AEDs [26]. Research on more individualized and efficient treatment plans for managing epilepsy is still needed to overcome these obstacles (see Table 7.1).

Table 7.1 The drawbacks and restrictions of current antiepileptic medications

Challenge/ Limitation	Description	References
Incomplete seizure control	Some people have seizures that are not well controlled by the antiepileptic medications (AEDs) that are now available, which makes controlling the disease difficult in the long run	[27]
Adverse effects	Dizziness, sleepiness, mood swings, and cognitive impairment are only a few of the adverse effects linked to AEDs. It is quite difficult to strike a balance between seizure control and manageable side effects	[27]
Drug interactions	There are interactions between several AEDs and other drugs that reduce their effectiveness or have unfavorable consequences. For those with comorbidities, controlling medication interactions is essential	[27]
Individual variability	Individuals might differ greatly in how they react to AEDs. It is often necessary to use a trial-and-error method to determine the appropriate medicine and dose, which causes delays in reaching the best seizure control	[27]
Development of tolerance	Some people are resistant to the effects of AEDs over time, which would need adjusting dosages or looking into alternate drugs to keep them effective	[27]
Cognitive and behavioral effects	Particularly in youngsters and the elderly, AEDs affect behavior and cognitive performance. It difficult to strike a balance between the advantages of seizure control and any possible cognitive and behavioral negative effects	[27]
Limited treatment options for specific types	Limited treatment choices exist for some epileptic syndromes or seizure types, and people with these disorders have challenges in locating appropriate AEDs to meet their unique requirements	[27]
Pregnancy considerations	Pregnancy poses hazards for some AEDs, which might result in congenital abnormalities. It is important to carefully weigh the necessity of managing seizures against the growing fetus's safety	[27]
Financial and access challenges	For some people, the cost and accessibility of AEDs barrier, making it difficult for them to regularly get and take prescribed drugs, which can influence seizure treatment	[27]

7.3 Previous Studies on Natural Antiepileptic Agents

The possibility of using different substances obtained from plants and other natural sources to treat epilepsy has been investigated in earlier research on natural antiepileptic drugs. Preclinical research has shown the antiepileptic qualities of extracts from medicinal plants, including *Ginkgo biloba*, *Bacopa monnieri*, and *Cannabis sativa* [28]. These organic substances often have neuroprotective and anti-inflammatory properties, regulating neurotransmitter function and reducing oxidative stress all of which are linked to illnesses related to seizures. Furthermore, the possible antiepileptic properties of essential oils including chamomile and lavender have been studied [29]. These oils might affect the central nervous system by having relaxing and anti-anxiety qualities. Cannabinoids derived from the cannabis plant have drawn attention for their antiepileptic qualities, especially cannabidiol (CBD). Cannabinoids have medicinal promise as shown by the approval of Epidiolex, a CBD-based medicine, for the treatment of certain kinds of epilepsy [30].

7.3.1 Citrus Reticulata Peels: Composition and Potential Health Benefits

Peels from Citrus reticulata are rich in minerals, vitamins, fiber, polyphenols, carotenoids, terpenes, limonoids, and citrus bioactive components in addition to essential oils [31]. The antioxidant, anti-inflammatory, and neuroprotective qualities of these constituents have positive effects on health [32]. The peels provide minerals for electrolyte balance, fiber for digestive health, and vitamins to help the immune system. The potential for cardiovascular and anticancer effects is suggested by the presence of flavonoids and limonoids [33]. Citrus reticulata peels have the potential to be a natural source of bioactive chemicals in a balanced diet; however, further study is necessary to completely understand and confirm these health benefits (see Table 7.2).

7.4 Utilization of Citrus to Combat Epilepsy

The utilization of citrus, particularly compounds found in Citrus aurantium and Citrus sinensis, has shown promise in combating epilepsy. Research indicates that flavonoid-rich extracts from orange juice (OJe) and essential oils from Citrus aurantium can exert anticonvulsant effects by interacting with NMDA and GABA receptors, increasing seizure latency in various experimental models. Additionally, Chinese herbal medicine, which includes components like Citrus aurantium, has been historically used to treat epilepsy, showing effectiveness in reducing seizure frequency and improving outcomes in clinical studies. These findings suggest that

Table 7.2 The possibility of using citrus reticulata peels in a balanced diet as a natural source of bioactive chemicals

Component	Composition in citrus reticulata peels	Potential health benefits	References
Essential oils	Limonene, myrcene, gamma-terpinene, etc.	Antioxidant, anti-inflammatory, and potential anti-seizure properties.	[34]
Flavonoids	Hesperidin, naringin, rutin, quercetin, etc.	Cardiovascular health, anti-inflammatory, and neuroprotective effects.	[34]
Vitamins	Vitamin C, folate, vitamin A, etc.	Immune support, skin health, and antioxidant properties.	[34]
Fiber	Pectin, hemicellulose, cellulose, etc.	Digestive health, blood sugar regulation, and potential weight management.	[34]
Polyphenols	Tangeretin, nobiletin, flavanones, etc.	Antioxidant, anti-inflammatory, and potential neuroprotective effects.	[35]
Minerals	Potassium, magnesium, calcium, etc.	Electrolyte balance, bone health, and support for cardiovascular function.	[36]
Carotenoids	Beta-carotene, lutein, zeaxanthin, etc.	Vision health, antioxidant properties, and potential anti-inflammatory effects.	[37]
Terpenes	Terpinolene, alpha-pinene, beta-pinene, etc.	Antioxidant, anti-inflammatory, and potential anti-anxiety properties.	[38]
Citrus bioactive compounds	Citrus flavonoids, coumarins, carotenoids, etc.	Anticancer potential, cardiovascular health support, and anti-inflammatory effects.	[39]
Limonoids	Limonin, nomilin, etc.	Antioxidant, potential anticancer effects, and liver health support.	[40]

citrus compounds may offer a natural and potentially beneficial approach to the management of epilepsy. Falero et al. (2015) investigated the aqueous extract from C. aurantium leaves increases seizure latency in zebrafish, suggesting potential anticonvulsant properties mediated by interactions with glutamate receptors, possibly offering new insights into natural epilepsy treatments [41].

7.4.1 Quantity and Composition of Citrus Reticulata Peels

The amount and makeup of Citrus reticulata peels, sometimes referred to as mandarin or tangerine peels, are important for several purposes, from cooking to possible medical usage. The distinct taste, aroma, and possible health advantages of these peels are attributed to their abundance of bioactive chemicals. Citrus reticulata peel yields vary in amount based on several variables, including citrus fruit type, growing techniques, and harvesting strategies. The peels make up a large amount of the fruit overall on average, thus their abundance makes them an intriguing source of important ingredients [42]. Essential oils, flavonoids, phenolic acids, and other bioactive substances are among the many components that make up Citrus reticulata peels. Remarkably, the essential oils isolated from the peels include chemicals like limonene, myrcene, and γ-terpinene that contribute to the distinct taste and scent. The food and fragrance industries have made use of these essential oils. Flavonoids with antioxidant qualities as rutin, hesperidin, and naringin are also present in the peels. Antioxidants are essential for scavenging free radicals from the body, which have positive effects on health by lowering inflammation and oxidative stress [43].

7.4.2 Environmental Concerns and Waste Management Practices

Biomass has been introduced as a sustainable renewable resource for biorefineries due to worries about the depletion of fossil fuels and the need to lessen the environmental impact [44]. A variety of valorization experiments have been conducted on citrus processing waste, a sustainable biowaste residue, to recover its matter and energy [45]. While green extraction techniques are employed to recover pectin, bioactive/nutraceuticals, and essential oils (mostly limonene), the waste is directly used to create biosorbents, nanomaterials, and organic fertilizer. Bioprocesses for energy valorization, such as those producing biogas or using sugar as a platform, often need a pre-treatment stage to get rid of limonene's inhibitory effects [46]. Integrated biorefineries are meant to speed up the process and are beneficial since they produce a wider range of chemical products using less energy, but they still have significant operating costs when done on a big scale [47]. The majority of these techniques have been verified in lab settings and sometimes in pilot settings. Since the industries addressing these problems are still in their infancy, they must step up their R&D efforts. Utilizing trash to extract valuable compounds and developing preventative measures against environmental disasters are the future directions of biowaste management [48]. This is especially crucial in the wake of the 2015 United Nations Climate Change Conference in Paris so that the most environmentally adaptive technology will endure into the future. Citrus reticulata peels, like those of other citrus fruits, have bioactive chemicals and essential oils that have potential health benefits. However, there are environmental issues with how citrus peels are disposed of and the waste management techniques used in their processing [49]. Flavonoids, other bioactive substances, and essential oils be found in citrus peels. Even while these substances find use in many sectors, incorrect disposal causes these substances to leak into the environment, where they harm the quality of soil and water. Peels are among the many waste products produced during the processing of citrus fruits. If not correctly handled, this trash leads to contamination in the environment [50]. Citrus peels and other organic wastes lead to the formation of methane, a powerful greenhouse gas, in landfills [51]. The use of landfills is not a sustainable option for the environment. Discovering new applications for citrus peels is a component of sustainable waste management techniques [52]. They are

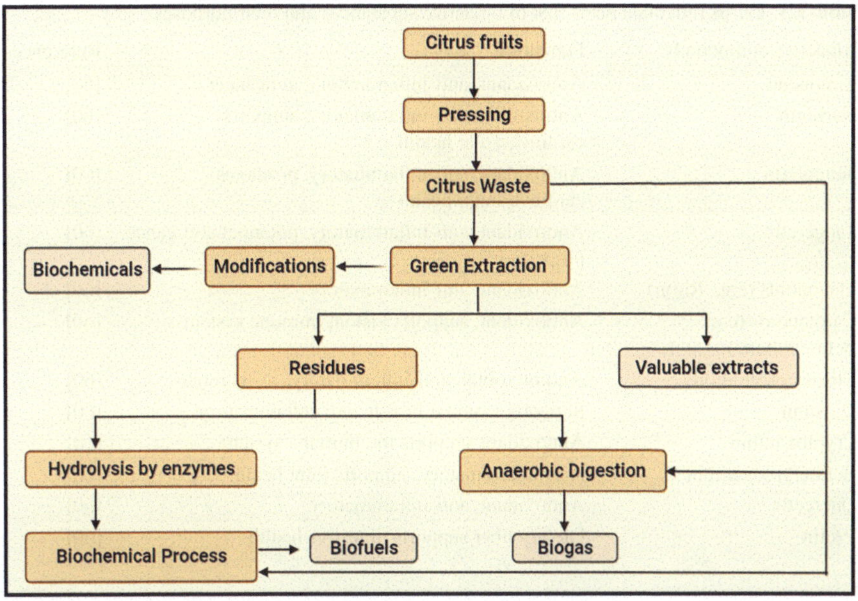

Fig. 7.1 Waste management techniques and environmental issues

used as a source of bioenergy via processes like anaerobic digestion, as well as to produce animal feed and natural cleaning products. Compost enriched with nutrients that be used in agriculture can be created by composting citrus peels [53]. Citrus peels contain essential oils that be extracted and used in cuisine, cosmetics, and aromatherapy, among other sectors. Both wastes are decreased and this technique can be profitable. Citrus peels decompose spontaneously in the correct environment and are biodegradable. Waste management is something that local companies and communities. Managing garbage from citrus peels requires strict adherence to environmental rules [54]. While citrus reticulata peels have potential uses and benefits, businesses involved in citrus processing should follow waste disposal guidelines and look into sustainable alternatives to reduce their environmental impact [55]. Waste management for citrus reticulata peels requires careful consideration to minimize environmental impact [56]. Composting, finding other applications for trash, and disposing of it responsibly are examples of sustainable activities that help lessen the environmental impact of orange peel waste (see Fig. 7.1).

7.5 Bioactive Compounds in Citrus Reticulata Peels

Citrus reticulata, a member of the Rutaceae family of citrus fruits, is sometimes referred to as tangerines or mandarin oranges [57]. Citrus reticulata peels include a wealth of different bioactive chemicals that add to their potential health advantages.

Table 7.3 Citrus reticulata peels' list of bioactive substances and their purposes

Bioactive compounds	Functions	References
Limonene	Antioxidant, anti-inflammatory, anticancer	[60]
Naringin	Antioxidant, anti-inflammatory, supports cardiovascular health	[60]
Tangeretin	Antioxidant, anti-inflammatory, promotes cardiovascular health	[60]
Tangeretin	Antioxidant, anti-inflammatory, potential anti-cancer properties	[60]
Flavonoids (e.g., Rutin)	Antioxidant, anti-inflammatory	[60]
Carotenoids (e.g., Beta-carotene)	Antioxidant, supports vision, immune system	[60]
Citrus essential oils	Antimicrobial, anti-inflammatory, stress relief	[60]
Diosmin	Supports vascular health, anti-inflammatory	[60]
Cryptoxanthin	Antioxidant supports the immune system	[60]
Beta-cryptoxanthin	Anti-inflammatory, supports joint health	[60]
Quercetin	Antioxidant, anti-inflammatory	[60]
Pectin	Dietary fiber supports digestive health	[60]

Bioactive substances are additional nutrients found in food, usually in trace amounts [58]. To assess their impact on health, they are being thoroughly investigated. To guarantee that the extract has therapeutic qualities, it is essential to measure the bioactive components. Table 7.3 lists the bioactive substances in citrus reticulata peels along with an explanation of each one's use. In addition to bioactive substances such as phenols, maximal amounts of flavonoids, protein, carbs, vitamin C, and tannins are present [59].

7.5.1 Identification and Characterization of Bioactive Compounds

The identification and characterization of bioactive chemicals require the process of dissecting, analyzing, and understanding the properties of chemical compounds that exhibit biological activity [61]. These substances be found in a wide range of natural sources, such as fungi, animals, plants, and microorganisms. First, collect biological material that could include bioactive compounds. Plant extracts, animal tissues, and microbiological cultures are a few examples of this material [62]. The acquired materials are processed to extract the bioactive compounds using the appropriate solvents or extraction procedures. A range of chromatographic techniques, including thin-layer chromatography, column chromatography, and high-performance liquid chromatography, used to separate certain bioactive compounds from the crude extract [63]. Apply analytical techniques to ascertain the chemical structure of the separated compounds [64]. This might include looking for antibacterial, anti-inflammatory, antioxidant, or other relevant qualities, depending on the intended usage. Use biochemical experiments to investigate the method of action of the bioactive compounds [65]. Understanding how the chemicals interact with biological systems is much easier as a result. Examine the safety and toxicity profiles of the bioactive compounds, especially if they are meant to be used in medications or other applications [66]. This might include both *in vitro* and *in vivo* toxicity studies. Optimize the production of bioactive molecules or modify their structures to improve efficacy, bioavailability, or other desirable properties [67]. Expand the manufacturing scale to generate larger quantities of bioactive chemicals after they have been identified, characterized, and polished. Examination of potential applications for the bioactive compounds in the domains of agriculture, cosmetics, medicine, and other industries [68].

7.5.2 Mechanisms of Action in Antiepileptic Properties

Drugs that target the abnormal neuronal excitability and synchronization characteristic of epileptic episodes have a variety of mechanisms of action that affect their antiepileptic efficacy [69]. One such method is ion channel modulation, particularly of voltage-gated sodium channels [70]. To prevent an excessive sodium ion influx and reduce neuronal excitability, antiepileptic drugs often block these channels. Gamma-aminobutyric acid neurotransmission modulation [71]. As the primary inhibitory neurotransmitter in the brain, GABA has an inhibitory influence on the excitability of individual neurons when its effects are increased. Some AEDs increase inhibitory tone overall by increasing GABA production or inhibiting its breakdown, which enhances GABAergic transmission. Medications that treat epilepsy also target glutamate, the primary excitatory neurotransmitter [72]. N-methyl-D-aspartate receptors are a subtype of glutamate receptors that can induce and propagate seizures. AEDs act as NMDA receptor antagonists, lowering the excitatory effects of glutamate and halting the progression of seizures. Calcium channel modulation is another way that antiepileptic drugs work. Reduced excitatory neurotransmission and neurotransmitter release result from blocking voltage-gated calcium channels [73].

7.5.3 Comparative Analysis with Conventional Antiepileptic Drugs

Citrus reticulata peels contain many bioactive chemicals, such as flavonoids, coumarins, and essential oils [74]. Examine the specific drugs present and any potential implications they have on epilepsy. Conventional AEDs often comprise drugs like phenytoin, carbamazepine, and valproic acid. Understand how these drugs work and how well they work to control seizures [75]. The mechanism of action explains how Citrus reticulata peels' active compounds affect neurotransmitter activity, ion channels, or other pathways linked to epilepsy by comparing them to those of conventional AEDs, which often involve modulating ion channels, neurotransmitters, or other targets. Evaluating available data, including clinical studies, to ascertain the efficacy of citrus reticulata peels in the treatment of epilepsy. Citrus reticulata peels used with conventional AEDs: safety features [76]. Factors including side effects, pharmacological interactions, and long-term effects on cognitive function [77]. The bioavailability of Citrus reticulata peels' active components, considering their metabolism, excretion, distribution, and absorption processes. Patients getting therapy based on Citrus reticulata peels should be compared to those receiving standard AEDs in terms of acceptability and adherence. A few elements that might influence compliance include patient preferences, taste, and ease of administration. Examine the economic feasibility of treatments with Citrus reticulata peels in comparison to conventional AEDs. Pricing and accessibility affect the feasibility of widespread usage. Analyzed the status of citrus reticulata peels as a regulated medical ingredient [78–85].

7.6 Antiepileptic Potential of Citrus Reticulata Peels

Due to their high bioactive chemical content, citrus reticulata peels have antiepileptic potential. Anti-inflammatory and antioxidant properties of certain essential oils, such as limonene and myrcene, suggest that they play a role in neuroprotection. Naringin and hesperidin are two examples of flavonoids with anti-inflammatory qualities that promote cardiovascular health and are helpful in the treatment of epilepsy. Terpenes and polyphenols also contribute to the overall antioxidant and anti-seizure properties of these peels. Studies have shown that some elements found in citrus peels, such as tangerine and nobiletin, control brain activity and have anticonvulsive effects. By targeting several molecular targets associated with epileptic processes, these chemicals have the potential to reduce both the frequency and severity of seizures. Moreover, citrus bioactive compounds including flavonoids and limonoids possess neuroprotective properties that preserve brain health and reduce the cellular harm caused by seizures [1].

7.7 *In vitro* Studies and *in vivo* Studies of Antiepileptic

In vitro and *in vivo* tests investigating antiepileptic activity are used by researchers to assess the effectiveness of various substances in preventing or reducing epileptic episodes [86]. Isolated cell culture or tissue is often used in *in vitro* experiments to assess the effects on neuronal excitability, neurotransmitter modulation, and ion channel function [87]. However, the total benefits of antiepileptic medications which include modifications to seizure frequency, cognitive function, and

Table 7.4 Mechanisms and possible medicinal advantages of drugs for treating epilepsy

Study	Methodology	Model system	Findings	References
1	*In vitro*	Neuronal cell culture (Rat)	Reduced neuronal excitability, and modulation of GABA receptors.	[89]
2	*In vivo*	Animal model (Mouse)	Decreased seizure frequency, and increased latency.	[89]
3	*In vitro*	Brain slice culture (Human)	Modulation of glutamate release, neuroprotective effects.	[89]
4	*In vivo*	Animal model (Rabbit)	Improved cognitive function, and reduced neuronal apoptosis.	[89]
5	*In vitro*	Hippocampal neurons (Mouse)	Inhibition of sodium channels, prevention of hyperexcitability.	[89]
6	*In vivo*	Animal model (Rat)	Suppression of inflammatory pathways reduced seizure severity.	[89]
7	*In vitro*	Cortical tissue culture (Human)	Enhanced GABAergic transmission, and decreased epileptiform activity.	[89]
8	*In vivo*	Animal model (Monkey)	Improved seizure control with minimal side effects.	[89]
9	*In vitro*	Neuronal cell line (Human)	Modulation of calcium channels, reduction in neurotransmitter release.	[89]
10	*In vivo*	Animal model (Dog)	Delayed onset of seizures, increased survival rate.	[89]

neuroprotective outcomes are seen *in vivo* studies using animal models [88]. The findings from both types of studies are combined by researchers to fully understand the mechanisms and potential therapeutic benefits of medications in the treatment of epilepsy (see Table 7.4).

7.8 Mechanistic Insights into Antiepileptic Activity

By controlling neuronal activity and balancing the electrical activity of the brain, antiepileptic drugs reduce or eliminate seizures [90]. The primary targets of these drugs'various mechanisms of action are ion channels and neurotransmitter systems [91]. AEDs improve the inhibitory activity of gamma-aminobutyric acid, the principal inhibitory neurotransmitter in the brain [92]. They either increase GABA synthesis, facilitate GABA release, or lengthen GABA's half-life at its receptors. This lowers the likelihood of excessive neuronal activity leading to seizures by enhancing the brain's overall inhibitory tone [93]. Sodium channels are examples of voltage-gated ion channels that are blocked by AEDs. By blocking these channels, the drugs reduce the quantity of sodium ions that enter the cell, stabilizing the neuron's membrane and putting an end to abnormal, excessive neuronal activity [94]. AEDs further alter glutamate receptors, potassium channels, and calcium channels to impact neurotransmission and neuronal excitability [95]. Different AEDs have different processes, which correspond to different types and origins of epileptic

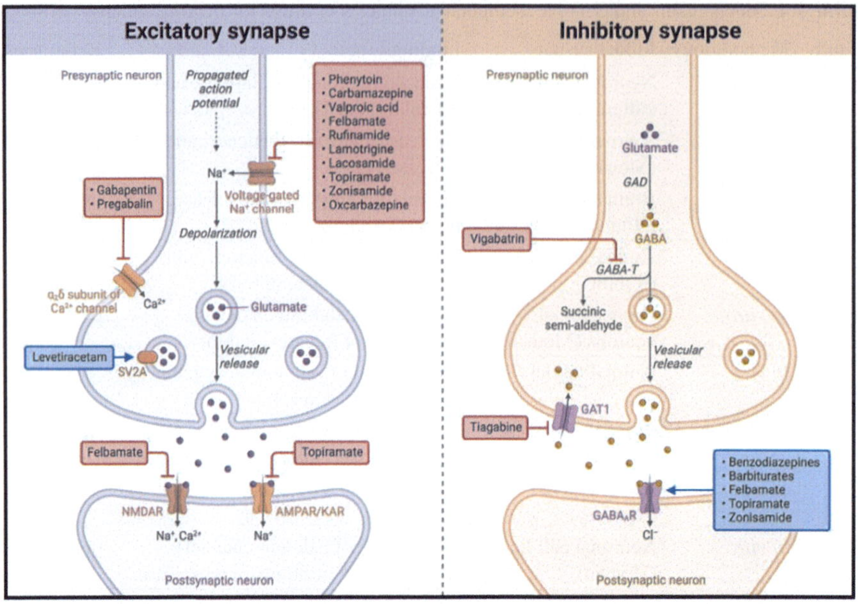

Fig. 7.2 Mechanism of antiepileptic drugs [101]

seizures. These drugs try to bring the brain's excitability and inhibition into balance, which reduces aberrant electrical activity and the frequency and intensity of seizures in epileptics [96]. *In vitro* studies are often used to investigate the intricate cellular and molecular mechanisms behind antiepileptic actions [97]. Researchers investigate the alteration of ion channels, including calcium and sodium channels, and the enhancement of inhibitory neurotransmission, particularly via gamma-aminobutyric acid receptors. The regulation of neuronal excitability and the avoidance of seizures' characteristic aberrant electrical discharges are functions of these systems [98]. Additionally, studies are being conducted on the neuroprotective properties, which include the possibility that compounds with antiepileptic properties might shield neurons from oxidative stress or excessive excitability. Anti-inflammatory qualities are especially crucial as inflammation plays a role in the pathogenesis of seizures [99]. *In vivo* studies provide a thorough understanding of the systemic effects of antiepileptic medications, which complement these in vitro findings. They assess changes in behavior, the intensity of seizures, and cognitive capacities in animal models to provide light on the substances' overall therapeutic potential and safety profile [100]. Moreover, considering the complex environment of the living body, these studies explore the possible impacts of drugs on neuronal network dynamics, synaptic plasticity, and neurotransmitter release see Fig. 7.2.

7.9 Conclusion and Future Perspective

How promising it is to exploit natural resources for medicinal applications is shown by the findings of this book chapter's inquiry into biowaste from Citrus reticulata peels as a potential source of potent antiepileptic medicines. Peels from citrus reticulata can alter many pathways associated with epileptic seizures based on their composition. These compounds include bioactive molecules, essential oils, and flavonoids. Research presented in this chapter shows that extracts from the peel of citrus reticulata have antiepileptic properties. The effects on ion channels, neurotransmitter systems, and anti-inflammatory pathways that have been documented help us understand the mechanisms by which these extracts reduce abnormal neuronal excitability. The fact that this research uses agricultural by products to make environmentally friendly antiepileptic medications makes it even more noteworthy. To fully understand the antiepileptic capabilities of Citrus reticulata peels, more research must concentrate on identifying and isolating specific bioactive compounds. Our search for novel antiepileptic drugs derived from natural sources will be aided by the ability to develop customized therapies with proven mechanisms of action. Clinical studies are necessary to verify Citrus reticulata peel extracts' safety, efficacy, and optimal dosage for use as an antiepileptic. Through these studies, the gap between scientific discoveries and therapeutic applications will be closed, ensuring that the natural compounds produced from citrus peels are developed into standardized and reliable medications for individuals with epilepsy. Further investigation is needed to examine potential interactions between Citrus reticulata peel extracts and antiepileptic drugs currently marketed. Understanding the interactions between these natural products and pharmaceutical medications can help guide combination treatment, perhaps leading to better outcomes for people with epilepsy. This book chapter lays the foundation for future research into Citrus reticulata peels' potential as an antiepileptic by emphasizing the importance of interdisciplinary collaboration between pharmacologists, clinicians, and botanists. Through the integration of scientific research and practical applications, orange peel biowaste be used for its medicinal characteristics, leading to the advancement of both eco-friendly practices and epilepsy treatment.

Acknowledgments The authors are thankful to the Principal, Department of Pharmacy, Teerthanker Mahaveer University, Moradabad for providing the necessary facilities.

Competing Interest Statement The authors declare no conflict of interest.

Disclaimer None

References

1. Sharma, P., Ruchika, D. P., Kumar, R., Saneja, A., & Singh, D. (2023). A solid dispersion of Citrus reticulata peel biowaste as an effective antiepileptic: Sustainable approach toward value addition and agro-industrial waste valorisation. *The Journal of Drug Delivery Science and Technology, 81*, 104238. Available from: https://www.sciencedirect.com/science/article/pii/S1773224723000904

2. Russo, C., Maugeri, A., Lombardo, G. E., Musumeci, L., Barreca, D., Rapisarda, A., et al. (2021). The second life of citrus fruit waste: A valuable source of bioactive compounds. *Molecules, 26*(19), 5991.
3. Sharma, P., Dhiman, P., & Singh, D. (2023). Dietary flavonoids-rich Citrus reticulata peel extract interacts with CREB signaling to suppress seizures and linked neurobehavioral impairments in a kindling mouse model. *Nutritional Neuroscience, 26*(7), 582–593.
4. Khan, U. M., Sameen, A., Aadil, R. M., Shahid, M., Sezen, S., Zarrabi, A., et al. (2021). Citrus genus and its waste utilization: A review on health-promoting activities and industrial application. *Evidence-Based Complementary and Alterntive Medicine, 2021*, 1–17.
5. Al Kamaly, O., Numan, O., Almrfadi, O. M. A., Alanazi, A. S., & Conte, R. (2022). Separation and evaluation of potential antioxidant, analgesic, and anti-inflammatory activities of limonene-rich essential oils from Citrus sinensis (L.). *Open Chemistry, 20*(1), 1517–1530.
6. Kamanna, K., & Amaregouda, Y. (2022). Synthesis of bioactive scaffolds catalyzed by agro-waste-based solvent medium. *Physical Sciences Reviews, 8*(10).
7. Lou, J., Babadi, M. R., Otadi, M., Tarahomi, M., Van Le, Q., Khonakdar, H. A., et al. (2023). Agricultural waste valorization towards (nano) catalysts for the production of chemicals and materials. *Fuel, 351*, 128935.
8. Hamidinasab, M., Ahadi, N., Bodaghifard, M. A., & Brahmachari, G. (2023). Sustainable and bio-based catalysts for multicomponent organic synthesis: An overview. *Polycyclic Aromatic Compounds, 43*(6), 5172–5226.
9. Jasuja, N. D., Saxena, R., Chandra, S., & Sharma, R. (2012). Pharmacological characterization and beneficial uses of Punica granatum. *Asian Journal of Plant Sciences, 11*(6), 251–267.
10. Ngo, T. C. Q., Van, T. L., Thao, N. P., Tran, T. K. N., & Hoang, N. B. (2022). An insight on Vietnamese bio-waste materials as activated carbon precursors for multiple applications in environmental protection. *Open Chemistry, 20*(1), 618–626.
11. Anal, A. K. (2013). Food processing by-products. In *Handbook of Plant Food Phytochemicals* (pp. 180–197).
12. Yamaguchi, K. K. L., Dias, D. S., Lamarão, C. V., Castelo, K. F. A., Lima, M. S., Antonio, A. S., et al. (2021). Amazonian bacuri (Platonia insignis Mart.) fruit waste valorisation using response surface methodology. *Biomolecules, 11*(12), 1767.
13. Veerapandian, M., Ramasundaram, S., Jerome, P., Chellasamy, G., Govindaraju, S., Yun, K., et al. (2023). Drug delivery application of functional nanomaterials synthesized using natural sources. *Journal of Functional Biomaterials, 14*(8), 426.
14. de Assis, R. C., Monteiro, G. R., Valentim, A. B., Maia, C. S. C., da Silva Felipe, S. M., Rabelo, C. A. F., et al. (2023). Biological properties of bioactive compounds from the fruit and leaves of the genipap tree (Genipa americana L.): A systematic review. *Food Bioscience, 53*, 102514.
15. Al-Nuaimy, M. N. M., Azizi, N., Nural, Y., & Yabalak, E. (2023). Recent advances in environmental and agricultural applications of hydrochars: A review. *Environmental Research, 250*, 117923.
16. Liu, H., Wu, J., Cai, Z., Deng, B., Liu, H., & Zhao, X. (2022). Extraction of red pigment from Chinese jujube peel and the antioxidant activity of the pigment extracts. *Open Chemistry, 20*(1), 849–862.
17. Das, N., Dhanawat, M., & Shrivastava, S. K. (2012). An overview on antiepileptic drugs. *Drug Discoveries & Therapeutics, 6*(4), 178–193.
18. Abraham, S., & Shaju, M. (2013). Innovations in epilepsy management–an overview. *Journal of Pharmacy & Pharmaceutical Sciences, 16*(4), 564–576.
19. Lasoń, W., Chlebicka, M., & Rejdak, K. (2013). Research advances in basic mechanisms of seizures and antiepileptic drug action. *Pharmacological Reports, 65*(4), 787–801.
20. Berkovic, S. F. (2005). Treatment with anti-epileptic drugs. *Australian Family Physician, 34*(12), 1017–1020.
21. Cavanna, A. E., Ali, F., Rickards, H. E., & McCorry, D. (2010). Behavioral and cognitive effects of anti-epileptic drugs. *Discovery Medicine, 9*(35), 138–144.

22. Siang, L. H., Arulsamy, A., Yoon, Y. K., & Shaikh, M. F. (2022, August 8). Fruits for seizures? A systematic review on the potential anti-Convulsant effects of fruits and their phytochemicals. *Current Neuropharmacology, 20*(10), 1925–1940.
23. Löscher, W. (2002). Basic pharmacology of valproate: A review after 35 years of clinical use for the treatment of epilepsy. *CNS Drugs, 16*, 669–694.
24. Barry, J. J., Ettinger, A. B., Friel, P., Gilliam, F. G., Harden, C. L., Hermann, B., et al. (2008). Consensus statement: The evaluation and treatment of people with epilepsy and affective disorders. *Epilepsy & Behavior, 13*, S1–S29.
25. Guerrini, R. (2006). Valproate as a mainstay of therapy for pediatric epilepsy. *Pediatric Drugs, 8*, 113–129.
26. Prentice, R. N., & Rizwan, S. B. (2023). Translational considerations in the development of intranasal treatments for epilepsy. *Pharmaceutics., 15*(1), 233.
27. Angus-Leppan, H., Sperling, M. R., & Villanueva, V. (2023). Antiseizure medications (antiepileptic drugs) in adults: Starting, monitoring and stopping. *Journal of Neurology, 270*(1), 573–581.
28. Yew, A. N. J., Schraagen, M., Otte, W. M., & van Diessen, E. (2023). Transforming epilepsy research: A systematic review on natural language processing applications. *Epilepsia, 64*(2), 292–305.
29. Salahinejad, A., Meuthen, D., Attaran, A., Chivers, D. P., & Ferrari, M. C. O. (2023). Effects of common antiepileptic drugs on teleost fishes. *Science of the Total Environment, 866*, 161324.
30. Tumani Üstdal, B., Soydan Çabuk, D., Coşgunarslan, A., Evlice, B., & Evlice, A. (2023). Evaluation of the effect of antiepileptic drugs on mandibular bone quality by fractal analysis. *Oral Radiology, 1–7*, 563–569.
31. Job, J. T., Visakh, N. U., Pathrose, B., Alfarhan, A., Rajagopal, R., Thayyullathil, J., et al. (2023). Chemical composition and biological activities of the essential oil from Citrus reticulata Blanco peels collected from Agrowastes. *Chemistry & Biodiversity, 21*, e202301223.
32. Rajput, S., Kaur, S., Panesar, P. S., & Thakur, A. (2023). Supercritical fluid extraction of essential oils from Citrus reticulata peels: Optimization and characterization studies. *Biomass Conversion and Biorefinery, 13*(16), 14605–14614.
33. Meryem, S., Mohamed, D., Nour-eddine, C., & Faouzi, E. (2023). Chemical composition, antibacterial and antioxidant properties of three Moroccan citrus peel essential oils. *Scientific African, 20*, e01592.
34. Šafranko, S., Šubarić, D., Jerković, I., & Jokić, S. (2023). Citrus by-products as a valuable source of biologically active compounds with promising pharmaceutical, biological and biomedical potential. *Pharmaceuticals., 16*(8), 1081.
35. Kaur, S., Panesar, P. S., & Chopra, H. K. (2023). Citrus processing by-products: An overlooked repository of bioactive compounds. *Critical Reviews in Food Science and Nutrition, 63*(1), 67–86.
36. Suri, S., Singh, A., & Nema, P. K. (2022). Current applications of citrus fruit processing waste: A scientific outlook. *Applied Food Research, 2*(1), 100050. Available from: https://www.sciencedirect.com/science/article/pii/S2772502222000105
37. Sarkar, R., Kundu, A., Dutta, A., Mandal, A., & Saha, S. (2022). Citrus Peel as a source for waste valorization and its greener processing. In: Melon breeding and genetics: Developments in food quality & safety. *American Chemical Society*. p. 11–147. (ACS Symposium Series; vol. 1415). Available from: https://doi.org/10.1021/bk-2022-1415.ch011.
38. Manthey, J. A., & Grohmann, K. (1996, January 1). Concentrations of hesperidin and other Orange Peel flavonoids in citrus processing byproducts. Journal of Agricultural and Food Chemistry 44(3):811–814. Available from https://doi.org/10.1021/jf950572g.
39. Mahato, N., Sharma, K., Sinha, M., Dhyani, A., Pathak, B., Jang, H., et al. (2021). Biotransformation of citrus waste-I: Production of biofuel and valuable compounds by fermentation. *Processes, 9*, 220.
40. Ferreira, S. S., Silva, A. M., & Nunes, F. M. (2018). Citrus reticulata Blanco peels as a source of antioxidant and anti-proliferative phenolic compounds. *Industrial Crops and*

Products, 111, 141–148. Available from: https://www.sciencedirect.com/science/article/pii/S0926669017306854
41. Rosa-Falero, C., Torres-Rodríguez, S., Jordán, C., Licier, R., Santiago, Y., Toledo, Z., Santiago, M., Serrano, K., Sosa, J., & Ortiz, J. G. (2015, February 13). Citrus aurantium increases seizure latency to PTZ induced seizures in zebrafish thru NMDA and mGluR's I and II. *Frontiers in Pharmacology., 5*, 284.
42. Tomás-Navarro, M., Vallejo, F., & Tomás-Barberán, F. A. (2014). Chapter 40 – Bioavailability and metabolism of citrus fruit beverage flavanones in humans. In R. R. Watson & V. R. Preedy (Eds.), *Zibadi SBTP in HH and D, editors* (pp. 537–551). Academic. Available from: https://www.sciencedirect.com/science/article/pii/B9780123984562000402
43. Said, N. S., Olawuyi, I. F., Cho, H. S., & Lee, W. Y. (2023). Novel edible films fabricated with HG-type pectin extracted from different types of hybrid citrus peels: Effects of pectin composition on film properties. *International Journal of Biological Macromolecules, 253*, 127238. Available from: https://www.sciencedirect.com/science/article/pii/S0141813023041351
44. Yadav, V., Sarker, A., Yadav, A., Miftah, A. O., Bilal, M., & Iqbal, H. M. N. (2022). Integrated biorefinery approach to valorize citrus waste: A sustainable solution for resource recovery and environmental management. *Chemosphere, 293*, 133459. Available from: https://www.sciencedirect.com/science/article/pii/S0045653521039333
45. Ferrara, A., Candia, V., & Ferranti, P. (2023). In P. B. T. S. F. S. A. C. A. Ferranti (Ed.), *4.22 – Technological processes to produce novel ingredients from Agri-food sources: Functional compounds from citrus wastes* (pp. 380–391). Elsevier. Available from: https://www.sciencedirect.com/science/article/pii/B9780128239605000421
46. da Costa, J. S., Maranduba, H. L., de Sousa, C. S., de Almeida Neto, J. A., & Rodrigues, L. B. (2022). Environmental performance of orange citrus waste as raw material for pectin and essential oil production. *Food and Bioproducts Processing, 135*, 165–177. Available from: https://www.sciencedirect.com/science/article/pii/S0960308522000839
47. Alzubi, E., Kassem, A., Melkonyan, A., & Noche, B. (2024). Enhancing economic-social sustainability through a closed-loop citrus supply chain: A life cycle cost analysis. *Resources, Conservation & Recycling Advances, 21*, 200199. Available from: https://www.sciencedirect.com/science/article/pii/S2667378923000718
48. Alzubi, E., Shbikat, N., & Noche, B. (2023). A system dynamics model to improving sustainable performance of the citrus farmers in Jordan Valley. *Cleaner Production Letters, 4*, 100034. Available from: https://www.sciencedirect.com/science/article/pii/S2666791623000076
49. Benny, N., Shams, R., Dash, K. K., Pandey, V. K., & Bashir, O. (2023). Recent trends in utilization of citrus fruits in production of eco-enzyme. *The Journal of Agriculture and Food Research, 13*, 100657. Available from: https://www.sciencedirect.com/science/article/pii/S2666154323001643
50. Falcone, G., Fazari, A., Vono, G., Gulisano, G., & Strano, A. (2024). Application of the LCA approach to the citrus production chain—A systematic review. *Cleaner Environmental Systems, 12*, 100156. Available from: https://www.sciencedirect.com/science/article/pii/S2666789423000508
51. Lucia, C., Laudicina, V. A., Badalucco, L., Galati, A., Palazzolo, E., Torregrossa, M., et al. (2022). Challenges and opportunities for citrus wastewater management and valorisation: A review. *The Journal of Environmental Management, 321*, 115924. Available from: https://www.sciencedirect.com/science/article/pii/S0301479722014979
52. Zema, D. A., Fòlino, A., Zappia, G., Calabrò, P. S., Tamburino, V., & Zimbone, S. M. (2018). Anaerobic digestion of orange peel in a semi-continuous pilot plant: An environmentally sound way of citrus waste management in agro-ecosystems. *Science of the Total Environment, 630*, 401–408. Available from: https://www.sciencedirect.com/science/article/pii/S0048969718305618
53. Cabot, M. I., Lado, J., Clemente, G., & Sanjuán, N. (2022). Towards harmonised and regionalised life cycle assessment of fruits: A review on citrus fruit. *Sustainable Production and*

Consumption, 33, 567–585. Available from: https://www.sciencedirect.com/science/article/pii/S2352550922001981
54. Phiri, R., Mavinkere Rangappa, S., & Siengchin, S. (2024). Agro-waste for renewable and sustainable green production: A review. *The Journal of Cleaner Production, 434*, 139989. Available from: https://www.sciencedirect.com/science/article/pii/S0959652623041471
55. Balkrishna, A., Srivastava, S., Srivastava, D., Sharma, N., Arya, V., & Gautam, A. K. (2023). Unleashing the potential of medicinal and aromatic plant wastes with particular consideration of vermicomposting: A comprehensive review of literature. *Journal of Applied Research on Medicinal and Aromatic Plants, 39*, 100527. Available from: https://www.sciencedirect.com/science/article/pii/S2214786123000712
56. Vásquez-Ibarra, L., Iriarte, A., Rebolledo-Leiva, R., Vásquez, M., Angulo-Meza, L., & González-Araya, M. C. (2021). Considering the influence of the variability in management practices on the environmental impacts of fruit production: A case study on raspberry production in Chile. *The Journal of Cleaner Production, 313*, 127609. Available from: https://www.sciencedirect.com/science/article/pii/S0959652621018278
57. Shang, N., Tong, P., Ye, P., Su, J., Bai, M., & Wu, H. (2023). Crc1,6RhaT is involved in the synthesis of hesperidin of the main bioactive substance in the Citrus reticulata 'Chachi' fruit. *Horticultural Plant Journal, 10*(6), 1321–1335. Available from: https://www.sciencedirect.com/science/article/pii/S2468014123001164
58. Kaur, S., Panesar, P. S., & Chopra, H. K. (2023). Extraction of dietary fiber from kinnow (Citrus reticulata) peels using sequential ultrasonic and enzymatic treatments and its application in development of cookies. *Food Bioscience, 54*, 102891. Available from: https://www.sciencedirect.com/science/article/pii/S2212429223005424
59. Shorbagi, M., Fayek, N. M., Shao, P., & Farag, M. A. (2022). Citrus reticulata Blanco (the common mandarin) fruit: An updated review of its bioactive, extraction types, food quality, therapeutic merits, and bio-waste valorization practices to maximize its economic value. *Food Bioscience, 47*, 101699. Available from: https://www.sciencedirect.com/science/article/pii/S2212429222001584
60. Shaw, D., Tripathi, A. D., Paul, V., Agarwal, A., Mishra, P. K., & Kumar, M. (2023). Valorization of essential oils from citrus peel powder using hydro-distillation. *Sustainable Chemistry and Pharmacy, 32*, 101036. Available from: https://www.sciencedirect.com/science/article/pii/S2352554123000700
61. Cai, X., Chen, J., Chen, J., Li, A., Li, Q., Bai, L., et al. (2023). Identification and quality evaluation of Citrus reticulata cv. Chachiensis varieties based on SNP markers. *The Arabian Journal of Chemistry, 16*(8), 104948. Available from: https://www.sciencedirect.com/science/article/pii/S1878535223004100
62. Zhang, Y., Zhao, C., Tian, G., Lu, C., Li, Y., He, L., et al. (2018). Simultaneous characterization of chemical structures and bioactivities of citrus-derived components using SERS barcodes. *Food Chemistry, 240*, 743–750. Available from: https://www.sciencedirect.com/science/article/pii/S0308814617312566
63. Cruz, A. G., Mtz-Enríquez, A. I., Díaz-Jiménez, L., Ramos-González, R., Valdés, J. A. A., Flores, M. E. C., et al. (2020). Production of fatty acid methyl esters and bioactive compounds from citrus wax. *Waste Management, 102*, 48–55. Available from: https://www.sciencedirect.com/science/article/pii/S0956053X19306518
64. Li, G., Rouseff, R., Cheng, Y., Zhou, Q., & Wu, H. (2021). Comprehensive identification and distribution pattern of 37 oxygenated heterocyclic compounds in commercially important citrus juices. *LWT, 152*, 112351. Available from: https://www.sciencedirect.com/science/article/pii/S0023643821015048
65. Rossi, R. C., da Rosa, S. R., Weimer, P., Lisbôa Moura, J. G., de Oliveira, V. R., & de Castilhos, J. (2020). Assessment of compounds and cytotoxicity of Citrus deliciosa Tenore essential oils: From an underexploited by-product to a rich source of high-value bioactive compounds. *Food Bioscience, 38*, 100779. Available from: https://www.sciencedirect.com/science/article/pii/S2212429220311172

66. Oliveira, A. M. B., Viganó, J., Sanches, V. L., Rostagno, M. A., & Martínez, J. (2022). Extraction of potential bioactive compounds from industrial Tahiti lime (citrus latifólia tan.) by-product using pressurized liquids and ultrasound-assisted extraction. *Food Research International, 157*, 111381. Available from: https://www.sciencedirect.com/science/article/pii/S0963996922004380
67. Gupta, A. K., Dhua, S., Pratiksha, K. V., Naik, B., Magwaza, L. S., et al. (2023). Current and emerging applications in detection and removal of bitter compounds in citrus fruit juice: A critical review. *Food Bioscience, 55*, 102995. Available from: https://www.sciencedirect.com/science/article/pii/S2212429223006466
68. Dadwal, V., Joshi, R., & Gupta, M. (2022). A comparative metabolomic investigation in fruit sections of Citrus medica L. and Citrus maxima L. detecting potential bioactive metabolites using UHPLC-QTOF-IMS. *Food Research International, 157*, 111486. Available from: https://www.sciencedirect.com/science/article/pii/S0963996922005440
69. Yang, H., Gu, X., Chen, H., Zeng, Q., Mao, Z., & Ge, Y. (2023). Omics techniques reveal the toxicity mechanisms of three antiepileptic drugs to juvenile zebrafish (Danio rerio) brain and liver. *Aquatic Toxicology, 262*, 106668. Available from: https://www.sciencedirect.com/science/article/pii/S0166445X23002709
70. Tsyvunin, V., Shtrygol', S., & Shtrygol, D. (2020). Digoxin enhances the effect of antiepileptic drugs with different mechanism of action in the pentylenetetrazole-induced seizures in mice. *Epilepsy Research, 167*, 106465. Available from: https://www.sciencedirect.com/science/article/pii/S0920121120305155
71. Ekici, M., Tuncay, H. O., Akalın, E., Yaman Bucak, A., & Üresin, A. Y. (2024). Evaluation of the efficacy, safety, and mechanism of action of plants traditionally used in the treatment of hypertension in Turkey. *Journal of Herbal Medicine, 43*, 100835. Available from: https://www.sciencedirect.com/science/article/pii/S2210803323002130
72. Gupta, G., Kazmi, I., Afzal, M., Rahman, M., Saleem, S., Ashraf, M. S., et al. (2012). Sedative, antiepileptic and antipsychotic effects of Viscum album L. (Loranthaceae) in mice and rats. *Journal of Ethnopharmacology, 141*(3), 810–816. Available from: https://www.sciencedirect.com/science/article/pii/S0378874112001717
73. Kwon, J. Y., Jeon, M. T., Jung, U. J., Kim, D. W., Moon, G. J., & Kim, S. R. (2019). Perspective: Therapeutic potential of flavonoids as alternative medicines in epilepsy. *Advances in Nutrition, 10*(5), 778–790. Available from: https://www.sciencedirect.com/science/article/pii/S2161831322004343
74. Viteva, E., & Zahariev, Z. (2018). Comparative effectiveness of add-on therapy with newer-generation antiepileptic drugs in Bulgarian patients with refractory epilepsy. *Epilepsy & Behavior, 87*, 137–145. Available from: https://www.sciencedirect.com/science/article/pii/S1525505018304803
75. Kumar, S., Sarangi, S. C., Tripathi, M., & Gupta, Y. K. (2020). Evaluation of adverse drug reaction profile of antiepileptic drugs in persons with epilepsy: A cross-sectional study. *Epilepsy & Behavior, 105*, 106947. Available from: https://www.sciencedirect.com/science/article/pii/S1525505019313812
76. Gunawan, C., Seneviratne, U., & D'Souza, W. (2019). The effect of antiepileptic drugs on epileptiform discharges in genetic generalized epilepsy: A systematic review. *Epilepsy & Behavior, 96*, 175–182. Available from: https://www.sciencedirect.com/science/article/pii/S1525505018310424
77. Higazy, I. M. (2020). Brain targeting stealth lipomers of combined antiepileptic-anti-inflammatory drugs as alternative therapy for conventional anti-Parkinson's. *The Saudi Pharmaceutical Journal, 28*(1), 33–57. Available from: https://www.sciencedirect.com/science/article/pii/S1319016419301380
78. Li, X., Wang, Q., Wang, M., Liu, Y., Chen, L., Wang, F., et al. (2024). Integrated metabolomics and network pharmacology revealed the key active ingredients for the treatment of ulcerative colitis in the Citrus reticulata 'Dahongpao' peel. *Journal of Pharmaceutical and*

Biomedical Analysis, 239, 115887. Available from: https://www.sciencedirect.com/science/article/pii/S0731708523006568

79. B'chir, F., & Arnaud, M. J. (2023). Chapter 4 – Chemical profile and extraction yield of essential oils from peel of Citrus limon, Citrus aurantium, and Citrus limetta: A review. In Atta-Ur-Rahman BTS in NPC (Ed.), (pp. 135–204). Elsevier. Available from: https://www.sciencedirect.com/science/article/pii/B9780443189616000093
80. Lee, G. J., Lee, S. Y., Kang, N. G., & Jin, M. H. (2022). A multi-faceted comparison of phytochemicals in seven citrus peels and improvement of chemical composition and antioxidant activity by steaming. *LWT, 160*, 113297. Available from: https://www.sciencedirect.com/science/article/pii/S0023643822002328
81. Zhao, J., Mao, X., Zhang, Q., Xiao, W., Yan, A., Hu, J., et al. (2024). A convenient and effective method for determining organophosphorus pesticides in citrus fruits based on a novel dispersive solid phase extraction using UiO-66/Alg bead as the sorbent. *Food Chemistry, 438*, 137991. Available from: https://www.sciencedirect.com/science/article/pii/S0308814623026092
82. Julaeha, E., Puspita, W. R., Permadi, N., Harja, A., Nurjanah, S., Wahyudi, T., et al. (2024). Optimization of Citrus aurantifolia peel extract encapsulation in alginate-gelatin hydrogel microbeads for antibacterial wound dressing applications. *Carbohydrate Polymer Technologies and Applications, 7*, 100406. Available from: https://www.sciencedirect.com/science/article/pii/S2666893923001263
83. Kulig, M., Galanty, A., Grabowska, K., & Podolak, I. (2022). Assessment of safety and health-benefits of Citrus hystrix DC. Peel essential oil, with regard to its bioactive constituents in an in vitro model of physiological and pathological skin conditions. *Biomedicine & Pharmacotherapy, 151*, 113151. Available from: https://www.sciencedirect.com/science/article/pii/S0753332222005406
84. Singhal, S., & Swami Hulle, N. R. (2022). Citrus pectins: Structural properties, extraction methods, modifications and applications in food systems—A review. *Applied Food Research, 2*(2), 100215. Available from: https://www.sciencedirect.com/science/article/pii/S2772502222001706
85. Wang, Y., Liu, J., Chen, L., Jin, S., An, C., Chen, L., et al. (2023). Effects of thermal treatments on the extraction and in vitro fermentation patterns of pectins from pomelo (Citrus grandis). *Food Hydrocolloids, 141*, 108755. Available from: https://www.sciencedirect.com/science/article/pii/S0268005X23003016
86. Zhang, J., Mu, K., Yang, P., Feng, X., Zhang, D., Fan, X., et al. (2021). Synthesis, antiepileptic effects, and structure-activity relationships of α-asarone derivatives: In vitro and in vivo neuroprotective effect of selected derivatives. *Bioorganic Chemistry, 115*, 105179. Available from: https://www.sciencedirect.com/science/article/pii/S0045206821005563
87. Jamal Gilani, S., Zaheen Hassan, M., Sarim Imam, S., Kala, C., & Prakash, D. S. (2019). Novel benzothiazole hydrazine carboxamide hybrid scaffolds as potential in vitro GABA AT enzyme inhibitors: Synthesis, molecular docking and antiepileptic evaluation. *Bioorganic & Medicinal Chemistry Letters, 29*(14), 1825–1830. Available from: https://www.sciencedirect.com/science/article/pii/S0960894X19302938
88. Yokoi, R., Shibata, M., Odawara, A., Ishibashi, Y., Nagafuku, N., Matsuda, N., et al. (2021). Analysis of signal components < 500 Hz in brain organoids coupled to microelectrode arrays: A reliable test-bed for preclinical seizure liability assessment of drugs and screening of antiepileptic drugs. *Biochemistry and Biophysics Reports, 28*, 101148. Available from: https://www.sciencedirect.com/science/article/pii/S2405580821002429
89. Ullah, M. I., Anwar, R., Zia, M., Gul, B., Kamran, S., & Kamran, S. H. (2023). Assessment of in vivo antiepileptic potential and phytochemical analysis of Cassia absus seed extracts. *Heliyon, 9*(4), e14660. Available from: https://www.sciencedirect.com/science/article/pii/S2405844023018674

90. Wang, Y., & Chen, Z. (2019). An update for epilepsy research and antiepileptic drug development: Toward precise circuit therapy. *Pharmacology & Therapeutics, 201*, 77–93. Available from: https://www.sciencedirect.com/science/article/pii/S0163725819300920
91. Boada, C. M., French, J. A., & Dumanis, S. B. (2020). Proceedings of the 15th antiepileptic drug and device trials meeting: State of the science. *Epilepsy & Behavior, 111*, 107189. Available from: https://www.sciencedirect.com/science/article/pii/S1525505020303681
92. Campos, G., Fortuna, A., Falcão, A., & Alves, G. (2018). In vitro and in vivo experimental models employed in the discovery and development of antiepileptic drugs for pharmacoresistant epilepsy. *Epilepsy Research, 146*, 63–86. Available from: https://www.sciencedirect.com/science/article/pii/S0920121118300639
93. Feldman, A. E., & Gidal, B. E. (2013). QTc prolongation by antiepileptic drugs and the risk of torsade de pointes in patients with epilepsy. *Epilepsy & Behavior, 26*(3), 421–426. Available from: https://www.sciencedirect.com/science/article/pii/S1525505012005835
94. Margineanu, D. G. (2014). Systems biology, complexity, and the impact on antiepileptic drug discovery. *Epilepsy & Behavior, 38*, 131–142. Available from: https://www.sciencedirect.com/science/article/pii/S1525505013004344
95. Upton, N. (1994). Mechanisms of action of new antiepileptic drugs: Rational design and serendipitous findings. *Trends in Pharmacological Sciences, 15*(12), 456–463. Available from: https://www.sciencedirect.com/science/article/pii/0165614794900590
96. Pereira, M., Costa, B., & Vale, N. (2023). Chapter 16 – Targeting DNA damage response pathways in glioblastoma: From mechanistic insights to advances in the clinic. In C. Vitorino, C. Balaña, & C. CBTNIIG (Eds.), (pp. 345–360). Academic. Available from: https://www.sciencedirect.com/science/article/pii/B9780323998734000177
97. Brodtkorb, E. (1998). Chapter 26 Antiepileptic drug treatment: Clinical considerations and concerns. In: Ottersen, O. P., Langmoen, I. A., Gjerstad, L. B. T. P. in BR, editors. Elsevier. p. 395–406. Available from: https://www.sciencedirect.com/science/article/pii/S0079612308604513
98. Stables, J. P., Bialer, M., Johannessen, S. I., Kupferberg, H. J., Levy, R. H., Loiseau, P., et al. (1995). Progress report on new antiepileptic drugs a summary of the second Eilat conference. *Epilepsy Research, 22*(3), 235–246. Available from: https://www.sciencedirect.com/science/article/pii/0920121195000542
99. Steinhäuser, C., Grunnet, M., & Carmignoto, G. (2016). Crucial role of astrocytes in temporal lobe epilepsy. *Neuroscience, 323*, 157–169. Available from: https://www.sciencedirect.com/science/article/pii/S0306452215000123
100. De Caro, C., Iannone, L. F., Citraro, R., Striano, P., De Sarro, G., Constanti, A., et al. (2019). Can we 'seize' the gut microbiota to treat epilepsy? *Neuroscience & Biobehavioral Reviews, 107*, 750–764. Available from: https://www.sciencedirect.com/science/article/pii/S0149763419306517
101. Bialer, M., & White, H. S. (2010). Key factors in the discovery and development of new antiepileptic drugs. *Nature Reviews Drug Discovery, 9*(1), 68–82. Available from: https://www.nature.com/articles/nrd2997

Chapter 8
Extraction and Valorization of Bioactive Essential Oils from Citrus Fruit Waste

Naduvilthara U. Visakh, Athira Shajin, Doppalapudi Vijaya Rani, and J. Reecha

8.1 Introduction

Citrus belongs to the Rutaceae family and is a commonly cultivated fruit worldwide. It is a staple fruit crop in human diets due to its widespread availability. The most prevalent types of citrus fruits grown commercially are *Citrus sinensis* (Orange), *Citrus reticulata* (tangerines), *Citrus aurantiifolia* (lime), *Citrus limon* (lemon), and *Citrus paradisi* (grapefruit) [1–3]. After World War II, the citrus sector has grown significantly over the past few decades due to increasing international trade and the steady rise in the intake of citrus fruits and their products. This includes quickly developing frozen concentrated orange juice processing technology and other citrus waste processing industries. Citrus fruits are primarily high in carbohydrates, such as sucrose, glucose, and fructose, and have very little protein and fat. Besides being an excellent dietary fiber source, fresh citrus fruits reduce blood cholesterol and may help avoid gastrointestinal diseases. The fruits provide phytochemicals, including carotenoids, flavonoids, and limonoids, as well as vitamin C and B complexes. Flavanones were predominant in the citrus peel and possess different properties, which were known to prevent oxidation, inhibit inflammation, and inhibit the effects of carcinogens [4]. China, India, Nigeria, Brazil, and Mexico are the major countries that grow citrus globally. In 2021, citrus fruits accounted for

N. U. Visakh (✉) · A. Shajin
Department of Agricultural Entomology, College of Agriculture, Kerala Agricultural University, Thrissur, Kerala, India

D. V. Rani
Department of Agronomy, Sardar Vallabhbhai Patel University of Agriculture & Technology, Meerut, Uttar Pradesh, India

J. Reecha
Department of Plant Pathology, Sardar Vallabhbhai Patel University of Agriculture and Technology, Meerut, Uttar Pradesh, India

A. Chauhan et al. (eds.), *Valorization of Citrus Food Waste*,
https://doi.org/10.1007/978-3-031-77999-2_8

161.8MT (million tons) of fruit grown in 10.2 Mha or more, making them the second-largest produced crop worldwide [5].

The waste from the citrus processing industry contains both solid and liquid waste. Depending on the technique used, the amount of waste produced during citrus processing ranges from roughly 50% to 70% of the fruits that have been processed [6]. Around 119.7MT of citrus fruit waste is produced globally yearly, whereas 7.8 million tons of waste from India. This waste comprises 20–40% seeds, peels 50–55%, pomace and wastewater [7]. The waste has environmental impacts and industrial values, leading to new industrial opportunities. This chapter discusses the composition and ecological effects of citrus fruit waste and the potential extraction of bioactive essential oils with various biological activities from citrus fruit peel waste.

8.2 Characterization and Composition of Citrus Fruit Waste

8.2.1 Citrus Peel

Approximately 30–34% of the peel from citrus fruits is obtained while making citrus juice. It contains highly fragrant compounds, pectin, dietary fibers, natural pigments and polyphenols [8]. Many phytochemicals and bioactive components can be found in citrus peels, such as polyphenol-flavonoids (nobiletin, naringin, hesperidin, tangerine), carotenoids, essential oils (limonene), carbohydrates (cellulose, hemicellulose, pectin, and dietary fibers), flavoring agents and pigments [9, 10]. As it comprises many phytochemicals, it is utilized in the drink, culinary, pharmaceutical, and cosmetic sectors.

Polysaccharides that are slightly esterified with methanol or acetic acid are the building blocks of pectin. It is made up of α-1,4 glycosidic linkages connecting units of D-galacturonic acid. It is a naturally occurring solidifying or gelling agent widely used for jam and jelly preparations as a texturizer, stabilizer, and thickening agent. It usually appears in color from white to light brown [11]. Dietary fibers (at least ten carbohydrate units) are non-starch polysaccharides, which are difficult for the intestines to digest and absorb. They can be found in soluble and insoluble forms, such as gum, pectin, cellulose, hemicellulose, and lignin [12]. Many fruits and vegetables produce pigments called carotenoids through biosynthesis. They are classified into two classes: hydrocarbon carotenoids (carotenes), which include lycopene and β-carotene, and oxygenated carotenoids (xanthophylls), which contain lutein and violaxanthin. These substances function as precursors to vitamin A, strengthening the immune system, encouraging good ocular health, and aiding the growth of epithelial tissues [13]. Plants produce flavonoids, a diverse class of secondary metabolites, to shield them from harmful UV rays and damage. They are separated into six classes based on their structure, which are flavones, flavanones, isoflavones, flavonols, and anthocyanidins. Flavanones are predominant in the citrus peel and possess different properties, which are known to prevent oxidation, inhibit inflammation, and inhibit the effects of carcinogens [14].

8.2.2 *Citrus Seed*

About 20–25% of citrus fruits have seeds, typically discarded once the fruit is processed. Citrus seeds could be a profitable product since they are a high protein and essential oils source. It has increased its culinary, cosmetic, and medical demand. Unhulled seed contains the following composition [15–17].

8.2.3 *Citrus Pomace*

After citrus juice processing, the remaining solid residue or other products is called citrus pomace. Citrus pomace is a valuable source of limonene, essential oils, molasses, and pectin. It is composed of pulp, membranes, and seeds. It also contains naturally occurring antioxidants and polyphenols, or flavonoids, which can be used in industries. The nutritive analysis of citrus pomace (Table 8.1) [18].

8.3 Environmental Impact

Citrus fruit waste can significantly impact the environment for several reasons. Citrus fruit cultivation requires various resources, including water, fertilizer, and pesticides. In addition to wasting the resources needed to grow them, throwing away these fruits will negatively impact the environment and diminish available resources. The main environmental issue with citrus waste is its high carbohydrate content, which can accelerate fermentation and attract pests and pathogens [19]. Conventional techniques like landfills and incineration were used to dispose of the waste. However, these techniques have become insufficient and have also caused environmental

Table 8.1 Nutritive analysis of citrus pomace

Nutrients	Composition (%)
Sugar	30–40%
Cellulose	13–17%
Pectin	14–25%
Moisture	10%
Fat	0.4–1.6%
Micronutrient	0.5%

impacts. As the waste from industry is thrown away in landfills, it decomposes to release methane without oxygen, a greenhouse gas that contributes to global warming [20]. Improper disposal of citrus waste on land can degrade soil quality and disrupt the soil nutrient balance. While some nutrients may benefit the soil, the excess can change soil acidity levels, making it unsuitable for plant growth and disrupting the ecosystem. The wastes also contain citric acid and some of the essential oils. When these compounds leach into water sources, harming aquatic habitats through eutrophication, algal blooms, and oxygen deficit. If the waste has been burned as a disposal method, high levels of nitrogen, carbon, sulfur oxides, particulate matter compounds (VOCs) and other pollutants are released into the environment, causing air pollution [21].

8.3.1 Addressing Waste Management Issues

As citrus fruits have a market value worldwide, they have been traded and consumed where they are not naturally grown. It has post-harvest stability that facilitates international trade. The industrialization of citrus fruits has become inescapable due to the continuous expansion of plantation areas and crop sizes. Citrus fruits are processed to prepare jam, jelly, and marmalade and to extract beneficial essential oils and flavonoid compounds [22, 23]. Approximately 18% of citrus fruits produced worldwide are utilized in industries mainly for juice production. With the increasing consumption of citrus fruits in industries, waste production is also comparatively increasing, which leads to environmental hazards. Discarding fruit pieces without proper treatment poses a risk to the environment [24]. Around half of the fresh fruit mass is wasted while producing citrus juice. This includes 50–55% of peel waste, 20–40% of seeds, others of pomace, and wastewater. Every year, citrus fruit processing produces waste materials of about 10 MT [25], leading to environmental issues. Investment is necessary to properly dispose of citrus waste since improper disposal pollutes the water and land and further damages the aquatic ecology.

Citrus fruit waste, such as seeds, pomace and peels, is reported to have numerous uses as a bio-resource. Many components found in citrus trash have the potential to be further investigated in the food industry. These components can be used as sources of pectin, colorants, flavoring agents or as additives to make candied goods in the confectionery industry [26]. Moreover, the pharmaceutical industry finds utility in it [27]. Citrus waste can be utilized for non-food applications such as packaging materials, biofuel, activated carbon synthesis, and bio-adsorbent [7, 28]. It also has a unique aroma and contains essential oils, making it suitable for cosmetic products [29].

8.3.2 Exploring the Potential of Bioactive Essential Oils

Bioactive essential oils derived from citrus fruit waste have many potential uses, from food and agriculture to cosmetics and pharmaceuticals. Citrus essential oils mainly consist of terpenes, particularly limonene, which can account for about 90% of the oil composition. The remaining terpenes are citral, geraniol, myrcene, β-pinene, β-pinene, and myrcene. These compounds contribute to citrus fruits' unique flavor and aroma and contain various biological properties. The chief essential oil limonene has been identified as the potential anticancer properties that prevent the proliferation of cancer cells. Essential oil has anti-microbial activity and has been widely used in disinfectants and other pesticide formulations. These oils contain antioxidant properties that prevent the cell from cellular damage and oxidative stress [30, 31]. The cosmetic industry uses essential oils from citrus in their skincare products, which act as cleansing and toning agents. They can also help to balance oily skin, reduce acne, and improve skin texture. These oils are also utilized in hair care products for their refreshing scent and potential benefits in promoting scalp health and stimulating hair growth. These essential oils are used as flavoring agents and preservatives in the food industry [32]. Some of the essential oils from citrus act as insect repellents and are also used in pesticide formulation to manage pest populations [33].

8.4 Citrus Essential Oils

The aromatic compound, which is a mixture of over two hundred compounds comprising mainly of hydrocarbons like mono and sesquiterpenes and their oxygenated counterparts such as ketones, aliphatic aldehydes, esters and alcohols, are essential oils. These are generally synthesized and present in aromatic plants. Most essential oils have a specific gravity less than that of water, and the specific gravity is higher in only a tiny fraction of essential oils. They can transfer their odor to the aqueous layer even though these are immiscible with water. These are also called volatile or ethereal oils [34].

Citrus, one of the essential genera belonging to the family of Rutaceae, is valued highly medicinally and economically. The essential oil from citrus is generally procured from flavedo, the fruit's rind. It is acquired from flowers and also the leaves of the plant. It is widely used in various industries, such as food, pharmaceuticals, cosmetics, and perfumery. Essential oils' volatile compounds and semi-volatile compounds constitute almost 86–98% of the whole oil fraction, which is of great importance [35].

8.4.1 Aromatic and Therapeutic Properties

The characteristic flavor and aroma of citrus can be attributed to essential oils, which have constituent components like aldehydes, alcohols, ketones, terpenes, and esters in their volatile parts [36, 37]. The aroma of the zest and the bitterness is due to the principal compound in the citrus fruit peel, limonoids [38, 39]. Due to its strong aroma, it is utilized as a flavoring agent in food. The existence of the bioactive phenolic compounds adds to its antimicrobial property, leading to its use as a potent food preservative, which can slow down food spoiling. In addition to that, it adds organoleptic qualities, flavor and aroma to the food products. With the same principal flavorings, essential oils can be used as a substitute for herbs and spices [40, 41]. It acts as a natural sanitizer, which can be safely used instead of synthetic chemicals with antimicrobial properties to prevent cross-contamination in the food industry [42–44]. Essential oils with pharmaceutical properties have been utilized in cosmetics and perfume manufacturing as flavoring agents. In soap, cosmetology and confectionaries, it plays the role of a skin-whitening agent [45, 46].

Aromatherapy, the use of fragrances in therapy, is one of the quickly growing therapies to mitigate or treat infections and diseases by inhaling the compounds of which essential oils are a part. The penetration of aromas into the bloodstream with the inhalation of the aromatic components affects the limbic system that controls memories and emotions, causing the body to be rejuvenated, uplifted and stimulated [47, 48]. It can be used as a relaxant and antidepressant to reduce stress-related issues like fear and anxiety, improving creativity and mood, as the oil mainly targets the nervous system of the body and can be administered via nebulization [49]. It can also be administered orally and topically applied on the skin. A reduction in the symptoms of anxiety has been observed in patients diagnosed with cancer [50].

Citrus peel essential oils exhibit anti-inflammatory, antiallergic properties and anticancer activities. It can be used as a disinfectant for the skin and minor wound healing. It helps in enhancing the memory in patients with Alzheimer's disease. It also shows hepatoprotective, hypoglycemic and neuroprotective effects. Respiratory tract infections can be treated with it. To some extent, it protects the kidneys from damage and neurodegenerative diseases [51, 52]. Antiulcerogenic and gastroprotective effects are exerted by the limonene component of the essential oil [53]. In essential oils, flavonoids such as apigenin, hesperidin, quercetin, rutin, and kaempferol impart many of these properties [54]. Anti-tumor, anti-allergic, anti-prostatitis, and anti-asthmatic properties have contributed to quercitin, one of the principally active ingredients [55, 56]. There are reports of many of its medicinal properties in American folk medicine, especially for controlling hypertension [37].

The anticancerous property is D-Limonene's contribution, one of the significant constituents with potential chemotherapeutic agents in the citrus essential oil [57, 58]. Malignant brain tumors could be clinically treated with enormous impact with the oxygenated monoterpene constituent of the oil, perillyl alcohol [59]. Moreover, it acts as an inhibitor of cough, helps enhance the secretion of digestive juice, promotes the motility of the gastrointestine, decreases inflammation, and helps dissolve gallstones in the gall bladder [60].

The premature aging of the skin can be prevented by using antioxidant components in the essential oil that fights the free radicle, thus making the essential oil a significant element in skin care products [61]. When administered individually or with other synthetic drugs, essential oil shows some anti-obesity properties [41, 62]. The synthetic antioxidants could be easily replaced by citrus peel oil as it shows similar antiradical potential as vitamin E, making it a safe substitute in medicinal and food materials where synthetic antioxidants are used [63]. With increasing beauty consciousness, the essential oil has gained more attention as a natural compound for formulating dermatological products for skin care [64].

8.4.2 Studies on Essential Oil Extraction from Citrus Waste

Studies on the extraction techniques of essential oils from the rind of citrus fruits have been high over the last decades, mainly due to their rising utilization in different industry sectors such as food, pharmaceuticals, perfumery, and cosmetics. These oils constitute many other compounds with a wide array of functions and can also be a substitute for certain synthetic additives in these industries. The studies regarding extraction and its composition focus mainly on studying the semivolatile and volatile fraction of the citrus rind or the fruit peel. Some studies also report on the composition and uses of oil from citrus flowers, that is, the neroli or oil from citrus leaves and little branches called the petitgrain and from the plant leaves [65]. Nutraceutical and economic importance have added to the interest in investigating the composition of various citrus waste essential oils through GC-MS analyses. Numerous studies have been performed on different cultivars and species of citrus to identify the significant components in their essential oil, and limonene and monoterpene hydrocarbon are the major components [66, 67].

Some of these studies include studies to provide additional information on four different cultivars of *Citrus deliciosa* var. *tangerine viz*. Spinosa Clementine, Fina Clementine, Thornless Clementine, and Nour Clementine. The hydrodistillation of peels and leaves was carried out, and the components were quantitatively and qualitatively analyzed for their components by GC-MS. Their medicinal and antioxidant properties were also studied [66]. To add some fundamental information to the food processing industry, many studies have been carried out on antimicrobial activity and antiradical potential, which can be widely utilized in the processing of citrus [63].

Apart from these, studies have also been done on the techniques used to extract oil. Many attempts have been made to compare various techniques with the traditional way of removing the oil so that an economically feasible method could be found and used on a large scale. In such studies, the main focus was given to the yield of the oil obtained during extraction, the time required for the process, the physical constants that are involved in the process, the effect of the process on the chemical composition of the oil, the effect on its property such as the antimicrobial property, antioxidant property and also the consumption of energy during the process. In most studies, the newer technologies are compared to conventional hydrodistillation [68].

The study also focused on comparing the hydrodistillation and cold pressing of lemon peels with supercritical CO2 extraction and its evaluation of the constituent components and the efficiency with which oil has been separated from the rind [69]. Innovative microwave-accelerated distillation was evaluated and compared for the efficiency of the extraction process with conventional hydrodistillation process and cold pressing [70].

Literature could also be found comparing various environmentally friendly extraction processes to the traditional methods for recovering essential oil from the citrus rind. One such is comparing five different approaches for *Citrus sinensis* peel essential oil—conventional hydro-distillation, Enzymes-assisted extraction by hydrodistillation, Salt-assisted extraction by hydrodistillation, Solvent-free micro-wave extraction and Ultrasound-assisted extraction by hydrodistillation. The time required for extraction, the composition of oil, the amount of oil obtained, the energy utilized during the process, and the antioxidant properties were critically analyzed and compared [71]. The study also found that the optimal extraction technique is Solvent-free micro-wave extraction among the five as it generated greater yield during a shorter period of the extraction process and was also found to be economical and eco-friendly in nature [72].

Studies emphasizing the biorefinery concept of oil extraction have also been found with the spike in environmentally friendlier technologies. Solar energy was used to extract oil from citrus rinds [73].

8.5 Extraction Methods

There are various extraction techniques for obtaining essential oil from the vegetative matter of the plant. Distillation and cold pressing are the traditionally used ones that consume much energy during the process and require much time to complete. With the need for faster and more economical extraction techniques, many have been discovered by scientists that are more environmentally friendly, too. Some innovative methods have been developed that reduce the cost of extraction and decrease the energy used in the process.

8.5.1 Eco-Friendly Techniques and Innovative Methods

A traditional extraction method is steam distillation, especially with a Clevenger-type hydrodistillation unit [74]. To prevent the thermal reactions of the constituent compounds that occur during the extraction process, a second condenser apparatus could be attached to it, giving rise to an improved and modified type of Clevenger apparatus that can also reduce the monoterpene compound oxidation [75]. The extraction method that leads to watery emulsion production, which needs further centrifuge to obtain the essential oil component, is cold pressing, which is also traditionally used [35].

A solid-liquid extraction method to obtain active ingredients such as essential oil is maceration. It is commonly used as it is cheap but requires appropriate solvents for immersing and grinding the plant material during the process for a specified period [76]. For the reduction in the volume of organic solvent, the use of Soxhlet can be considered as it has the benefit of using the same volume of solvent repeatedly and the extract thus obtained will be much more concentrated, efficiently using the time [77].

There are various modern techniques for the distillation process. Some of them are DIC—Instant Controlled Pressure Drop Technique, SFE—Supercritical fluid extraction, MHG—Microwave Hydrodiffusion and Gravity and MSD -Microwave Steam Distillation. In an instant controlled pressure drop technique of extracting the essential oil, the matrix expands, thus significantly improving the oil extraction from the vegetative matter [78].

One of the economical and surging techniques is Supercritical fluid extraction. This technique does not utilize harmful and expensive organic solvents, avoiding the clean-up steps and making the process speedy and efficient even at room temperatures. Carbon dioxide is a nontoxic gas and readily available, and due to its non-explosive nature, it is considered an appropriate solvent in this method. The ease with which it can be eliminated after the extraction also adds to using the gas in the extraction technique [79, 80].

The extraction techniques that facilitate the production of essential oil without any difference in composition and have a speed of several folds are microwave hydrodiffusion and gravity. Microwave Steam Distillation breaks down the plant's cell wall quickly with the application of microwave radiation, which results in the bursting of the cell cytoplasm that releases essential oil quickly [81].

Ultrasonic-assisted extraction in the UAE is a highly efficient and eco-friendly method with minimum energy usage and a faster extraction process. It avoids the generation of chemical residues and contaminants and operates at a lower temperature, which makes it eco-friendly. The principal behind the method is acoustic cavitation, which increases the mass transfer rates to several folds by the captivating phenomenon of implosion of a gas vapour-filled bubble by creating turbulence and liquid currents. With the use of this method, the particle size is reduced, solvent usage is minimized, the time of extraction also shortens, and thereby the efficiency of extraction is enhanced [82, 83].

A cost-effective and efficient extraction technique widely utilized in industries is ultrasound-assisted hydrodistillation. Pressurized hot water extraction, called subcritical or superheated water, is a green technology for extracting essential oil that uses water at high temperatures and pressure [84]. A solar energy-dependent hydrodistillation unit constitutes the biorefinery concept, a greener and zero-energy-consuming process [73].

A method patented for essential oil extraction is microwave 'dry' distillation or fast microwave-accelerated distillation, which combines natural distillation with microwave energy [70]. The oil glands in the vegetal matter burst, and the volatile matter evaporates when placed in a microwave reactor, which is further condensed and collected. This method gives a higher yield in a shorter period with less energy consumption, making it economically feasible. This is a cleaner technology as no solvent or water is used in the extraction process, and thus, residue is generated [70].

8.5.2 Optimization for Enhanced Oil Yield

The oil yield that can be obtained from citrus peel depends on many factors. The thickness of the fruit's peel is one of the significant factors determining the amount of oil. Compared to thin-peeled citrus fruits, those with thicker peels will contain more oil glands; thus, oil isolated from them will also be higher. Grapefruit and sour orange are thick-peeled and yield more essential oil than thin-peeled citrus species.

The oil yield also depends on the maturity stage at which the fruit is harvested. When the extraction of oil from lemon, bitter orange, orange, and mandarin was done in various maturity stages, it was revealed that the citrus peel showed different oil yields, and the composition also varied depending on the fruiting stage [85]. Lemon showed maximum oil yield and high limonene content at its immature stage, while orange was at the maturity stage. The semimature stages were the best for extracting oil for the other two to obtain a higher oil yield [85].

The oil extraction method also determines essential oil yield from citrus peels. When compared to extraction with the Soxhlet apparatus in lime peel, the maceration method gives a comparatively higher yield. This can be attributed to the fact that there is direct contact between the peel and the organic solvent for the whole duration of the extraction process and that the same volume of the solvent circulates continuously, which results in a higher amount of oil [86]. When the citrus peel essential oil was extracted with ultrasound-assisted hydrodistillation, the oil yield was 0.48%, while that from hydrodistillation was only 0.22%. The higher yield was obtained because the oil was released from the oil glands due to the disrupting action of ultrasound on oil glands [87].

A high yield was obtained with hydrodistillation and a lower yield with supercritical fluid extraction when essential oil yield was compared between cold pressing, supercritical fluid extraction, and hydrodistillation from the rinds of lemon [84]. While seven different methods were compared, namely hydrodistillation, Ohmic-assisted hydro distillation, solvent free microwave extraction, commercial hydrodistillation, steam distillation, microwave-assisted hydrodistillation and solvent-less microwave extraction, there was a significant difference in the yield from each method [71]. The yield was highest in the traditional hydro distillation method. At the same time, the oil accumulation rate was higher in the case of microwave-assisted extraction, and the extraction time was less in solventless microwave extraction [71, 88].

The other factor that affects the yield of essential oil is the duration of extraction. In general, as the duration of the process of extraction increases, the amount of oil obtained will also be higher. The amount of time the peels remain in contact with the solvent increases as the time of extraction increases, and this results in the transferring of more amount of essential oil mass in the citrus peels into the organic solvent, thereby increasing the oil yield from peels [84, 89, 90].

The character and properties of the solvent used and the duration of its contact with the peel are key determining elements in the essential oil yield. Thus, selecting the appropriate solvent is critical to obtaining an optimum yield. The solvent will show higher extraction efficiency if the polarity value is close to the solute's. The other factors to be considered are the solubility, cost, selectivity of the solvent used, and handling safety. When the lime peel was extracted with three different solvents, viz., water, hexane and ethanol, the highest yield was obtained when extracted with hexane compared to others, and the yield increased as the extraction period was extended [91].

The biomass and solvent ratio is also a determining factor for obtaining an optimum yield. During the UAE from sweet lime, the ratio of 1:10 showed the highest yield. In contrast, the yield gradually decreased by increasing the ratio to 1:30. As the properties of the solvent, such as solubility, viscosity and diffusivity, are significantly associated with the temperature of the solvent, temperature becomes a major factor for maximizing yield during the extraction process. Generally, an increase in the efficiency of the extraction process is found with an increase in temperature [84].

8.6 Yield and Chemical Composition

Essential oils are a mixture of over two hundred compounds at concentrations. It consists of different classes of chemicals, which can be generally classified into two fractions. The volatile fraction makes up almost 85–99% of the oil, with limonene being the highest fraction of 32–98% and the non-volatile fraction of 1–15% with fatty acids, waxes, polyethoxylated flavonoids, sterols etc. [49].

These compounds arise from two different biosynthetic pathways. Shikimic acid and phenylpropanoids are the parent compounds for the aromatic compounds, while terpenoids are synthesized from the acetate–malonic acid in their intermediate form. The essential oil constituents can also be classified according to their functional group, like hydrocarbons, aldehydes, ketones, alcohols, phenols, oxides, esters, peroxides and phenolic ethers. The oxygenated terpenes are called the terpenoids, and the terpenes are a significant part of the essential oil. It is of different types based on their structure. The isoprene units in the terpene structure form the basis for classifying the terpenes into hemiterpenes, monoterpenes, sesquiterpenes etc. The monoterpene with a single loop present in the citrus peel is limonene.

8.6.1 Analysis of Essential Oil Yield and Components

One of the widely used tools for analyzing is gas chromatography-mass spectrometry (GC-MS). At the same time, the non-volatile residue could be studied with liquid chromatography, a thin layer or high-performance liquid chromatography [46]. More complicated techniques for authenticating the volatile compounds in citrus exist in the field, and they depend on the ratios of the carbon isotope. This is mainly utilized to keep in check the adulteration of citrus essential oil, from adulterants like citral from lemon grass oil and synthetic substitutes. This technique is called gas chromatography hyphenated to carbon isotope ratio mass spectrometry (GC-C-IRMS). This has been utilized to distinguish the citrus species and determine its geographical origin, eliminating environmental influences [92].

An extensively used method for analyzing essential oil is headspace solid-phase microextraction, which is more straightforward and sensitive. With the combination of nanotechnology, the accuracy of the analysis for determining and isolating the compounds can be enhanced [87]. Table 8.2 explains the essential oils extracted from different citrus species peel waste with yield and primary chemical composition.

Molecular distillation is a liquid-liquid separation technology that separates essential oil into two phases under high vacuum and low temperature. It is used for the thermolabile substances in the food industry and is also exploited for refining and concentrating essential oils [93].

8.6.2 Impact of Extraction Methods on the Chemical Composition

The method by which the oil is being extracted often determines its quality. Some bioactive and natural compounds could get destroyed during extraction if the procedures are not done correctly, resulting in an ill-quality oil [106]. An improved and modified Clevenger apparatus, when used to extract orange peel essential oil, resulted in significantly higher component yield, and it also decreased the conversion of β-myrcene after thermally reacting to β-thujene and oxidation of compounds such as β and α-pinene in comparison to the traditionally used Clevenger apparatus [75]. No difference was observed when the chromatographic analysis was done for oils obtained through conventional hydrodistillation and solar-mediated hydrodistillation, but the latter was faster [73]. Steam distillation, if carried out for a very long time, will convert the esters present in the essential oils into alcohol and acids upon hydration. To prevent hydration, extraction time for steam distillation should be limited [107].

There was quite a difference in the quantity and quality of the essential oils of lemon, berkane, navel and bitter orange when oil extracted by hydrodistillation was compared to the oil extracted with microwave-assisted hydrodistillation. When

Table 8.2 Yield and composition of citrus peel oil

Citrus species	Oil yield (%)	Components	Percent (%)	References
Sweet Lime - *C. limetta*	.54	Limonene	91.8	[52, 94]
		Trans-carveol	0.2	
		β-myrcene	1.6	
		Cis-linalool oxide	0.3	
		n-octanol	0.3	
		linalool	0.9	
		α-pinene	0.4	
		camphene	0.8	
Sweet orange - *Citrus sinensis*	1.48 (w/w)	D-Limonene	92.74	[95, 96]
		b-Myrcene	2.23	
		α -Pinene	0.59	
		Sabine	1.64	
		Linalool	.21	
Lime - *C. aurantifolia*	3 (w/w)	citronellal	0.1	[37, 65, 97, 98, 99]
		p-pinene	28.4	
		geranial	2.1	
		geraniol	7.5	
		limonene	39.3	
		neral	5.3	
Kaffir lime - *C. hystrix*	1.8 (w/w)	p-pinene and	39.3	[37]
		limonene	14.2	
		citronellal	11.7	
		terpinene-4-ol	8.9	
		α-terpineol α-	3.0	
		terpinene	2.4	
		pinene	2.0	
		linalool	1.9	
		furanoid cis-linalool oxide	1.9	
Bitter/sour orange – *C. aurantium*	10.36 (dw)	D-Limonene	93.38	[95]
		b-Myrcene	2.41	
		α -Pinene	.5	
		Sabine	.82	
		Linalool	.37	
		Capric aldehyde	.44	
		b-Ocimene	.49	
Sour lime - *C. acida*	1.7 (w/w)	α-pinene	0.27	[52, 98]
		(E)- β-ocimene	0.31	
		verbenone	0.12	
		sabinene	0.28	
		Limonene	96.86	
		β-pinene	1.37	

(continued)

Table 8.2 (continued)

Citrus species	Oil yield (%)	Components	Percent (%)	References
Lemon – *C. limon*	.41 (w/w)	limonene	55.40	[34, 100, 101]
		trans-carveol	1.33	
		ethyl p-methoxycinnamate	2.2	
		α-terpineol	1.07	
		neral	10.39	
		linalool	1.16	
		geraniol	1.48	
		Nonanal	1.19	
		trans-verbenol	6.43	
		cis-α-bergamotene	1.60	
		decanal	3.25	
		ethyl cinnamate	2.21	
Berkane – *C. clementina*	5.9 (dw)	D-Limonene	90.6	[95]
		β -Myrcene	3.9	
		α -Pinene	1.22	
		Sabine	2.04	
		Linalool	.4	
		(+)-4Carene	1.22	
Citron – *C. medica*	0.75 w/w	limonene	33.60	[102]
		myristicin	24.36	
		nerol	1.30	
		neral	1.50	
		geranial	2.53	
		carvacrol	8.10	
		α-bergamotene	2.67	
		β- bisabolene	4.40	
		apiol	5.34	
		elemicin	2.46	
Grapefruit - *C. paradisi*	.5 (w/w)	linalool	0.4	[52]
		β-Phellandrene	4.2	
		decanal	0.6	
		D-Limonene	82.9	
		o-cymene	1.2	
		β-Pinene	0.8	
		β-myrcene	2.5	
Pomelo - *C. grandis*	1.06	α-phellandrene	0.79	[103, 104]
		α-pinene	0.68	
		α-limonene	96.49	
		β-cis-ocimene	0.138	
		β-pinene	0.24	
		β-myrcene	1.64	

(continued)

Table 8.2 (continued)

Citrus species	Oil yield (%)	Components	Percent (%)	References
Mandarin - *C. reticulata*	.9 (w/w)	octanal	0.2	[52, 105]
		sabinene	0.3	
		α-pinene	0.5	
		myrcene	2.1	
		γ-terpinene	3.9	
		Limonene	90.7	

microwave-assisted hydrodistillation was done. The percentage of limonene in lemon and geraniol in berkane increased in the essential oil. Microwave-assisted hydro distillation of the peels of navel and berkane showcased a higher number of constituents when the essential oil was analyzed for its components, wherein there was an increase in the components when traditional hydrodistillation was performed to extract the oil from lemon and orange [95].

Conventional hydro-distillation, ultrasound-assisted hydrodistillation, salt-assisted hydrodistillation, solvent-free microwave extraction and enzymes-assisted hydro-distillation methods when compared for the oil extraction from the *Citrus sinensis* peel, limonene was seen to be dominant, but its concentration varied slightly with the extraction technique [108]. The enzyme-assisted hydrodistillation showed the highest percentage of limonene content at 88.3%, while the lowest was found in solvent-free microwave extraction at 85.4%. However, the proportion of oxygenated monoterpenes, especially the linalool, was 4.8%, and the natural aroma was high in the Solvent-free microwave extraction method compared to other methods where the linalool content ranged from 2% to 3%. This was because the thermal effects were low, and the water used was also comparatively less [72].

Upon comparison of microwave-accelerated distillation, cold pressing and traditional hydrodistillation, a higher number of compounds were detected in microwave-accelerated distillation with 57 compounds, while the least number of compounds with cold pressing with 44 compounds and 46 compounds with hydrodistillation. The proportion of the compounds also varied with the technique used. The amount of oxygenated compounds was higher in essential oils obtained from microwave-assisted hydrodistillation when compared with the other two. However, the limonene content was less in microwave-assisted hydrodistillation [70].

The extraction technique changed the proportion and constituent compounds of oil obtained from bergamot. When extracted by steam distillation, it was found that alcohols (24.73%), olefins (39.75%) and aldehydes (26.38%) were found to be significant components. In comparison, acid (44.3%), ester (7.89%) and olefin (26.67%) were found to be the significant compounds when a low-temperature continuous phase change extraction technique was carried out [109–111].

The study on essential oil extraction by microwave-assisted hydrodistillation process from orange rinds shows that with the increase in the power, the constituent compounds of the oil get affected. The quality of the oil deteriorates with increasing power as inappropriate reactions take place. The chemical nature of the constituents changes, and the recovery efficiency also decreases with increased power. The presence of alcohol, aldehydes, sesquiterpene, ethylphenol, and phenol was found to be almost nil in the oil extracted with microwave-assisted hydrodistillation, and limonene was found to be predominant [107].

The number of oxygenated products and cyclic oxygenated monoterpenoids is less in essential oils extracted by vacuum distillation. The essential oils from this extraction method have distinctive flavors as the components that provide that characteristic flavor maintained during the extraction process [112]. Microwave-assisted hydrodistillation, compared to solvent-free microwave extraction, showed that the oil composition was similar when the GC-MS analysis was done, and only a little difference in their quantity was found. The volatile compounds deteriorate less when extracted with microwave-assisted hydrodistillation than conventional hydrodistillation [113].

8.7 Biological Activities of Citrus Peel Waste Essential Oils

Citrus is an essential part of the tropical and subtropical regions; its fruits produce over 100 million tons annually [114]. Citrus fruits are notable for their tangible properties and healthy benefits and thus play a vital role in the worldwide food systems [115]. In addition to the known healthy benefits, due to the presence of bioactive essential oils, they can be used in various sectors [116]. Indeed, the traditional medicine of Asia has long coupled the therapeutic benefits of citrus-derived components. Yet, modern science is only beginning to identify and discover bioactive components in citrus wastes. Consecutively, citrus byproducts (peels and seeds) contain many essential oils, alkaloids, coumarins, and extra bioactive compounds with potential human medical advantages [117]. Citrus contains vitamins C, A, and E-rich as well as a complete set of minerals and antioxidants essential for well-being.

8.7.1 Citrus Bioactive Compounds

Peels and seeds, along with the pulp, account for a large part of fruit processing waste, which is citrus fruits. They are the products that usually end up as waste materials. Mainly considered a waste disposal issue, it is now proven to be an underrated source of wealth. Many active essential oils are contained in that waste. These compressed compounds are very versatile and thus hold immense potential in various sectors. They are a sustainable way to handle waste while investing in exciting and promising areas. These insights revealed that these waste products were no longer considered useless as their deep reservoir of valuable components was highlighted, mainly pectin and antioxidants. However, there are two purposes for which contemporary chemical extraction techniques might be helpful: citrus waste's value creation by converting it into a sustainable source of biologically active substances [118, 119]. These important compounds will be the core of the lecture's focus, and we will highlight their usage in various fields such as insecticides, medicine, pharmaceuticals, cosmetics and many more. Starting from the position that they benefit health and wellbeing and make possible the manufacturing of industrial processes as a better alternative, essential oils provide sustainable and eco-friendly methodology.

8.7.1.1 Super Powerful Antioxidants—Flavonoids

Flavonoids are a heterogeneous class of natural polyphenolics present in plants' tissues. They are citruses constituting the primary sources of these flavonoids. Thus, they are the primary natural products that contribute significantly to the antioxidant profiles in the human diet. From a chemical perspective, they are composed of a special 15-atom carbon skeleton with two phenyl rings connected with a fundamental cyclic ring centered on oxygen. Flavonoids have subspecies caused by the different chemical modifications within the univalent cycle. It is a matter of fact that beta-cryptoxanthin, lutein, alpha-carotene, beta-carotene, lycopene, and xanthophylls make up the common subgroups. Citrus fruits are generally richer in many different types of flavonoids, such as anthocyanins (cyanidin, glucosides), flavones (nobiletin, tangeritin) and flavanone-7-O- glycosides (naringin, hesperidin) [120].

8.7.1.2 Powers of Carotenoids and Apocarotenoids as Color and Function Agents

Carotenoids are pigments present in all plants, for instance, in citrus. They are indispensable parts of these essential functions in plants and cells. Chemically, they can be categorized into two main groups: carotenoids (carbon-based) and xanthophyll (oxygen-rich). The α-carotene, β-carotene, lycopene, lutein and β-cryptoxanthin are among the most well-known xanthophylls. One crucial difference between them is that xanthophylls are always known to carry fatty acids (being esterified). At the same time, carotenes have a more straightforward structure, and therefore, they are fundamentally unable to be esterified. Case in citrus fruits, xanthophylls often occur as esters of different fatty acids, namely, lauric, capric, myristic and oleic acids [121].

8.7.2 Essential Oils

The peeling part of citrus contains essential oil (called flavedo in scientific terms), an economical product important for health and other benefits. All these oils are made from terpenes, limonoids, flavonoids, carotenoids, and coumarins, partaking in a vast bioactivity [122]. Their aromas, usually pleasant and natural, strengthen their demand as the main ingredients for food, beauty, perfume, and pharmaceutical products [123]. Not only do they serve their beautiful fragrances, but citrus essential oils have a broad spectrum of therapeutic advantages. They got an appreciation for their anti-oxidant, analgesic, anxiolytic, neuroprotective and antimicrobial effects [124]. The bacteria-inhibiting efficacy is crucial because it can damage the bacterial cell membrane, which makes a bacteria cell's components leak out and finally makes the cellular functions inactive. The acute efficacy of this organism against the microbes has made their commercialization quite a consequential move of recent times that may lead to their use during fruit, vegetable, meat and preserving processed foods [125].

8.7.2.1 Insecticidal Activity of Citrus Essential Oils

8.7.2.1.1 Evaluation of Citrus Oils as Fumigant Toxicants

The essential oil extracted from sweet orange and other citrus spp. peels and its identified components as a fumigant were evaluated to test the toxicity against the cowpea weevil, *Callosobruchus* spp. and red flour beetle, *Tribolium castaneum* which affects storage products like seeds and commodities, causing specific symptoms (for cowpea weevils; presence of eggs on seed surface, seed with holes, and seed infection and for red flour beetles; turning products greyish, releasing gaseous quinines pungent odour and tunnels inside the flour). Generally, an average of adult weevils in the test group will be kept in a sealed vial. As to the filter strips, which were soaked with the different concentrations of the essential oil, their components were loaded into the vials. Insect mortality was recorded every specific time after exposure [126, 127].

8.7.2.1.2 Assessment of Citrus Essential Oils as Contact Toxicants

The *C. sinensis* (sweet orange) essential oil and its compound toxicity through contact was evaluated against red flour beetles and pulse beetles through the method established, which was by initial dilution with acetone plus essential oil and its components applied in different concentrations to the insect dorsal thorax using a calibrated pipette. The testing doses are determined by studying various concentrations used beforehand. The treated and controlled insects were observed within a controlled environment for 24 hours. The pulse beetles, *Callosobruchus* spp. and red flour beetles were considered dead if they displayed zero movement when gently prodded. Observations of mortality were recorded after the 24-hour observation period [115, 128–130].

8.7.2.2 Antimicrobial Properties

The antifungal and antibacterial effect of citrus oils was assessed. Essential oils from peels of orange and mosambi showed different levels of inhibitory activity against various fungi and bacteria, etc.; Peel oil of lemon demonstrated the best efficiency, with the lowest Minimum Inhibitory Concentration (MIC) (against *A. flavus* 0.3%, Penicillium and Fusarium 0.2% respectively) values as an antifungal and most substantial zone of inhibition as antibacterial for most test substances [129].

Highlighting the bactericidal activity, lemon peel oil uniquely displayed the highest inhibition zones against all bacteria involved. Significantly, only Lemon peel oil demonstrated immense inhibition against *S. aureus* and *S. enterica*, the leading species in the study group. It outperformed other essences of oil herb. The more profound analysis identified MIC values for lemon peel oil, with MIC being zero for *S. aureus*, *Escherichia coli*, and *Erwinia*, while 0.7% for *P. aeruginosa* inhibited it [129].

8.7.3 Essential Oils in Aromatherapy

Stress alerts the system, and we start breathing differently, activating the brain's emotional center (the limbic system). It can cause emotional flutters, which can turn out to be dangerous. Studies show that some respiratory diseases can be connected not only to conditions like anxiety and depression [130]. But also to other psychological health problems. Aromatherapy using natural oils, which may possess antioxidant properties, may be considered by patients with the disease either alone or in combination with other treatments as part of their medical management. As respiration of essential oils activates several systems inside a body, the quality of an overall mental and physical being can be boosted. Tests have suggested that citrus essential oils may be efficient for insomnia, maybe sidestepping the drugs' side effects for treating these symptoms. Today, individuals select aromatherapy because it is perceived as safe and has no side effects. However, research on aromatherapy safety is still needed. Essential oil ingestion (pills, drops, food) proceeds through breathing or application directly [131].

8.7.4 Citrus Peel Waste Application in the Food Industry

8.7.4.1 Citrus as a Dietary Fiber Option in Bakery Foods

Citrus fiber added fiber content in bakery products like muffins. Baked goods with a higher fiber content, replacing some wheat flour with citrus fiber concentrate, can provide good health benefits that satisfy various dietary needs.

8.7.4.2 Pectin from Citrus as a Stabilizer

The food industry uses pectin from citrus as a thickener, a gelling agent and a stabilizer. Occasionally, this pectin is from the piths of citrus peels to add firmness to yogurts or milk drinks that may be acidic. These drinks are loved for their taste and nutritive contents, but without the stabilizers, they may have the proteins coagulating, or ultimately, the beverages are layered. Citrus pectin is a safety guard that solves the abovementioned problems and allows us to make a product of outstanding quality [132].

8.7.4.3 Citrus Waste as Beverages

Citrus processing produces a lot of waste, although the sidestream has many benefits for the beverage sector. Its bio-functions create opportunities to improve food attributes and even benefit consumers healthwise. Citrus sub-products can be turned into hydrating and let curriculum, which is rich in vitamin C [133]. Essential oils extracted from citrus waste add flavor and aroma to drinks and other products, including food and cosmetics. For this purpose, phenols and carotenoids in these byproducts are suitable for biofertilizers, increasing food storage life [134].

8.7.4.4 Essential Oils from Citrus as a Natural Approach to Food Safety

Essential oils have been in the food industry for a long time. Through the years, essential oils have been one of the most used and well-known food preservatives and one of the best options in the food industry. Researchers are scanning their assorted properties for use as alternative natural preservatives to synthetic preservatives. Recent studies pointed out that essential oils are used as antimicrobials in packaging, films, edibles and nanoplates to extend the storage life of fruits, veggies and meats. Essential oils, high-quality phytochemicals, have an edge over lower-grade chemical cut oils. Research has proven that the benefits of essential oils such as tea tree, cinnamon, and lemon oils offer antioxidants and antibacterial agents [135] that help in the safety of food and preserve the shelf life of many foods. Essential oils preserve food safety as the constituents present among terpenes and aromatic compounds, such as health protection, flavor, and aroma. Together, essential oils' factored antibacterial and antioxidant activities suggest that they are up to being used as natural alternatives to some chemical preservatives so that product freshness found in breakfast cereals and salads, for instance, remains uncompromised [136].

8.7.5 Citrus Peel Waste Application in the Pharmaceutical World

Peels from citrus contain rich bioactive components, which makes them valuable ingredients for medicine formulation in the pharmaceutical industry.

8.7.5.1 Flavonoids with Diverse Benefits

Besides natural oils, citrus peels offer flavonoids like polymethoxylated flavones such as nobiletin and tangerine. The bioactive compounds offer significant medical therapeutic effects against cardiovascular diseases, cancer and inflammation. Research has been on the health-improving features of citrus peels that might have anti-bacterial, anti-defying as well and liver health-promoting effects [137].

8.7.5.2 Antimicrobial Potential

Researchers suggest the capability of essential oil extracted from the citrus peel to combat microbes. Such a study showed oil's potency against some bacteria, which can be used as a preservative in the pharmaceutical industry [138–141].

8.7.5.3 Gastrointestinal Applications

Citrus sinensis peel extracts were used in reducing gastric ulcers in rats, where researchers observed that gastric pH increased significantly, accompanied by a decrease in gastric volume, number of lesions, and gastric acid secretion [141, 142].

8.7.5.4 Citrus Uses in Weight Control

Unarguably researched how citrus peels can prevent obesity. The study in mice indicates that citrus peel extract could help control weight and fight against fat build-up by supporting intestinal health and suppressing fat accumulation [142].

8.8 The Ecological Expense and Challenge of Citrus Production and Waste Management

The cultivation and processing of citrus fruits profoundly impact the environment. One of the main concerns is energy, as revealed by LCA (Life Cycle Assessment) studies that show considerable variation depending on the citrus type and cultivation area [143]. Citrus processing plants are notorious for their high energy consumption, contributing to greenhouse gas emissions. Another environmental issue is the wastewater from the processing. The high levels of organic matter increase the biological oxygen demand (BOD) and chemical oxygen demand (COD) of the process wastewater. However, the latter is not entirely harmful as it contains several valuable compounds, such as byproduct recovery fibers, phenolics, sugars and organic acids.

The environmentally friendly control of citrus production activity is manifested through the white mountains of wastes generated from citrus peels, seeds and leftover leaves. They are strongly organic materials that release carbon richly quickly, making waste disposal and management difficult. In contrast, classical technologies such as burning or landfilling do not save the environment as well as they don't use energy efficiently. But, despite this, the mirror of hope is still there. Fermentation and physical treatments can convert Citrus peels into fuels or chemicals. Hence, an alternative for sustainable production can be provided by the use of fermentation and physical conversion techniques. In addition to this, the European Union (EU), South America and Asia policies restricted the landfills of organic waste such as citrus. For this, the recovery of energy and resources had to be done first [144].

8.9 Conclusion

In conclusion, the extraction and valorization of bioactive essential oils from citrus fruit waste present significant opportunities for sustainable resource utilization and economic value creation. Despite containing a variety of bio-valuable phytochemicals, citrus processing produces a substantial amount of unwanted materials. Citrus peels and seeds are responsible for most of the waste generated in citrus processing. The present chapter deals with an excellent opportunity present within citrus waste, namely, bioactive essential oils. As high concentrations of bioactive compounds are found, citrus essential oils provide a wealth of opportunities, both in their own right and in the transformation of current waste disposal protocols within the citrus sector.

In the future, the integration of biorefinery concepts and the utilization of waste-to-energy technologies could further enhance the sustainability and economic viability of citrus waste valorization processes. By adopting a holistic approach that considers environmental, economic, and social aspects, the extraction and valorization of bioactive essential oils from citrus fruit waste can contribute to the transition towards a more sustainable and circular bioeconomy. In summary, the exploration of citrus waste as a valuable source of bioactive compounds offers promising prospects for sustainable resource management, waste reduction, and value creation across various industries. With continued research and innovation, the extraction and valorization of bioactive essential oils from citrus fruit waste have the potential to drive positive environmental and socio-economic impacts while meeting the growing demand for natural and sustainable products in the global market.

References

1. Baldwin, E. (1993). *Citrus fruit* (pp. 107–149). Springer.
2. Gmitter, F. G., Chen, C., Nageswara Rao, M., & Soneji, J. R. (2007). *Citrus fruits. Fruits and nuts* (pp. 265–279).
3. Rashid, U., Ibrahim, M., Yasin, S., Yunus, R., Taufiq-Yap, Y., & Knothe, G. (2013). Biodiesel from Citrus reticulata (mandarin orange) seed oil, a potential non-food feedstock. *Industrial Crops and Products, 45*, 355–359.
4. Kundu, D., Das, M., Mahle, R., Biswas, P., Karmakar, S., & Banerjee, R. (2020). *Citrus fruits. Valorization of fruit processing by-products* (pp. 145–166). Elsevier.
5. Gonzatto, M. P., & Santos, J. S. (2023). *Introductory chapter: World citrus production and research. Citrus research-horticultural and human health aspects*. IntechOpen.
6. Varmie, E. B., & Thakur, M. (2021). Utilization of citrus processing waste: A review. *Pharma Innovation Journal, 10*, 682–697.
7. Suri, S., Singh, A., & Nema, P. K. (2022). Current applications of citrus fruit processing waste: A scientific outlook. *Applied Food Research, 2*(1), 100050.
8. Rafiq, S., Kaul, R., Sofi, S., Bashir, N., Nazir, F., & Nayik, G. A. (2018). Citrus peel as a source of functional ingredient: A review. *Journal of the Saudi Society of Agricultural Sciences, 17*(4), 351–358.
9. Singh, B., Singh, J. P., Kaur, A., & Singh, N. (2020). Phenolic composition, antioxidant potential and health benefits of citrus peel. *Food Research International, 132*, 109114.
10. Liew, S. S., Ho, W. Y., Yeap, S. K., & Sharifudin, S. A. B. (2018). Phytochemical composition and in vitro antioxidant activities of Citrus sinensis peel extracts. *PeerJ, 6*, e5331.
11. Wang, C., Qiu, W.-Y., Chen, T.-T., & Yan, J.-K. (2021). Effects of structural and conformational characteristics of citrus pectin on its functional properties. *Food Chemistry, 339*, 128064.
12. Huang, J.-y., Liao, J.-s., Qi, J.-r., Jiang, W.-x., & Yang, X.-q. (2021). Structural and physicochemical properties of pectin-rich dietary fiber prepared from citrus peel. *Food Hydrocolloids, 110*, 106140.
13. Agócs, A., Nagy, V., Szabó, Z., Márk, L., Ohmacht, R., & Deli, J. (2007). Comparative study on the carotenoid composition of the peel and the pulp of different citrus species. *Innovative Food Science & Emerging Technologies, 8*(3), 390–394.
14. Tripoli, E., La Guardia, M., Giammanco, S., Di Majo, D., & Giammanco, M. (2007). Citrus flavonoids: Molecular structure, biological activity and nutritional properties: A review. *Food Chemistry, 104*(2), 466–479.
15. Juhaimi, F. A., Matthäus, B., Özcan, M. M., & Ghafoor, K. (2016). The physico-chemical properties of some citrus seeds and seed oils. *Zeitschrift für Naturforschung C., 71*(3–4), 79–85.
16. Anwar, F., Naseer, R., Bhanger, M., Ashraf, S., Talpur, F. N., & Aladedunye, F. A. (2008). Physico-chemical characteristics of citrus seeds and seed oils from Pakistan. *Journal of the American Oil Chemists' Society, 85*(4), 321–330.
17. Malacrida, C. R., Kimura, M., & Jorge, N. (2012). Phytochemicals and antioxidant activity of citrus seed oils. *Food Science and Technology Research, 18*(3), 399–404.
18. Widmer, W., Zhou, W., & Grohmann, K. (2010). Pretreatment effects on orange processing waste for making ethanol by simultaneous saccharification and fermentation. *Bioresource Technology, 101*(14), 5242–5249.
19. Lin, C. S. K., Pfaltzgraff, L. A., Herrero-Davila, L., Mubofu, E. B., Abderrahim, S., Clark, J. H., et al. (2013). Food waste as a valuable resource for the production of chemicals, materials and fuels. Current situation and global perspective. *Energy & Environmental Science, 6*(2), 426–464.
20. Wei, Y., Li, J., Shi, D., Liu, G., Zhao, Y., & Shimaoka, T. (2017). Environmental challenges impeding the composting of biodegradable municipal solid waste: A critical review. *Resources, Conservation and Recycling, 122*, 51–65.

21. Russo, C., Maugeri, A., Lombardo, G. E., Musumeci, L., Barreca, D., Rapisarda, A., et al. (2021). The second life of Citrus fruit waste: A valuable source of bioactive compounds. *Molecules, 26*(19), 5991.
22. Richa, R., Kohli, D., Vishwakarma, D., Mishra, A., Kabdal, B., Kothakota, A., et al. (2023). Citrus fruit: Classification, value addition, nutritional and medicinal values, and relation with pandemic and hidden hunger. *Journal of Agriculture and Food Research, 14*, 100718.
23. Satari, B., & Karimi, K. (2018). Citrus processing wastes: Environmental impacts, recent advances, and future perspectives in total valorization. *Resources, Conservation and Recycling, 129*, 153–167.
24. Mahawar, M. K., Jalgaonkar, K., Bibwe, B., Bhushan, B., Meena, V. S., & Sonkar, R. K. (2020). Post-harvest processing and valorization of Kinnow mandarin (Citrus reticulate L.): A review. *Journal of Food Science and Technology, 57*, 799–815.
25. Zema, D., Calabrò, P., Folino, A., Tamburino, V., Zappia, G., & Zimbone, S. (2018). Valorisation of citrus processing waste: A review. *Waste Management, 80*, 252–273.
26. Wedamulla, N. E., Fan, M., Choi, Y.-J., & Kim, E.-K. (2022). Citrus peel as a renewable bioresource: Transforming waste to food additives. *Journal of Functional Foods, 95*, 105163.
27. Chavan, P., Singh, A. K., & Kaur, G. (2018). Recent progress in the utilization of industrial waste and by-products of citrus fruits: A review. *Journal of Food Process Engineering, 41*(8), e12895.
28. Bisht, B., Gururani, P., Aman, J., Vlaskin, M. S., Anna, K., Joshi, S., et al. (2023). A review on holistic approaches for fruits and vegetables biowastes valorization. *Materials Today: Proceedings, 73*, 54–63.
29. Sharmeen, J. B., Mahomoodally, F. M., Zengin, G., & Maggi, F. (2021). Essential oils as natural sources of fragrance compounds for cosmetics and cosmeceuticals. *Molecules, 26*(3), 666.
30. Andrade, M. A., Barbosa, C. H., Shah, M. A., Ahmad, N., Vilarinho, F., Khwaldia, K., et al. (2022). Citrus by-products: Valuable source of bioactive compounds for food applications. *Antioxidants, 12*(1), 38.
31. Bora, H., Kamle, M., Mahato, D. K., Tiwari, P., & Kumar, P. (2020). Citrus essential oils (CEOs) and their applications in food: An overview. *Plants, 9*(3), 357.
32. Mahato, N., Sharma, K., Koteswararao, R., Sinha, M., Baral, E., & Cho, M. H. (2019). Citrus essential oils: Extraction, authentication and application in food preservation. *Critical Reviews in Food Science and Nutrition., 59*(4), 611–625.
33. Jing, L., Lei, Z., Li, L., Xie, R., Xi, W., Guan, Y., et al. (2014). Antifungal activity of citrus essential oils. *Journal of Agricultural and Food Chemistry, 62*(14), 3011–3033.
34. Kamaliroosta, L., Zolfaghari, M., Shafiee, S., Larijani, K., & Zojaji, M. (2016). Chemical identifications of citrus peels essential oils. *Journal of Food Science and Technology, 6*(2), 69–76.
35. Tranchida, P. Q., Bonaccorsi, I., Dugo, P., Mondello, L., & Dugo, G. (2012). Analysis of Citrus essential oils: State of the art and future perspectives. *Flavour and Fragrance Journal, 27*(2), 98–123.
36. Kelebek, H., & Selli, S. (2011). Determination of volatile, phenolic, organic acid and sugar components in a Turkish cv. Dortyol (*Citrus sinensis* L. Osbeck) orange juice. *Journal of the Science of Food and Agriculture, 91*(10), 1855–1862.
37. Jain, S., Arora, P., & Popli, H. (2020). A comprehensive review on Citrus aurantifolia essential oil: Its phytochemistry and pharmacological aspects. *Brazilian Journal of Natural Sciences, 3*(2), 354–362.
38. Rafiq, S., Kaul, R., Sofi, S. A., Bashir, N., Nazir, F., & Ahmad, N. G. (2018). Citrus peel as a source of functional ingredient: A review. *Journal of the Saudi Society of Agricultural Sciences, 17*(4), 351–358.
39. Ahmad, A., Dubey, P., Younis, K., & Yousuf, O. (2022). Mosambi (*Citrus limetta*) peel and Sago based biodegradable film: Development and characterization of physical, water barrier and biodegradation properties. *Bioresource Technology Reports, 18*, 101016.

40. Velázquez-Nuñez, M. J., Avila-Sosa, R., Palou, E., & López-Malo, A. (2013). Antifungal activity of orange (*Citrus sinensis* var. *Valencia*) peel essential oil applied by direct addition or vapor contact. *Food Control, 31*(1), 1–4.
41. Asnaashari, S., Delazar, A., Habibi, B., Vasfi, R., Nahar, L., Hamedeyazdan, S., et al. (2010). Essential oil from *Citrus aurantifolia* prevents ketotifen-induced weight-gain in mice. *Phytotherapy Research, 24*(12), 1893–1897.
42. Bajpai, V. K., Baek, K. H., & Kang, S. C. (2012). Control of Salmonella in foods by using essential oils: A review. *International Food Research Journal, 45*(2), 722–734.
43. Fisher, K., & Phillips, C. (2008). Potential antimicrobial uses of essential oils in food: Is citrus the answer? *Trends in Food Science and Technology, 19*(3), 156–164.
44. Lin, C. M., Sheu, S. R., Hsu, S. C., & Tsai, Y. H. (2010). Determination of bactericidal efficacy of essential oil extracted from orange peel on the food contact surfaces. *Food Control, 21*(12), 1710–1715.
45. Matsuura, R., Ukeda, H., & Sawamura, M. (2006). Tyrosinase inhibitory activity of citrus essential oils. *Journal of Agricultural and Food Chemistry, 54*(6), 2309–2313.
46. Palazzolo, E., Laudicina, V. A., & Germanà, M. A. (2013). Current and potential use of citrus essential oils. *Current Organic Chemistry, 17*(24), 3042–3049.
47. Gökkaya, İ., Koçer, G. G., & Renda, G. (2024). What does a community think about aromatherapy? *Holistic Nursing Practice, 38*(2), 73–84.
48. Lahlou, M. (2004). Methods to study the phytochemistry and bioactivity of essential oils. *Phytotherapy Research, 18*(6), 435–448.
49. Dugo, P., Mondello, L., Dugo, L., Stancanelli, R., & Dugo, G. (2000). LC-MS for the identification of oxygen heterocyclic compounds in citrus essential oils. *Journal of Pharmaceutical and Biomedical Analysis, 24*(1), 147–154.
50. Wilkinson, S., Love, S., Westcombe, A., Gambles, M., Burgess, C., Cargill, A., et al. (2007). Effectiveness of aromatherapy massage in the management of anxiety and depression in patients with cancer: A multicenter randomized controlled trial. *Clinical Oncology, 25*(5), 532–539.
51. Palma, C. E., Cruz, P. S., Cruz, D. T. C., Bugayong, A. M. S., & Castillo, A. L. (2019). Chemical composition and cytotoxicity of Philippine calamansi essential oil. *Industrial Crops and Products, 128*, 108–114.
52. Singh, B., Singh, J. P., Kaur, A., & Yadav, M. P. (2021). Insights into the chemical composition and bioactivities of citrus peel essential oils. *International Food Research Journal, 143*, 110231.
53. Moraes, T. M., Kushima, H., Moleiro, F. C., Santos, R. C., Machado Rocha, L. R., Marques, M. O., et al. (2009). Effects of limonene and essential oil from *Citrus aurantium* on gastric mucosa: Role of prostaglandins and gastric mucus secretion. *Chemico-Biological Interactions, 180*(3), 499–505.
54. Zeb, A. (2021). *Phenolic antioxidants in foods: Chemistry, biochemistry and analysis. Phenolic antioxidants in foods: Chemistry, biochemistry and analysis* (pp. 385–411). Springer.
55. Yu, X., Lin, H., Wang, Y., Lv, W., Zhang, S., Qian, Y., et al. (2018). D-limonene exhibits antitumor activity by inducing autophagy and apoptosis in lung cancer. *Oncotargets and Therapy, 11*, 1833–1847.
56. Lv, X., Zhao, S., Ning, Z., Zeng, H., Shu, Y., Tao, O., et al. (2015). Citrus fruits as a treasure trove of active natural metabolites that potentially provide benefits for human health. *Chemistry Central Journal, 9*, 1–14.
57. Jia, S. S., Xi, G. P., Zhang, M., Chen, Y. B., Lei, B., Dong, X. S., et al. (2013). Induction of apoptosis by D-limonene is mediated by inactivation of Akt in LS174T human colon cancer cells. *Oncology Reports, 29*(1), 349–354.
58. Shi, Y., Huang, S., He, Y., Wu, J., & Yang, Y. (2018). Navel orange peel essential oil to control food spoilage molds in potato slices. *Journal of Food Protection, 81*(9), 1496–1502.

59. Chen, T. C., da Fonseca, C. O., Levin, D., & Schönthal, A. H. (2021). The Monoterpenoid Perillyl alcohol: Anticancer agent and medium to overcome biological barriers. *Pharmaceutics, 13*(12), 1580–1593.
60. Bustamante, J., van Stempvoort, S., García-Gallarreta, M., Houghton, J. A., Briers, H. K., Budarin, V. L., et al. (2016). Microwave assisted hydro-distillation of essential oils from wet citrus peel waste. *Journal of Cleaner Production, 137*, 598–605.
61. Bungau, S., Agarwal, P., Sebghatollahi, Z., Kamal, M., Dhyani, A., Shrivastava, A., et al. (2022). Citrus essential oils in aromatherapy: Therapeutic effects and mechanisms. *Antioxidants, 11*(12), 2374.
62. Ho, T. M., Yoshii, H., Terao, K., & Bhandari, B. R. (2021). *Functionality of cyclodextrins in encapsulation for food applications. Functionality of Cyclodextrins in encapsulation for food applications* (pp. 255–276). Springer.
63. Abd-Elwahab, S. M., El-Tanbouly, N. D., Moussa, M. Y., Abdel-Monem, A. R., & Fayek, N. M. (2016). Antimicrobial and antiradical potential of four agro-waste citrus peels cultivars. *Journal of Essential Oil Bearing Plants, 19*(8), 1932–1942.
64. Clavaud, C., Jourdain, R., Bar-Hen, A., Tichit, M., Bouchier, C., Pouradier, F., et al. (2013). Dandruff is associated with disequilibrium in the proportion of the major bacterial and fungal populations colonizing the scalp. *PLoS One, 8*(3), e58203.
65. González-Mas, M. C., Rambla, J. L., López-Gresa, M. P., Amparo Blázquez, M., & Granell, A. (2019). Volatile compounds in citrus essential oils: A comprehensive review. *Frontiers in Plant Science, 10*, 433929.
66. El-hawary, S., Taha, K., Abdel-Monem, A., Kirillos, F., & Mohamed, A. (2013). Chemical composition and biological activities of peels and leaves essential oils of four cultivars of *Citrus deliciosa* var. *tangarina*. *American Journal of Essential Oils, 1*(2), 1–6.
67. Fanciullino, A.-L., Tomi, F., Luro, F., Desjobert, J. M., & Casanova, J. (2006). Chemical variability of peel and leaf oils of mandarins. *Flavour and Fragrance Journal, 21*(2), 359–367.
68. Golmakani, M., & Moayyedi, M. (2015). Comparison of heat and mass transfer of different microwave-assisted extraction methods of essential oil from *Citrus limon* (Lisbon variety) peel. *Food Science & Nutrition, 3*(6), 506–518.
69. Gök, A., İsmail Kirbaşlar, Ş., & Gülay, K. F. (2014). Comparison of lemon oil composition after using different extraction methods. *Journal of Essential Oil Research, 27*(1), 17–22.
70. Ferhat, M. A., Meklati, B. Y., & Chemat, F. (2007). Comparison of different isolation methods of essential oil from Citrus fruits: Cold pressing, hydrodistillation and microwave "dry" distillation. *Flavour and Fragrance Journal, 22*(6), 494–504.
71. Mohagheghniapour, A., Saharkhiz, M., Golmakani, M., & Niakousari, M. (2018). Variations in chemical compositions of essential oil from sour orange (*Citrus aurantium* L.) blossoms by different isolation methods. *Sustainable Chemistry and Pharmacy, 10*, 118–124.
72. Taktak, O., Ben Youssef, S., Abert Vian, M., Chemat, F., & Allouche, N. (2021). Physical and chemical influences of different extraction techniques for essential oil recovery from Citrus sinensis peels. *Journal of Essential Oil Bearing Plants, 24*(2), 290–303.
73. Hilali, S., Fabiano-Tixier, A. S., Ruiz, K., Hejjaj, A., Ait Nouh, F., Idlimam, A., et al. (2019). Green extraction of essential oils, polyphenols, and Pectins from Orange Peel employing solar energy: Toward a zero-waste biorefinery. *ACS Sustainable Chemistry & Engineering, 7*(13), 11815–11822.
74. Ben Hsouna, A., Ben Halima, N., Smaoui, S., & Hamdi, N. (2017). Citrus lemon essential oil: Chemical composition, antioxidant and antimicrobial activities with its preservative effect against Listeria monocytogenes inoculated in minced beef meat. *Lipids in Health and Disease, 16*(1), 1–11.
75. Chen, Y., Wu, J., Xu, Y., Fu, M., & Xiao, G. (2014). Effect of second cooling on the chemical components of essential oils from orange peel (*Citrus sinensis*). *Journal of Agricultural and Food Chemistry, 62*(35), 8786–8790.

76. Njila, N., Mahdi, E., Lembe, M., Nde, Z., & Nyonseu, D. (2017). Review on extraction and isolation of plant secondary metabolites. In *7th Int'l conference on agricultural, chemical, biological and environmental sciences* (pp. 67–72).
77. Ghaderi, A., & Bahram, E. (2015). Soxhlet extraction and gas chromatography mass spectrometry analysis of extracted oil from *Pistacia atlantica* kurdica nuts and optimization of process using factorial design of experiments. *Journal of Analytical Chemistry, 3*(6), 122–126.
78. Allaf, K., Besombes, C., Berka, B., Kristiawan, M., Sobolik, V., & Allaf, T. (2011). Instant controlled pressure drop technology in plant extraction processes. In *Enhancing extraction processes in the food industry* (pp. 255–303).
79. Yousefi, M., Rahimi-Nasrabadi, M., Pourmortazavi, S. M., Wysokowski, M., Jesionowski, T., Ehrlich, H., et al. (2019). Supercritical fluid extraction of essential oils. *Trends in Analytical Chemistry, 118*, 182–193.
80. Aresta, A., Cotugno, P., De Vietro, N., Massari, F., & Zambonin, C. (2020). Determination of polyphenols and vitamins in wine-making by-products by supercritical fluid extraction (SFE). *Analytical Letters, 53*(16), 2585–2595.
81. Bousbia, N., Vian, M. A., Ferhat, M. A., Meklati, B. Y., & Chemat, F. (2009). A new process for extraction of essential oil from Citrus peels: Microwave hydrodiffusion and gravity. *Journal of Food Engineering, 90*(3), 409–413.
82. Varo, M. A., Jacotet-Navarro, M., Serratosa, M. P., Mérida, J., Fabiano-Tixier, A. S., Bily, A., et al. (2019). Green ultrasound-assisted extraction of antioxidant phenolic compounds determined by high performance liquid chromatography from bilberry (*Vaccinium Myrtillus* L.) juice by-products. *Waste & Biomass Valorization, 10*(7), 1945–1955.
83. Wang, Y., Zhang, X., Ma, X., Zhang, K., Li, S., Wang, X., et al. (2019). Study on the kinetic model, thermodynamic and physicochemical properties of Glycyrrhiza polysaccharide by ultrasonic assisted extraction. *Ultrasonics Sonochemistry, 51*, 249–257.
84. Khandare, R. D., Tomke, P. D., & Rathod, V. K. (2021). Kinetic modeling and process intensification of ultrasound-assisted extraction of d-limonene using citrus industry waste. *Chemical Engineering and Processing, 159*, 108181.
85. Bourgou, S., Zohra Rahali, F., Ourghemmi, I., Sa, M., & Tounsi ıdani. (2012). Changes of Peel essential oil composition of four Tunisian citrus during fruit maturation. *Scientific World Journal, 2012*, e528593.
86. Zhang, Q. W., Lin, L. G., & Ye, W. C. (2018). Techniques for extraction and isolation of natural products: A comprehensive review. *Chinese Medicine, 13*(1), 1–26.
87. Xing, C., Qin, C., Li, X., Zhang, F., Linhardt, R., Sun, P., et al. (2019). Chemical composition and biological activities of essential oil isolated by HS-SPME and UAHD from fruits of bergamot. *LWT - Food Science and Technology, 104*, 38–44.
88. Kazemi, M., Niazi, A., & Yazdanipour, A. (2022). Extraction of Satureja Rechingeri volatile components through ultrasound-assisted and microwave-assisted extractions and comparison of the chemical composition with headspace solid-phase microextraction. *Journal of Essential Oil Research, 34*(1), 21–35.
89. Grover, S., Aggarwal, P., Kaur, S., Yadav, R., & Kumar, A. (2023). Physicochemical assessment, characterization, antioxidant and antimicrobial potential study of essential oil extracted from the Peel of different Galgal (*Citrus Pseudolimon*) cultivars. *Waste & Biomass Valorization, 15*, 1–11.
90. Tambunan, A. P., Bahtiar, A., & Raymond, R. T. (2017). Influence of extraction parameters on the yield, phytochemical, TLC-densitometric quantification of quercetin, and LC-MS profile, and how to standardize different batches for long term from *Ageratum conyoides* L. leaves. *Pharmacognosy Journal, 9*(6), 39–42.
91. Hasibuan, R., & Gultom, E. (2021). The effect of method, type of solvent and extraction time towards the yield of oil on essential oil extraction from lime peel (*Citrus aurantifolia*). *IOP Conference Series: Materials Science and Engineering, 1122*(1), 012108.

92. Schipilliti, L., Bonaccorsi, I. L., Occhiuto, C., Dugo, P., & Mondello, L. (2018). Authentication of citrus volatiles based on carbon isotope ratios. *Journal of Essential Oil Research, 30*(1), 1–15.
93. Deng, W., Liu, K., Cao, S., Sun, J., Zhong, B., Molecules, J. C., et al. (2020). Chemical composition, antimicrobial, antioxidant, and antiproliferative properties of grapefruit essential oil prepared by molecular distillation. *Molecules, 25(5)*, 127.
94. Maurya, A. K., Mohanty, S., Pal, A., Chanotiya, C. S., & Bawankule, D. U. (2018). The essential oil from *Citrus limetta* Risso peels alleviates skin inflammation: In-vitro and in-vivo study. *Journal of Ethnopharmacology, 212*, 86–94.
95. Brahmi, F., Mokhtari, O., Legssyer, B., Hamdani, I., Asehraou, A., Hasnaoui, I., et al. (2021). Chemical and biological characterization of essential oils extracted from citrus fruits peels. *Materials Today Proceedings, 45*(8), 7794–7799.
96. Yang, C., Chen, H., Chen, H., Zhong, B., Luo, X., & Chun, J. (2017). Antioxidant and anticancer activities of essential oil from Gannan navel orange peel. *Molecules, 8*, 1391.
97. Lemes, R. S., Alves, C. C. F., Estevam, E. B. B., Santiago, M. B., Martins, C. H. G., Dos Santos, T. C. L., et al. (2018). Chemical composition and antibacterial activity of essential oils from Citrus aurantifolia leaves and fruit peel against oral pathogenic bacteria. *Anais da Academia Brasileira de Ciências, 90*(2), 1285–1292.
98. Buathong, R., & Duangsrisai, S. (2023). Plant ingredients in Thai food: A well-rounded diet for natural bioactive associated with medicinal properties. *Peer Journal, 11*, e14568.
99. Hosni, K., Zahed, N., Chrif, R., Abid, I., Medfei, W., Kallel, M., et al. (2010). Composition of peel essential oils from four selected Tunisian Citrus species: Evidence for the genotypic influence. *Food Chemistry, 123*(4), 1098–1104.
100. Paw, M., Begum, T., Gogoi, R., Pandey, S. K., & Lal, M. (2020). Chemical composition of Citrus limon L. Burmf Peel essential oil from north East India. *Journal of Essential Oil Bearing Plants, 23*(2), 337–344.
101. Jiang, H., Zhang, W., Xu, Y., Chen, L., Cao, J., & Jiang, W. (2022). An advance on nutritional profile, phytochemical profile, nutraceutical properties, and potential industrial applications of lemon peels: A comprehensive review. *Trends in Food Science and Technology, 124*, 219–236.
102. Vahidi, R., Pourahmad, R., & Mahmoudi, R. (2019). Chemical compounds and antibacterial and antioxidant properties of citron (*Citrus medica* L.) peel essential oil. *Journal of Food and Bioprocess Engineering, 2*(1), 71–76.
103. Ngan, T., Muoi, N., Quan, P., & Cang, M. (2020). Evaluation of physical and chemical properties of pomelo (*Citrus grandis* L.) essential oil using steam distillation process. *Asian Journal of Chemistry, 6*, 1433–1436.
104. Li, L. J., Hong, P., Chen, F., Sun, H., Yang, Y. F., Yu, X., et al. (2016). Characterization of the aldehydes and their transformations induced by UV irradiation and air exposure of white Guanxi honey Pummelo (*Citrus Grandis* (L.) Osbeck) essential oil. *Journal of Agricultural and Food Chemistry, 64*(24), 5000–5010.
105. Kirbaşlar, F., Tavman, A., Dülger, B., & Türker, G. (2009). Antimicrobial activity of Turkish citrus peel oils. *Pakistan Journal of Botany, 41*(6), 3207–3212.
106. Tongnuanchan, P., & Benjakul, S. (2014). Essential oils: Extraction, bioactivities, and their uses for food preservation. *Journal of Food Science, 79*(7), 1231–1249.
107. Toan, T., Truc, T., Le, X., Quyen, N., & Tran, T. (2020). Study on extraction process and analysis of components in essential oils of Vietnamese orange peel (*Citrus sinensis*) by microwave assisted hydrodistillation extraction. *IOP Conference Series: Materials Science and Engineering, 991*(1), 012125.
108. Golmakani, M. T., Farahmand, M., Ghassemi, A., Eskandari, M. H., & Niakousari, M. (2017). Enrichment of citral isomers in different microwave-assisted extraction of essential oil from fresh and dried lemon verbena (*Aloysia citridora*) leaves. *Journal of Food Processing & Preservation, 41*(6), e13215.

109. Lin MinHao, L., Zhou AiMei, Z., Yang Hui, Y., Xie Wei, X., Li XueNan, L., Liu ZhongHao, L., et al. (2015). Research on the extraction of bergamot essential oil by steam distillation and component analysis. *Journal of Food Safety, 2*, 619–625.
110. Yang, H., Zhou, A., Xia, X., Gou, B., Liu, X., & Chen, S. (2015). Study on the extract technology and component analysis of bergamot essential oil by low-temperature continuous phase transition. *Science and Technology of Food Industry, 35*, 289–293.
111. Xing, C., Qin, C., Li, X., Zhang, F., Linhardt, R. J., Sun, P., et al. (2019). Chemical composition and biological activities of essential oil isolated by HS-SPME and UAHD from fruits of bergamot. *Food Science and Technology, 104*, 38–44.
112. Deng, G., Craft, J., Steinberg, K., Li, P., Pokharel, S. K., & Setzer, W. N. (2017). Influence of different isolation methods on chemical composition and bioactivities of the fruit Peel oil of *Citrus medica* L. var. *sarcodactylis* (Noot.) Swingle. *Medicine, 4(1)*, 1.
113. Golmakani, M., & nutrition MM-F science. (2015). Comparison of heat and mass transfer of different microwave-assisted extraction methods of essential oil from *Citrus limon* (Lisbon variety) peel. *Food Science & Nutrition, 3*(6), 506–518.
114. Zou, Z., Xi, W., Hu, Y., Nie, C., & Zhou, Z. (2016). Antioxidant activity of Citrus fruits. *Food Chemistry, 196*, 885–896. https://doi.org/10.1016/j.foodchem.2015.09.072
115. Liu, Y., Heying, E., & Tanumihardjo, S. A. (2012). History, global distribution, and nutritional importance of citrus fruits. *Comprehensive Reviews in Food Science and Food Safety, 11*(6), 530–545. https://doi.org/10.1111/j.1541-4337.2012.00201.x
116. Rawat, N., Kiran, S. P., Du, D., Gmitter, F. G., & Deng, Z. (2015). Comprehensive meta-analysis, co-expression, and miRNA nested network analysis identifies gene candidates in citrus against Huanglongbing disease. *BMC Plant Biology, 15*(1), 184. https://doi.org/10.1186/s12870-015-0568-4
117. Anwar, F., Naseer, R., Bhanger, M. I., Ashraf, S., Talpur, F. N., & Aladedunye, F. A. (2008). Physico-chemical characteristics of citrus seeds and seed oils from Pakistan. *Journal of the American Oil Chemists' Society, 85*(4), 321–330. https://doi.org/10.1007/s11746-008-1204-3
118. Maqbool, Z., et al. (2023). Citrus waste as source of bioactive compounds: Extraction and utilization in health and food industry. *Molecules, 28*(4), MDPI. https://doi.org/10.3390/molecules28041636
119. Putnik, P., et al. (2017). Innovative 'Green' and novel strategies for the extraction of bioactive added value compounds from citrus wastes—A review. *Molecules, 22*(5), 680. https://doi.org/10.3390/molecules22050680
120. Chen, X.-M., Tait, A. R., & Kitts, D. D. (2017). Flavonoid composition of orange peel and its association with antioxidant and anti-inflammatory activities. *Food Chemistry, 218*, 15–21. https://doi.org/10.1016/j.foodchem.2016.09.016
121. Walia, A., Gupta, A. K., & Sharma, V. (2019). Role of bioactive compounds in human health. *Acta Scientific Medical, 3*, 25–33.
122. Teixeira, B., et al. (2013). Chemical composition and antibacterial and antioxidant properties of commercial essential oils. *Industrial Crops and Products, 43*, 587–595. https://doi.org/10.1016/j.indcrop.2012.07.069
123. Lemes, R. S., et al. (2018). Chemical composition and antibacterial activity of essential oils from Citrus aurantifolia leaves and fruit peel against oral pathogenic bacteria. *Anais da Academia Brasileira de Ciências, 90*(2), 1285–1292. https://doi.org/10.1590/0001-3765201820170847
124. Bayala, B., Bassole, I. H. N., Scifo, R., Gnoula, C., Morel, L., Lobaccaro, J.-M. A., & Simpore, J. (2014). Anticancer activity of essential oils and their chemical components—A review. *American Journal of Cancer Research, 4*, 591.
125. Javed, S., et al. (2014). Phytochemistry, GC-MS analysis, antioxidant and antimicrobial potential of essential oil from five citrus species. *Journal of Agricultural Science, 6*(3). https://doi.org/10.5539/jas.v6n3p201
126. Oyedeji, A. O., Okunowo, W. O., Osuntoki, A. A., Olabode, T. B., & Ayo-folorunso, F. (2020). Insecticidal and biochemical activity of essential oil from Citrus sinensis peel and constituents on Callosobrunchus maculatus and Sitophilus zeamais. *Pesticide Biochemistry and Physiology, 168*, 104643. https://doi.org/10.1016/j.pestbp.2020.104643

127. Visakh, N. U., Pathrose, B., Chellappan, M., Ranjith, M. T., Sindhu, P. V., & Mathew, D. (2022a). Chemical characterisation, insecticidal and antioxidant activities of essential oils from four Citrus spp. fruit peel waste. Food. *Bioscience, 50*, 102163.
128. Visakh, N. U., Pathrose, B., Narayanankutty, A., Alfarhan, A., & Ramesh, V. (2022b). Utilization of Pomelo (Citrus maxima) Peel waste into bioactive essential oils: Chemical composition and insecticidal properties. *Insects, 13*(5), 480.
129. Aruna, T., Hemalatha, G., Kumutha, K., Kanchana, S., & Vellaikumar, S. (2022). Physicochemical, antioxidant and antimicrobial properties of citrus peel essential oils. *Journal of Applied and Natural Science, 14*(2), 640–646. https://doi.org/10.31018/jans.v14i2.3484
130. Farrar, A. J., & Farrar, F. C. (2020). Clinical aromatherapy. *Nursing Clinics of North America, 55*(4), 489–504. https://doi.org/10.1016/j.cnur.2020.06.015
131. Shiina, Y., et al. (2008). Relaxation effects of lavender aromatherapy improve coronary flow velocity reserve in healthy men evaluated by transthoracic Doppler echocardiography. *International Journal of Cardiology, 129*(2), 193–197. https://doi.org/10.1016/j.ijcard.2007.06.064
132. Nasseri, A. T., Thibault, J.-F., & Ralet, M.-C. (2008). Citrus pectin: Structure and application in acid dairy drinks. In *Tree and forestry science and biotechnology; global science books*.
133. Varmie, E. B., & Thakur, M. (2021). Utilization of citrus processing waste: A review. *Pharmaceutical Innovation Journal, 10*, 682–697.
134. Okino Delgado, C. H., & Fleuri, L. F. (2016). Orange and mango by-products: Agro-industrial waste as source of bioactive compounds and botanical versus commercial description—A review. *Food Reviews International, 32*(1), 1–14. https://doi.org/10.1080/87559129.2015.1041183
135. Khalid, et al. (2023). Chia seeds (*Salvia hispanica* L.): A therapeutic weapon in metabolic disorders. *Food Science & Nutrition, 11*(1), 3–16. https://doi.org/10.1002/fsn3.3035
136. Bhavaniramya, S., Vishnupriya, M. S., Al-Aboody, R. V., & Baskaran, D. (2019). Role of essential oils in food safety: Antimicrobial and antioxidant applications. *Grain & Oil Science and Technology, 2*(2), 49–55. https://doi.org/10.1016/j.gaost.2019.03.001
137. Tripoli, E., La Guardia, M., Giammanco, S., Di Majo, D., & Giammanco, M. (2007). Citrus flavonoids: Molecular structure, biological activity and nutritional properties: A review. *Food Chemistry, 104*(2), 466–479. https://doi.org/10.1016/j.foodchem.2006.11.054
138. Job, J. T., Visakh, N. U., Pathrose, B., Alfarhan, A., Rajagopal, R., Thayyullathil, J., et al. (2024). Chemical composition and biological activities of the essential oil from Citrus reticulata Blanco peels collected from Agrowastes. *Chemistry & Biodiversity, 21*(3), e202301223.
139. Narayanankutty, A., Visakh, N. U., Sasidharan, A., Pathrose, B., Olatunji, O. J., Al-Ansari, A., et al. (2022). Chemical composition, antioxidant, anti-bacterial, and anti-cancer activities of essential oils extracted from Citrus limetta Risso Peel waste remains after commercial use. *Molecules, 27*(23), 8329.
140. Matsuzaki, K., Nakajima, A., Guo, Y., & Ohizumi, Y. (2022). A narrative review of the effects of citrus peels and extracts on human brain health and metabolism. *Nutrients, 14*(9), 1847. https://doi.org/10.3390/nu14091847
141. Ishfaq, M., et al. (2021). Antioxidant and wound healing potential of essential oil from Citrus reticulata Peel and its chemical characterization. *Current Pharmaceutical Biotechnology, 22*(8), 1114–1121. https://doi.org/10.2174/1389201021999200918102123
142. Jiang, H., Zhang, W., Xu, Y., Chen, L., Cao, J., & Jiang, W. (2022). An advance on nutritional profile, phytochemical profile, nutraceutical properties, and potential industrial applications of lemon peels: A comprehensive review. *Trends in Food Science and Technology, 124*, 219–236. https://doi.org/10.1016/j.tifs.2022.04.019
143. Satari, B., & Karimi, K. (2018. Elsevier B.V). Citrus processing wastes: Environmental impacts, recent advances, and future perspectives in total valorization. *Resources, Conservation and Recycling, 129*, 153–167. https://doi.org/10.1016/j.resconrec.2017.10.032
144. Suri, S., Singh, A., & Nema, P. K. (2022). Current applications of citrus fruit processing waste: A scientific outlook. *Applied Food Research, 2*(1), 100050. https://doi.org/10.1016/j.afres.2022.100050

Chapter 9
Citrus Food Waste is a Valuable Source of Bioactive Components and Application in the Food Industry

Mouandhe Imamou Hassani, Shreya Saha, Laura Jūrienė, Doli Saikia, and Hakki Bilgin

9.1 Introduction

In modern times of sustainability and resource conservation, food waste utilization has become an indispensable area of study and innovation. According to FAO's estimation, over 32% of the world's food supply went to waste in year 2009. Citrus fruits contribute a significant amount of food waste, including seeds and peels. As an example, the amount of pure juice yielded by a single citrus fruit is just half its weight; the other half consists of leftovers such as peel, pulp, pomace, and seeds. The disposal or incineration of large quantities of these leftovers pollutes the environment and reduces the amount of oxygen in polluted water. Nevertheless, there is an opportunity to lessen the environmental impact of food waste by tapping into this under-researched source of bioactive chemicals, which also happens to be rich in compounds with numerous health-promoting applications.

M. Imamou Hassani (✉)
Kaunas University of Technology, Department of Food Science and Technology, Kaunas, Lithuania

Institut National de Recherche Pour L'Agriculture, La Pêche et L'Environnement (INRAPE), Moroni, Comoros
e-mail: mouandhe.imamou@ktu.edu

S. Saha
Department of Dairy Microbiology, ICAR-National Dairy Research Institute, Karnal, Haryana, India

L. Jūrienė · H. Bilgin
Kaunas University of Technology, Department of Food Science and Technology, Kaunas, Lithuania

D. Saikia
Department of Allied Sciences, School of Health Science and Technology, UPES, Bidholi, Dehradun, Uttarakhand, India

A. Chauhan et al. (eds.), *Valorization of Citrus Food Waste*,
https://doi.org/10.1007/978-3-031-77999-2_9

With a growing demand, global increase in food production and processing has given rise to a significant increase in food waste or by-product generation which brings about concern related to environmental, economic, and social aspects worldwide. Estimations have shown that globally around 1.3–1.6 Gt of food is wasted per year, resulting in the emission of 3.3 Gt of CO2 into the atmosphere and causing 750 billion US dollars in direct economic losses annually [57]. Besides these impacts, these food waste materials have been shown as a crucial source of valuable bioactive compounds [15]. Therefore, to decrease the generated waste and increase the volarization of these by-products, waste management techniques and technologies have been gaining significant attention all over the world.

Citrus fruits are wasted in large amounts throughout their conversion from cultivation to consumption. Lime (Citrus aurantifolia), sweet orange (Citrus sinensis), mandarin (Citrus reticulata), lemon (Citrus limonum), and grapefruit (Citrus vitis) are among the commonly consumed citrus fruits. They are an abundant supply of vitamins, minerals, and different types of dietary fiber that are crucial for human growth and nourishment [34]. Citrus food waste contains bioactive components that have been shown to have numerous health benefits. These benefits include antimicrobial activities that prevent the growth of pathogenic bacteria and fungi, antioxidant qualities that can fight oxidative stress and inflammation, and potential anti-cancer effects due to their ability to modify cellular pathways [63]. Furthermore, these compounds have demonstrated potential in a range of applications, including nutraceuticals, functional foods. Medicines, and cosmetics. This chapter examines the diverse range of bioactive elements found in citrus food waste, including their extraction, characterization, and possible uses in different food sectors.

9.2 Sources of Citrus food Waste

Citrus processing industries, juice extraction operations, fruit markets, commercial kitchens and households. Citrus processing plants, where fruits are juiced or processed into products like jams, marmalades, and concentrates, are the main sources of citrus waste [73]. The primary sources of citrus waste are the citrus processing industries, which include fruit processing facilities, oil extraction facilities, and juice factories [13]. During these processing operations, a lot of peels, pulp, seeds, and other residuals are generated. It is obvious that the outer peel is one of the major byproducts discarded after juice production [9].

Citrus fruit juice extraction operations produce a significant quantity of waste, whether the juice is used for industrial or fresh consumption [27]. Peels are removed and discarded in large quantities during the juice extraction process, which can be a major source of citrus waste [55]. Furthermore, pulp and fibrous debris remain after juice extraction. Although this pulp could have nutrients and bioactive elements, it is frequently thrown away as waste.

In fruit markets and distribution facilities, where spoiled or overripe fruits are thrown away, citrus trash is also produced. Vendors and marketplaces frequently reject overripe, bruised, or unsightly citrus fruits in order comply with standards for product quality. Citrus fruits can deteriorate more quickly when improperly handled and stored, which can result in spoiling and waste [61]. Fruit deterioration and increased waste can be caused by variables such temperature swings, poor ventilation, and physical damage during shipping or storage [1]. The total citrus waste stream is impacted by these wasted fruits [22]. Fruits that fail to adhere to grading norms can be rejected or degraded, which would result in waste. Throughout the supply chain, from manufacturers to wholesalers to retailers, this can happen at several points. Because citrus fruits develop in different seasons, there are often excesses when harvest quantities surpass demand. During peak seasons, fruit markets could find it challenging to sell every item of produce, which could lead to waste if an accumulation of extra fruit occurs.

In both residential and commercial kitchens, citrus waste is produced when fruits are consumed and the peels and other waste are thrown away [5]. When preparing meals, citrus fruit seeds are also removed and thrown away. Although citrus trash produced by individual households may be relatively insignificant but when added to the entire waste stream, it can have a substantial impact.

9.3 Composition of Citrus Waste

Among different food products, fruits are one of the most widely consumed groups which causes a significant amount of waste production during processing [30]. Among different fruit industries, one of the highest production rates belongs to citrus species worldwide [17]. Recently, global manufacturing and utilization of these species in the juice industry have been increasing with a growing demand. The increase in production has been causing an increase in waste generation, mainly membranes, pulp, peels, and seeds. Peel alone covers around 50% of the raw product weight. Even though produced waste materials were considered as animal feed or discarded previously, research has shown that a lot of valuable compounds are present in citrus waste and they can be used for further usage [58].

The term citrus comes from the fruit genus Citrus, and it consists of different species such as limes, lemons, grapefruit, sweet oranges, sour or bitter oranges, and mandarins. In food processing, each can be used for different purposes and products due to different compositions. While in the juice industry, mostly sweet oranges are preferred, mandarins and grapefruits are used during the production of marmalades, jams, and flavoring agents [73]. During citrus fruit processing, around 40–60% of waste product, also known as pomace, is generated [7]. The composition of waste may change based on factors such as fruit type, processing methods, and extraction techniques. However, on average, citrus waste contains citrus peels which consist of surface layer flavedo, includes oil glands and pigments, and inner soft layer albedo, pulp, seeds, and membranes. Among these constitutes, citrus waste consists mainly of peels, around 60–65%, which is followed by internal tissues and seeds by weight [58].

9.4 Bioactive Components from Citrus Waste and Their Extraction

Waste generated by citrus processing has high-value-added compounds that can be used in the production of functional foods, nutritional supplements, and medications that aim to promote general well-being due to its broad array of bioactive compounds. In addition to decreasing waste, successful exploitation and application of these compounds can aid in the production of environmentally friendly and health-enhancing goods. Every part of the citrus is rich in important minerals and phytochemicals, from the peels to the seeds. The majority of citrus waste, about 50–55% of the fruit weight, is made up of citrus peels which are rich in nutritious and valuable compounds such as dietary fibers, pigments, essential oils, and phenolics with high bioactivity. [69]. D-limonene which is derived from citrus peels, is in high demand since they are used in food products, drinks, perfumery, and the cosmetics industry. Additionally, soluble sugars, preferred for bioethanol production, including sucrose, fructose, and glucose are present in significant amounts. Peel's cellular component consists mainly of arabinose, galactose, cellulose, and pectin. The abundance of soluble and insoluble carbohydrates implies that peel can be effectively used to produce value-added products through ensuing metabolic processes [52]. Citrus seeds, recognized as the inedible components of citrus fruits, are notable for their substantial oil content, comprising approximately 20–40% of the total mass [54] and in terms of fatty acid, seeds show high-quality content with polyunsaturated fatty acids [43].

Generated waste during processing contains a significant amount of valuable substances with numerous health advantages. Flavonoids found in abundance in peels may lower the risk of chronic illnesses. With their potential to prevent cancer, limonoids are mostly present in citrus peels and seeds. Citronene and citral, two substances with antibacterial and mood-boosting qualities, are found in essential oils made from citrus peels. Furthermore, pectin—a soluble fiber good for blood sugar management and digestive health [48].

9.4.1 Phenolic Compounds

Phenolics are found abundantly in plants, fruits, and vegetables and they are known for their antioxidant properties provided by aromatic rings present in the structure. Some of those compounds are flavanols, flavanones, flavonols, anthocyanidins, and isoflavanones. In citrus fruits, phenolic compounds contribute to various aspects of fruit quality, including color, flavor, and nutritional value. According to studies, plant phenolics are present in all portions of the plant, and they have been linked to a variety of biological impacts [48].

Citrus waste is rich in high-value-added substances such as phenolics, which possess antioxidant properties by effectively scavenging the free radicals, neutralizing their harmful effects, and reducing oxidative stress in the body. According to scavenging characteristics for suppressing the activity of free radicals, citrus by-products have two sub-classes which are known as hydroxybenzoic and hydrocinnamic acids [24].

Flavonoids, constituting a subgroup within the phenolic compounds exist plentifully in citrus species [58]. Distinguished by their multifaceted chemical compositions encompassing flavones, flavonols, flavanones, and flavanonols, these compounds serve important functions in both the growth and defense mechanisms of plants. Moreover, they significantly influence the pigmentation, taste, and nutritional profile of various fruits and vegetables. Despite the fact that flavonoids are typically thought of as non-nutritive substances, many researchers are interested in learning more about their potential to prevent serious chronic diseases. According to a study, peels of citrus fruits were found to have the largest concentrations of polymethoxylated flavones when compared to the edible sections [38, 68].

Incorporating waste into dietary supplements, functional foods, and cosmetic formulations may offer a sustainable and effective way to harness its scavenging activity for promoting health and well-being. For this purpose, different extraction techniques are widely used to obtain the best efficiency and high-quality end product. Microwave is a highly used technique with short time, low energy and solvent consumption, and high-efficiency features [3]. The research investigated the phenolics extracted from peel by this technique and found a significant amount of narirutin and hesperidin [25]. As another novel extraction technique, ultrasound was used for phenolic extraction in lime and mandarine waste peels [59]. In another study, it was revealed that low power ultrasound technique showed significant yield for hesperidin and total phenol recovery. Additionally, it was reported that the ultrasound method was 1.77 times more efficient than the microwave technique [44]. The impact of a pulsed electric field for flavonoids and total phenols recovery in peel was examined by Luengo et al. who observed that the technique improved the yield by an average of 140%. Furthermore, the pulsed electric field sample displayed higher levels of naringin, hesperidin, and overall antioxidant activity in comparison to the untreated sample.

9.4.2 Dietary Fiber

Carbohydrates such as lignin, b-glucans, cellulose, hemicellulose, pectins, gums, b-glucans, and lignin make up the majority of dietary fiber composition. It is usually classified as insoluble and soluble fibers as two primary categories. The cell wall of cereals, vegetables, and fruits is the main source of dietary fiber consumed. The advantageous physiological benefits of dietary fiber have been gaining attention recently. According to research, diets rich in dietary fiber help prevention or treatment of coronary diseases [67]. Besides assisting in avoiding hydrolysis, absorption, and digestion in the human body, dietary fiber also sustains insulin levels, lowers preprandial cholesterol, increases colonic fermentation, and improves fecal bulking efficiency [10, 20]. Dietary fibers act differently during processing in the gastrointestinal tract depending on the techniques they are obtained, physicochemical characteristics, and composition [11]. In this aspect, compared to other alternative sources, including cereals, orange juice extraction residues have a larger proportion of soluble dietary fibers and are readily available in large quantities, making them potentially an ideal source of dietary fibers. Research has shown that 1.5% orange fiber can be a good alternative to add as a fat replacer to fermented sausages [21]. Besides that, fiber from citrus products shows the ability to inhibit the oxidation of lipids resulting in an improvement of oxidative stability followed by increasing shelf life [18, 56]. In addition, it was reported that fiber from the citrus peel by-product improves intestinal activity and balance [12]. Research has indicated a process for making use of orange trash that included extracting the essential oil and then using the leftover cellulolytic fibers to produce dietary fiber. It was observed that assuming a 40% efficiency in essential oil extraction, 8000 tons of total solid residues produced during the manufacture of orange juice can give 5200 tons of fiber and about 12.8 tons of essential oils [50].

9.4.3 Essential Oils

Essential oils are concentrated liquids derived from plant-based products through distillation or cold pressing, containing natural aromatic compounds, giving them characteristic fragrances and therapeutic properties. Citrus fruit peels undergo cold pressing primarily to extract volatile elements, such as essential oils. This physical extraction method yields a watery emulsion which is processed by centrifugation to obtain oil. Besides these methods, new techniques such as microwave extraction, controlled pressure drop process, subcritical water, supercritical fluid, and ultrasound methods are also used to obtain essential oils. Among these techniques, supercritical fluid extraction has been developed to handle difficult separation processes involving small amounts of valuable chemicals. It works by using a gas, like CO_2, above its critical temperature and pressure. Together with quick extraction, great selectivity, and low solvent consumption, its main benefits also include no extract residue and high concentration. This method uses CO_2, an odorless, nonflammable gas that is created by both human and natural processes, as the solvent instead of more conventional organic solvents like hexane. Supercritical CO_2 extraction is used in different sectors, including pharmaceuticals, cosmetics, and food. It is especially useful for extracting weakly polar chemicals including carotenoids, triglycerides, fatty acids, and smells. Essential oils have also been extracted successfully using ultrasound waves because of their cavitational impact, which breaks down plant cell walls and speeds up heat and mass transport while facilitating the release of extractable chemicals. With the use of ultrasounds, extraction can be finished quickly, reproducibly, and with little solvent usage. Besides oils, the technique also helps extraction of phenolics [33]. The controlled pressure drop process was used to obtain volatile oil from a solid matrix depending on the thermo-mechanical process. It is triggered quickly lowering the product's steam pressure to a vacuum. Research has shown that the method is able to isolate oil from peels of orange [51]. The concentration of peel oil varies depending on the type of fruit, ranging from 0.05 to 5.0% [36].

Essential oils can be found in significant quantities in citrus fruits. Their distinctive flavor and scent are attributed to the unique oil glands found in their peels, which house these aromatic chemicals. Waste from citrus fruits, especially peels, is a major source of essential oils. Much of the fruit peel is usually thrown away as garbage after it is processed for juice or other goods. However, creative extraction techniques were devised for effective recovery. Solvent extraction, steam distillation, and cold pressing are some of these techniques widely used to obtain oil. Essential oils derived from citrus species are utilized in beverages and food, aromatherapy, cosmetics, and cleaning goods, among other industries. They are used as natural substitutes for artificial fragrances and cleaning agents, to enhance the flavor of food and drinks, and to give aroma to skincare items and perfumes.

In addition to their aromatic properties, their health benefits, have been gaining attention recently [6]. Oil extracted from orange peel includes mainly, D-limonene, and cyclic terpene which are abundant in essential oils and responsible for their characteristic aromas and it is the most abundant high-value added compound in citrus essential oils, comprising up to 95% of the oil extracted from citrus peels. [2]. Even though it has another optical isomer, which is L-limonene, they show differences in fragrances. While D-limonene smells like orange, L-limonene smells like turpentine. D-limonene is known for its distinct citrus scent and flavor, which makes it a popular ingredient in different product formulations such as cosmetics, fragrances, beverages, food, and cleaning agents. It is one of the most widely utilized agent for flavoring of beverages, and confectionery goods, providing a natural citrus taste and aroma. In addition to its culinary uses, D-limonene has gained attention by showing effective activity on cancer, diabetic, and tumoral activity treatment [41]. Besides health aspects, it can also be widely used as a natural pesticide instead of traditional synthetic ones which causes unwanted results in environmental aspect [49].

9.4.4 Pectin

Plant cell walls, particularly those of citrus fruits, contain the complex polymer pectin. Fruits get their texture and firmness from this structural feature, which gives the cell walls their strength and rigidity. Alpha-1,4 glycosidic links, which hold long chains of galacturonic acid molecules together, are the building blocks of pectin. As byproducts of processing citrus, huge quantities of waste that is high in pectin are produced. These consist of membranes, pomace, and citrus peels. As a functional food ingredient obtained from citrus fruits and apples, it is one of the most widely used compounds with an EU code of E440, worldwide. Pectin is mostly employed in the industry for its gelling qualities, which are used in fruit juice, pastry fillings, jams and jellies, and confections. However, pectin is also used to stabilize yogurts and acidified milk beverages. Pectins are well known for their antidiarrheal properties and are utilized as detoxifying agents in the pharmaceutical industry. They are also employed as gels and pastes in cosmetics [71].

The primary source materials used in industrial pectin production are the leftovers from fruit juice manufacturing. Strong acids are used in industrial processes to chemically extract pectins; this process is known as traditional acid extraction. Although the method provides waste volarization for pectin, environmental concerns should be considered. To solve such weakness, the enzymatic method combination of proteolytic, hemicellulolytic, pectinolytic, and cellulolytic has been investigated in studies for pectin extraction [71].

9.5 Applications in the Food Industry

Due to the valuable functional components present in citrus waste and the growing concern over waste management, researchers are exploring various ways to utilize citrus waste, thereby reducing environmental damage caused by improper disposal. Peels, pulp, pomace, and seeds are all parts of citrus waste. These contain many beneficial phytochemicals, such as pectin, essential oils, polyphenolic flavonoids, carotenoids, dietary fibre, limonoids, organic acids, and vitamins (Fig. 9.1). In fact, the bioactive compounds that can be extracted from orange trash can be used to develop functional foods and nutraceuticals that are beneficial for human health such as protection against pathogen, allergens, and prevention of several illnesses [62].

Moreover, citrus waste finds applications in animal feed, compost preparation, biofuels, and bioethanol production. Through various processes, solid and concentrated liquid citrus waste can be converted into citrus waste-based molasses, lactic acid, feed yeast, industrial alcohol/ethanol, and vinegar. Recently, researchers have proposed utilizing citrus waste for the synthesis of iron and silver nanoparticles with cytotoxic properties, as well as nanocellulose and enzymes such as pectinase and larvicides [29].

Citrus peels are rich in essential oils that have diverse applications in food preservation, food safety, and the nutraceutical industry. Several research have investigated several aspects of essential oils, including their antibacterial mechanisms, spray applications, microencapsulation with biodegradable polymers, and thin films based on essential oils. Citrus essential oils are widely used in the cosmetics, pharmaceutical, food, and packaging industries because to their usefulness in ensuring the safety of food throughout storage, transportation, and preparation.

Fig. 9.1 Applications of Citrus food Waste in the food industry

9.5.1 Dairy Industry

Citrus pectin, a plant cell wall polymer, is essential to the preparation of food. The key functions are stabilization, thickening, and gelling. Pectin, made from fruit peels and byproducts like citrus peels and apple pomace, is widely utilized in the food business. It stabilizes acidified milk beverages and yogurts, which are popular for their natural flavor and nutritional value. Research shows that acidified milk beverages without stabilizers commonly cause protein flocculation and whey separation. To effectively reduce these difficulties and ensure the stability of milk drinks, citrus pectin is added as a protective colloid [42].

9.5.2 Beverage Industry

The bio-functional characteristics of citrus waste make its utilisation in the food industry critical for improving food quality and customer health. It helps in the creation of high-quality foods and drinks that are rich in nutrients. Juices and other nutrient-rich beverages can make good use of citrus waste due to its high vitamin C concentration [66]. Additionally, the essential oils (EOs) present in citrus waste find applications in various beverages, food additives, pharmaceuticals, and cosmetic products, owing to their flavor-enhancing and aromatic properties. In addition, citrus processing wastes are rich in carotenoids and phenols, which make them biofertilizers that can prolong the freshness and longevity of foods and drinks [46].

The favorable effects of incorporating Citrus natsudaidai (CN) peel extracts into aqueous solutions. These solutions, flavored with CN extracts, exhibited superior odor compared to commercial citrus-flavored beverages in electronic nose tests [4]. Furthermore, they exhibited a blend of acidity, astringency, and a flavour reminiscent of oranges, with a level of palatability similar to that of commercially available citrus-flavored drinks. The use of citrus essential oils in diverse culinary items such as ice cream, marmalade, and jams has also been extensively recorded [64].

Citrus peel pectin, when hydrolyzed, yields hydrolysates with enhanced solubility and potential applications as dietary fiber sources in liquid food products, particularly beverages. Additionally, natural colorants extracted from citrus peel are extensively utilized in food products as eco-friendly alternatives to synthetic colorants, further promoting the sustainable utilization of waste materials [29].

9.5.3 Bakery Industry

The incorporation of citrus fiber into muffins, significantly increasing their fiber content by 40–63% compared to control muffins. This substitution of wheat flour with citrus fiber allowed for the creation of baked goods with elevated levels of dietary fiber, potentially beneficial for individuals with diverse nutritional needs [53].

The use of citrus powder or flour in bread and confectionery items to improve their performance and effectiveness. For example, a research project created low-fat brioches that were enhanced with debittered orange fibres, leading to enhanced stability and nutritional characteristics. The use of citrus albedo in bread also increased its shelf life as a result of its high pectin and fibre content. In addition, it was discovered that citrus peel is useful for improving the nutritional and functional characteristics of cakes [8].

Additionally, bioactive chemicals extracted from citrus waste, noted for their antioxidant properties, are employed as food additives, including in candied goods for the confectionery and baking industries. Enriching muffins with dietary-fiber-rich orange bagasse extract reduced their predicted glycemic index without compromising sensory attributes. Similarly, biscuits formulated with mandarin peel powder exhibited enhanced nutritional content and antioxidant activity [4].

Citrus peel and rind are valued for their distinct flavor, color, and nutritional qualities, commonly used as food additives in candied items for the confectionery and bakery sectors. There has been tremendous expansion in the worldwide market for bakery products, with biscuits leading the pack and bread and rolls, cakes and pastries, and rusk following closely behind [29].

9.5.4 Meat Industry

Citrus extracts have emerged as promising natural additives for enhancing the quality and extending the shelf life of meat products and their effectiveness in controlling lipid oxidation in meat balls during frozen storage [45]. These extracts contain natural antioxidants derived from citrus peels, which inhibit enzymatic reactions responsible for oxidative damage. Furthermore, their incorporation into meat products has shown positive effects on color, flavor, and overall sensory properties. Additionally, by-products such as mosambi peel have been utilized to enhance fiber content and improve storage stability in sausages and patties [72]. The utilization of citrus waste not only adds value to the food industry but also offers sustainable solutions for food preservation. Moreover, citrus seed waste, known for its high protein, mineral, and fiber content, can be utilized as a functional additive in various food formulations, including pancakes, doughnuts, and meat-based items, owing to its excellent emulsion action and stability. Furthermore, advancements in processing techniques have allowed for the development of citrus flour-based seasonings, which not only preserve the nutritional integrity of orange peel but also enhance the flavor and nutritional properties of meat products without the need for solvent extraction. Overall, the incorporation of citrus extracts and by-products presents promising avenues for improving the quality and extending the shelf life of meat products in a natural and sustainable manner.

9.6 Functional Properties of Citrus Waste

Citrus waste exhibits remarkable functional properties due to its rich content of bioactive compounds like flavonoids and phenolic acids, making it a valuable resource in various industries. Its high antioxidant capacity renders it useful for extending the shelf life of food products and preventing lipid oxidation. Additionally, citrus waste can serve as a natural source of dietary fiber, contributing to improved digestion and overall health when incorporated into food formulations [35]. Apart from antioxidant and antimicrobial properties, few functional properties of citrus food waste include prebiotic, encapsulating agent, as edible packaging material.

9.6.1 Prebiotic

Although pectin from citrus peels is commonly used to make jam, jellies, and marmalades, it really has strong prebiotic benefits. Oligosaccharides produced from citrous pectin are known to be an excellent prebiotic [62]. Prebiotics are indigestible dietary components that stimulate the development and function of beneficial microbes in the colon, resulting in enhanced host health. They are becoming ever more important in everyday life. Recent studies have shown that discarded orange peels may be used to produce potent prebiotic food that exhibits improved bio-functionality. Pectic oligosaccharides (POS) are a kind of prebiotic that is derived from agricultural waste, namely orange peel waste. They are produced by the partial breakdown of pectin, as described by Foti et al. [19] and Suri et al. [62]. Korean Citrous hallabong peel pectic polysaccharides HBE-III were the subject of an investigation into their potential anti-cancer effects [31]. The study demonstrated that HBE-III suppresses the spread of cancer cells to the lungs in a manner that is influenced by the dosage administered. The HBE-III greatly increased the ability of NK cells to kill Yac-1 tumour cells. Vinegar was derived from the waste of bergamot (Citrous bergamia) [31]. At a pH of 2.9, vinegar was made with the Acetobacter aceti strain. The study showed that bergamot can be used to make citrous vinegar with high amounts of bioactive ingredients like brutieridin and melitidin, as long as the right microbial starters are used [16]. A recent study used an in vitro model to look at how citrous waste food fibre affects the gut system of piglets in order to learn more about its prebiotic effects. The study found that pigs had a higher acetate ratio and better immune systems in their guts [65]. Foti et al. [19] did study that suggested using orange peel fibre to make oligosaccharides with extra value. Pectin made from enzymes breaking down orange peel fibre had better prebiotic effects than pectin that is sold in stores. Because of this, pectic oligosaccharides made from citrous waste could be used as new prebiotics that work better.

9.6.2 Encapsulating Agent

In recent years, due to advancements in nanotechnology research, pectin extracted from citrus peel has been used as an encapsulating agent. To improve the stability and efficacy of encapsulation, polymer blends are being studied as encapsulating agents [47]. An oil emulsion was created using citrus peel waste, pectin, citric acid, Ca2+ ion, and ascorbic acid. The research found that the emulsion formed was stable with a decreased droplet size. Nanoparticles containing a high concentration of resveratrol were synthesized using zein and citrus peel pectin. With a salt concentration of less than 50 mM, the nanoparticles remained stable over a pH range of 2–7, even after being heated to 80 °C for 1 hour. Results from in-vitro studies show that bioaccessibility is higher for encapsulated resveratrol than for free resveratrol. According to Huang et al. [23], resveratrol's antioxidant activities were found to be increased.

9.6.3 Packaging Material

A new area of study is looking into how to use used citrous peels to make recyclable packing materials. Citrous limon and citrous aurantifolia peels were used to make coverings for fresh strawberries that can be eaten [40]. A new kind of packing film was made by mixing pectin from lemon peel with starch from sweet potatoes in a

3:1 ratio. Titanium oxide nanoparticles (0.5–2.0% by weight) were also added. Through the casting process, pomelo peel flour and tea polyphenols were used to make a functional edible wrapping film. The research found that a mixed packing material made of pomelo peel and tea polyphenol, with 10% tea polyphenol, was very good at keeping out moisture and had good antibacterial, antioxidant, and mechanical properties. If you want to store soybean oil for 30 days in an eco-friendly way, this package could work [39, 60]. Pectin from blood orange peel and fish gelatine were recently mixed to make a tasty film. A hybrid film made of equal parts pectin from orange peel and gelatine from fish had great tensile strength and also had antibacterial and protective properties. Using the wrapping method, an edible layer was put on top of the ricotta cheese. This made the cheese more stable in terms of its physical, chemical, textural, and microbial properties [14, 26].

The utilisation of citrus wastes demonstrates the potential use of the newly developed edible GFPec-based films as antimicrobial food packaging. These films can help fresh-cut fruits and vegetables last longer and are better for the environment than fake food packingmaterials [28].

9.7 Concluding Remarks

Environmental concern and food insecurity across the globe has prompted researchers and industry leaders to the biorefining of citrus fruit waste. In the last decade, citrus waste has emerged as a promising source of food additives and functional food ingredients. It is linked to several bioactivities such as antimicrobial, anti-inflammatory, and antioxidant properties due to the containment of compounds such as polyphenols, pectin, and essential oils. The abundance of various bioactive components has paved the way for several applications in the food industry. For instance, the chemical composition of dairy, meat, or bakery products could be enhanced by enrichment with bioactive compounds extracted from citrus food waste. Moreover, several studies have demonstrated the possible use of refined waste as packaging material, encapsulation agent, or prebiotic. Nevertheless, to unleash the full potential of citrus fruit waste, more research on sustainable extraction of bioactive compounds and bioavailability of its bioactive compounds is needed.

References

1. Ahmad, M. S., Siddiqui, M. W., Ahmad, M. S., & Siddiqui, M. W. (2015). Factors affecting postharvest quality of fresh fruits. In *Postharvest quality assurance of fruits: Practical approaches for developing countries* (pp. 7–32). Springer.
2. Anagnostopoulou, M. A., Kefalas, P., Papageorgiou, V. P., Assimopoulou, A. N., & Boskou, D. (2006). Radical scavenging activity of various extracts and fractions of sweet orange peel (citrus sinensis). *Food Chemistry, 94*(1), 19–25.

3. Anticona, M., Blesa, J., Frigola, A., & Esteve, M. J. (2020). High biological value compounds extraction from citrus waste with non-conventional methods. *Foods, 9*(6), 811.
4. Ben Hsouna, A., Sadaka, C., Generalić Mekinić, I., Garzoli, S., Švarc-Gajić, J., Rodrigues, F., Morais, S., Moreira, M. M., Ferreira, E., Spigno, G., Brezo-Borjan, T., Akacha, B. B., Saad, R. B., Delerue-Matos, C., & Mnif, W. (2023). The chemical variability, nutraceutical value, and food-industry and cosmetic applications of citrus plants: A critical review. *Antioxidants (Basel, Switzerland), 12*(2), 481. https://doi.org/10.3390/antiox12020481
5. Benny, N., Shams, R., Dash, K. K., Pandey, V. K., & Bashir, O. (2023). Recent trends in utilization of citrus fruits in production of eco-enzyme. *Journal of Agriculture and Food Research, 13*, 100657.
6. Braddock, R. J., Temelli, F., & Cadwallader, K. R. (1986). *Citrus essential oils: A dossier for material safety data sheets.*
7. Caballero, S., Li, Y. O., McClements, D. J., & Davidov-Pardo, G. (2022). Encapsulation and delivery of bioactive citrus pomace polyphenols: A review. *Critical Reviews in Food Science and Nutrition, 62*(29), 8028–8044.
8. Caggia, C., Palmeri, R., Russo, N., Timpone, R., Randazzo, C. L., Todaro, A., & Barbagallo, S. (2020). Employ of citrus by-product as fat replacer ingredient for bakery confectionery products. *Frontiers in Nutrition, 7*, 46. https://doi.org/10.3389/fnut.2020.00046
9. de Castro, L. A., Lizi, J. M., das Chagas, E. G. L., de Carvalho, R. A., & Vanin, F. M. (2020). From orange juice by-product in the food industry to a functional ingredient: Application in the circular economy. *Foods, 9*(5), 593.
10. Champ, M., Langkilde, A. M., Brouns, F., Kettlitz, B., & Collet, Y. L. B. (2003). Advances in dietary fibre characterisation. 1. Definition of dietary fibre, physiological relevance, health benefits and analytical aspects [review]. *Nutrition Research Reviews, 16*(1), 71–82.
11. Chau, C. F., & Huang, Y. L. (2003). Comparison of the chemical composition and physicochemical properties of different fibers prepared from the peel of citrus sinensis L. Cv. Liucheng. *Journal of Agricultural and Food Chemistry, 51*(9), 2615–2618.
12. Chau, C.-F., Sheu, F., Huang, Y.-L., & Su, L.-H. (2005). Improvement in intestinal function and health by the peel fibre derived from Citrus Sinensis L Cv Liucheng. *Journal of the Science of Food and Agriculture, 85*(7), 1211–1216.
13. Chavan, P., Singh, A. K., & Kaur, G. (2018). Recent progress in the utilization of industrial waste and by-products of citrus fruits: A review. *Journal of Food Process Engineering, 41*(8), e12895.
14. Costa, M. J., Maciel, L. C., Teixeira, J. A., Vicente, A. A., & Cerqueira, M. A. (2018). Use of edible films and coatings in cheese preservation: Opportunities and challenges. *Food Research International, 107*, 84–92.
15. Cristóbal, J., Caldeira, C., Corrado, S., & Sala, S. (2018). Techno-economic and profitability analysis of food waste biorefineries at European level. *Bioresource Technology, 259*, 244–252.
16. Di Donna, L., Bartella, L., De Vero, L., Gullo, M., Giuffrè, A. M., Zappia, C., Capocasale, M., Poiana, M., D'Urso, S., & Caridi, A. (2020). Vinegar production from Citrus bergamia by-products and preservation of bioactive compounds. *European Food Research and Technology, 246*(10), 1981–1990. https://doi.org/10.1007/s00217-020-03549-1
17. FAO. (2017). *Citrus fruit fresh and processed—Statistical bulletin.*
18. Fernández-Ginés, J. M., Fernández-López, J., Sayas-Barberá, E., Sendra, E., & Pérez-Alvarez, J. A. (2003). Effect of storage conditions on quality characteristics of Bologna Sausages made with citrus fiber. *Journal of Food Science, 68*(2), 710–714.
19. Foti, P., Ballistreri, G., Timpanaro, N., Rapisarda, P., & Romeo, F. V. (2022). Prebiotic effects of citrus pectic oligosaccharides. *Natural Product Research, 36*(12), 3173–3176. https://doi.org/10.1080/14786419.2021.1948845
20. Fuentes-Zaragoza, E., Riquelme-Navarrete, M. J., Sánchez-Zapata, E., & Pérez-Álvarez, J. A. (2010). Resistant starch as functional ingredient: A review. *Food Research International, 43*(4), 931–942.

21. García, M. L., Dominguez, R., Galvez, M. D., Casas, C., & Selgas, M. D. (2002). Utilization of cereal and fruit fibres in low fat dry fermented sausages. *Meat Science, 60*(3), 227–236.
22. Haldar, D., Shabbirahmed, A. M., Singhania, R. R., Chen, C.-W., Dong, C.-D., Ponnusamy, V. K., & Patel, A. K. (2022). Understanding the management of household food waste and its engineering for sustainable valorization-A state-of-the-art review. *Bioresource Technology, 358*, 127390.
23. Huang, X., Liu, Y., Zou, Y., Liang, X., Peng, Y., McClements, D. J., & Hu, K. (2019). Encapsulation of resveratrol in zein/pectin core-shell nanoparticles: Stability, bioaccessibility, and antioxidant capacity after simulated gastrointestinal digestion. *Food Hydrocolloids, 93*, 261–269.
24. Ignat, I., Volf, I., & Popa, V. I. (2011). A critical review of methods for characterisation of polyphenolic compounds in fruits and vegetables. *Food Chemistry, 126*(4), 1821–1835.
25. Inoue, T., Tsubaki, S., Ogawa, K., Onishi, K., & Azuma, J.-I. (2010). Isolation of hesperidin from peels of thinned citrus Unshiu fruits by microwave-assisted extraction. *Food Chemistry, 123*(2), 542–547.
26. Jridi, M., Abdelhedi, O., Salem, A., Kechaou, H., Nasri, M., & Menchari, Y. (2020). Physicochemical, antioxidant and antibacterial properties of fish gelatin-based edible films enriched with orange peel pectin: Wrapping application. *Food Hydrocolloids, 103*, 105688.
27. Kandemir, K., Piskin, E., Xiao, J., Tomas, M., & Capanoglu, E. (2022). Fruit juice industry wastes as a source of bioactives. *Journal of Agricultural and Food Chemistry, 70*(23), 6805–6832.
28. Khalil, R. K., Sharaby, M. R., & Abdelrahim, D. S. (2023). Novel active edible food packaging films based entirely on citrus peel wastes. *Food Hydrocolloids, 134*, 107961.
29. Khan, U. M., Sameen, A., Aadil, R. M., Shahid, M., Sezen, S., Zarrabi, A., Ozdemir, B., Sevindik, M., Kaplan, D. N., Selamoglu, Z., Ydyrys, A., Anitha, T., Kumar, M., Sharifi-Rad, J., & Butnariu, M. (2021). Citrus genus and its waste utilization: A review on health-promoting activities and industrial application. *Evidence-based Complementary and Alternative Medicine: Ecam, 2021*, 2488804. https://doi.org/10.1155/2021/2488804. (Retraction published Evid Based Complement Alternat Med. 2023 Dec 13;2023:9849716).
30. Kosseva, M. R. (2013). Chapter 3 – Sources, characterization, and composition of food industry wastes. In M. R. Kosseva & C. Webb (Eds.), *Food industry wastes* (pp. 37–60). Academic Press.
31. Lee, E. H., Park, H.-R., Shin, M.-S., Cho, S. Y., Choi, H.-J., & Shin, K.-S. (2014). Antitumor metastasis activity of pectic polysaccharide purified from the peels of Korean Citrus Hallabong. *Carbohydrate Polymers, 111*, 72–79.
32. Li, S., Wang, H., Guo, L., Zhao, H., & Ho, C.-T. (2014). Chemistry and bioactivity of nobiletin and its metabolites. *Journal of Functional Foods, 6*, 2–10.
33. Ma, Y.-Q., Chen, J.-C., Liu, D.-H., & Ye, X.-Q. (2009). Simultaneous extraction of phenolic compounds of citrus peel extracts: Effect of ultrasound. *Ultrasonics Sonochemistry, 16*(1), 57–62.
34. Mahajan, M., & Sadana, R. (2023). Citrus based food products and their shelf life. In *Recent advances in citrus fruits* (pp. 367–393). Springer.
35. Mahato, N., Sharma, K., Sinha, M., & Cho, M. H. (2018). Citrus waste derived nutra–/pharmaceuticals for health benefits: Current trends and future perspectives. *Journal of Functional Foods, 40*, 307–316.
36. Mamma, D., & Christakopoulos, P. (2014). Biotransformation of citrus by-products into value added products. *Waste and Biomass Valorization, 5*(4), 529–549.
37. Mandalari, G., Bennett, R. N., Bisignano, G., Saija, A., Dugo, G., Lo Curto, R. B., et al. (2006). Characterization of flavonoids and pectins from bergamot (Citrus Bergamia Risso) peel, a major byproduct of essential oil extraction. *Journal of Agricultural and Food Chemistry, 54*(1), 197–203.

38. Manthey, J. A., & Grohmann, K. (2001). Phenols in citrus peel byproducts. Concentrations of hydroxycinnamates and polymethoxylated flavones in citrus peel molasses [article]. *Journal of Agricultural and Food Chemistry, 49*(7), 3268–3273.
39. Moustafa, H., Youssef, A. M., Darwish, N. A., & Abou-Kandil, A. I. (2019). Eco-friendly polymer composites for green packaging: Future vision and challenges. *Composites Part B: Engineering, 172*, 16–25.
40. Muñoz-Labrador, A., Moreno, R., Villamiel, M., & Montilla, A. (2018). Preparation of citrus pectin gels by power ultrasound and its application as an edible coating in strawberries. *Journal of the Science of Food and Agriculture, 98*(13), 4866–4875. https://doi.org/10.1002/jsfa.9018
41. Murali, R., Karthikeyan, A., & Saravanan, R. (2013). Protective effects of D-limonene on lipid peroxidation and antioxidant enzymes in streptozotocin-induced diabetic rats. *Basic & Clinical Pharmacology & Toxicology, 112*(3), 175–181.
42. Naseri, A. T., Thibault, J. F., & Ralet-Renard, M. C. (2008). Citrus pectin: Structure and application in acid dairy drinks. In *Tree and forestry science and biotechnology*. Global Science Books.
43. Ndayishimiye, J., & Chun, B. S. (2017). Optimization of carotenoids and antioxidant activity of oils obtained from a co-extraction of citrus (Yuzu Ichandrin) by-products using supercritical carbon dioxide. *Biomass and Bioenergy, 106*, 1–7.
44. Nipornram, S., Tochampa, W., Rattanatraiwong, P., & Singanusong, R. (2018). Optimization of low power ultrasound-assisted extraction of phenolic compounds from mandarin (Citrus Reticulata Blanco Cv. Sainampueng) peel. *Food Chemistry, 241*, 338–345.
45. Nishad, J., Koley, T. K., Varghese, E., & Kaur, C. (2018). Synergistic effects of nutmeg and citrus peel extracts in imparting oxidative stability in meat balls. *Food Research International (Ottawa Ont.), 106*, 1026–1036. https://doi.org/10.1016/j.foodres.2018.01.075
46. Okino Delgado, C. H., & Fleuri, L. F. (2016). Orange and mango by-products: Agro-industrial waste as source of bioactive compounds and botanical versus commercial description—A review. *Food Reviews International, 32*, 1–14.
47. Panwar, D., Saini, A., Panesar, P. S., & Chopra, H. K. (2021). Unraveling the scientific perspectives of citrus by-products utilization: Progress towards circular economy. *Trends in Food Science & Technology, 111*, 549–562.
48. Rafiq, S., Kaul, R., Sofi, S. A., Bashir, N., Nazir, F., & Ahmad Nayik, G. (2018). Citrus peel as a source of functional ingredient: A review. *Journal of the Saudi Society of Agricultural Sciences, 17*(4), 351–358.
49. Reddy, S. G., Kirti Dolma, S., Koundal, R., & Singh, B. (2016). Chemical composition and insecticidal activities of essential oils against diamondback moth, Plutella Xylostella (L.) (Lepidoptera: Yponomeutidae). *Natural Product Research, 30*(16), 1834–1838.
50. Rezzadori, K., Benedetti, S., & Amante, E. R. (2012). Proposals for the residues recovery: Orange waste as raw material for new products. *Food and Bioproducts Processing, 90*(4), 606–614.
51. Rezzoug, S. A., & Louka, N. (2009). Thermomechanical process intensification for oil extraction from orange peels. *Innovative Food Science & Emerging Technologies, 10*(4), 530–536.
52. Rivas, B., Torrado, A., Torre, P., Converti, A., & Domínguez, J. M. (2008). Submerged citric acid fermentation on orange peel autohydrolysate. *Journal of Agricultural and Food Chemistry, 56*(7), 2380–2387.
53. Romero-Lopez, M. R., Osorio-Diaz, P., Bello-Perez, L. A., Tovar, J., & Bernardino-Nicanor, A. (2011). Fiber concentrate from orange (Citrus sinensis L.) bagase: Characterization and application as bakery product ingredient. *International Journal of Molecular Sciences, 12*(4), 2174–2186. https://doi.org/10.3390/ijms12042174
54. Rosa, A., Era, B., Masala, C., Nieddu, M., Scano, P., Fais, A., et al. (2019). Supercritical Co2 extraction of waste citrus seeds: Chemical composition. *Nutritional and Biological Properties of Edible Fixed Oils, 121*(7), 1800502.
55. Satari, B., & Karimi, K. (2018). Citrus processing wastes: Environmental impacts, recent advances, and future perspectives in total valorization. *Resources, Conservation and Recycling, 129*, 153–167.

56. Sáyago-Ayerdi, S. G., Brenes, A., & Goñi, I. (2009). Effect of grape antioxidant dietary Fiber on the lipid oxidation of raw and cooked chicken hamburgers. *LWT – Food Science and Technology, 42*(5), 971–976.
57. Scialabba, N., Jan, O., Tostivint, C., Turbé, A., O'Connor, C., Lavelle, P., et al. (2013). *Food wastage footprint: Impacts on natural resources*. Summary Report.
58. Sharma, K., Mahato, N., Cho, M. H., & Lee, Y. R. (2017). Converting citrus wastes into value-added products: Economic and environmently friendly approaches. *Nutrition, 34*, 29–46.
59. Singanusong, R., Nipornram, S., Tochampa, W., & Rattanatraiwong, P. (2015). Low power ultrasound-assisted extraction of phenolic compounds from mandarin (Citrus Reticulata Blanco Cv. Sainampueng) and lime (Citrus Aurantifolia) peels and the antioxidant. *Food Analytical Methods, 8*(5), 1112–1123.
60. Srinivasa, P. C., & Tharanathan, R. N. (2007). Chitin/chitosan—Safe, ecofriendly packaging materials with multiple potential uses. *Food Reviews International, 23*(1), 53–72. https://doi.org/10.1080/87559120600998163
61. Strano, M. C., Altieri, G., Allegra, M., Di Renzo, G. C., Paterna, G., Matera, A., & Genovese, F. (2022). Postharvest technologies of fresh citrus fruit: Advances and recent developments for the loss reduction during handling and storage. *Horticulturae, 8*(7), 612.
62. Suri, S., Singh, A., & Nema, P. K. (2022). Current applications of citrus fruit processing waste: A scientific outlook. *Applied Food Research, 2*(1), 100050.
63. Swamy, C. T., Sivadurga, K., Reddy, M. P., Marimuthu, G., Prashantkumar, C. S., Premkumar, C., & Purewal, S. S. (2023). Citrus waste: A treasure of promised phytochemicals and its nutritional-nutraceutical value in health promotion and industrial applications. In *Recent advances in citrus fruits* (pp. 395–426). Springer.
64. Tekgül, Y., & Baysal, T. (2018). Comparative evaluation of quality properties and volatile profiles of lemon peels subjected to different drying techniques. *Journal of Food Process Engineering, 41*, 12902.
65. Uerlings, J., Arevalo Sureda, E., Schroyen, M., Kroeske, K., Tanghe, S., De Vos, M., Bruggeman, G., Wavreille, J., Bindelle, J., & Purcaro, G. (2021). Impact of citrus pulp or inulin on intestinal microbiota and metabolites, barrier, and immune function of weaned piglets. *Frontiers in Nutrition, 8*, 650211.
66. Varmie, E. B., & Mamta Thakur, M. (2021). Utilization of citrus processing waste: A review. *Pharma Innovation, 10*(5), 682–697.
67. Villanueva-Suárez, M. J., Redondo-Cuenca, A., Rodríguez-Sevilla, M. D., & de las Heras Martínez, M. (2003). Characterization of nonstarch polysaccharides content from different edible organs of some vegetables, determined by Gc and Hplc: Comparative study. *Journal of Agricultural and Food Chemistry, 51*(20), 5950–5955.
68. Wang, L., Wang, J., Fang, L., Zheng, Z., Zhi, D., Wang, S., et al. (2014). Anticancer activities of citrus peel polymethoxyflavones related to angiogenesis and others. *BioMed Research International, 2014*, 453972.
69. Wang, L., Xu, H., Yuan, F., Fan, R., & Gao, Y. (2015). Preparation and physicochemical properties of soluble dietary fiber from orange peel assisted by steam explosion and dilute acid soaking. *Food Chemistry, 185*, 90–98.
70. Widmer, W., Zhou, W., & Grohmann, K. (2010). Pretreatment effects on orange processing waste for making ethanol by simultaneous saccharification and fermentation. *Bioresource Technology, 101*(14), 5242–5249.
71. Willats, W. G. T., Knox, J. P., & Mikkelsen, J. D. (2006). Pectin: New insights into an old polymer are starting to gel. *Trends in Food Science & Technology, 17*(3), 97–104.
72. Younis, K., Ahmad, S., & Malik, M. A. (2019). WITHDRAWN: Physicochemical properties and shelf life stability of meat products: Effect of mosambi peel powder. *Asian-Australasian Journal of Animal Sciences*. https://doi.org/10.5713/ajas.19.0451. Advance online publication.
73. Zema, D. A., Calabrò, P. S., Folino, A., Tamburino, V., Zappia, G., & Zimbone, S. M. (2018). Valorisation of citrus processing waste: A review. *Waste Management, 80*, 252–273.

Chapter 10
Valorization of Citrus Waste for the Synthesis of Value Added Products

Kenganora Mruthunjaya, Santhepete Nanjundiah Manjula, Himanshu Sharma, Amit Anand, and Madhuchandra Kenchegowda

10.1 Introduction

One of the most significant fruit crops in the world is citrus. Citrus fruit production has skyrocketed in the previous three decades, rising from an average of 62 million tons annually to around 100 million tons. Around the world, citrus is cultivated in over 100 nations, mostly in tropical and subtropical regions with ideal soil and climate for citrus growing [1]. The two major ways that citrus fruits are sold are as processed juice or as fresh fruit. One of the fruits that is grown, processed, and eaten the most in the world is citrus [2]. Citrus processing enterprises produce a great deal of by-products, which, if not properly treated and disposed of, seriously pollute the environment [3]. Nevertheless, these by-products represent a rich source of chemicals with additional value, such carotenoids, pectin, polyphenols, and essential oils. These compounds may be recovered and used in the culinary, cosmetic, and pharmaceutical industries to create products that have a variety of health advantages

K. Mruthunjaya (✉) · A. Anand
Department of Pharmacognosy, JSS College of Pharmacy,
JSS Academy of Higher Education and Research, Mysuru, Karnataka, India
e-mail: kmruthunjaya@jssuni.edu.in

S. N. Manjula
Department of Pharmacology, JSS College of Pharmacy, JSS Academy of Higher Education and Research, Mysuru, Karnataka, India
e-mail: snmanjula@jssuni.edu.in

H. Sharma
Teerthanker Mahaveer College of Pharmacy, Teerthanker Mahaveer University, Moradabad, Uttar Pradesh, India

M. Kenchegowda
Department of Pharmaceutics, JSS College of Pharmacy, JSS Academy of Higher Education and Research, Mysuru, Karnataka, India

A. Chauhan et al. (eds.), *Valorization of Citrus Food Waste*,
https://doi.org/10.1007/978-3-031-77999-2_10

[4]. Furthermore, by biotechnological means, these citrus by-products may also be recycled into other bio-products including enzymes, biofuels, and bio-polymers [5]. Citrus by-product valorisation solutions have therefore been created to address the growing demand from society for functional meals, fuels, and energy while also minimizing the adverse environmental effect [6]. Peels are generated in a huge amount as a by-product of processing citrus fruits; these peels do not contribute to the product's value since they are thrown away or disposed of. Because citrus peels contain a variety of physiologically active substances, including naturally occurring antioxidants such phenolic compounds, their potential usage as value-added products has been extensively researched [7]. Vitamin C, dietary fibre, potassium, folate, calcium, thiamine, niacin, vitamin B6, phosphorus, magnesium, copper, riboflavin, and pantothenic acid may all be found in abundance in citrus fruits [8]. Moreover, flavonoids, alkaloids, coumarins, limonoids, carotenoids, phenol acids, essential oils, and phytochemicals are among the secondary metabolites found in these fruits [9]. Currently, 92,088 billion tons of citrus are produced worldwide; 68.44% of this amount is consumed, and the remaining 21.51% is utilized to make other goods. After China and Brazil, India is the third-largest producer of citrus fruits [10]. The largest concentration of various citrus is found in North-East India. The citrus business generates a massive quantity of waste every year, usually in the shape of peels, seeds, and pulpy leftovers, all over the world [11]. The expanding demand for citrus goods is creating a need for ecologically acceptable and sustainable methods of disposing of citrus by-products. Although citrus trash was always considered a by-product, there has been a paradigm shift in recent years that sees it as a valuable resource with untapped potential. Citrus garbage is full of bioactive compounds that have several health-promoting properties, including as flavonoids, essential oils, and terpenes [12]. This technique, called "value-added product conversion," is a viable and sustainable way to transform discarded citrus into useful products using cutting-edge technologies. This tactic addresses the environmental problems linked to the disposal of citrus waste while also promoting a circular economy. Recycling citrus trash into value-added products follows the green chemistry maximization of resource efficiency and mitigation of environmental impact [13]. Researchers and industry companies are adopting a range of technical techniques, including fermentation, extraction, and biotechnological processes, to harness the bioactive components contained in citrus waste in order to develop high-value products. Furthermore, the value-adding of citrus garbage advances the global environmental agenda by promoting the use of renewable resources and reducing reliance on fossil fuels [14]. The development of novel extraction techniques and biorefinery technology has opened up new avenues for the manufacture of natural flavours, nutritional supplements, biofuels, and antioxidants from citrus waste [10]. In addition to the environmental benefits, citrus-producing nations can benefit economically from the value-adding of their waste. By converting trash into valuable commodities, stakeholders in the citrus industry have the potential to boost overall profitability, diversify revenue sources, and facilitate the development of a more resilient and sustainable agricultural sector [15]. In conclusion, the process of repurposing citrus trash provides a workable and all-encompassing way to deal with the problems

associated with citrus by products from an economic and environmental standpoint [16]. This approach creates a more circular and sustainable way to produce citrus by utilizing scientific, technical, and sustainable breakthroughs to transform garbage into money [17]. As research in this field continues to increase, the synthesis of value-added products from citrus garbage is projected to play an important role in determining the future of the citrus sector and sustainable waste management approaches [18]. This book chapter's main goal is to thoroughly examine and clarify the value-adding process of citrus trash as a sustainable and profitable source for producing high-quality products. The goal of this chapter is to provide readers with a comprehensive grasp of the techniques used in the extraction, transformation, and use of citrus waste elements for the production of high-value goods by means of a careful analysis of creative and environmentally responsible methods. This chapter aims to add to the body of knowledge surrounding citrus waste valorization by synthesising recent research findings and emerging trends [19].

10.1.1 Significance of Citrus Waste Valorization

The process of turning citrus trash-peels, pulp, seeds, and other by products from processing citrus fruit into valuable goods is known as "citrus waste valorization." By repurposing by-products that would otherwise be thrown away as garbage, citrus waste valorization reduces the environmental effect of the citrus processing industries. Because valorization repurposes these materials, it offers an environmentally benign alternative to landfill congestion caused by citrus waste [20]. Pectin, essential oils, bioactive substances, and dietary fibres are among the useful components found in citrus trash. In order to promote resource efficiency, valorization techniques seek to extract and use these components for a variety of uses. Through the process of transforming citrus trash into commodities like animal feed, pectin, essential oils, and biofuels, industries may now make more money from something that was previously regarded as a waste [21]. By lowering waste disposal costs and even generating new revenue streams, citrus waste valorization can save costs for the citrus processing sectors. Waste citrus can be used to produce bioenergy in the form of biofuels like ethanol or biomethane. This offers a green energy source in addition to lowering reliance on fossil fuels. Bioactive chemicals and dietary fibre with potential health advantages can be found in citrus trash. These ingredients can be extracted using valorization procedures and used in medications, supplements, or functional foods [1]. Essential oils containing aromatic chemicals are abundant in citrus peels. These oils can be extracted by the valorisation process and employed in the fragrance and taste industries, helping to make food additives, cosmetics, and perfumes. Citrus waste valorization closes the resource use loop, which is consistent with the tenets of the circular economy. It turns garbage into useful goods, transforming industrial processes into ones that are more circular and sustainable [22]. Agricultural methods can also profit from the value-adding of citrus waste. For instance, using leftover orange peels as animal feed supports sustainable agricultural methods. Research and innovation in the creation of new technology, procedures, and goods are stimulated by the investigation of citrus waste valorization, leading to improvements in waste management and sustainability. In conclusion, citrus waste valorization converts trash into useful resources, promoting a more resource-efficient and circular economy. This provides a comprehensive solution to environmental, economic, and social sustainability [23]. By doing so, it hopes to provide insights that can help practitioners, policymakers, and researchers find sustainable and resource-efficient solutions for waste management and product development [24].

10.2 Citrus Waste Composition and Generation

10.2.1 Types of Citrus Waste

The most widely grown citrus fruits worldwide are oranges, lemons, grapefruits, mandarins, and limes. Production increases annually as a result of rising consumer needs. The citrus processing industries generate a lot of garbage every year; just the peel waste accounts for almost half of the wet fruit mass. Citrus fruit processing produces a variety of wastes, such as solid, liquid, and distillery effluents. Solid trash includes things like residue, rags, peels, and sludge [25]. Because citrus is mostly employed in the food manufacturing process to make fresh juice or drinks with citrus flavours, peels, pulp, and seeds make up a sizable portion of the trash citrus produces annually. Nevertheless, a sizable amount of bioactive substances may be present in the seeds, pulp, and skins of many fruits and vegetables [26]. The recovery of vitamin C, phenolic compounds, and antioxidant activity from citrus fruit waste (grapefruit, orange, and lemon) was investigated. Ethanol extracts were made from the peel, entire fruit, and pulp containing seeds of each citrus fruit. The findings indicated that in each citrus variety, the peels exhibited greater levels of phenolic compounds, vitamin C, flavonoids, and antioxidant activity in comparison to their internal, discarded components (seeds and pulp) [27].

10.2.1.1 Citrus Peel

Due to its rich vitamin content and fragrant qualities, orange peel often regarded as citrus waste-can be used in a variety of ways. The colourful outer layer of a citrus peel, known as zest, is rich in essential oils and can enhance the flavour of food. It's frequently utilised for cooking, baking, and garnishing. Simmering citrus peel in a sugar syrup allows it to become candied. Cakes, confections, and sweets frequently use candied peel. It is possible to add flavour to beverages like teas, syrups, and spirits by using citrus peels. It contains natural oils that are suitable for use in DIY cleaning solutions. The oils leave a pleasant smell and have antibacterial qualities [28]. Vinegar and citrus peels can be used to make an all-purpose cleanser. Citrus peels contain essential oils that can be extracted and used in aromatherapy. Scents of citrus are frequently linked to feelings of energising and elevating. Citrus peels can be dried and used as a natural air freshener in sachets or potpourri. Because citrus peels have a strong aroma, you can spread them about plants or in gardens to keep animals and pests away [29]. Because they are high in nutrients, citrus peels can be put to compost bins to increase the nutritional content of the compost. It can be incorporated into DIY face washes and masks [30]. Citrus peels' natural acids have the ability to exfoliate. Because citrus peel extracts brighten skin, they are occasionally utilised in cosmetic goods. Because citrus peels have a strong scent, you can arrange them near windows or other entry points to help repel insects. Because of their high nutritional value, dried citrus peels have been the subject of some research on their usage in animal feed [31].

10.2.1.2 Citrus Seeds

To extract phytoconstituents from the seeds, a number of downstream processes must be optimised once the seeds are collected, washed, dried, and ground. Some of the steps in downstream processes include solvents, time, temperature, pressure, and particle size. Citrus seeds are extracted using a variety of methods, including as solvent extraction, cold pressing, ultrasonic-assisted extraction, and supercritical CO_2 extraction. Because they are considered to be useless, citrus seeds are usually thrown away as waste. Ammerman and Arrington state that 4.8% of dried citrus pulp typically contains seeds. Because these seeds are rich in protein, it is profitable to utilise them as a supplement for cattle. Citrus seeds' 30% oil content (by weight) further increases their potential for usage as biodiesel [32].

10.2.1.3 Citrus Pomace

The surplus material left over after processing citrus fruit to make juice or other products is called citrus pomace. Almost half of the citrus fruit's volume is lost during the industrial processing of citrus juice, and the massive amount of citrus pomace created poses major ecological issues. The pomace produced by the agro-fruit business when insufficient citrus is used has a detrimental effect on the environment and will cause large financial losses. Fruit pomace contains a multitude of health-promoting bioactive ingredients, such as phenolic compounds, polysaccharides, dietary fibre, natural antioxidants, carbs, and a host of other nutrients [33].

Table 10.1 List of citrus waste components with their chemical composition and uses

Components	Chemical composition	Uses	References
Citrus pulp	Cellulose, hemicellulose, pectin	Dietary fiber, bulking agent in food products	[35]
Citrus peel	Essential oils, flavonoids	Flavouring, antioxidant properties	[36]
Citrus seeds	Oil, protein, fiber	Source of oil for industrial use, protein	[37]
Citrus juice residue	Water, sugars, citric acid	Water retention, flavouring in animal feed	[38]
Citrus essential oils	Limonene, citral, linalool	Flavouring, fragrance in cosmetics	[39]
Citrus waxes	Citrus wax, fatty acids	Food coating, protective layer	[40]
Citrus pectin	Pectic substances	Gelling agent in food, thickening agent	[41]
Citrus bioflavonoids	Hesperidin, quercetin	Antioxidant, potential health benefits	[42]

10.2.2 Chemical Composition

The by-products produced during the processing of citrus fruits, including oranges, lemons, limes, and grapefruits, are referred to as citrus trash. The peels, pulp, seeds, and membranes that remain after juice or essential oils are extracted are examples of these by-products. Citrus by-products, with their complex chemical makeup, have drawn attention to their potential usefulness rather than being thrown away as garbage. Citrus waste has a different chemical makeup depending on the type of fruit, how it is processed, and how it is extracted [34]. However, citrus waste typically contains a range of useful bioactive compounds that is illustrated in the Table 10.1.

10.2.3 Quantities Produced in the Citrus Industry

The group of plants that provide citrus fruits is made up of several species from the genera citrus, fortunella, and poncirus. Lemons (Citrus limon), limes (e.g., *Citrus latifolia*, *Citrus aurantifolia*, and *Citrus limettioides*), grapefruits (*Citrus paradisi*), pummelos (*Citrus maxima*), kumquats (Fortunella spp.), trifoliate oranges (*Poncirus trifoliata*), and sweet and bitter oranges (*Citrus sinensis* and *Aurantium*) are among the main species. The citrus family produces the vast majority of commercially valuable fruits for human consumption. In small quantities, kumquats, or the Fortunella fruit species, are also useful. Besides, pomelos and grapefruits (*C. paradisi* and *C. máxima*) and other fruits (*Citrus medica*, *Citrus bergamia*, *Citrus myrtifolia* and Fortunella spp.) are also notorious [27]. The production of other citrus fruits is 13.90 million tonnes, whereas that of melons and grapefruits is 9.56 million tonnes. You can name China, Brazil, and India as the top three generating nations.

China produces a lot more tangerines and other citrus fruits, which is largely responsible for its high productivity. The world's largest producer of limes and lemons is India. Furthermore, pomelos and grapefruits (*Citrus paradisi* and *C. máxima*) are also well-known fruits, as are *Citrus medica*, *Citrus bergamia*, *Citrus myrtifolia*, and Fortunella spp. Melons and grapefruits are produced in 9.56 million tonnes, while other citrus fruits are produced in 13.90 million tonnes [43].

10.3 Challenges and Environmental Impact of Citrus Waste Disposal

10.3.1 Environmental Concerns

The waste that is produced during the preparation of citrus fruits presents many environmental issues. Peels, seeds, pulp, and other by products from the extraction and processing of citrus fruit juice are referred to as citrus waste. Environmental problems may arise if citrus trash is improperly disposed of in landfills. Due to its bulky nature and slow decomposition, citrus trash contributes to the production of methane, a powerful greenhouse gas. In addition to taking up valuable property, landfilling can produce leachate, which can contaminate nearby water and soil. Citrus waste may contaminate water sources if it is not properly managed. Citrus waste's high organic content can influence water quality by causing fertiliser runoff. Runoff has the potential to cause eutrophication, which can harm aquatic habitats by resulting in algal blooms [44]. Carbon dioxide and other greenhouse gases are released when citrus debris breaks down. The climate is changing as a result of these gases. Reducing these emissions can be achieved by using sustainable disposal techniques or by discovering new use for citrus trash. Incineration is one of the traditional disposal techniques that uses energy and can pollute the air. Some of these environmental effects can be mitigated by processing citrus trash in an energy-efficient manner or using it to generate energy [45]. Essential oils and bioactive compounds are among the beneficial components found in citrus waste that are frequently lost when the waste is improperly treated. Utilising cutting-edge technologies to recover these minerals can have a positive economic and environmental impact. The health of the soil may be impacted by improper citrus waste disposal or land application [46]. Citrus waste's acidity can change the pH of the soil, which can affect microbial activity and nutrient availability. Ecosystem health and agricultural productivity may suffer as a result of this. Waste citrus can be used as a source of bioactive compounds, to produce biofuels, to extract pectin, or as animal feed. Reducing the environmental impact of citrus waste can be achieved by finding economically viable applications for it. One eco-friendly way to handle citrus debris is to compost it. Citrus waste that has been properly composted can be applied as a soil conditioner to enhance soil fertility and structure. Citrus trash can be transformed from a waste product to a resource by extracting valuable components using advanced technology like biorefineries. Promoting environmentally friendly waste management techniques and encouraging industry adoption can be accomplished by governments and regulatory agencies. Citrus trash may be made into a useful resource and its environmental impact reduced by addressing these issues and putting sustainable policies in place [47].

10.3.2 Economic Implications

Orange, lemon, and grapefruit peels, pulp, and seeds are examples of citrus waste, which can have a variety of negative economic effects. The way garbage is handled and used can have a beneficial or negative effect on the economy. Municipalities and businesses may incur additional charges for waste management if they dispose of citrus waste in landfills. The improper handling of organic waste disposal might result in higher costs and environmental problems. Waste citrus can be used for a number of things, like making compost, biofuels, and animal feed. Anaerobic digestion and fermentation are two methods for producing bioenergy from the organic materials found in citrus trash. Citrus waste can be used to produce bioenergy, which will boost the renewable energy industry and have financial advantages. It is possible to extract useful substances from citrus trash, including pectin, essential oils, and bioactive chemicals [48]. Waste citrus fruit can be used as a soil conditioner or organic fertiliser in agriculture. Farmers may save money as a result, and sustainable farming methods may benefit. Research & development initiatives aimed at coming up with creative uses for leftover citrus peel have the potential to produce whole new markets and goods. Job creation and economic growth may be fuelled by this innovation [49]. Reducing the overall ecological footprint and promoting environmental sustainability can be achieved through the efficient use of citrus waste. Consequently, companies who implement eco-friendly policies may see an improvement in their reputation. Determining commercially feasible methods for handling citrus trash can improve the supply chain's overall effectiveness in citrus-producing areas. Stakeholders in the food processing, waste management, and agriculture industries may need to work together on this. Businesses may experience negative economic effects from increased environmental restrictions due to the expense of compliance. Utilising and managing citrus waste properly can help businesses comply with regulations and stay out of trouble. Market demand for companies that actively manage and use citrus waste in an eco-friendly manner may be driven by consumer preferences for sustainable and eco-friendly products. This may have an effect on market competitiveness and brand reputation [50].

10.3.3 Regulatory Framework

Environmental restrictions in several nations control the disposal of organic waste, including trash from citrus fruits. The goals of these laws are to safeguard ecosystems and stop pollution. Regarding the management, transportation, and disposal of citrus trash, there can be certain laws. These rules might specify how to manage waste in a way that has the least negative effects on the environment. Regulations concerning the use of citrus trash in agriculture may exist in some areas. Regulations may define acceptable practices and amounts for using citrus trash as an organic fertiliser or soil conditioner [51]. Regulations pertaining to the handling and elimination of biowaste, which encompasses organic waste items such as citrus peels, may be special to some jurisdictions. A circular economy may be supported by laws or efforts in some nations or areas. This might promote the use of citrus trash for bioenergy production, value-added product creation, or other sustainable uses. to get the most recent and accurate information possible. These authorities could be departments of agriculture, environmental agencies, or other pertinent organisations [52]. In India, different federal and state government agencies are likely to be in charge of enforcing waste management laws, which include those pertaining to citrus trash. These rules might address things like recycling garbage for sustainable development and economic gain, as well as environmental protection and waste disposal. India's waste management procedures are governed by environmental laws and regulations [53]. Guidelines for the secure disposal of waste from agriculture and food production may be among them. In India, FSSAI is in charge of policing and overseeing food safety. They may be responsible for regulations pertaining to trash and by products from food manufacturing, especially citrus waste. In India, environmental rules and regulations are implemented by Pollution Control Boards in each state. SPCBs have the authority to establish rules and recommendations for the handling of agricultural waste, including citrus trash. Guidelines for the handling of agricultural by products, such as citrus waste, may be found under the Ministry of Agriculture's NHB. Rules and regulations are in place in India for the handling of biodegradable trash. Because citrus garbage is organic, it may be governed by these regulations [54].

10.4 Valorization Approaches

10.4.1 Overview of Valorization

The process of obtaining value from waste materials often by turning them into energy or valuable products is referred to as valorization. Orange, lemon, and grapefruit waste are examples of citrus waste that can be economically and environmentally valued through a variety of valorization techniques. Essential oils found in citrus peels can be extracted and utilised in a variety of products, including food,

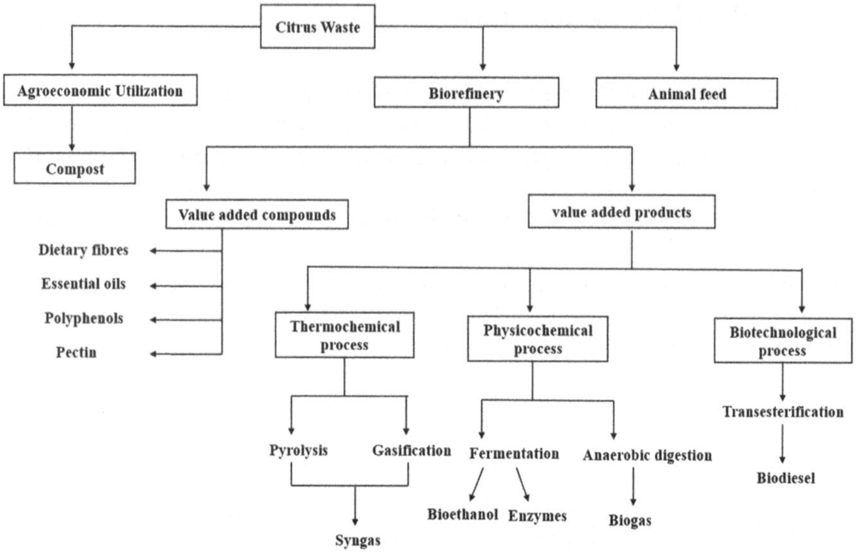

Fig. 10.1 A diagrammatic representation for the valorization of citrus waste

cosmetics, and pharmaceuticals. In addition to lowering trash and its negative effects on the environment, the value-adding of citrus waste helps businesses generate new sources of income. This is an illustration of how waste may be utilised creatively and sustainably to provide an invaluable resource. Any industrial processing operation that aims to reuse, recycle, or compost wastes, residues, usable products, or energy sources is referred to as "waste valorization." Generally, it manifests itself in one of the following ways: turning waste materials into raw materials; using finished or semi-finished products as a source of energy or as raw materials; using waste materials at various stages of the manufacturing process; and, finally, incorporating waste materials into finished products. Environmental regulations state that waste kinds utilised in valuation are generally not harmful and are not the same as residential trash. This is because waste valorization is much more homogeneous and involves a significant amount of production-related wastes and by-products' processing [55]. A flow diagram depicting the valorization of citrus waste into useful value added products and compounds can be seen in the Fig. 10.1.

10.4.2 Biochemical Conversion

Biochemical conversion is the process of using biological agents like bacteria or enzymes to convert biomass or organic resources into bio-based goods, fuels, or energy. Usually, biological reactions are used in this procedure to break down complex organic compounds into simpler ones. Harvesting the energy contained in biomass and transforming it into more beneficial forms, like biofuels or biochemicals, is the main objective of biochemical conversion. Citrus waste can be biochemically converted to produce useful goods like organic acids, biofuels, enzymes, and other bio-based commodities by using biological processes to convert its organic components. Peels, pulp, seeds, and other by products produced during citrus processing make up the majority of citrus trash [56].

10.4.2.1 Citrus Waste as Feedstock

The by-products produced during the processing of citrus fruits, including oranges, lemons, limes, and grapefruits, are referred to as citrus trash. Peels, seeds, pulp, and other remnants that are usually thrown away as waste can be considered by-products. But there's been a rise in interest in using citrus trash as a useful feedstock for a number of uses, from bioenergy to agriculture and beyond. A by-product of processing citrus, citrus pulp is frequently fed to livestock as an additional feed. It has a lot of fibre, carbs, and several important elements. Animal feed can also contain dried and processed citrus peel [57]. Citrus peel's strong essential oil content, however, might prevent some cattle from using it in big amounts. Pectin, a complex carbohydrate utilised as a gelling agent and stabiliser in the food industry, is found in substantial amounts in citrus peels. Pectin extraction from citrus peels has the potential to be a profitable technology. Citrus peels can be used to extract essential oils, which are then used in a variety of industries such food flavouring, cosmetics, and perfumery. Waste citrus, especially the pulpy leftovers, can be used to produce bioenergy. Anaerobic digestion and pyrolysis are two methods that can be used to turn citrus waste into biogas, bioethanol, or charcoal. Composting or processing citrus trash can result in the production of organic fertilisers [58]. Citrus trash is an excellent source of nitrogen and phosphorus due to its nutrient-rich composition. Packaging materials that degrade naturally can be made using pectin that is produced from citrus peels. These substances can lessen the harm that traditional plastic packaging causes to the environment. Bioactive substances found in citrus trash, such polyphenols and flavonoids, may be beneficial to health. Citrus trash is a valuable feedstock that helps minimise waste and reduce environmental impact. It also enables many businesses establish economically viable and sustainable procedures. Researchers and businesses are still looking for creative methods to integrate citrus trash into various applications and extract value from it [59].

10.4.2.2 Enzymatic Processes

The biological reactions and transformations that enzymes catalyse are referred to as enzymatic processes. Biological molecules called enzymes mostly proteins work as catalysts to quicken chemical reactions without getting consumed in the process. Enzymes are vital for many biological processes, and the correct operation of living things depends on their participation in metabolic pathways. Citrus trash is treated

Table 10.2 List of some enzymatic processes used in the valorization of citrus waste

Enzyme	Substrate	Process	Products	References
Pectinase	Pectin in citrus	Pectin hydrolysis	Release of sugars, improvement in juice extraction	[61]
Cellulase	Cellulose in peels	Cellulose hydrolysis	Production of glucose, potential for bioethanol	[62]
Amylase	Starch in citrus	Starch hydrolysis	Release of glucose and other sugars	[63]
Protease	Proteins in citrus	Protein hydrolysis	Production of amino acids, peptides	[64]
Lipase	Lipids in citrus	Lipid hydrolysis	Release of fatty acids and glycerol	[65]
Xylanase	Xylan in citrus	Xylan hydrolysis	Breakdown of hemicellulose, potential for bioethanol	[66]
Polygalacturonase	Polygalacturonic acid in citrus	Polygalacturonic acid hydrolysis	Breakdown of pectin, improved processing	[67]

enzymatically, which breaks down complex molecules in citrus by-products like peels and pulp into simpler and more valuable parts. The extraction of essential oils, bioactive compounds, and other valuable items from citrus trash is a common application of this procedure [60]. A summarised enzymatic process involved in the valorisation of citrus waste is represented in the Table 10.2.

10.4.3 Pyrolysis and Gasification

Two thermal processes that are used to transform organic materials into fuels, energy, and other useful products are pyrolysis and gasification. Despite their similarities, they are different in their qualities and uses. The process of heat breakdown without oxygen is called pyrolysis. Organic materials, like trash or biomass, are heated in an oxygen-free or low-oxygen environment to high temperatures (usually between 400 and 800 °C). Biochar, bio-oil, and syngas are the main outcomes of pyrolysis. Syngas, also known as synthesis gas, is a mixture of gases including carbon monoxide, hydrogen, and other gases. Biochar is a solid residue that is rich in carbon [68]. Bio-oil is a liquid mixture of organic molecules. Biofuels, soil-improving biochar, and other useful compounds can be made via pyrolysis. It is regarded as a flexible technology having uses in agriculture, waste management, and energy generation. Gasification is a thermochemical process that produces syngas, a combination of gases, from carbonaceous sources like coal, biomass, and organic waste. In contrast to pyrolysis, gasification needs controlled air or oxygen concentrations. The by-product of gasification is called syngas, and it is mostly made up of hydrogen, carbon monoxide, methane, and other gases. The composition of the syngas is determined by the feedstock and process parameters. Syngas is produced during gasification and is useful for a number of industrial processes. It may also be used as a fuel source for the production of electricity and as a precursor to create chemicals and synthetic fuels [69]. Energy generation and waste-to-energy conversion are two areas where gasification finds use. Pyrolysis occurs without the presence of oxygen, whereas gasification involves the carefully controlled addition of air or oxygen. The primary by product of gasification is syngas; pyrolysis yields biochar, bio-oil, and syngas. Pyrolysis often happens at lower temperatures (400–800 °C), whereas gasification typically happens at higher temperatures (800–1600 °C). While both methods can be applied to the production of energy and the disposal of waste, the specific products and their uses may vary [70]. In order to recover resources and produce sustainable energy, gasification and pyrolysis are viewed as promising methods that can reduce greenhouse gas emissions and effectively utilise organic materials. The thermochemical techniques of pyrolysis and gasification can be utilised to transform citrus waste into valuable commodities including biochar, syngas, and bio-oil. These procedures, which fall under the larger category of biomass conversion, can provide important bioenergy and bio-based products as well as aid in waste management. By heating biomass without oxygen, a process known as pyrolysis, organic materials are broken down into syngas, biochar, and bio-oil. This solid carbon-rich substance can be added to soil as a soil amendment to improve soil fertility and sequester carbon. a liquid that can be utilised as a precursor for other compounds or processed further to create biofuels. a hydrogen and carbon monoxide combination that can be used to produce electricity and heat. The partial burning of biomass in the presence of regulated air or oxygen content is known as gasification. It generates a syngas that has a multitude of uses [71]. Citrus waste may be reduced considerably via gasification and pyrolysis, both of which provide a sustainable waste management option. Carbon can be sequestered and soil quality can be enhanced by the biochar created. The production of syngas can be utilised to generate electricity and heat, which lessens the need for fossil fuels. It can be difficult to regulate the process parameters in order to maximise product yields. The feedstock and process parameters can affect the quality and content of the biochar, bio-oil, and syngas that are generated [72].

10.4.4 Chemical Extraction

Chemical extraction methods mainly include the use of solvents or other chemical operations to extract usable components from citrus waste. Citrus fruit peels, seeds, and pulp contain a range of bioactive compounds, such as flavonoids, essential oils, and pectin, which can be isolated and applied to a number of industries [73]. Some of the extensively used methods for the chemical extraction of citrus waste is shown in the Table 10.3.

Table 10.3 A summary of various methods of extraction of component from citrus waste with their applications

Methods of Extraction	Process	Applications	References
Steam distillation	Essential oils, which are abundant in citrus peels, can be extracted using steam distillation. In this method, steam is passed through the citrus peels, causing the essential oils to evaporate. The steam and essential oil vapor are then condensed back into liquid form.	The extracted essential oils can be used in the food, fragrance, and cosmetic industries.	[74]
Solvent extraction	Organic solvents, such as hexane or ethanol, can be used to dissolve and extract various compounds from citrus waste. The solvent is then evaporated, leaving behind the concentrated extract	This method is suitable for extracting lipophilic compounds like flavonoids and certain essential oils.	[75]
Supercritical fluid extraction (SFE)	Carbon dioxide in a supercritical state (high pressure and temperature) is used as a solvent to extract compounds from citrus waste. The supercritical CO_2 has both gas and liquid-like properties, making it an efficient solvent.	SFE is often used to extract essential oils and other bioactive compounds from citrus peels.	[76]
Acid hydrolysis for pectin extraction	Citrus peels are rich in pectin, a polysaccharide. Acid hydrolysis involves treating the peels with acids to break down the pectin into soluble fragments, which can then be separated and recovered	Pectin extracted through this method is commonly used as a gelling agent in the food industry.	[77]
Enzymatic extraction	Enzymes are used to break down the complex structures of citrus waste, facilitating the release of bioactive compounds. Enzymatic hydrolysis is often employed for extracting specific compounds without harsh chemical treatments	This method is suitable for obtaining specific compounds like flavonoids and can be applied in the food and pharmaceutical industries	[78]

10.4.5 Emerging Technologies in Citrus Waste Valorization

The goal of emerging technologies in citrus waste valorization is to make the most use of peels, pulp, and seeds-by products produced during the citrus processing process. These technologies seek to improve resource efficiency, lessen the impact on the environment, and turn citrus waste into products with added value. Essential oils from orange peels are extracted using sophisticated methods including microwave-assisted extraction and supercritical fluid extraction. In order to turn citrus waste into biofuels like ethanol and biogas, advanced bioconversion technologies, such as enzymatic hydrolysis and microbial fermentation, are being investigated. By separating these fibres for use in the food and pharmaceutical industries, novel extraction techniques hope to advance the creation of functional foods and nutritional supplements [79]. By maximising resource efficiency, reducing trash output, and effectively employing all parts of citrus waste, these strategies seek to establish a circular economy. Novel technologies are being researched to turn citrus waste into energy, such as pyrolysis and anaerobic digestion. These procedures can produce thermal energy, biogas, or charcoal, all of which support sustainable energy sources. Technologies to turn citrus trash into premium animal feed are being explored. This not only minimises waste but also gives the livestock and aquaculture sectors a reasonably priced, nutrient-dense substitute. There is hope for a more sustainable and profitable citrus processing sector thanks to the continuous research and development in these cutting-edge technologies for citrus waste valorization. As these technologies advance, they may be able to address environmental issues and provide value from materials that were previously thought of as waste [80].

10.5 Synthesis of Value-Added Products

10.5.1 Citrus Essential Oils

α-/β-pinene, sabinene, β-myrcene, d-limonene, linalool, α-humulene, and α-terpineol, which belong to the monoterpenes, monoterpene aldehyde/alcohol, and sesquiterpenes groups, respectively, are some of the main biologically active chemicals found in citrus essential oils (CEOs) [81]. These chemicals have antibacterial and other health-promoting qualities, such as anti-inflammatory, anti-cancer, and antioxidant, which offer enormous potential for use in food applications [82]. Essential oils derived from citrus fruits, including oranges, lemons, limes, and grapefruits, are highly valued natural goods. These oils have a variety of aromatic and functional qualities and are abundant in volatile molecules, mostly terpenes. Citrus essential oils are used to create value-added products by extracting, isolating, and converting these molecules into fresh, functional materials. The food and beverage industry make considerable use of citrus essential oils to add organic tastes and scents. Citrus essential oils are also used by perfume and fragrance manufacturers as important ingredients to create a broad variety of smells for personal care items and perfumes [83]. Because of their invigorating and revitalising aromas, citrus essential oils are widely used in aromatherapy. To encourage relaxation and stress alleviation, they can be added to candles, diffusers, massage oils, and bath products. Citrus essential oils have antibacterial qualities that make them appropriate for use in home cleansers. These oils can be added to natural cleaning solutions to enhance their antibacterial qualities and pleasing aroma. Because of their scent and possible skin advantages, citrus essential oils are frequently used in cosmetics and personal care products like lotions, creams, shampoos, and soaps [84]. Certain citrus

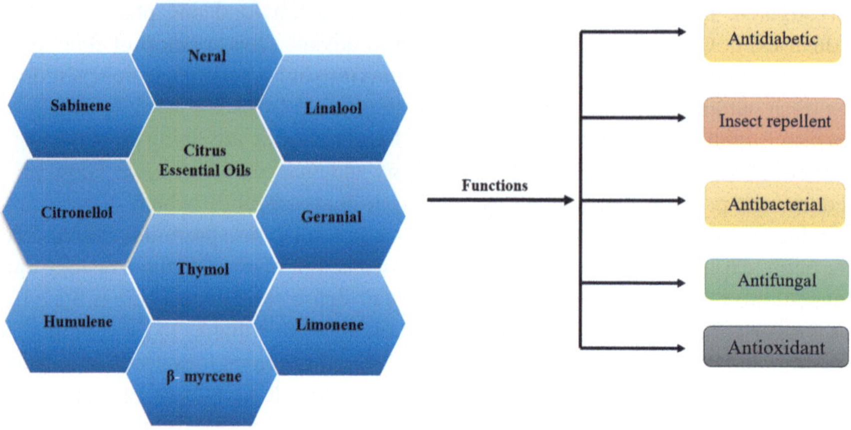

Fig. 10.2 Components of citrus essential oil and their functions

essential oils, such orange and lemon oils, have bioactive ingredients that may be advantageous to health. These oils have the flavour and possibly even medicinal qualities that make them suitable for use as natural flavourings in dietary supplements. Citrus essential oils can be utilised to create biopesticides and insect repellents because of their insecticidal qualities. When compared to their synthetic chemical counterparts, these natural substitutes may be less hazardous and more environmentally friendly. Citrus essential oils contain certain chemicals that can be isolated and processed to produce bio-based products and polymers. These materials could be used in coatings, packaging, and other industrial processes, resulting in more environmentally friendly and sustainable options [85]. A representation of various components present in the citrus essential oils with their function can be seen in the Fig. 10.2.

10.5.1.1 Extraction Methods

The techniques used to extract different chemicals, as well as essential oils from citrus and citrus trash, are associated with energy use and carbon dioxide emissions, which have an effect on the environment and expenses. As a result, using effective extraction techniques can maximise extraction, control waste, and increase the value of the by-product [86]. Conventional extraction techniques, such as Soxhlet extraction, maceration, infusion, solid-liquid extraction (SLE), and liquid-liquid extraction (LLE), necessitate longer extraction durations and considerable energy consumption. Furthermore, the heat-labile active chemicals in the extract may be harmed by the high working temperature used in the typical extraction procedure. As a result, the emphasis and specifications around extraction techniques are currently moving towards green extraction, which is environmentally benign and requires less energy and CO_2 emissions. For example, in liquid-liquid extraction (LLE), the use of bio-renewable solvents based on different alkane diols has been evaluated as environmentally friendly alternatives to separate bioactive terpenoids from terpenes to obtain EOs during downstream processing [87]. The evaluation of greenhouse gas (GHG) emissions associated with the biorefinery of citrus processing waste (CPW) found that GHG emissions were reduced by 134% and 77%, respectively. In order to operate the extraction process, modern and unconventional techniques like critical pressure, supercritical fluid (liquefied CO_2), temperature, continuous electricity supply, pumps, pressure containers, and highly sealed vessels, as well as high maintenance, microwave, and ultrasound generating equipment, may need a lot of energy. Nevertheless, these techniques use less solvent, produce an extract of higher quality, and increase extraction yield in a shorter amount of time [88]. Supercritical fluid extraction (SFE), steam explosion, ultrasound-assisted extraction (UAE), and microwave-assisted extraction (MAE) are a few of the green extraction techniques. The most popular extraction techniques among them for essential oils and other natural products derived from citrus species are UAE and MAE. These techniques need less energy and produce less CO_2 emissions when used quickly and easily. In a similar vein, supercritical fluids function at low to moderate temperatures and pressures without producing chemical residue and are harmless and non-flammable [89]. Furthermore, it was discovered that the most promising hybrid techniques were the pairings of UAE with a few cutting-edge methods as the microwave technique, instant-controlled pressure drop technique (DIC), and SFE. Aside from this, wet citrus peel waste has also been subjected to microwave-assisted hydro-distillation (MAHD) in order to extract essential oils. Furthermore, microwave hydro-diffusion and gravity (MHG) is an improved technique over traditional methods since it doesn't require the use of solvents, has a quick extraction time, and utilises the combined forces of Earth's gravity and microwave radiation [90]. Utilising microwave and ultrasonic techniques, solvent-free extraction methods have been effectively utilised to extract essential oils (EOs) via hydrothermal processing. Moreover, cavitation-based extraction (CE) techniques shorten extraction times while improving the quantity and quality of the extracted material. UAE, negative pressure cavitation (NPC) extraction, and hydrodynamic cavitation extraction (HCE) are examples of cavitation-based extraction (CE) techniques. In order to create HCE, a liquid is often pumped via Venturi tubes or orifice plates, creating a pressure gradient in the flow that encourages the formation, expansion, and collapse of microbubbles. Orange peel trash is managed and other food waste is processed using large-scale HCE technologies. Improved yields, scalability, and performance have all been demonstrated by HC-based methods. Additionally, solar hydro-distillation (SSD), a low-energy, environmentally friendly method based on solar energy, has been developed to extract essential oils [91].

10.5.1.2 Applications

CEOs are used in a variety of industries, including cosmetics, medicines, food formulation, and packaging. Nonetheless, the applications of CEOs in relation to food safety, packaging, and preservation were the main emphasis of this study. Commercial antimicrobial compounds are employed in the management of food spoilage and contamination. Growing health concerns have focused emphasis on natural antimicrobials such essential oils derived from plants. Antimicrobial and food preservation qualities of EOs and their constituents protect against a wide range of infections [92]. Various dangerous insects can be effectively controlled using conventional chemical pesticides. On the other hand, persistent and unchecked pesticide usage may cause these insects to adapt and develop pesticide resistance. Chemical-based pesticides usually affect the ecosystem in a perishable way, endangering not just the nontargeted creatures but also the entire food chain. Since citrus plants have therapeutic qualities, they could take the place of traditional chemical pesticides in this situation. Therefore, bioactive compounds of citrus, particularly CEOs, can be employed as an alternative to insecticides generated from synthetic and inorganic chemicals [93]. Because germs are become more resistant to antimicrobial medications, the effectiveness of antibiotics in treating human pathogenic diseases is currently declining. Studies on the antibacterial activity of CEOs have produced encouraging results about the development of plant-based antibiotics [79]. The antibacterial efficacy of the peel extract of citrus fruits (EO) from both matured and un ripened citrus fruits was examined against four human pathogenic bacteria: *Bacillus subtilis*, *Salmonella typhimurium*, *Staphylococcus aureus*, and *Escherichia coli*. Fresh fruits and vegetables, cereals, processed foods, and other food items can all become contaminated by mould, which can result in both quantitative and qualitative losses [94]. Consequently, CEOs have the potential to be utilised as a natural antifungal agent to reduce fungal growth and contamination while prolonging the shelf life of different foods. CEOs are fungicidal agents with a wide range of effects. It was discovered that *C. sinensis's* EO was effective against *Aspergillus niger*, Eutypa sp., *Botryospaeria dothidea*, and *Fomitiporia mediterranea*-three fungal plant diseases that infect grapevines—were controlled by lemon essential oil (*Citrus limon*) [95].

10.5.2 Biofuels

10.5.2.1 Biodiesel

Transesterification is a method that can be used to turn citrus waste into biodiesel. Peels, pulps, and other by products from processing citrus fruits are referred to as citrus trash, because they contain a lot of oil. Biodiesel, an environmentally acceptable and renewable substitute for regular diesel fuel, can be made from this oil. After the oil from the waste citrus is extracted, it is reacted with an alcohol (usually methanol or ethanol) in the presence of a catalyst. As a by-product of this chemical process, called transesterification, glycerol and biodiesel are produced. In a variety of applications, including power generation and transportation, conventional diesel fuel can be replaced with biodiesel made from citrus waste. The generation of biodiesel from citrus trash has various advantages. It offers an environmentally friendly and waste-free method of using agricultural by products. Furthermore, compared to conventional diesel, biodiesel made from citrus waste often emits fewer greenhouse gases, which helps with efforts to combat climate change [96].

10.5.2.2 Bioethanol

The trash that is produced during the processing of citrus fruits can be used as a feedstock to produce bioethanol. As a substitute for conventional fossil fuels, bioethanol is an environmentally acceptable and sustainable fuel. Peels, pulp, and seeds from citrus fruits like oranges, lemons, and grapefruits are considered citrus trash. Because these by products are frequently thrown away as waste, environmental problems arise. Pre-treatment procedures are used to the gathered citrus trash in order to reduce complex carbs to simpler sugars. This may entail the use of chemical, physical, or enzymatic techniques to speed up the fermentation process that follows. After the citrus waste has been pre-treated, it is fermented, a process that turns the sugars into carbon dioxide and ethanol by means of microorganisms called yeast. Because it transforms the biomass into alcohol, this stage is essential to the synthesis of bioethanol [97]. A combination of water and ethanol can be found in the fermentation broth that results. Ethanol is concentrated for use as fuel by distillation, which separates it from water. Molecular sieves or other dehydration methods could be used as further steps to eliminate any leftover water and improve the ethanol's purity. The end result is bioethanol, a clean-burning, renewable fuel that may be used in flexible fuel cars either as a stand-alone fuel or as an addition to petrol. Citrus processing facilities can generate additional cash, reduce waste disposal concerns, and contribute to a more environmentally friendly and sustainable energy sector by using citrus waste for bioethanol production. Furthermore, as compared to fossil fuels, bioethanol made from citrus waste helps reduce greenhouse gas emissions [98].

10.5.3 High-Value Chemicals

10.5.3.1 Limonene

The presence of limonene in CWs is the primary barrier to their use as a substrate for the production of biogas. This ingredient lowers the biogas yield and is extremely harmful to microorganisms that break down food. Strong inhibition of ethanol synthesis by microbes is exhibited by limonene. As a result, before beginning the fermentation or digesting process, this component needs to be isolated from the CWs [99]. It should be noted that although hydrolysis by chemicals or enzymes is required to convert these polymers to sugars, which are subsequently fermented by bacteria, such as baker's yeast, to ethanol, digesting bacteria can hydrolyse the carbohydrate polymers in CWs and ultimately convert them to biogas. Prior research has established two different enzymatic hydrolysis-based methods for producing ethanol from CWs. In the first method, cellulose, pectinase, and β-glucosidase were used to hydrolyse CWs. Next, fermentable sugars were fermented to produce ethanol, and limonene was extracted from the hydrolysate by filtering it out. In the second option, steam stripping was used to partially liberate the limonene. Then, a simultaneous saccharification and fermentation (SSF) technique was used to hydrolyse and ferment carbohydrate polymers into ethanol. Non-fermentable sugars and solid polymer waste were dried and used as calf feed in both approaches. The sluggish rate of hydrolysis reactions and the high cost of enzymes make it difficult to apply these alternatives [100]. Based on the fact that steam condensation caused the slurry of CW to dilute by 65.30% during hydrolysis, it can be computed that the flashing process in the flash drum eliminated 99% of the limonene content of the slurry. The recovered limonene can be obtained by condensing the flash drum's vapour output. The hydrolysate's residual limonene concentration (0.0035% w/v) is less than the *S. cerevisiae* minimum inhibitory limonene concentration of 0.01% w/v [101].

10.5.3.2 Pectin

Citrus peel contains three types of soluble sugars: glucose, fructose, and sucrose. The polysaccharides found in the cell walls of all citrus peels are pectin, cellulose, and hemicellulose. The main ingredients of pectin and hemicelluloses are galactose, arabinose, and galacturonic acid; rhamnose, glucose, and xylose are far less common. The two main parts of citrus peels are called flavedo and albedo. The primary source of pectin is the albedo. Methoxylated polygalacturonic acid makes up pectin. Mature fruit typically contains 3–7% pectin on a dry weight basis and 0.1–1.1% pectin on a fresh weight basis [102]. Citrus fruits are a good source of soluble dietary fibre that can be used in milk beverages, juices, desserts, and other fillings due to their moderately high pectin concentration and low-calorie value. Under the right circumstances, pectin can readily create gels with sugar and acid. It is appropriate for use in the making of jelly. Pectin has been applied as coating and bulking agents, emulsifiers, chelators, water-soluble biodegradable films, and viscosity modifiers. In the commercial setting, pectin is extracted by heating diluted acid to a pH of 2 for 2–4 hours. It is necessary to precipitate the extracted pectin with ethanol or isopropyl alcohol [103]. The production of bioethanol and some value added products are represented by a flowchart as shown in the Fig. 10.3.

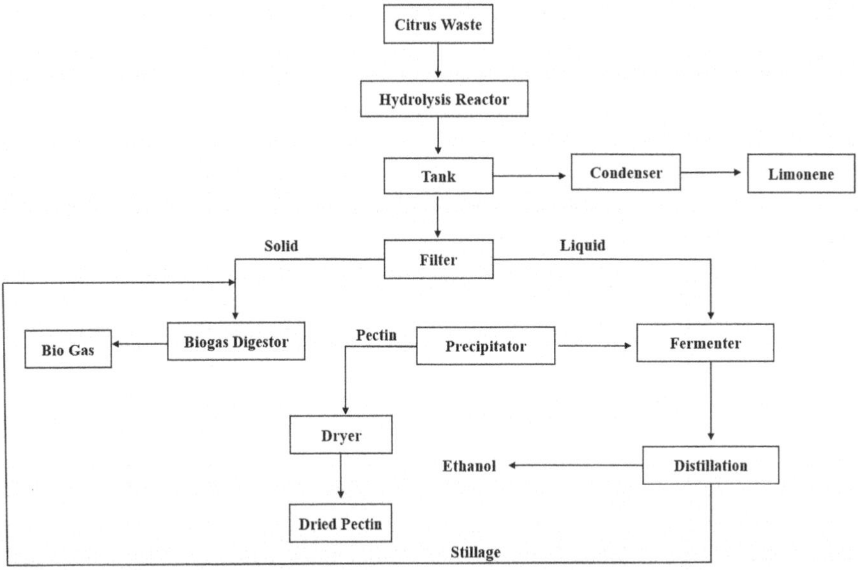

Fig. 10.3 A Process flow diagram for the production of bioethanol, biogas and limonene from citrus waste

10.5.3.3 Citric Acid

This tricarboxylic acid has six carbons and is produced as an intermediary in the tricarboxylic acid cycle (TCA). Initially, it was separated from lemon juice. About 60% of the total soluble solids in the edible portion of a lemon are made up of citric acid. Citrus juice contains free form of citric acid. By adding calcium oxide or lime, it is possible to extract citric acid from juice or pressed juice made from pith and pulp waste. The latter precipitates out as calcium citrate after reacting with citric acid. This is gathered by filtering, and then sulfuric acid is added to extract citric acid from its calcium salt. With a growing trend in demand and consumption, the yearly output of citric acid worldwide has surpassed 1.4 million tonnes in recent years [104]. Since *Aspergillus niger* is used in fermentation, citric acid is produced industrially. Ninety percent of the entire supply of citric acid worldwide is produced with a fermentation process. Solid state fermentation (SSF) and submerged liquid surface fermentation are the two most often used fermentation processes for producing citric acid from citrus waste. The latter method, which uses *A. niger*, has been used more frequently. There has been evidence in recent years that the submerged fermentation process is not as effective as the SSF technology. The primary benefits of the former are increased aeration, which facilitates O_2 and CO_2 exchange between the gas and substrate matrix, and reduced water content, which results in lower levels of water activity [105]. In addition, the SSF approach demonstrated its versatility by requiring only sterilisation and no additional nutrition or treatments. To recover refined citric acid, lime and then sulfuric acid are added to the crude citric acid that is created during fermentation. Because citric acid does not undergo phase shift, a sizable volume of pure citric acid can be produced using the streamlined production methods. The production of highly purified citric acid necessitates the application of further extraction and purification processes, such as membrane separation, solvent extraction, adsorption and ion exchange, and spray techniques. An advantage of solvent extraction methods over chemical procedures is their ability to reduce the need for sulfuric and lime acids. The advantages of adsorption and ion exchange processes are minimal regeneration consumption, high capacity, specificity, and rapid recovery. Furthermore, no by products, such as calcium sulphate, are produced using these methods [106]. However, using these techniques has certain drawbacks, such as the expensive cost of analytical grade chemicals due to the significant amount of desorbing solvent required, which dilutes the recovered citric acid and produces large amounts of waste liquid [107]. Membrane separation techniques make use of a thin membrane that permits solvent and solute molecules to move selectively over it. Among the frequently used membrane separation techniques are reverse osmosis, nanofiltration, ultrafiltration, and electrodialysis. It is thought that the spray extraction method purifies the citric acid that is collected as a by-product of fermentation. To get rid of contaminants, mostly sugars and other organic acid by products like malic and oxalic acid, it uses organic solvents like ethanol or acetone. Fractional precipitation of the residual sugars at a low pressure of about 20 bars is another step in the purification process. Whereas oxalic acid stays in the solution because it can only precipitate at pressures more than 120 bars, both malic and citric acids precipitate in the same pressure range. By catalysing oxidation through its chelating activity towards metal ions, citric acid demonstrates antioxidant capabilities. Moreover, food deterioration is reduced by the chelation of metal ions, which prevents bacteria from growing on the substrate [108].

10.5.4 Animal Feed and Fertilizers

Citrus waste is produced during the processing of citrus fruits, such as grapefruits, oranges, and lemons. It can be used in a variety of ways, such as fertiliser and animal feed. By minimising waste, this strategy lessens its negative effects on the environment and advances the adoption of sustainable agricultural techniques. After the juice is extracted, the pulp that is left behind can be dried and added to animal feed as a high-fiber ingredient. It can be incorporated into ruminant diets because it is high in carbs. For animal feed, dried citrus peel can be an excellent source of fibre and energy. It has essential oils in it that might be antibacterial and good for the health of animals. To make handling and feeding citrus trash easier, it can be pelletized. Additionally, pelleting might lengthen the product's shelf life [109]. Composting citrus waste, such as peels and pulps, can produce organic fertiliser. The resulting compost is high in organic matter, which helps to improve soil structure and gives plants nourishment. Mulch made from finely chopped citrus trash can be applied to plants. It distributes nutrients into the soil, inhibits the growth of weeds, and helps hold on to soil moisture as it breaks down. Anaerobic digestion of citrus trash can yield biogas, a sustainable energy source. Digestate, a by-product of this process, is still usable as a fertiliser [110]. Waste citrus can also be utilised to make biofuels like ethanol. This offers another way to lessen the negative effects of citrus waste on the environment. Citrus waste contains some components, like flavonoids, that can be added to animal feed additives and act as natural antioxidants. It should be emphasised that suitable processing techniques should be used to eliminate any potential pollutants, such as pesticides or residues from post-harvest treatments, before employing citrus trash in any application [111].

10.5.5 Other Bioproducts

10.5.5.1 Citrus Molasses

10–15% of the soluble particles in the citrus trash are found in the liquid portion. These soluble solids include between 50% and 70% fermentable or otherwise perishable carbohydrates. Because of this, the liquor has a biological oxygen demand (BOD) of 40,000–100,000 ppm. If left untreated, this can seriously pollute the environment as it can seep through the ground, reach lakes and the subterranean water table, and weaken the aquatic ecosystem [98]. Molasses can be made from this liquor portion of the leftover citrus trash. The thick, viscous liquid known as molasses is produced as a by-product of extracting citrus juice and has a huge ranging from dark brown to nearly black. It is a significant by product of processing citrus trash. It tastes almost indigestible and is extremely bitter. It is a desirable place to find limonin glucosides. It is utilised as a fermentation substrate and has a high sugar content (60–75%). It is either added back into dried citrus pulp or used as a raw material in distilleries. Other uses for it include combining it with citrus pulp to make cow feed, adding it to grass silage to make animal feed, or combining it with other nutrients to make mixed feed products. The extraction and purification of molasses from citrus wastes collected from juice processing factories could boost its commercial worth because it is not artificially synthesised or commercially available [112].

10.5.5.2 Dietary Fibers, Carbohydrates, and Sugars

Citrus waste is a significant source of dietary fibres, which are divided into two main categories: soluble and insoluble fibres. Soluble fibres make up between 50% and 60% of cellulose and hemicelluloses. Pectin, gum, mucus, and some cellulose are examples of soluble fibres. Conversely, lignin, hemicelluloses, and cellulose are examples of insoluble fibres. Complex colloidal carbohydrates known as pectic compounds are found in the spaces between neighbouring cell walls in the middle lamellae of plant tissues. Galactose, free acid radicals of galacturonic acid, and galacturonic acid itself, particularly those with large molecular weights. The polygalacturonic acids' free carboxyl groups can be partially neutralised by one or more bases, or partially esterified by –CH3 groups [113]. Citrus fruits contain a variety of pectic substances, such as pectic acid (colloidal polygalacturonic acid without methyl ester groups), pectin (water-soluble pectinic acids with varying amounts of methyl ester), and protopectin (water-insoluble pectic substances) [114]. A heterogeneous polysaccharide structure of linear α (1→4)-linked D-galacturonic acid units makes up pectin. These hydrolyse to produce pectinic acid. The latter are methyl ester-containing colloidal polygalacturonic acids. Colloidal polygalacturonic acids devoid of methyl ester groups are known as pectic acids. Alcohol precipitation and weighing, base-induced demethylation, and precipitation using calcium salts are the methods used to quantify pectic acids. The amount of a coloured complex produced with carbazol and sulfuric acid is used to quantify pectic compounds. One of the best sources of pectic compounds is citrus fruits [115].

10.5.5.3 Xanthan Gum

Xanthan gum is a type of biological polymer that has a molecular weight of several million Dalton and is obtained from plants, seeds, and seaweeds. It is commonly referred to as gum. It is a hetero-polysaccharide in structure, made by the bacterium *Xanthomonas compestris*. Since it was initially made available for purchase in the 1960s, over 30,000 tonnes are produced year worldwide. Mainly used as a stabiliser and enhancer of viscosities, xanthan gum is used in the culinary, pharmaceutical, and petrochemical industries. Depending on the growth conditions and the microorganism utilised, the amount of acetyl and pyruvic contents in xanthan gum can change, resulting in polymer solutions with varying molecular weights, compositions, and rheological behaviours. Depending on the Xanthomonas species, additional components include glucose-glucuronic acid, mannose, pyruvate, acetate, galactose, etc. The media composition, which is influenced by temperature, pH, dissolved oxygen, and other factors, determines the pyruvate content [116]. Xanthan gum with a high pyruvate percentage (about 4–4.8%) thickens more readily than gum with a lower pyruvate content (2.5–5%). However, enhanced oil recovery (EOR) makes use of xanthan that is free of pyruvate. According to reports, they conducted a comparative analysis of four distinct fractions of citrus waste—whole citrus waste, pectic extract, hemi cellulosic extract, and cellulosic extract—as substrate material for the fermentation process of producing xanthan gum. Instead of using glucose or sucrose, citrus waste is thought to be a cheap and inexpensive substrate to generate the final xanthan product at a minimal expanse. It was found that the water-soluble components of the citrus waste, such as pectin, organic acids, and simple carbohydrates, could be easily transformed into xanthan. However, because complex carbohydrates like hemicellulose and cellulose are less biodegradable than whole citrus trash, the amount of xanthan they produced was reduced by 36% and 60%, respectively [117].

10.5.5.4 Organic Acids

Because of the high concentration of soluble organic acids and sugars in their juice, citrus fruits are often referred to as acid fruits. Citric and malic acids are the primary sources of acidity. Apart from these two, there are other acids as well, such as tartaric, lactic, succinic, benzoic, oxalic, and formic. The combination of citric and malic acids, along with the salts found in citrus fruits, creates a buffering system in juice that prevents the acidity (pH) from changing much when diluted [118].

10.6 Techno-Economic Analysis and Feasibility

10.6.1 Cost-Benefit Analysis

When it comes to citrus trash, a cost-benefit analysis (CBA) compares the expenses of disposal or treatment against the possible gains from using the garbage for different purposes in order to determine whether controlling and using it is economically feasible. The expenses related to gathering and moving citrus trash from processing plants to disposal or utilisation centres is calculated. Analysis of the expenses is done which is associated with handling or processing citrus waste, which may incorporate techniques like anaerobic digestion, composting, or chemical extraction [106]. Determine the potential energy output and economic value of using citrus trash for bioenergy production through procedures such as anaerobic digestion or combustion. The advantages of applying citrus trash as a fertiliser or soil amendment, considering increases in crop yields and soil quality analysis is recommended. This could entail projecting energy, extracted chemical, or cost savings from increased soil fertility market pricing [107]. Establish the time frame for evaluating the costs and benefits. To decide whether to move further with the citrus waste management or utilisation project, a specified decision criterion is applied. It is important to keep in mind that a thorough cost-benefit analysis should consider both qualitative and quantitative aspects. It gives stakeholders a methodical way to assess if managing citrus waste is economically feasible and supports them in making defensible decisions [119].

10.6.2 Economic Viability

Citrus waste utilisation is dependent on a number of factors, including market demand, technological advancements, operational size, and regional concerns. Peels, pulp, and seeds are examples of citrus waste that can be recycled in a variety of ways to provide value and support sustainability. Essential oils found in citrus peels are extensively utilised in the food, beverage, and fragrance sectors. It can be economical to extract essential oils from citrus trash. Animal feed can contain both citrus peels and pulp. Citrus trash is a good source of nutrition for poultry and cattle feed. Waste from citrus fruits can be fermented or anaerobically digested to produce biofuel. This may aid in the creation of renewable energy [120]. Citrus peels can be used to extract pectin, a natural gelling ingredient that is employed in the food business. This has the potential to be profitable, particularly when used as a thickening agent for jams and jellies. Additional revenue streams can be generated by the production of goods utilising citrus extracts, such as cleaning products, dietary supplements, and cosmetics. Composting citrus waste can produce organic fertiliser. This can improve the health of the soil by being sold or utilised for on-site farming. Citrus-based products can be utilised to create biodegradable packaging or plastics, meeting the increasing need for environmentally acceptable substitutes. An extra source of income can be generated by using citrus waste to make juices, smoothies, or flavoured beverages [121].

10.6.3 Market Trends and Potential

Globally, there is an increasing focus on circular economy and sustainability measures. Waste from citrus fruits, such as peels and pulps, can be recycled to make goods with additional value, reducing waste and maximising resource use. Citrus waste is increasingly being used in the creation of dietary supplements, natural extracts, and functional food ingredients as consumer knowledge of nutraceuticals and functional foods grows. Packaging that degrades naturally can be made using materials derived from citrus, such as pectin that is produced from citrus peels. Citrus waste-derived materials provide a viable replacement for conventional packaging in light of the growing need for environmentally friendly packaging options. Anaerobic digestion and bioethanol synthesis are two methods that can be used to produce bioenergy from citrus trash. This is in line with the need to discover sustainable alternatives to conventional energy sources and the growing interest in renewable energy sources. Animal feed can benefit from the addition of citrus trash as a source of nutrients and fibre. It can also be applied to agriculture as a fertiliser or soil conditioner. These applications help to improve agricultural practices' circularity and lessen their negative effects on the environment. The personal care and cosmetics industries employ citrus essential oils and discarded citrus extracts. These organic components are prized for their pleasant scents and possible cosmetic advantages [122].

10.7 Challenges and Future Perspectives

Opportunities and difficulties arise when citrus waste is valued for the creation of products with additional value. It is necessary to develop effective techniques for removing valuable chemicals like flavonoids, pectin, and essential oils from citrus trash. It is essential to increase extraction efficiency while reducing the use of energy and resources. It can be difficult to guarantee the consistency and calibre of goods that are synthesised from citrus waste. It is crucial to develop technologies that will improve the generated products' overall quality and shelf life. The content of citrus waste differs according on the type of citrus fruit, how it is processed, and seasonal changes. To ensure that the manufacturing of value-added products is profitable, cost-effective techniques for valorization must be developed [123]. The main goals of research should be to increase productivity and optimise operations. It might be difficult to find and develop a market for products made from citrus trash. To promote the value of such items, effective marketing methods and customer education are required. Citrus waste-derived goods might compete with well-known brands. To get market recognition, citrus waste-derived products must either carve out a specialised market or exhibit excellent qualities [124]. Value maximisation seeks to make use of citrus waste, but waste production from citrus fruit processing should also be kept to a minimum. This implies maximising the extraction procedures and determining what to do with the leftover waste. It is imperative to evaluate and reduce the carbon footprint linked to the process of valorization. This entails maximising the use of energy, transportation, and other elements that raise greenhouse gas emissions. Complying with food safety, quality, and environmental regulations is essential to the effective implementation of citrus waste valorisation methods [125]. Future perspectives anticipate that these difficulties will be addressed by continued research and innovation. Working together, government agencies, business, and academics can hasten the creation of long-term, financially sound strategies for monetizing citrus waste. Technological developments combined with a heightened consciousness of environmental sustainability could lead to a more effective and extensive use of citrus waste in the creation of valuable products [126].

10.8 Conclusion

In conclusion, the valorization of citrus waste for the synthesis of value-added products represents a promising and sustainable approach to address both environmental and economic challenges. This innovative utilization of citrus by-products not only reduces the environmental impact associated with waste disposal but also taps into the untapped potential of these resources. The synthesis of value-added products from citrus waste not only adds economic value to what was once considered as waste but also contributes to the development of a circular economy. Through various extraction and processing techniques, researchers and industries have demonstrated the ability to transform citrus waste into valuable compounds such as essential oils, bioactive compounds, pectin, and biofuels. These products find applications in various industries, including food and beverage, pharmaceuticals, cosmetics, and bioenergy. The valorization of citrus waste not only offers economic benefits by creating new revenue streams but also aligns with the growing demand for sustainable and eco-friendly practices in production and consumption. However, successful implementation of citrus waste valorization requires collaborative efforts from researchers, industries, and policymakers. Technological advancements, coupled with supportive regulatory frameworks, can further enhance the efficiency and scalability of citrus waste valorization processes. Additionally, raising awareness among consumers about the environmental and economic benefits of utilizing citrus waste can drive demand for products derived from these sustainable practices. In essence, the valorization of citrus waste embodies a holistic approach towards sustainability, promoting the efficient use of resources and minimizing environmental impact. As we continue to explore innovative solutions to address global challenges, the valorization of citrus waste stands out as a tangible example of turning a waste stream into a valuable resource, contributing to a more sustainable and circular future.

Acknowledgments The authors express heartfelt gratitude towards the leadership and management of JSS College of Pharmacy, JSS Academy of Higher Education & Research (JSS AHER), Mysuru, Karnataka, India.

Competing Interest Statement The authors declare no conflict of interest.

Disclaimer None.

References

1. Liu, Y., Heying, E., & Tanumihardjo, S. A. (2012). History, global distribution, and nutritional importance of citrus fruits. *Comprehensive Reviews in Food Science and Food Safety, 11*(6), 530–545.
2. Ladaniya, M. S. (2008). Commercial fresh citrus cultivars and producing countries. In *Citrus fruit: Biology, technology and evaluation* (pp. 13–65). Academic Press.
3. Mahato, N., Agarwal, P., Mohapatra, D., Sinha, M., Dhyani, A., Pathak, B., Tripathi, M. K., & Angaiah, S. (2021). Biotransformation of citrus waste-II: Bio-sorbent materials for removal of dyes, heavy metals and toxic chemicals from polluted water. *Processes, 9*(9), 1544.
4. Reguengo, L. M., Salgaço, M. K., Sivieri, K., & Júnior, M. R. (2022). Agro-industrial by-products: Valuable sources of bioactive compounds. *Food Research International, 152*, 110871.
5. Panwar, D., Panesar, P. S., & Chopra, H. K. (2021). Recent trends on the valorization strategies for the management of citrus by-products. *Food Reviews International, 37*(1), 91–120.
6. Suri, S., Singh, A., & Nema, P. K. (2022). Current applications of citrus fruit processing waste: A scientific outlook. *Applied Food Research, 2*(1), 100050.
7. Mahato, N., Sharma, K., Sinha, M., & Cho, M. H. (2018). Citrus waste derived nutra-/pharmaceuticals for health benefits: Current trends and future perspectives. *Journal of Functional Foods, 40*, 307–316.
8. Vincente, A. R., Manganaris, G. A., Ortiz, C. M., Sozzi, G. O., & Crisosto, C. H. (2014). Nutritional quality of fruits and vegetables. In *Postharvest handling* (pp. 69–122). Academic Press.
9. Ahmed, E., Arshad, M., Khan, M. Z., Amjad, M. S., Sadaf, H. M., Riaz, I., Sabir, S., & Ahmad, N. (2017). Secondary metabolites and their multidimensional prospective in plant life. *Journal of Pharmacognosy and Phytochemistry, 6*(2), 205–214.

10. Kundu, D., Banerjee, S., Karmakar, S., & Banerjee, R. (2021). Valorization of citrus lemon wastes through biorefinery approach: An industrial symbiosis. *Bioresource Technology Reports, 15*, 100717.
11. Sarao, L. K., Kaur, S., Kaur, P., Ankita, & Bakala, H. S. (2022). Production of bioethanol from fruit wastes: Recent advances. In *Food waste to green fuel: Trend & development* (pp. 213–253). Springer.
12. Köse, M. D., & Bayraktar, O. (2018). Valorization of citrus peel waste. *Natural Volatiles and Essential Oils, 5*(2), 10–18.
13. Gómez-Mejía, E., Rosales-Conrado, N., León-González, M. E., & Madrid, Y. (2019). Citrus peels waste as a source of value-added compounds: Extraction and quantification of bioactive polyphenols. *Food Chemistry, 295*, 289–299.
14. Mamma, D., & Christakopoulos, P. (2014). Biotransformation of citrus by-products into value added products. *Waste and Biomass Valorization, 5*, 529–549.
15. Satari, B., & Karimi, K. (2018). Citrus processing wastes: Environmental impacts, recent advances, and future perspectives in total valorization. *Resources, Conservation and Recycling, 129*, 153–167.
16. Lamine, M., Hamdi, Z., Zemni, H., Rahali, F. Z., Melki, I., Mliki, A., & Gargouri, M. (2024). From residue to resource: The recovery of high-added values compounds through an integral green valorization of citrus residual biomass. *Sustainable Chemistry and Pharmacy, 37*, 101379.
17. Magare, M. E., Sahu, N., Kanade, G. S., Chanotiya, C. S., & Thul, S. T. (2020). An integrated process of value addition to citrus waste and performance of Fenton process for its conversion to biogas. *Waste and Biomass Valorization, 11*(1), 165–172.
18. Sharma, P., Vishvakarma, R., Gautam, K., Vimal, A., Gaur, V. K., Farooqui, A., Varjani, S., & Younis, K. (2022). Valorization of citrus peel waste for the sustainable production of value-added products. *Bioresource Technology, 351*, 127064.
19. Lucarini, M., Durazzo, A., Nazhand, A., Kiefer, J., Bernini, R., Romani, A., Souto, E. B., & Santini, A. (2022). Lemon (Citrus limon) bio-waste: Chemistry, functionality and technological applications. In *Mediterranean fruits bio-wastes: Chemistry, functionality and technological applications* (pp. 303–322). Springer International Publishing.
20. Negro, V., Mancini, G., Ruggeri, B., & Fino, D. (2016). Citrus waste as feedstock for bio-based products recovery: Review on limonene case study and energy valorization. *Bioresource Technology, 214*, 806–815.
21. Shrestha, S., Khatiwada, J. R., Sharma, H. K., & Qin, W. (2021). Bioconversion of fruits and vegetables wastes into value-added products. In *Sustainable bioconversion of waste to value added products*. (pp. 145–163). Springer.
22. Chavan, P., Singh, A. K., & Kaur, G. (2018). Recent progress in the utilization of industrial waste and by-products of citrus fruits: A review. *Journal of Food Process Engineering, 41*(8), e12895.
23. El-gengaihi, S. E., Mohammed, M. A., Aboubaker, D. H., Shoaib, R. M., Asker, M. S., Abdelhamid, S. A., & Hassan, E. M. (2020). Chemical, biological, and molecular studies on different citrus species wastes. *Plant Archives, 20*(1), 2773–2782.
24. Mohsin, A., Hussain, M. H., Zaman, W. Q., Mohsin, M. Z., Zhang, J., Liu, Z., Tian, X., Salim-ur-Rehman, Khan, I. M., Niazi, S., & Zhuang, Y. (2022). Advances in sustainable approaches utilizing orange peel waste to produce highly value-added bioproducts. *Critical Reviews in Biotechnology, 42*(8), 1284–1303.
25. Liu, N., Li, X., Zhao, P., Zhang, X., Qiao, O., Huang, L., Guo, L., & Gao, W. (2021). A review of chemical constituents and health-promoting effects of citrus peels. *Food Chemistry, 365*, 130585.
26. Rafiq, S., Kaul, R., Sofi, S. A., Bashir, N., Nazir, F., & Nayik, G. A. (2018). Citrus peel as a source of functional ingredient: A review. *Journal of the Saudi Society of Agricultural Sciences, 17*(4), 351–358.

27. Zayed, A., Badawy, M. T., & Farag, M. A. (2021). Valorization and extraction optimization of Citrus seeds for food and functional food applications. *Food Chemistry, 355*, 129609.
28. Zannini, D., Dal Poggetto, G., Malinconico, M., Santagata, G., & Immirzi, B. (2021). Citrus pomace biomass as a source of pectin and lignocellulose fibers: From waste to upgraded biocomposites for mulching applications. *Polymers, 13*(8), 1280.
29. Singh, B., Singh, J. P., Kaur, A., & Yadav, M. P. (2021). Insights into the chemical composition and bioactivities of citrus peel essential oils. *Food Research International, 143*, 110231.
30. Schalow, S., Baloufaud, M., Cottancin, T., Fischer, J., & Drusch, S. (2018). Orange pulp and peel fibres: Pectin-rich by-products from citrus processing for water binding and gelling in foods. *European Food Research and Technology, 244*, 235–244.
31. Vahidi, R., Pourahmad, R., & Mahmoudi, R. (2019). Chemical compounds and antibacterial and antioxidant properties of citron (*Citrus medica* L.) peel essential oil. *Journal of Food and Bioprocess Engineering, 2*(1), 71–76.
32. El-Adawy, T. A., El-Bedawy, A. A., Rahma, E. H., & Gafar, A. M. (1999). Properties of some citrus seeds. Part 3. Evaluation as a new source of protein and oil. *Food/Nahrung, 43*(6), 385–391.
33. Chinapongtitiwat, V., Jongaroontaprangsee, S., Chiewchan, N., & Devahastin, S. (2013). Important flavonoids and limonin in selected Thai citrus residues. *Journal of Functional Foods, 5*(3), 1151–1158.
34. Sharmeen, J. B., Mahomoodally, F. M., Zengin, G., & Maggi, F. (2021). Essential oils as natural sources of fragrance compounds for cosmetics and cosmeceuticals. *Molecules, 26*(3), 666.
35. Baker, E. A., Procopiou, J., & Hunt, G. M. (1975). The cuticles of Citrus species. Composition of leaf and fruit waxes. *Journal of the Science of Food and Agriculture, 26*(8), 1093–1101.
36. Diaz, J. V., Anthon, G. E., & Barrett, D. M. (2007). Nonenzymatic degradation of citrus pectin and pectate during prolonged heating: Effects of pH, temperature, and degree of methyl esterification. *Journal of Agricultural and Food Chemistry, 55*(13), 5131–5136.
37. Dadwal, V., & Gupta, M. (2023). Recent developments in citrus bioflavonoid encapsulation to reinforce controlled antioxidant delivery and generate therapeutic uses. *Critical Reviews in Food Science and Nutrition, 63*(9), 1187–1207.
38. Wilson, C. P. (1928). Relation of chemistry to the citrus products industry. *Industrial & Engineering Chemistry, 20*(12), 1302–1307.
39. Chinnici, G., Zarbà, C., Hamam, M., Pecorino, B., & D'Amico, M. (2019). A model of circular economy of citrus industry. *International Multidisciplinary Scientific Geo Conference: SGEM, 19*(4.2), 19–26.
40. Joglekar, S. N., Pathak, P. D., Mandavgane, S. A., & Kulkarni, B. D. (2019). Process of fruit peel waste biorefinery: A case study of citrus waste biorefinery, its environmental impacts and recommendations. *Environmental Science and Pollution Research, 26*, 34713–34722.
41. Yadav, V., Sarker, A., Yadav, A., Miftah, A. O., Bilal, M., & Iqbal, H. M. (2022). Integrated biorefinery approach to valorize citrus waste: A sustainable solution for resource recovery and environmental management. *Chemosphere, 293*, 133459.
42. Rivas-Cantu, R. C., Jones, K. D., & Mills, P. L. (2013). A citrus waste-based biorefinery as a source of renewable energy: Technical advances and analysis of engineering challenges. *Waste Management & Research, 31*(4), 413–420.
43. Chen, W., Zhou, Y., & Chen, Y. (2021). The environmental impacts of citrus residue management in China: A case study in the three gorges reservoir region. *Waste Management, 133*, 80–88.
44. Lanfranchi, M. (2012). Economic analysis on the enhancement of citrus waste for energy production. *Journal of Essential Oil Research, 24*(6), 583–591.
45. Sharma, K., Mahato, N., Cho, M. H., & Lee, Y. R. (2017). Converting citrus wastes into value-added products: Economic and environmently friendly approaches. *Nutrition, 34*, 29–46.
46. Raimondo, M., Caracciolo, F., Cembalo, L., Chinnici, G., Pecorino, B., & D'Amico, M. (2018). Making virtue out of necessity: Managing the citrus waste supply chain for bioeconomy applications. *Sustainability, 10*(12), 4821.

47. Ferrari, A., Morone, P., & Tartiu, V. E. (2016). Tackling uncertainty through business plan analysis—A case study on citrus waste valorisation in the south of Italy. *Agriculture, 6*(1), 5.
48. Valenti, F., Porto, S. M., Chinnici, G., Selvaggi, R., Cascone, G., Arcidiacono, C., & Pecorino, B. (2017). Use of citrus pulp for biogas production: A GIS analysis of citrus-growing areas and processing industries in South Italy. *Land Use Policy, 66*, 151–161.
49. Boccia, F., Di Pietro, B., & Covino, D. (2021). Food waste and environmental-sustainable innovation: A scenario for the Italian citrus market. *Quality-Access to Success, 22*(182), 145.
50. Cecilia, J. A., García-Sancho, C., Maireles-Torres, P. J., & Luque, R. (2019). Industrial food waste valorization: A general overview. In *Biorefinery: Integrated sustainable processes for biomass conversion to biomaterials, biofuels, and fertilizers* (pp. 253–277). Springer.
51. Nayak, A., & Bhushan, B. (2019). An overview of the recent trends on the waste valorization techniques for food wastes. *Journal of Environmental Management, 233*, 352–370.
52. Srivastava, A. K., & Singh, S. (2006). Biochemical markers and nutrient constraints diagnosis in citrus: A perspective. *Journal of Plant Nutrition, 29*(5), 827–855.
53. Patsalou, M., Samanides, C. G., Protopapa, E., Stavrinou, S., Vyrides, I., & Koutinas, M. (2019). A citrus peel waste biorefinery for ethanol and methane production. *Molecules, 24*(13), 2451.
54. Jeong, D., Park, H., Jang, B. K., Ju, Y., Shin, M. H., Oh, E. J., Lee, E. J., & Kim, S. R. (2021). Recent advances in the biological valorization of citrus peel waste into fuels and chemicals. *Bioresource Technology, 323*, 124603.
55. Njokweni, S. G., Weimer, P. J., Botes, M., Cruywagen, C. W., & van Zyl, W. H. (2021). Extraruminal fermentation of citrus, grape and apple pomaces: Assessing the potential to serve as feedstock for production of volatile fatty acids. *Waste and Biomass Valorization, 12*, 3671–3681.
56. Choi, I. S., Lee, Y. G., Khanal, S. K., Park, B. J., & Bae, H. J. (2015). A low-energy, cost-effective approach to fruit and citrus peel waste processing for bioethanol production. *Applied Energy, 140*, 65–74.
57. Ni, H., Yang, Y. F., Chen, F., Ji, H. F., Yang, H., Ling, W., & Cai, H. N. (2014). Pectinase and naringinase help to improve juice production and quality from pummelo (*Citrus grandis*) fruit. *Food Science and Biotechnology, 23*, 739–746.
58. Güzel, M., & Akpınar, Ö. (2019). Production and characterization of bacterial cellulose from citrus peels. *Waste and Biomass Valorization, 10*, 2165–2175.
59. Mohamed, S. A., Drees, E. A., El-Badry, M. O., & Fahmy, A. S. (2010). Biochemical properties of α-amylase from peel of Citrus sinensis cv. Abosora. *Applied Biochemistry and Biotechnology, 160*, 2054–2065.
60. Cantú, M. D., Mariano, A. G., Palma, M. S., Carrilho, E., & Wulff, N. A. (2008). Proteomic analysis reveals suppression of bark chitinases and proteinase inhibitors in citrus plants affected by the citrus sudden death disease. *Phytopathology, 98*(10), 1084–1092.
61. Huang, R., Zhang, Y., Shen, S., Zhi, Z., Cheng, H., Chen, S., & Ye, X. (2020). Antioxidant and pancreatic lipase inhibitory effects of flavonoids from different citrus peel extracts: An in vitro study. *Food Chemistry, 326*, 126785.
62. Song, L. W., Qi, J. R., Liao, J. S., & Yang, X. Q. (2021). Enzymatic and enzyme-physical modification of citrus fiber by xylanase and planetary ball milling treatment. *Food Hydrocolloids, 121*, 107015.
63. Mohamed, S. A., Al-Malki, A. L., & Kumosani, T. A. (2009). Characterization of a polygalacturonase from Trichoderma harzianum grown on citrus peel with application for apple juice. *Australian Journal of Basic and Applied Sciences, 3*(3), 2770–2777.
64. Gądek, W., Mlonka-Mędrala, M., Prestipino, M., Evangelopoulos, P., Kalisz, S., & Yang, W. (2016). Gasification and pyrolysis of different biomasses in lab scale system: A comparative study. In *E3S web of conferences* (Vol. 10, p. 00024). EDP Sciences.
65. Prestipino, M., Chiodo, V., Maisano, S., Zafarana, G., Urbani, F., & Galvagno, A. (2017). Hydrogen rich syngas production by air-steam gasification of citrus peel residues from cit-

rus juice manufacturing: Experimental and simulation activities. *International Journal of Hydrogen Energy, 42*(43), 26816–26827.

66. Volpe, M., Panno, D., Volpe, R., & Messineo, A. (2015). Upgrade of citrus waste as a biofuel via slow pyrolysis. *Journal of Analytical and Applied Pyrolysis, 115*, 66–76.
67. Chiodo, V., Urbani, F., Zafarana, G., Prestipino, M., Galvagno, A., & Maisano, S. (2017). Syngas production by catalytic steam gasification of citrus residues. *International Journal of Hydrogen Energy, 42*(46), 28048–28055.
68. Alvarez, J., Hooshdaran, B., Cortazar, M., Amutio, M., Lopez, G., Freire, F. B., Haghshenasfard, M., Hosseini, S. H., & Olazar, M. (2018). Valorization of citrus wastes by fast pyrolysis in a conical spouted bed reactor. *Fuel, 224*, 111–120.
69. Maqbool, Z., Khalid, W., Atiq, H. T., Koraqi, H., Javaid, Z., Alhag, S. K., Al-Shuraym, L. A., Bader, D. M., Almarzuq, M., Afifi, M., & Al-Farga, A. (2023). Citrus waste as source of bioactive compounds: Extraction and utilization in health and food industry. *Molecules, 28*(4), 1636.
70. Shakir, I. K., & Salih, S. J. (2015). Extraction of essential oils from citrus by-products using microwave steam distillation. *Iraqi Journal of Chemical and Petroleum Engineering, 16*(3), 11–22.
71. Sikdar, D. C., Menon, R., Duseja, K., Kumar, P., & Swami, P. (2016). Extraction of citrus oil from orange (Citrus sinensis) peels by steam distillation and its characterizations. *International Journal of Technical Research and Applications, 4*(3), 341–346.
72. Khaw, K. Y., Parat, M. O., Shaw, P. N., & Falconer, J. R. (2017). Solvent supercritical fluid technologies to extract bioactive compounds from natural sources: A review. *Molecules, 22*(7), 1186.
73. Kaya, M., Sousa, A. G., Crépeau, M. J., Sørensen, S. O., & Ralet, M. C. (2014). Characterization of citrus pectin samples extracted under different conditions: Influence of acid type and pH of extraction. *Annals of Botany, 114*(6), 1319–1326.
74. Nadar, S. S., Rao, P., & Rathod, V. K. (2018). Enzyme assisted extraction of biomolecules as an approach to novel extraction technology: A review. *Food Research International, 108*, 309–330.
75. Suri, S., Singh, A., & Nema, P. K. (2021). Recent advances in valorization of citrus fruits processing waste: A way forward towards environmental sustainability. *Food Science and Biotechnology, 30*, 1601–1626.
76. González-Miquel, M., & Díaz, I. (2020). Valorization of citrus waste through sustainable extraction processes. In *Food industry wastes* (pp. 113–133). Academic Press.
77. Palazzolo, E., Laudicina, V. A., & Germanà, M. A. (2013). Current and potential use of citrus essential oils. *Current Organic Chemistry, 17*(24), 3042–3049.
78. Mustafa, N. E. M. (2015). Citrus essential oils: Current and prospective uses in the food industry. *Recent Patents on Food, Nutrition & Agriculture, 7*(2), 115–127.
79. Viuda-Martos, M., Ruiz-Navajas, Y., Fernández-López, J., & Pérez-Álvarez, J. A. (2010). Effect of added citrus fibre and spice essential oils on quality characteristics and shelf-life of mortadella. *Meat Science, 85*(3), 568–576.
80. González-Mas, M. C., Rambla, J. L., López-Gresa, M. P., Blázquez, M. A., & Granell, A. (2019). Volatile compounds in citrus essential oils: A comprehensive review. *Frontiers in Plant Science, 10*, 12.
81. Haokip, S. W., Sheikh, K. A., Das, S., Devi, O. B., Singh, Y. D., Wangchu, L., & Heisnam, P. (2023). Unraveling physicochemical profiles and bioactivities of citrus peel essential oils: A comprehensive review. *European Food Research and Technology, 249*(11), 2821–2834.
82. Mahato, N., Sharma, K., Koteswararao, R., Sinha, M., Baral, E., & Cho, M. H. (2019). Citrus essential oils: Extraction, authentication and application in food preservation. *Critical Reviews in Food Science and Nutrition, 59*(4), 611–625.
83. Manhongo, T. T., Chimphango, A. F., Thornley, P., & Röder, M. (2022). Current status and opportunities for fruit processing waste biorefineries. *Renewable and Sustainable Energy Reviews, 155*, 111823.

84. Agrawal, B. J. (2015). Supercritical carbon-dioxide assisted dyeing of textiles: An environmental benign waterless dyeing process. *International Journal of Innovative Research and Creative Technology, 1*(2), 201–205.
85. Asiri, A. M., & Isloor, A. M. (Eds.). (2019). *Green sustainable process for chemical and environmental engineering and science: Supercritical carbon dioxide as green solvent*. Elsevier.
86. Panda, D., & Manickam, S. (2019). Cavitation technology—The future of greener extraction method: A review on the extraction of natural products and process intensification mechanism and perspectives. *Applied Sciences, 9*(4), 766.
87. Bora, H., Kamle, M., Mahato, D. K., Tiwari, P., & Kumar, P. (2020). Citrus essential oils (CEOs) and their applications in food: An overview. *Plants, 9*(3), 357.
88. Brah, A. S., Armah, F. A., Obuah, C., Akwetey, S. A., & Adokoh, C. K. (2023). Toxicity and therapeutic applications of citrus essential oils (CEOs): A review. *International Journal of Food Properties, 26*(1), 301–326.
89. Tranchida, P. Q., Bonaccorsi, I., Dugo, P., Mondello, L., & Dugo, G. (2012). Analysis of Citrus essential oils: State of the art and future perspectives. A review. *Flavour and Fragrance Journal, 27*(2), 98–123.
90. Regnier, T., Combrinck, S., Veldman, W., & Du Plooy, W. (2014). Application of essential oils as multi-target fungicides for the control of Geotrichum citri-aurantii and other postharvest pathogens of citrus. *Industrial Crops and Products, 61*, 151–159.
91. Kumar, N., Upadhyay, A., & Shukla, S. (2023). Effects of essential oils and ultrasonic treatments on properties of edible coatings and their application on citrus fruits. *Starch-Stärke, 76*, 2300104.
92. Swaine, R. L., & Swaine, R. L., Jr. (1988). Citrus oils: Processing, technology, and applications. *Perfume & Flavorist, 13*, 1–20.
93. Taghizadeh-Alisaraei, A., Hosseini, S. H., Ghobadian, B., & Motevali, A. (2017). Biofuel production from citrus wastes: A feasibility study in Iran. *Renewable and Sustainable Energy Reviews, 69*, 1100–1112.
94. Choi, I. S., Kim, J. H., Wi, S. G., Kim, K. H., & Bae, H. J. (2013). Bioethanol production from mandarin (Citrus unshiu) peel waste using popping pretreatment. *Applied Energy, 102*, 204–210.
95. Boluda-Aguilar, M., & López-Gómez, A. (2013). Production of bioethanol by fermentation of lemon (Citrus limon L.) peel wastes pretreated with steam explosion. *Industrial Crops and Products, 41*, 188–197.
96. Santiago, B., Moreira, M. T., Feijoo, G., & González-García, S. (2020). Identification of environmental aspects of citrus waste valorization into D-limonene from a biorefinery approach. *Biomass and Bioenergy, 143*, 105844.
97. Ruiz, B., & Flotats, X. (2016). Effect of limonene on batch anaerobic digestion of citrus peel waste. *Biochemical Engineering Journal, 109*, 9–18.
98. Ozturk, B., Winterburn, J., & Gonzalez-Miquel, M. (2019). Orange peel waste valorisation through limonene extraction using bio-based solvents. *Biochemical Engineering Journal, 151*, 107298.
99. Pourbafrani, M., Forgács, G., Horváth, I. S., Niklasson, C., & Taherzadeh, M. J. (2010). Production of biofuels, limonene and pectin from citrus wastes. *Bioresource Technology, 101*(11), 4246–4250.
100. Devi, W. E., Kumar, R. S., & Mishra, A. A. (2014). Extraction of pectin from citrus fruit peel and its utilization in preparation of jelly. *International Journal of Engineering Research & Technology, 3*(5), 1925–1932.
101. Torrado, A. M., Cortés, S., Salgado, J. M., Max, B., Rodríguez, N., Bibbins, B. P., Converti, A., & Domínguez, J. M. (2011). Citric acid production from orange peel wastes by solid-state fermentation. *Brazilian Journal of Microbiology, 42*, 394–409.
102. Soccol, C. R., Vandenberghe, L. P., Rodrigues, C., & Pandey, A. (2006). New perspectives for citric acid production and application. *Food Technology and Biotechnology, 44*(2), 141–149.

103. Satheeshkumar, S., Sivagurunathan, P., Muthulakshmi, K., & Uma, C. (2019). Utilization of fruit waste for the production of citric acid by using Aspergillus niger. *Journal of Drug Delivery and Therapeutics, 9*(4-A), 9–14.
104. Aravantinos-Zafiris, G., Tzia, C., Oreopoulou, V., & Thomopoulos, C. D. (1994). Fermentation of orange processing wastes for citric acid production. *Journal of the Science of Food and Agriculture, 65*(1), 117–120.
105. Gaden, E. L., Pelciavas, D. N., & Winoker, J. (1954). Citrus waste utilization, microbiological production of riboflavin and citric acid from citrus molasses. *Journal of Agricultural and Food Chemistry, 2*(12), 632–638.
106. Alnaimy, A., Gad, A. E., Mustafa, M. M., Atta, M. A., & Basuony, H. A. (2017). Using of citrus by-products in farm animals feeding. *Open Access Journal of Science, 1*(3), 58–67.
107. Tayengwa, T., & Mapiye, C. (2018). Citrus and winery wastes: Promising dietary supplements for sustainable ruminant animal nutrition, health, production, and meat quality. *Sustainability, 10*(10), 3718.
108. Manju Wadhwa, M. W., Bakshi, M. P., & Makkar, H. P. (2015). Waste to worth: Fruit wastes and by-products as animal feed. *CABI Reviews, 15*, 1–26.
109. Wing, J. M., Van Horn, H. H., Sklare, S. D., & Harris, B., Jr. (1988). Effects of citrus molasses distillers solubles and molasses on rumen parameters and lactation. *Journal of Dairy Science, 71*(2), 414–420.
110. Chen, M. C., Ammerman, C. B., Henry, P. R., Palmer, A. Z., & Long, S. K. (1981). Citrus condensed molasses solubles as an energy source for ruminants. *Journal of Animal Science, 53*(1), 253–259.
111. Chen, L., Wu, Y., Jiang, X., Gan, D., Fan, J., Sun, Y., Liu, W., & Li, X. (2023). Dietary fiber extraction from citrus peel pomace: Yield optimization and evaluation of its functionality, rheological behavior, and microstructure properties. *Journal of Food Science, 88*(8), 3507–3523.
112. Kaur, S., Panesar, P. S., & Chopra, H. K. (2023). Extraction of dietary fiber from kinnow (Citrus reticulata) peels using sequential ultrasonic and enzymatic treatments and its application in development of cookies. *Food Bioscience, 54*, 102891.
113. Dranca, F., & Oroian, M. (2018). Extraction, purification and characterization of pectin from alternative sources with potential technological applications. *Food Research International, 113*, 327–350.
114. Bilanovic, D., Shelef, G., & Green, M. (1994). Xanthan fermentation of citrus waste. *Bioresource Technology, 48*(2), 169–172.
115. dos Santosa, F. P., Jra, A. M., Nunesa, T. P., de Farias Silvab, C. E., & de Souza Abud, A. K. (2016). Bioconversion of agro-industrial wastes into xanthan gum. *Chemical Engineer, 49*, 145–150.
116. Peng, M., Liu, J., Liu, Z., Fu, B., Hu, Y., Zhou, M., Fu, C., Gao, B., Wang, C., Li, D., & Xu, N. (2018). Effect of citrus peel on phenolic compounds, organic acids and antioxidant activity of soy sauce. *LWT, 90*, 627–635.
117. Zhou, W., Widmer, W., & Grohmann, K. (2007). Economic analysis of ethanol production from citrus peel waste. *Proceedings of the Florida State Horticultural Society, 120*, 310–315.
118. Alcon, F., Martin-Ortega, J., Pedrero, F., Alarcon, J. J., & de Miguel, M. D. (2013). Incorporating non-market benefits of reclaimed water into cost-benefit analysis: A case study of irrigated mandarin crops in southern Spain. *Water Resources Management, 27*, 1809–1820.
119. Alzubi, E., Kassem, A., Melkonyan, A., & Noche, B. (2024). Enhancing economic-social sustainability through a closed-loop citrus supply chain: A life cycle cost analysis. *Resources, Conservation & Recycling Advances, 21*, 200199.
120. Durkin, A., Taptygin, I., Kong, Q., Gunam Resul, M. F., Rehman, A., Fernández, A. M., Harvey, A. P., Shah, N., & Guo, M. (2019). Scale-up and sustainability evaluation of biopolymer production from citrus waste offering carbon capture and utilisation pathway. *ChemistryOpen, 8*(6), 668–688.

121. Lucia, C., Laudicina, V. A., Badalucco, L., Galati, A., Palazzolo, E., Torregrossa, M., Viviani, G., & Corsino, S. F. (2022). Challenges and opportunities for citrus wastewater management and valorisation: A review. *Journal of Environmental Management, 321*, 115924.
122. Putnik, P., Bursać Kovačević, D., Režek Jambrak, A., Barba, F. J., Cravotto, G., Binello, A., Lorenzo, J. M., & Shpigelman, A. (2017). Innovative "green" and novel strategies for the extraction of bioactive added value compounds from citrus wastes—A review. *Molecules, 22*(5), 680.
123. Khalil, M. N., Farghal, H. H., & Farag, M. A. (2022). Outgoing and potential trends of composition, health benefits, juice production and waste management of the multi-faceted Grapefruit Citrus X paradisi: A comprehensive review for maximizing its value. *Critical Reviews in Food Science and Nutrition, 62*(4), 935–956.
124. Berenguer, C. V., Andrade, C., Pereira, J. A., Perestrelo, R., & Câmara, J. S. (2022). Current challenges in the sustainable valorisation of agri-food wastes: A review. *Processes, 11*(1), 20.
125. Nirmal, N. P., Khanashyam, A. C., Mundanat, A. S., Shah, K., Babu, K. S., Thorakkattu, P., Al-Asmari, F., & Pandiselvam, R. (2023). Valorization of fruit waste for bioactive compounds and their applications in the food industry. *Food, 12*(3), 556.
126. Panwar, D., Saini, A., Panesar, P. S., & Chopra, H. K. (2021). Unraveling the scientific perspectives of citrus by-products utilization: Progress towards circular economy. *Trends in Food Science & Technology, 111*, 549–562.

Chapter 11
Extraction of Bioactive Compounds from Citrus Waste Through Green Technologies

Sakhawat Riaz, Moazam Hyder, Liu Xian, Zhu Kai Yue, Ghazanfer Abbas, Muhammad Waqar, Fatima Irfan, Danish Kamal, Huma Tabassum, and Anwar Ali

11.1 Introduction

Citrus fruits, which belong to the family Rutaceae, are the natural commodity that is subject to the greatest amount of marketing on earth. In addition to being recognized by the Romance loanword agrumes, which translates to "sour fruits," citrus is one of the most widely cultivated plants in the world. It is notable for its widespread use and acceptance in the food service sector [1]. Grown mostly in tropical and subtropical climates, citrus is one of the most important and biggest crops. More than 140 countries fill more than 150 million tons of citrus-based natural goods annually. The Food and Agriculture Organization (FAO) states that the top exporters of organic citrus and its products are Israel, Brazil, the United States, Costa Rica, Mexico, Cuba, and Italy, Argentina, Greece, and Spain are foremost exporters [2].

S. Riaz
The State Key Laboratory of Tea Plant Biology and Utilization, School of Tea & Food Science, Anhui Agricultural University, Hefei, China

Food and Nutrition Society, Gilgit Baltistan, Pakistan

M. Hyder
Guangdong Engineering Research Centre for Mineral Oil Pesticides, Institute of Zoology, Guangdong Academy of Science, Guangzhou, China

L. Xian · Z. K. Yue · D. Kamal
The State Key Laboratory of Tea Plant Biology and Utilization, School of Tea & Food Science, Anhui Agricultural University, Hefei, China

G. Abbas
Department of Chemistry, Ghazi University, Dera Ghazi Khan, Pakistan

M. Waqar
Department of Agricultural Economics, College of Agriculture, University of Sargodha, Sargodha, Pakistan

A. Chauhan et al. (eds.), *Valorization of Citrus Food Waste*,
https://doi.org/10.1007/978-3-031-77999-2_11

Many Citrus species, such as oranges (Citrus *sinensis*), mandarins (Citrus *reticulata*), grapefruits (Citrus *paradisi*), lemons (Citrus *limon*), Citrus *clementina*, and Citrus *unshiu* are consumed as juice or used for the development of new products all over the planet [3]. The world's most processed citrus fruits produce a lot of leftovers that are rich in soluble sugars, hemicellulose, cellulose, essential oils, pectin, and several other bioactive components [4]. The utilization of leftovers and residues from global production in the creation of new food products will be beneficial to the government of the area [5]. Food debris and wastes may be employed as food additives or raw materials, which might benefit the business in the marketplace, reduce nutritional deficiencies, encourage beneficial health treatments, and limit the damaging environmental consequences of waste treatment [6].

Food waste is produced at every step of the food life cycle, involving cultivation, processing in factories, storage, and delivery. Fruit and vegetable losses during the agricultural stages are substantial in underdeveloped countries, but they are primarily specified by the processing period, which accounts for 25% of losses [7]. 42% of wasted food is created during domestic activities, 39% is lost during food production, 14% is eliminated during food service (catering, and lodging), and 5% is compromised during delivery. Because food leftover contains a significant quantity of both micro and macronutrients, it is important [8]. Food processing, as in many developing nations, produces a lot of leftovers, and food byproducts, which may negatively affect the worldwide economy and biodiversity. Nonetheless, food wastes or leftovers have the potential to produce biochemically active substances that may improve human well-being and eliminate hunger [9]. Up to 50% of fresh fruit is made up of fruit byproducts, such as skins, seeds, husks, and bran, which contain multiple times the amount of nutrients compared to the finished product [10].

The manufacturing of citrus juice results in the creation of waste, which accounts for over half of the total mass of the fresh fruit crop. Pomace, peels (which account for 50–55% of the total fruit), seeds (which account for 20–40% mass), and wastewater are all included in the leftovers. Fruit pulp, and peels that have gone bad are all components that are included in citrus wastewater. Each year, the processing of citrus fruits throughout the world results in the generation of around ten million

F. Irfan
Sustainable Development Study Centre, Government College University, Lahore, Pakistan

H. Tabassum
Institute of Social and Cultural Studies, Department of Public Health, University of the Punjab, Lahore, Pakistan

A. Ali (✉)
Food and Nutrition Society, Gilgit Baltistan, Pakistan

Institute of Human Nutrition Sciences, Warsaw University of Life Sciences, Nowoursynowska 159 St., Warsaw, Poland

Fig. 11.1 Different types of Citrus fruits

(MT) of trash [11, 12]. It's a concern because of its huge amount and perishable nature. The trash rapidly attracts mold, insects, bacteria, and generates toxins. Citrus waste and byproducts must be handled since they have an adverse influence on biodiversity and contain significant bioactive substances (Fig. 11.1).

11.2 Bioactive Compounds in Citrus Waste

Fruit wastes gathered by the food agro-industries are usually composed of residual biomasses that have not been fully exploited and are rich in a number of bioactive substances [13]. Because fruit wastes are rich in phytochemicals, there has been a study done on extracting phenolic substances, fibers, and other beneficial components from them. Most citrus by-products consist of peels, pomace, and seed parts, which may be an excellent source of high-value bioactive substances such as dietary fibers, polysaccharides, flavor substances, and phytonutrients [7]. One of the most popular fruit genera for cultivation and consumption is the citrus genus. When compared to the edible part, the processing industries' leftovers are said to have higher concentrations of polyphenols, namely flavonoids and phenolic acids [14]. Higher concentrations of phenolic substances, flavonoids, vitamin C, and antioxidants may be found in the peel, entire fruit, and pulp extracts of citrus fruits than in their interior, discarded components (seeds and pulp) [15]. Approximately 30–34% of the peel from citrus fruits is obtained during the process of making citrus juice, making it the primary waste product [16]. Numerous phytochemicals and bioactive components can be found in citrus peels, including essential oils (D-limonene), carotenoids, polyphenols-flavonoids (polymethoxylated flavones: hesperidin, nobiletin, naringin, and tangeretin), sugars (cellulose, pectin, & hemicellulose), pigments and flavoring agents [17–20]. Citrus peels may be used to recover bioactive chemicals, fortify/incorporate them into functional meals, or add value to products [12]. Citrus fruits include around 20–25% seeds, which are often wasted when the fruit is processed. Citrus seeds may have high economic significance as a source of essential oil and protein [21]. Citrus seeds contain highly useful components such as essential oils, limonoids, carotenoids, polyphenols, and tocopherols, which may be used extensively for a variety of purposes [22]. Citrus pomace is the solid remains that remain after citrus fruits are processed to make juice or other goods. Citrus pomace is composed of seeds, pulp, and rags. Pomace is an excellent source of pectin, molasses, aromatic oils, and limonene. Furthermore, it includes polyphenols, flavonoids and antioxidants that may be used for commercial uses [23, 24].

11.3 Significance of Citrus Waste Valorization

Growing amounts of food waste are emerging as one of the biggest worldwide food production concerns, threatening the food system's sustainability and putting an impact on the already insecure food system [25]. The production of industrialized citrus goods creates huge volumes of trash between 50% and 70% of total citrus fruit is allocated for this purpose [26]. Approximately 15,000–25,000 tons of citrus waste are generated each year, which is an appealing source for many beneficial applications [27]. On the other hand, the eventual disposal of Citrus waste creates pollution challenges, particularly in poor nations, and handling it can have a large economic and environmental consequence [28]. Citrus wastes have a low pH (3–4), a high concentration of water (80–90%), and organic matter (95% of total solids) [29]. For years, the major application of Citrus leftovers has been pectin extraction or animal feed, compost manufacture, and fuel generation, which in certain situations may be rather expensive [30]. As a result, solutions for the most suitable and environmentally sustainable treatment of this sort of waste have been presented, such as its use for the creation of nanocellulose, larvicides, pectinase, and bioethanol, biofuels, containing nanoparticles of iron and silver with cytotoxic action [26]. Citrus waste has great biological value and possible health advantages [31], Due to its pectin content [32], high-quality fiber [24], and the occurrence of bioactive substances like polyphenols (phenolic acids & flavonoids), essential oils (EOs) [33] and carotenoids [34]. This may be utilized in the culinary, pharmaceutical, and cosmetic sectors [35].

All treatment/valorization solutions for citrus waste are dependent on a variety of criteria that determine economic feasibility. Citrus waste may be utilized directly or after basic processing as animal feed or soil conditioners (for a long time the most frequent destination), or processed further in the biorefinery industries [28]. Because the economic burden of waste management comprises a substantial portion of the yearly budget of the citrus processing business, its competitiveness may be greatly boosted by economically viable citrus residue management solutions [36]. The best approach must combine the most effective and ecologically friendly treatment/

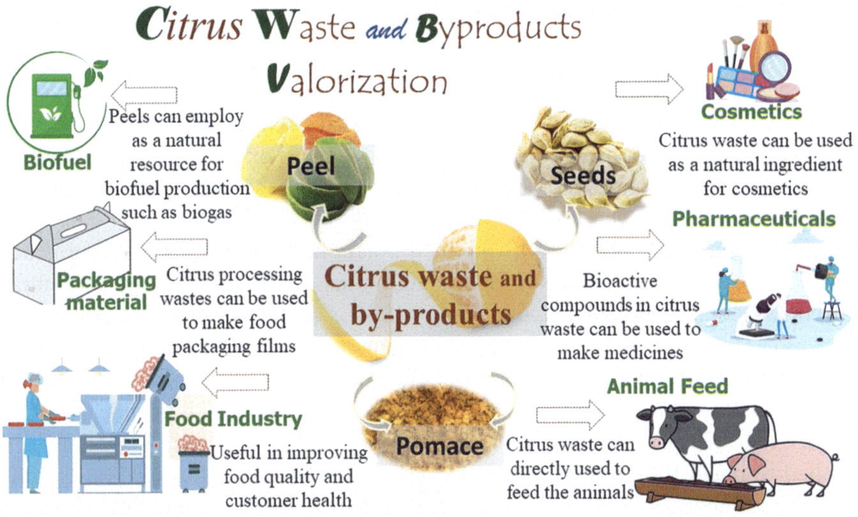

Fig. 11.2 Valorization of citrus fruit waste and byproducts

valorization technology with unique economic limitations based on local market actualities. Over the years, scientists worldwide have focused on creating new processing technologies for the full utilization of citrus waste [14, 37].

The most popular and easiest approach to handling the raw material is to use the whole citrus peel without separating specific ingredients (such as feed for animals, organic fertilizer, or the basis for compost) [38] (Fig. 11.2).

11.4 Direct Use of Citrus Waste

As technology advanced, new ideas and materials emerged. Citrus trash may be used in agriculture in two ways: directly on the soil or after composting. However, the enormous levels of EOs and other bioactive compounds in citrus waste are well recognized for their antibacterial properties, which may be highly hazardous to soil microorganisms [39, 40]. Furthermore, the soil utilization of non-stabilized organic matter might induce oxygen depletion of soil and odor emission. Proper land spreading procedures, composting, or suitable treatment of the bioactive components must be applied before the agronomic usage of citrus waste [28]. Agronomic exploitation of citrus residue by direct land application is a good approach to enhance the organic matter content of soil and hence improve its fertility [41]. Citrus waste has the potential to generate high-quality compost. Citrus waste compost, like biomass from other organic leftovers, is regarded as a partial or entire alternative for nonrenewable things [42]. Animal feed is the most common use of citrus waste in various regions of the Mediterranean basin, such as Southern Italy, accounting for up to 70–80% of output [43].

11.5 Food Industry Applications

Citrus peels, seeds, and pomace contain bioactive compounds that may be employed in the food business for various reasons such as food preservation, food safety, food additives, and nutraceuticals [31]. The capacity of citrus extracts to prevent the formation of germs, increase the cooked meatballs' storage shelf life, and have antioxidant properties [44]. The use of citrus phytonutrients to enhance food quality in terms of texture and color. Apart from their potent biological properties, citrus by-products are also an excellent source of dietary fiber. Because of its high concentration of beneficial components, this dietary fiber is preferred above other sources, such as cereal grains [45]. Dietary fiber is often utilized in meat and meat products to enhance cooking yield and texture because of its well-known ability to bind water and fat [14, 28]. One popular dietary fiber made from citrus peel is pectin. The main applications of pectin are as a thickener, emulsifier, and stabilizer. As a result, it's applied to make fruit juice, confections, jams, marmalade, jellies, and pastry fillings. Additionally, yogurts and other acidified milk beverages are stabilized with it [46]. Numerous instances exist of the direct utilization of citrus waste product essential oils and extracts, along with their potential for antibacterial and antioxidant purposes. The safety intake boundaries and their lasting effects on human wellness, remain unclear and need more investigation due to several factors, such as the problems surrounding synthetic chemicals. However, using natural ingredients and reducing the content of synthetic additives may be accomplished temporarily with the help of active food packaging [47] (Fig. 11.3).

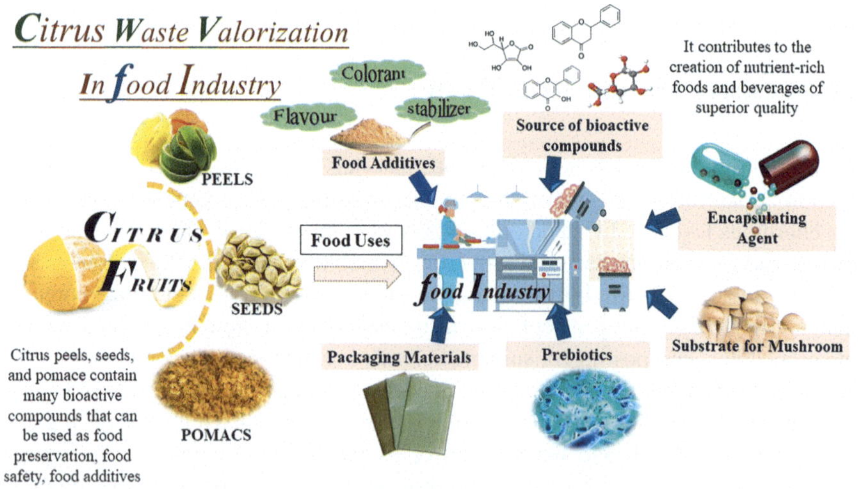

Fig. 11.3 Utilization of Citrus waste and byproduct in food sector

11.6 Extraction of Bioactive Compounds

The kind of extraction technique used to extract bioactive components from most fruit wastes and peels determines how well those chemicals may be recovered. The majority of investigations on fruit waste valorization conducted in recent years have concentrated on identifying appropriate extraction techniques and obtaining bioactive components [48]. The food, medicine, and chemical industries, among others, need low-cost extraction techniques with improved extraction efficiency [7]. Conventionally, typical techniques such as the soxhlet method, maceration, and hydro distillation, procedures were used to extract bioactive components from the material [49]. Because of some of these traditional extraction processes' disadvantages, they are no longer recommended. Common issues that must be resolved for an efficient extraction process include a longer extraction time, a slower extraction percentage, excessive solvent costs and consumption, low-purity extracts, and solvent impacts [50]. Depending on which bioactive compounds are being extracted, several extraction techniques may be used. The separation processes may be affected by many factors, including temperature, plant materials, pressure, and fluid type [51]. Valorization of citrus waste boosts the world economy in addition to lowering its negative effects on the ecosystem. Additionally, in terms of improved extraction yields and shorter extraction times, novel methods have been shown to be a more environmentally friendly and sustainable alternative to traditional extraction techniques [52]. These days, it's much simpler to extract bioactive chemicals, which are then used in the food and pharmaceutical sectors to improve the quality of food as well as in the pharmaceutical sectors to produce innovative medications against serious illnesses.

11.6.1 Conventional Extraction Techniques

Classical processes are referred to as usual ways since they have been in use for a very long period. These procedures are based on the provided energy and/or solvent extraction capability, or both. "Traditional" extraction techniques included hydro-distillation, Soxhlet extraction, and maceration [53]. Although it was originally developed just for lipid extraction, soxhlet extraction has long been employed as a conventional technique for removing significant bioactive components from a variety of plant parts [54]. Hydro-distillation is a recognized technique that serves as a benchmark for essential oil extraction [55]. The general method of extracting medicinal plants that are frequently used for herbal remedies is called maceration [56].

11.6.2 Green Technologies for Citrus Waste Extraction

To address the shortcomings of the traditional techniques, a number of innovative and environmentally friendly extraction technologies were tested. Some of the prospective methods with favorable results include microwave-assisted extraction, supercritical fluid extractions using greener solvents, pulsed electric field extraction, pressured liquid extraction, and ultrasound-assisted extraction [57].

11.6.2.1 Enzyme-Assisted Extraction (EAE)

Bioactive materials are often extracted from plant waste using enzymes. Plant cell walls include three polysaccharides called pectin, hemicellulose, and cellulose that act as barriers to the release of intracellular compounds. Enzymes include xylanase, pectinase, β-glucosidase and β-gluconase hydrolyze and depolymerize plant cell wall polysaccharides, releasing related chemicals in the process [58]. Since enzyme extraction uses water as a solvent instead of organic solvents, it is thought to be a more sustainable method of extracting oil and bioactive chemicals [59]. Enzyme arrangements, individually or in combination, may enhance the process of extracting and breaking down plant cells to liberate bioactive compounds. EAE is a viable alternative to conventional solvent-based extraction methods. It is predicated on the capacity of enzymes to catalyze interactions in aqueous solutions with mild processing settings [60].

When correlated to other extraction methods, the enzyme-assisted extraction approach is far more productive since it requires less time and yields higher-quality extracted chemicals. It necessitates familiarity with the fundamentals of the straightforward extraction process. During and after the extraction procedure, the extracted compounds will maintain their structural, physiochemical, and configured stability [61]. By using this method, we can precisely identify the intracellular chemical we need and extract it with great precision, improving the derived component's purity. Enzymes may effectively break down the cellular barrier without compromising the bioactive substances [62, 63]. The problematic aspect of this approach is that maintaining and storing significant quantities of these enzymes is challenging, making the separation and filtration of enzymes for the treatment procedure an essential and challenging task [64].

11.6.2.2 Microwave-Assisted Extraction

The food business, which is expanding quickly, produces a significant quantity of solid waste. Using this waste, which can be converted into highly valued goods, is one of the most crucial stages in the process of sustainable growth [65]. In the present scenario, it is evident that the isolation of phenolic substances from fruit waste requires the application of ecologically appropriate, durable, and possibly inexpensive equipment to accomplish the goals in a green manner. For example, microwave-aided extraction (MAE) is a novel method of extracting various phytochemicals using microwave energy [7]. Given that it uses less solvent and expedites the extraction process, it is a green extraction technique. The foundation of MAE is the dielectric warming idea, which describes how polar matrix molecules' molecular dipoles rotate in response to microwave ionizing waves to warm a dielectric material [66]. MAE may be carried out in open vessels that operate at lower ambient pressure or in sealed extraction vessels that operate at severe pressures and temperatures, permitting bigger product outputs. The latter method has the advantage of just requiring inexpensive equipment that can handle greater quantities, making it ideal for thermolabile compounds. Devices that operate in nitrogen atmospheres or vacuums have also recently been established [67].

11.6.2.3 Pulsed Electric Field Extraction

A relatively novel and creative method for extracting valuable compounds from fruit leftovers is the pulsed electric field (PEF). It's a special extraction method where a material placed between two electrodes is exposed to high microsecond pulses [68, 69]. It has been claimed that the PEF technique works well for refining the operations of clenching, dissolution, drying out, and extracting. A steam explosion is another well-liked and safe processing technique for vegetable products. It is associated with increased vapor lysis (between 160 and 280 °C) and a rapid discharge of high pressure (between 0.7 and 4.8 MPa) within a short contact time (between a few seconds and a few minutes). Following processing, the materials are released via narrow orifices, causing an explosive decomposition of biomass [70]. A typical pumpable fluid processing setup consists of a PEF generating unit, an appropriate product processing apparatus, a treatment container, and surveillance and monitoring devices [71]. PEF has no effect on the quality of the compounds recovered, but it increases the final products and yields of different chemicals. In the PEF device, the pulse amplitude ranges from 0.1–0.3 to 20–80 kV/cm. Cell membrane deterioration with little warming pattern is caused by mild electric fields (0.5 and 1 kV/cm; lasting 104–102 s). PEF is thus a great way to separate substances that are sensitive to heat. The electric field strength and treatment duration are the two main variables that define PEF therapy. The kind of energy and amount of pulse that are given determines the processing characteristics [72]. As a pre-treatment, it is non-thermal and more effective than heat-based pre-treatments in terms of recovery. Owing to the precise extraction of intracellular molecules, the energy required is negligible and does not affect the cell's overall design. Due to the higher extracts, fewer further purification steps are needed, which lowers total capital expenditure [73]. With a processing time of 60 ms and a pressing time of 30 min, the maximum polyphenol extraction was obtained at a field level of 7 kV/cm, while the maximum output of naringin and hesperidin was obtained at a field strength of 5 kV/cm. The highest outputs of polyphenol and anthocyanin were obtained at temperatures of 50 °C and 35 °C, respectively [74].

11.6.2.4 Ultrasound-Assisted Extraction

Phenolic compounds may be extracted from food waste with less time and solvent when using ultrasound-assisted extraction (UAE). UAE is regarded as one of the easiest extraction methods since it just needs a few simple lab tools, namely an ultrasonic bath [75]. The bioactive substances used in the UAE are obtained from by-products of fruits and vegetables and have a frequency ranging from 20 to 120 kHz. While high-frequency low-energy densities ultrasound produces a large amount of reactive radicals, low-frequency high intensity ultrasound offers tremendous shear and mechanical force, which is desired in the extraction process. There are significant time, temperature, energy, and chemical benefits associated with the extraction of beneficial components from fruit and vegetable waste [76–78].

11.6.2.5 Supercritical Fluid Extraction (SFE)

It has also been predicted that a different sustainable method based on supercritical CO2 (scCO2) would address the environmental issues related to more conventional methods. ScCO2 is superior to conventional solvent-based methods. Because it doesn't need concentration phases, it is perfect for lipophilic compounds like lipids because it allows for the preferred separation of molecules soluble in scCO2 [79]. The polarity of the scCO2 may be altered by adding a co-solvent, such as ethanol, which is well-regarded by many industrial companies, making it possible to separate more polar chemicals [80]. Moreover, it is possible to adequately modify the operating heat to stop the degradation of thermolabile chemicals. According to recent research, there is a noticeable improvement over conventional extraction in terms of collection, specificity, length, chemical durability, and overall power savings [81]. One key element in SFE that affects the velocity of isolation is the target compounds' solubility in scCO2. Targeting chemical solubility is mostly influenced by heat. Specifically, supercritical fluid density and solvation power rise with pressure. To maintain the right conditions throughout the extraction process, the raw materials are placed in an extraction vessel equipped with heat controls. The liquid is then pumped into the extraction vessel. After the fluid and dissolved chemicals are transported to the separators, the output is collected by a tap that is mounted at the bottom of the separators. In the end, the fluid is collected, recycled, or released into the surroundings. This method employs a variety of compounds as solvents, but the selection of the supercritical fluid is crucial to the process's success [82].

11.7 Challenges and Future Perspectives

Citrus waste has enormous potential for use in biotechnological invention and exploitation for product manufacturing, as well as in the isolation of value-added biomolecules via green extraction procedures and subsequent processing for food and other applications [83]. CW is becoming more and more popular as an affordable biomaterial for biodegradable food packaging and as a source of nutritional fiber in food compositions. Grape seed extracts are used in the textile dye industry to color cotton garments [13]. A number of other uses are being considered, mostly in the textile and agro-based sectors. Under this situation, significant scientific advancements should be directed toward the creation of sophisticated biorefineries based on citrus waste, with an emphasis on integrated, multi-product processes. Grape pomace may be easily converted into useful chemicals, fuels, and food items for the biochemical, energy, food, and feed industries [84]. To design sustainable processes that may spur economic development, engineering parameters, which are now insufficient, should be combined with research on life-cycle analysis and system analysis. Process integration and scale-up should also be investigated [85]. Pomace, peel, and seeds have a high organic content that should be considered when developing additional chemical or biotechnological processing methods. This is something that integrated biorefineries based on citrus waste must consider. Additionally, the price of CW and its seasonal availability need to be taken into consideration for total industrial output [86].

11.8 Conclusion

The use of citrus waste in the agro-food supply chain is outlined in this chapter. The inadequate pre- and post-harvest treatment and management methods result in citrus waste, but the investigation has shown that these remains, or by-products, are rich in phytochemical compounds like antioxidants, phenolic compounds, dietary fibers, and enzymes that have a lot of potential applications in the food and pharmaceutical industries. Bioactive components may be extracted from citrus waste and byproducts using cutting-edge extraction techniques including microwave-assisted extraction, ultrasound-assisted extraction, and other environmentally friendly techniques. Appropriate methods are used to bio-transform these wastes into useful products with excellent nutritional potential and low cost. Without a doubt, incorporating leftovers removes pollution-related problems in addition to disposal problems. Therefore, more government regulations and main financing are needed to get these advantageous products into the market.

Conflict of Interest The author declare no conflict of interest.

References

1. Riaz, S., Ahmad, A., Farooq, R., Hussain, N., Riaz, T., Hussain, K., et al. (2022). Citrus*: An overview of food uses and health benefits*. IntechOpen.
2. Liu, P. (2003). *World markets for organic citrus and citrus juices*. Food and Agriculture Organization of the United Nations (FAO).

3. Ma, G., Zhang, L., Sugiura, M., & Kato, M. (2020). Citrus and health. In *The genus citrus* (pp. 495–511). Elsevier.
4. Chavan, P., Singh, A. K., & Kaur, G. (2018). Recent progress in the utilization of industrial waste and by-products of citrus fruits: A review. *Journal of Food Process Engineering, 41*(8), e12895.
5. Rațu, R. N., Veleşcu, I. D., Stoica, F., Usturoi, A., Arsenoaia, V. N., Crivei, I. C., et al. (2023). Application of agri-food by-products in the food industry. *Agriculture, 13*(8), 1559.
6. Faustino, M., Veiga, M., Sousa, P., Costa, E. M., Silva, S., & Pintado, M. (2019). Agro-food byproducts as a new source of natural food additives. *Molecules, 24*(6), 1056.
7. Ahmad, N., Riaz, S., & Ali, A. (2023). Ingredients for food products. In *Palm trees and fruits residues* (pp. 115–153). Elsevier.
8. Ayala-Zavala, J., Vega-Vega, V., Rosas-Domínguez, C., Palafox-Carlos, H., Villa-Rodriguez, J., Siddiqui, M. W., et al. (2011). Agro-industrial potential of exotic fruit byproducts as a source of food additives. *Food Research International, 44*(7), 1866–1874.
9. Ali, A., Riaz, S., Sameen, A., Naumovski, N., Iqbal, M. W., Rehman, A., et al. (2022). The disposition of bioactive compounds from fruit waste, their extraction, and analysis using novel technologies: A review. *Processes, 10*(10), 2014.
10. Tlais, A. Z. A., Fiorino, G. M., Polo, A., Filannino, P., & Di Cagno, R. (2020). High-value compounds in fruit, vegetable and cereal byproducts: An overview of potential sustainable reuse and exploitation. *Molecules, 25*(13), 2987.
11. Kuna, A., Sowmya, M., Sahoo, M. R., Mayengbam, P. D., Dasgupta, M., & Sreedhar, M. (2018). Value addition and sensory evaluation of products made from underutilized Kachai lemon (Citrus jambhiri) Lush. fruits. *Journal of Pharmacognosy and Phytochemistry, 7*(5), 3032–3036.
12. Suri, S., Singh, A., & Nema, P. K. (2022). Current applications of citrus fruit processing waste: A scientific outlook. *Applied Food Research, 2*(1), 100050.
13. Sharma, M., Usmani, Z., Gupta, V. K., & Bhat, R. (2021). Valorization of fruits and vegetable wastes and by-products to produce natural pigments. *Critical Reviews in Biotechnology, 41*(4), 535–563.
14. Sharma, K., Mahato, N., Cho, M. H., & Lee, Y. R. (2017). Converting citrus wastes into value-added products: Economic and environmently friendly approaches. *Nutrition, 34*, 29–46.
15. Elkhatim, K. A. S., Elagib, R. A. A., & Hassan, A. B. (2018). Content of phenolic compounds and vitamin C and antioxidant activity in wasted parts of Sudanese citrus fruits. *Food Science & Nutrition, 6*(5), 1214–1219.
16. Rafiq, S., Kaul, R., Sofi, S., Bashir, N., Nazir, F., & Nayik, G. A. (2018). Citrus peel as a source of functional ingredient: A review. *Journal of the Saudi Society of Agricultural Sciences, 17*(4), 351–358.
17. Oyeleye, S. I., Ajayi, O. E., Ademosun, A. O., & Oboh, G. (2022). GC characterization and erectogenic enzyme inhibitory effect of essential oils from tangerine and lemon peels: A comparative study. *Flavour and Fragrance Journal, 37*(1), 33–42.
18. Lee, G. J., Lee, S. Y., Kang, N.-G., & Jin, M. H. (2022). A multi-faceted comparison of phytochemicals in seven citrus peels and improvement of chemical composition and antioxidant activity by steaming. *LWT, 160*, 113297.
19. Liu, N., Li, X., Zhao, P., Zhang, X., Qiao, O., Huang, L., et al. (2021). A review of chemical constituents and health-promoting effects of citrus peels. *Food Chemistry, 365*, 130585.
20. Saleem, M., Durani, A. I., Asari, A., Ahmed, M., Ahmad, M., Yousaf, N., et al. (2023). Investigation of antioxidant and antibacterial effects of citrus fruits peels extracts using different extracting agents: Phytochemical analysis with in silico studies. *Heliyon, 9*(4), e15433.
21. Berk, Z. (2016). By-products of the citrus processing industry. In *Citrus fruit processing* (pp. 219–233). Elsevier.
22. Rosa, A., Era, B., Masala, C., Nieddu, M., Scano, P., Fais, A., et al. (2019). Supercritical CO2 extraction of waste citrus seeds: Chemical composition, nutritional and biological properties of edible fixed oils. *European Journal of Lipid Science and Technology, 121*(7), 1800502.

23. Asif, M., Khan, M. K. I., Khan, M. I., Maan, A. A., Helmick, H., & Kokini, J. L. (2023). Effects of citrus pomace on mechanical, sensory, phenolic, antioxidant, and gastrointestinal index properties of corn extrudates. *Food Bioscience, 55*, 103012.
24. Chen, L., Wu, Y., Jiang, X., Gan, D., Fan, J., Sun, Y., et al. (2023). Dietary fiber extraction from citrus peel pomace: Yield optimization and evaluation of its functionality, rheological behavior, and microstructure properties. *Journal of Food Science, 88*(8), 3507–3523.
25. Kavitha, S., Kannah, R. Y., Kumar, G., Gunasekaran, M., & Banu, J. R. (2020). Introduction: Sources and characterization of food waste and food industry wastes. In *Food waste to valuable resources* (pp. 1–13). Elsevier.
26. Anticona, M., Blesa, J., Frigola, A., & Esteve, M. J. (2020). High biological value compounds extraction from citrus waste with non-conventional methods. *Food, 9*(6), 811.
27. Taghizadeh-Alisaraei, A., Hosseini, S. H., Ghobadian, B., & Motevali, A. (2017). Biofuel production from citrus wastes: A feasibility study in Iran. *Renewable and Sustainable Energy Reviews, 69*, 1100–1112.
28. Zema, D., Calabrò, P., Folino, A., Tamburino, V., Zappia, G., & Zimbone, S. (2018). Valorisation of citrus processing waste: A review. *Waste Management, 80*, 252–273.
29. Negro, V., Mancini, G., Ruggeri, B., & Fino, D. (2016). Citrus waste as feedstock for bio-based products recovery: Review on limonene case study and energy valorization. *Bioresource Technology, 214*, 806–815.
30. Siles, J., Vargas, F., Gutiérrez, M., Chica, A., & Martín, M. (2016). Integral valorisation of waste orange peel using combustion, biomethanisation and co-composting technologies. *Bioresource Technology, 211*, 173–182.
31. Maqbool, Z., Khalid, W., Atiq, H. T., Koraqi, H., Javaid, Z., Alhag, S. K., et al. (2023). Citrus waste as source of bioactive compounds: Extraction and utilization in health and food industry. *Molecules, 28*(4), 1636.
32. Twinomuhwezi, H., Godswill, A. C., & Kahunde, D. (2023). Extraction and characterization of pectin from orange (Citrus sinensis), lemon (Citrus limon) and tangerine (Citrus tangerina). *American Journal of Physical Sciences, 1*(1), 17–30.
33. Grover, S., Aggarwal, P., Kumar, A., Kaur, S., Yadav, R., & Babbar, N. (2024). Utilizing citrus peel waste: A review of essential oil extraction, characterization, and food-industry potential. *Biomass Conversion and Biorefinery, 15*, 1–22.
34. Kaur, S., Panesar, P. S., & Chopra, H. K. (2023). Citrus processing by-products: An overlooked repository of bioactive compounds. *Critical Reviews in Food Science and Nutrition, 63*(1), 67–86.
35. Mahato, N., Sharma, K., Sinha, M., & Cho, M. H. (2018). Citrus waste derived nutra-/pharmaceuticals for health benefits: Current trends and future perspectives. *Journal of Functional Foods, 40*, 307–316.
36. Zema, D. A. (2017). Planning the optimal site, size, and feed of biogas plants in agricultural districts. *Biofuels, Bioproducts and Biorefining, 11*(3), 454–471.
37. Satari, B., & Karimi, K. (2018). Citrus processing wastes: Environmental impacts, recent advances, and future perspectives in total valorization. *Resources, Conservation and Recycling, 129*, 153–167.
38. José, Á. S. L., Li, Q., & Thompson, I. P. (2010). Biorefinery of waste orange peel. *Critical Reviews in Biotechnology, 30*(1), 63–69.
39. Andiloro, S., Bombino, G., Tamburino, V., Zema, D. A., & Zimbone, S. M. (2013). Aerated lagooning of agro-industrial wastewater: Depuration performance and energy requirements. *Journal of Agricultural Engineering, 44*(s2), 827–832.
40. Zema, D. A., Andiloro, S., Bombino, G., Tamburino, V., Sidari, R., & Caridi, A. (2012). Depuration in aerated ponds of citrus processing wastewater with a high concentration of essential oils. *Environmental Technology, 33*(11), 1255–1260.
41. Varmie, E. B., & Thakur, M. (2021). Utilization of citrus processing waste: A review. *Pharma Innovation Journal, 10*, 682–697.

42. Bernal, V., Ros, M., Tittarelli, F., Intrigliolo, F., & Pascual, J. (2008). Citrus compost and its water extract for cultivation of melon plants in greenhouse nurseries. Evaluation of nutriactive and biocontrol effects. *Bioresource Technology, 99*(18), 8722–8728.
43. Tamburino, V., Zema, D., & Zimbone, S. (2007). Orange peel utilizations in Southern Italy. In *Proceedings of the CIGR section VI international symposium on food and agricultural products: Processing and innovations, Naples, Italy* (pp. 24–26).
44. Fernandez-Lopez, J., Zhi, N., Aleson-Carbonell, L., Pérez-Alvarez, J. A., & Kuri, V. (2005). Antioxidant and antibacterial activities of natural extracts: Application in beef meatballs. *Meat Science, 69*(3), 371–380.
45. Angulo-López, J. E., Flores-Gallegos, A. C., Ascacio-Valdes, J. A., Contreras Esquivel, J. C., Torres-León, C., Rúelas-Chácon, X., et al. (2022). Antioxidant dietary fiber sourced from agroindustrial byproducts and its applications. *Food, 12*(1), 159.
46. Belkheiri, A., Forouhar, A., Ursu, A. V., Dubessay, P., Pierre, G., Delattre, C., et al. (2021). Extraction, characterization, and applications of pectins from plant by-products. *Applied Sciences, 11*(14), 6596.
47. Andrade, M. A., Barbosa, C. H., Shah, M. A., Ahmad, N., Vilarinho, F., Khwaldia, K., et al. (2023). Citrus by-products: Valuable source of bioactive compounds for food applications. *Antioxidants, 12*(1), 38.
48. Anoopkumar, A., Aneesh, E. M., Sirohi, R., Tarafdar, A., Kuriakose, L. L., Surendhar, A., et al. (2023). Bioactives from citrus food waste: Types, extraction technologies and application. *Journal of Food Science and Technology, 61*, 444–458.
49. Rifna, E., Misra, N., & Dwivedi, M. (2023). Recent advances in extraction technologies for recovery of bioactive compounds derived from fruit and vegetable waste peels: A review. *Critical Reviews in Food Science and Nutrition, 63*(6), 719–752.
50. Wen, L., Zhang, Z., Sun, D.-W., Sivagnanam, S. P., & Tiwari, B. K. (2020). Combination of emerging technologies for the extraction of bioactive compounds. *Critical Reviews in Food Science and Nutrition, 60*(11), 1826–1841.
51. Milovanovic, S., Lukic, I., Stamenic, M., Kamiński, P., Florkowski, G., Tyśkiewicz, K., et al. (2022). The effect of equipment design and process scale-up on supercritical CO2 extraction: Case study for Silybum marianum seeds. *The Journal of Supercritical Fluids, 188*, 105676.
52. Gadi, S., Pérez-Vega, S., Minjares-Fuentes, R., Morales-Oyervides, L., Contreras-Esquivel, J. C., & Montañez, J. (2024). Novel extraction technologies for the recovery of bioactive compounds from citrus by-products: Recent findings. In *Bioresources and bioprocess in biotechnology for a sustainable future* (pp. 201–226). CRC Press.
53. Khoddami, A., Wilkes, M. A., & Roberts, T. H. (2013). Techniques for analysis of plant phenolic compounds. *Molecules, 18*(2), 2328–2375.
54. Sirisompong, W., Jirapakkul, W., & Klinkesorn, U. (2011). Response surface optimization and characteristics of rambutan (Nephelium lappaceum L.) kernel fat by hexane extraction. *LWT-Food Science and Technology, 44*(9), 1946–1951.
55. Chenni, M., El Abed, D., Rakotomanomana, N., Fernandez, X., & Chemat, F. (2016). Comparative study of essential oils extracted from Egyptian basil leaves (Ocimum basilicum L.) using hydro-distillation and solvent-free microwave extraction. *Molecules, 21*(1), 113.
56. Calvi, L., Pavlovic, R., Panseri, S., Giupponi, L., Leoni, V., & Giorgi, A. (2018). Quality traits of medical Cannabis sativa L. inflorescences and derived products based on comprehensive mass-spectrometry analytical investigation. In *Recent advances in cannabinoid research* (pp. 55–78). IntechOpen.
57. Rashid, R., Wani, S. M., Manzoor, S., Masoodi, F., & Dar, M. M. (2023). Green extraction of bioactive compounds from apple pomace by ultrasound assisted natural deep eutectic solvent extraction: Optimisation, comparison and bioactivity. *Food Chemistry, 398*, 133871.
58. Singh, G., Verma, A., & Kumar, V. (2016). Catalytic properties, functional attributes and industrial applications of β-glucosidases. *3 Biotech, 6*, 1–14.
59. Puri, M., Sharma, D., & Barrow, C. J. (2012). Enzyme-assisted extraction of bioactives from plants. *Trends in Biotechnology, 30*(1), 37–44.

60. Nadar, S. S., Rao, P., & Rathod, V. K. (2018). Enzyme assisted extraction of biomolecules as an approach to novel extraction technology: A review. *Food Research International, 108*, 309–330.
61. Maled, S. B., Bhat, A. R., Hegde, S., Sivamani, Y., Muthuraman, A., & Elayaperumal, S. (2024). Enzyme-assisted extraction. In *Bioactive extraction and application in food and nutraceutical industries* (pp. 173–200). Springer.
62. Wang, L., Wu, Y., Liu, Y., & Wu, Z. (2017). Complex enzyme-assisted extraction releases antioxidative phenolic compositions from guava leaves. *Molecules, 22*(10), 1648.
63. Cheng, X., Bi, L., Zhao, Z., & Chen, Y. (2015). Advances in enzyme assisted extraction of natural products. In *3rd International conference on material, mechanical and manufacturing engineering (IC3ME 2015)* (pp. 371–375). Atlantis Press.
64. Zhang, M., Ma, W., Wang, C., Yang, X., Lou, Y., Xia, X., et al. (2021). Optimization of enzyme-assisted extraction and purification of flavonoids from Pinus koraiensis nut-coated film and antioxidant activity evaluation. *Molecules, 26*(7), 1950.
65. Turan, O., Isci, A., Yılmaz, M. S., Tolun, A., & Sakiyan, O. (2024). Microwave-assisted extraction of pectin from orange peel using deep eutectic solvents. *Sustainable Chemistry and Pharmacy, 37*, 101352.
66. Ran, X.-l., Zhang, M., Wang, Y., & Adhikari, B. (2019). Novel technologies applied for recovery and value addition of high value compounds from plant byproducts: A review. *Critical Reviews in Food Science and Nutrition, 59*(3), 450–461.
67. Talmaciu, A. I., Volf, I., & Popa, V. I. (2015). A comparative analysis of the "green" techniques applied for polyphenols extraction from bioresources. *Chemistry & Biodiversity, 12*(11), 1635–1651.
68. Manzoor, M. F., Zeng, X. A., Ahmad, N., Ahmed, Z., Rehman, A., Aadil, R. M., et al. (2020). Effect of pulsed electric field and thermal treatments on the bioactive compounds, enzymes, microbial, and physical stability of almond milk during storage. *Journal of Food Processing and Preservation, 44*(7), e14541.
69. Manzoor, M. F., Ahmed, Z., Ahmad, N., Aadil, R. M., Rahaman, A., Roobab, U., et al. (2020). Novel processing techniques and spinach juice: Quality and safety improvements. *Journal of Food Science, 85*(4), 1018–1026.
70. Wang, K., Chen, J., Sun, S.-N., & Sun, R.-C. (2015). Steam explosion. In *Pretreatment of biomass* (pp. 75–104). Elsevier.
71. Barba, F. J., Parniakov, O., Pereira, S. A., Wiktor, A., Grimi, N., Boussetta, N., et al. (2015). Current applications and new opportunities for the use of pulsed electric fields in food science and industry. *Food Research International, 77*, 773–798.
72. Martínez, J. M., Delso, C., Angulo, J., Álvarez, I., & Raso, J. (2018). Pulsed electric field-assisted extraction of carotenoids from fresh biomass of Rhodotorula glutinis. *Innovative Food Science & Emerging Technologies, 47*, 421–427.
73. Dhua, S., Kumar, K., Sharanagat, V. S., & Nema, P. K. (2022). Bioactive compounds and its optimization from food waste: Review on novel extraction techniques. *Nutrition & Food Science, 52*(8), 1270–1288.
74. Brianceau, S., Turk, M., Vitrac, X., & Vorobiev, E. (2015). Combined densification and pulsed electric field treatment for selective polyphenols recovery from fermented grape pomace. *Innovative Food Science & Emerging Technologies, 29*, 2–8.
75. Garcia-Salas, P., Morales-Soto, A., Segura-Carretero, A., & Fernández-Gutiérrez, A. (2010). Phenolic-compound-extraction systems for fruit and vegetable samples. *Molecules, 15*(12), 8813–8826.
76. Paniwnyk, L., Alarcon-Rojo, A., Rodriguez-Figueroa, J. C., & Toma, M. (2017). The use of ultrasound as an enhancement aid to food extraction. In *Ingredients extraction by physicochemical methods in food* (pp. 399–440). Elsevier.
77. Chen, X., Fang, D., Zhao, R., Gao, J., Kimatu, B. M., Hu, Q., et al. (2019). Effects of ultrasound-assisted extraction on antioxidant activity and bidirectional immunomodulatory activity of

Flammulina velutipes polysaccharide. *International Journal of Biological Macromolecules, 140*, 505–514.
78. Dzah, C. S., & Dzigbor, A. (2023). Ultrasound assisted extraction: A relook at solvent to material ratio, its effects on process efficiency and how it can be exploited for different uses. *Journal of Food Process Engineering, 46*(7), e14339.
79. Da Silva, R. P., Rocha-Santos, T. A., & Duarte, A. C. (2016). Supercritical fluid extraction of bioactive compounds. *TrAC Trends in Analytical Chemistry, 76*, 40–51.
80. De Melo, M., Silvestre, A., & Silva, C. (2014). Supercritical fluid extraction of vegetable matrices: Applications, trends and future perspectives of a convincing green technology. *The Journal of Supercritical Fluids, 92*, 115–176.
81. Blicharski, T., & Oniszczuk, A. (2017). Extraction methods for the isolation of isoflavonoids from plant material. *Open Chemistry, 15*(1), 34–45.
82. Leiper, K. A., & Miedl, M. (2006). Brewhouse technology. In *Handbook of brewing* (pp. 398–461). CRC Press.
83. Sirohi, R., Tarafdar, A., Singh, S., Negi, T., Gaur, V. K., Gnansounou, E., et al. (2020). Green processing and biotechnological potential of grape pomace: Current trends and opportunities for sustainable biorefinery. *Bioresource Technology, 314*, 123771.
84. Yadav, V., Sarker, A., Yadav, A., Miftah, A. O., Bilal, M., & Iqbal, H. M. (2022). Integrated biorefinery approach to valorize citrus waste: A sustainable solution for resource recovery and environmental management. *Chemosphere, 293*, 133459.
85. Ali, A., Yu, L., Kousar, S., Khalid, W., Maqbool, Z., Aziz, A., et al. (2022). Crocin: Functional characteristics, extraction, food applications and efficacy against brain related disorders. *Frontiers in Nutrition, 9*, 1009807.
86. Benny, N., Shams, R., Dash, K. K., Pandey, V. K., & Bashir, O. (2023). Recent trends in utilization of citrus fruits in production of eco-enzyme. *Journal of Agriculture and Food Research, 13*, 100657.

Chapter 12
Citrus Waste as a Potential Source of Nutraceuticals

C. C. Ezemba, Victor O Etikudike, O. J. Osuala, A. S. Ezemba, P. I. Ezemba, V. E. Anyaogu, J. A. Mmaduekwe, I. W. Udoye, P. S. Obiasogu, A. Salvagain, and C. Ifeanyi-ofoka

12.1 Introduction

Citrus fruits, renowned for their vibrant colors, refreshing flavors, and rich nutritional content, have long been cherished as dietary staples and culinary delights across the globe. From the succulent sweetness of oranges to the tangy tartness of lemons and the zesty aroma of limes, citrus fruits have found their way into diverse culinary traditions, beverages, desserts, and health-promoting products [1]. However, with the increasing demand for citrus products, there comes a consequential challenge: the generation of substantial amounts of citrus waste during processing. This book chapter aims to explore the untapped potential of citrus waste as a valuable source of nutraceuticals, unlocking opportunities for sustainable utilization and value addition within the citrus industry.

In recent years, there has been a growing interest in the concept of nutraceuticals—products derived from food sources with purported health benefits beyond basic nutritional value. This burgeoning industry bridges the gap between food and

C. C. Ezemba (✉) · V. O. Etikudike · A. S. Ezemba · P. I. Ezemba · J. A. Mmaduekwe
I. W. Udoye
ChyChy Gilgal Laboratory and Consultancy Services, Ichida Nnewi, Anambra State, Nigeria
e-mail: cc.ezemba@coou.edu.ng

O. J. Osuala
Department of Pharmaceutical Microbiology and Biotechnology, Madonna University, Elele, Rivers State, Nigeria

V. E. Anyaogu · P. S. Obiasogu · A. Salvagain
Department of Microbiology, Chukwuemeka Odumegwu Ojukwu University, Uli, Anambra State, Nigeria

C. Ifeanyi-ofoka
Department of Biochemistry, Chukwuemeka Odumegwu Ojukwu University, Uli, Anambra State, Nigeria

A. Chauhan et al. (eds.), *Valorization of Citrus Food Waste*,
https://doi.org/10.1007/978-3-031-77999-2_12

pharmaceuticals, offering functional foods, dietary supplements, and natural health products formulated to promote health, prevent disease, and enhance overall well-being. Citrus waste, once viewed as a mere byproduct of fruit processing, is now gaining recognition as a promising reservoir of bioactive compounds with potential nutraceutical properties [1].

As the global demand for citrus products continues to soar, so does the volume of citrus waste generated by fruit processing facilities, juice extraction plants, and citrus packinghouses. Citrus waste comprises various components, including peels, pulp, seeds, and membranes, which are typically discarded or utilized for low-value purposes such as animal feed, composting, or landfilling. However, the disposal of citrus waste poses environmental challenges, including waste management issues, land degradation, greenhouse gas emissions, and potential pollution of soil and water resources [1, 2]. In light of these concerns, there is a pressing need to explore innovative approaches for valorizing citrus waste and extracting maximum value from this abundant resource.

Citrus waste contains a diverse array of bioactive compounds, including flavonoids, polyphenols, vitamins, minerals, dietary fiber, and essential oils, which exhibit various health-promoting properties such as antioxidant, anti-inflammatory, antimicrobial, cardioprotective, and anti-obesity effects. These bioactive compounds, once extracted and concentrated from citrus waste, can be incorporated into nutraceutical formulations, functional foods, dietary supplements, and pharmaceutical preparations, offering a natural and sustainable alternative to synthetic ingredients and chemical additives [3].

The extraction of bioactive compounds from citrus waste presents both opportunities and challenges, necessitating the adoption of efficient and eco-friendly extraction technologies. Various extraction methods, including solvent extraction, supercritical fluid extraction (SFE), microwave-assisted extraction (MAE), enzyme-assisted extraction (EAE), and ultrasound-assisted extraction (UAE), offer distinct advantages in terms of yield, selectivity, efficiency, and sustainability [2, 3]. By optimizing extraction parameters and selecting appropriate techniques, researchers can maximize the recovery of bioactive compounds from citrus waste while minimizing environmental impact and energy consumption.

The utilization of citrus waste-derived nutraceuticals extends beyond the realm of health supplements, encompassing a wide range of applications in functional foods, beverages, cosmetics, pharmaceuticals, and agrifood industries. Citrus waste extracts can be incorporated into formulations for antioxidant-rich beverages, fortified foods, natural food colorants, skincare products, pharmaceutical formulations, and agricultural inputs such as biopesticides and plant growth promoters [4]. Furthermore, the growing consumer demand for natural, plant-based ingredients presents lucrative market opportunities for citrus waste-derived nutraceuticals, driving innovation and market expansion within the nutraceutical industry.

The exploration of citrus waste as a potential source of nutraceuticals represents a paradigm shift in the utilization of agricultural byproducts, transforming waste into wealth and creating value-added opportunities within the citrus industry. By harnessing the bioactive potential of citrus waste and leveraging advanced extraction technologies, researchers and industry stakeholders can pave the way for sustainable solutions to waste management challenges while promoting human health and well-being through the development of functional, natural products derived from citrus waste [5–7]. This chapter endeavors to review the various citrus waste utilization, offering insights, innovations, and inspirations for unlocking the untapped potential of citrus waste as a valuable resource for nutraceutical discovery and development.

12.2 Nutritional Composition of Citrus Waste

12.2.1 Overview of Citrus Fruits

Citrus fruits are a diverse group of fruits belonging to the Rutaceae family, primarily cultivated for their edible fruit, juice, and peel. They are characterized by their refreshing flavor, vibrant colors, and high nutritional value, making them popular worldwide. This section provides an elaborate overview of citrus fruits and their production, including their botanical classification, global distribution, cultivation practices, and economic significance.

Botanical Classification Citrus fruits belong to the genus Citrus and are classified within the family Rutaceae. The genus Citrus comprises various species, hybrids, and cultivars, each exhibiting unique characteristics in terms of fruit size, shape, color, flavor, and nutritional composition. Common citrus species include oranges (Citrus sinensis), lemons (Citrus limon), limes (Citrus aurantiifolia), grapefruits (Citrus paradisi), and mandarins (Citrus reticulata) [1–3].

Global Distribution Citrus fruits are cultivated in various regions across the world, with major production hubs located in tropical and subtropical climates. Leading citrus-producing countries include Brazil, China, India, the United States, Spain, and Mexico. These countries benefit from favorable environmental conditions such as ample sunlight, moderate temperatures, well-drained soils, and sufficient water availability, which are conducive to citrus cultivation [2, 3].

Cultivation Practices The cultivation of citrus fruits typically involves several key practices, including site selection, soil preparation, planting, irrigation, fertilization, pest and disease management, and harvesting. Citrus trees are generally propagated through seeds, cuttings, or grafting onto rootstocks, depending on the desired characteristics of the cultivar. They require well-drained soils with adequate organic matter content and pH levels ranging from 6 to 7. Citrus trees are generally spaced apart to allow for optimal growth, air circulation, and light penetration [1, 4].

Economic Significance Citrus fruits hold significant economic importance due to their widespread consumption and commercial value. They are cultivated for various purposes, including fresh fruit consumption, juice extraction, processing into canned or frozen products, and extraction of essential oils and bioactive compounds for pharmaceutical and cosmetic industries. Citrus cultivation contributes to the livelihoods of millions of people worldwide, providing employment opportunities, income generation, and export earnings for producing countries [5].

12.2.2 Overview of Citrus Waste

Citrus waste, comprising various components such as peels, pulp, seeds, and membranes, is generated during the processing of citrus fruits in juice extraction plants, fruit processing facilities, and citrus packinghouses. Historically viewed as a byproduct of fruit processing, citrus waste presents both challenges and opportunities in terms of waste management, environmental sustainability, and value addition within the citrus industry [7].

Composition of Citrus Waste

Citrus waste encompasses different parts of the citrus fruit that are not utilized for primary product manufacturing. The primary constituents of citrus waste include:

- *Peels*: The outer rind or peel of citrus fruits, comprising the zest and flavedo layers rich in essential oils, flavonoids, and pectin.
- *Pulp*: The fleshy interior of citrus fruits, containing juice sacs, fibers, and soluble solids such as sugars, organic acids, and vitamins.
- *Seeds*: The seeds or kernels found within citrus fruits, which may contain oils, proteins, and bioactive compounds.
- *Membranes*: The thin, fibrous membranes that separate citrus fruit segments, containing pectin, fiber, and bioactive compounds.

Generation of Citrus Waste

Citrus waste is generated as a byproduct of citrus fruit processing operations, including juice extraction, fruit pulping, and citrus peeling. During these processes, citrus fruits undergo mechanical squeezing, pressing, or cutting to extract juice or pulp, resulting in the separation of the edible portion from the non-edible components. The discarded peels, pulp residues, seeds, and membranes constitute citrus waste, which is typically collected and disposed of through various methods such as composting, landfilling, or animal feed supplementation [8].

Disposal Methods and Environmental Impact

Traditionally, citrus waste has been disposed of through landfilling, composting, or animal feed supplementation. However, these disposal methods pose environmental challenges and sustainability concerns. Landfilling of citrus waste contributes to solid waste accumulation, methane emissions, and soil pollution, while composting may require significant time and resources for decomposition. Moreover, utilizing citrus waste as animal feed may not fully utilize its nutritional potential and may not be economically viable in all contexts [9].

Potential Utilization Avenues for Citrus Waste

In recent years, there has been growing interest in exploring alternative uses for citrus waste, driven by concerns about waste management, environmental sustainability, and resource optimization. Several potential utilization avenues for citrus waste have emerged, including:

- *Nutraceutical Extraction*: Citrus waste contains bioactive compounds such as flavonoids, polyphenols, vitamins, and dietary fiber, which exhibit various health-promoting properties. Extracting these bioactive compounds from citrus waste for use in nutraceuticals, dietary supplements, and functional foods offers a sustainable approach to value addition and waste reduction within the citrus industry [1].
- *Biofuel Production*: Citrus waste can be utilized as feedstock for biofuel production through processes such as anaerobic digestion, fermentation, or pyrolysis. The organic matter present in citrus waste can be converted into biogas, bioethanol, or bio-oil, offering renewable energy sources and reducing reliance on fossil fuels [1, 2].
- *Animal Feed Supplementation*: Citrus waste can be processed and incorporated into animal feed formulations as a source of dietary fiber, energy, and essential nutrients for livestock and poultry. However, proper processing and supplementation strategies are required to mitigate potential anti-nutritional factors and ensure animal health and performance.
- *Agricultural and Soil Amendment*: Citrus waste-derived products such as compost, biochar, or organic fertilizers can be utilized as soil amendments to improve soil structure, fertility, and microbial activity. These products can enhance nutrient cycling, water retention, and crop productivity in agricultural systems, promoting sustainable agriculture and soil health.
- *Cosmetic and Pharmaceutical Applications*: Citrus waste extracts containing essential oils, flavonoids, and antioxidants can be utilized in cosmetic and pharmaceutical formulations for skincare, haircare, and therapeutic purposes. Citrus essential oils are valued for their aromatic properties, antimicrobial activity, and potential health benefits in aromatherapy and personal care products [3].

Citrus waste represents a valuable yet underutilized resource with diverse potential applications in various sectors, including food, agriculture, energy, cosmetics, and pharmaceuticals. By exploring innovative utilization avenues for citrus waste and implementing sustainable waste management practices, stakeholders in the citrus industry can unlock value, reduce environmental impact, and promote circular economy principles. Further research, collaboration, and investment are needed to maximize the utilization of citrus waste and harness its full potential as a valuable resource for sustainable development and economic growth.

12.3 Bioactive Compounds Found in Citrus Waste

Citrus waste, comprising peels, pulp, seeds, and membranes discarded during fruit processing, contains a rich array of bioactive compounds that contribute to its nutritional and functional properties. These bioactive compounds include flavonoids, polyphenols, vitamins, minerals, dietary fiber, essential oils, and other phytochemicals, each offering various health-promoting benefits and potential applications in nutraceuticals, functional foods, cosmetics, and pharmaceuticals [11].

Flavonoids

Flavonoids are a class of polyphenolic compounds abundant in citrus fruits and their byproducts, including citrus peels and seeds. These compounds, including flavanones (e.g., hesperidin, narirutin), flavones (e.g., naringenin), and flavonols (e.g., quercetin, kaempferol), exhibit antioxidant, anti-inflammatory, anti-cancer, and cardioprotective properties. Hesperidin, one of the predominant flavonoids found in citrus peels, has been studied for its potential benefits in reducing oxidative stress, lowering cholesterol levels, and improving cardiovascular health [11, 12].

Polyphenols

Polyphenols are secondary metabolites abundant in citrus fruits and their byproducts, contributing to their antioxidant and anti-inflammatory properties. Citrus polyphenols, including phenolic acids (e.g., ferulic acid, caffeic acid), flavonoids, and tannins, exhibit diverse biological activities such as free radical scavenging, metal chelation, and enzyme inhibition. These compounds play a crucial role in protecting cells from oxidative damage, reducing the risk of chronic diseases, and promoting overall health and well-being [12].

Vitamins

Citrus fruits are renowned for their high vitamin content, particularly vitamin C (ascorbic acid), which is concentrated in the juice and peels of citrus fruits. Vitamin C serves as a potent antioxidant, supporting immune function, collagen synthesis, and iron absorption. Citrus peels also contain significant amounts of vitamin A (beta-carotene), vitamin E (tocopherols), and vitamin B complex (e.g., thiamine, riboflavin, niacin), which contribute to their nutritional value and health benefits [7–9].

Minerals

Citrus waste contains essential minerals such as potassium, calcium, magnesium, phosphorus, and trace elements like zinc, copper, and manganese. These minerals play vital roles in various physiological processes, including bone health, muscle function, nerve transmission, and enzyme activation. Citrus peels, in particular, are rich in calcium and potassium, making them valuable sources of these essential minerals for human nutrition and health.

Dietary Fiber

Citrus peels and pulp residues are rich sources of dietary fiber, including soluble and insoluble fibers such as pectin, cellulose, and hemicellulose. Dietary fiber contributes to digestive health, weight management, and blood sugar control by promoting satiety, regulating bowel movements, and modulating gut microbiota composition. Incorporating citrus waste-derived dietary fiber into functional foods and supplements can enhance their nutritional value and promote gastrointestinal health [8, 10].

Essential Oils

Citrus peels contain volatile essential oils rich in terpenes, limonene, and other aromatic compounds responsible for their characteristic aroma and flavor. Citrus essential oils possess antimicrobial, antifungal, and anti-inflammatory properties, making them valuable ingredients in aromatherapy, personal care products, and household cleaners. These oils are extracted from citrus peels through processes such as cold pressing, steam distillation, or solvent extraction and can be utilized in various applications, including food flavoring, fragrance formulation, and therapeutic use [11].

Other Phytochemicals

In addition to the well-known bioactive compounds such as flavonoids, polyphenols, vitamins, minerals, dietary fiber, and essential oils, citrus waste also contains other phytochemicals that contribute to its nutritional and functional properties. These phytochemicals include carotenoids, coumarins, alkaloids, and lignans, each offering unique health benefits and potential applications [9–11].

I. *Carotenoids*: Carotenoids are pigments responsible for the yellow, orange, and red colors in citrus fruits and their peels. Examples of carotenoids found in citrus waste include beta-carotene, lutein, and zeaxanthin. These compounds possess antioxidant properties and are precursors of vitamin A, playing essential roles in eye health, immune function, and skin health. Carotenoids in citrus waste contribute to its nutritional value and antioxidant activity, offering potential benefits for human health and well-being.

II. *Coumarins*: Coumarins are aromatic compounds found in citrus peels and seeds, known for their antioxidant, anti-inflammatory, and antimicrobial properties. Citrus coumarins, such as umbelliferone and scoparone, exhibit various biological activities, including anti-coagulant, vasodilatory, and hepatoprotective effects. These compounds have been studied for their potential therapeutic applications in cardiovascular health, skin disorders, and metabolic diseases, highlighting their value as functional ingredients derived from citrus waste.

III. *Alkaloids*: these are nitrogen-containing compounds found in citrus seeds and certain varieties of citrus fruits. While present in relatively low concentrations, citrus alkaloids such as synephrine and octopamine have been studied for their physiological effects, including appetite suppression, metabolism regulation, and energy expenditure enhancement. Citrus alkaloids are being investigated for their potential role in weight management, sports performance, and metabolic health, offering opportunities for further research and development.

IV. *Lignans*: Lignans are polyphenolic compounds found in citrus seeds and pulp residues, known for their antioxidant, estrogenic, and anti-cancer properties.

Citrus lignans, such as naringenin and hesperetin, exhibit estrogenic activity and have been studied for their potential role in hormone balance, bone health, and menopausal symptoms. These compounds may also have anti-inflammatory effects and contribute to the overall health-promoting properties of citrus waste-derived products.

These phytochemicals, along with flavonoids, polyphenols, vitamins, minerals, dietary fiber, and essential oils, collectively contribute to the nutritional and functional properties of citrus waste. By harnessing the bioactive potential of these phytochemicals through extraction, purification, and formulation processes, researchers and industry stakeholders can unlock new avenues for utilization and value addition within the citrus industry and beyond [12, 13]. Further research into the bioactivities, mechanisms of action, and potential health benefits of these phytochemicals is warranted to fully understand their impact on human health and well-being.

Citrus waste is a valuable source of bioactive compounds with diverse health-promoting properties and potential applications in nutraceuticals, functional foods, cosmetics, and pharmaceuticals. The bioactive compounds found in citrus waste, including flavonoids, polyphenols, vitamins, minerals, dietary fiber, essential oils, and other phytochemicals, contribute to its antioxidant, anti-inflammatory, antimicrobial, and cardioprotective effects, supporting human health and well-being [13]. Harnessing these bioactive compounds from citrus waste through extraction, purification, and formulation processes offers opportunities for sustainable utilization, waste reduction, and value addition within the citrus industry and beyond. Further research, innovation, and collaboration are needed to unlock the full potential of citrus waste-derived bioactive compounds and promote their integration into functional products and applications for the benefit of consumers and society.

12.4 Extraction Techniques for Nutraceuticals from Citrus Waste

The extraction of nutraceuticals from citrus waste involves the separation and concentration of bioactive compounds, including flavonoids, polyphenols, vitamins, and essential oils, from the solid matrix of citrus peels, pulp residues, seeds, and membranes. Various extraction techniques have been developed to maximize the recovery of bioactive compounds while minimizing solvent usage, energy consumption, and environmental impact [13]. This section provides a detailed exploration of the extraction techniques commonly used for nutraceutical extraction from citrus waste, including their principles, advantages, limitations, and applications.

1. Solvent Extraction Methods
Solvent extraction is one of the most widely used techniques for nutraceutical extraction from citrus waste. It involves the use of organic solvents such as ethanol, methanol, acetone, or ethyl acetate to dissolve and extract bioactive compounds

from the solid matrix of citrus waste. The solvent-extraction process can be performed using conventional techniques such as maceration, Soxhlet extraction, or reflux extraction, as well as modern methods such as accelerated solvent extraction (ASE) or pressurized liquid extraction (PLE). Solvent extraction offers high extraction efficiency, versatility, and scalability, making it suitable for large-scale production of citrus waste-derived nutraceuticals [14].

Examples of Solvent extraction

(a) Maceration

Maceration is a simple and conventional solvent extraction method commonly used for extracting bioactive compounds from citrus waste. The process involves soaking citrus waste (e.g., peels, pulp residues) in a suitable organic solvent (e.g., ethanol, methanol) for a specified period to allow the solvents to penetrate the solid matrix and dissolve the target compounds. The mixture is then agitated or stirred intermittently to enhance mass transfer and extraction efficiency. After the desired extraction time, the solvent extract is separated from the solid residue by filtration or centrifugation. The extract can be further concentrated by evaporation under reduced pressure or rotary evaporation to obtain a concentrated extract rich in bioactive compounds [13, 14].

(b) Accelerated Solvent Extraction (ASE)

Accelerated solvent extraction (ASE), also known as pressurized solvent extraction (PSE) or pressurized liquid extraction (PLE), is a modern extraction technique that utilizes elevated temperature and pressure to accelerate the extraction process. ASE involves placing citrus waste in an extraction cell or cartridge and adding a solvent to the system. The extraction cell is then sealed and pressurized, allowing the solvent to penetrate the solid matrix and dissolve the target compounds rapidly. The system is heated to the desired temperature, and extraction occurs under controlled conditions. After extraction, the solvent extract is collected, and the extract can be concentrated by evaporation or solvent removal. ASE offers advantages such as rapid extraction, improved efficiency, and reduced solvent usage compared to conventional methods [14].

(c) Soxhlet Extraction

Soxhlet extraction is a classical extraction technique used for extracting bioactive compounds from citrus waste using continuous refluxing with a solvent. The process involves placing citrus waste in a porous extraction thimble and inserting it into a Soxhlet extractor, which is connected to a round-bottom flask containing the solvent. The solvent is heated to its boiling point, causing it to vaporize and condense repeatedly, cycling through the citrus waste in the extraction thimble. As the solvent passes through the solid matrix, it dissolves the target compounds, and the resulting extract is collected in the round-bottom flask. Soxhlet extraction allows for prolonged contact between the solvent and the citrus waste, leading to efficient extraction of bioactive compounds. However, this method requires a large volume of solvent and is time-consuming [15].

Solvent extraction methods offer versatile and effective approaches for extracting bioactive compounds from citrus waste, providing valuable ingredients for nutraceuticals, functional foods, and natural health products. By optimizing extraction parameters and selecting appropriate solvents, researchers and industry stakeholders can maximize the recovery of bioactive compounds while minimizing environmental impact and resource consumption. Further research into solvent extraction techniques, including process optimization, solvent recycling, and green solvent development, is essential to advancing sustainable extraction practices and harnessing the full potential of citrus waste as a valuable resource for nutraceutical discovery and development [15, 16].

2. Supercritical Fluid Extraction (SFE)
Supercritical fluid extraction (SFE) utilizes supercritical fluids such as carbon dioxide (CO2) as solvents to extract bioactive compounds from citrus waste. Under specific temperature and pressure conditions, CO2 becomes a supercritical fluid with properties of both liquids and gases, enabling efficient extraction of lipophilic compounds such as essential oils and certain polyphenols. SFE offers several advantages, including high selectivity, minimal solvent residue, and environmentally friendly operation. However, SFE equipment can be costly, and optimization of extraction parameters is required to achieve optimal yields [17].

3. Microwave-Assisted Extraction (MAE)
Microwave-assisted extraction (MAE) employs microwave radiation to enhance the extraction of bioactive compounds from citrus waste. Microwave energy penetrates the solid matrix of citrus waste, causing rapid heating and localized extraction of target compounds. MAE offers several advantages, including reduced extraction time, improved efficiency, and lower solvent consumption compared to conventional extraction methods. However, careful optimization of microwave power, extraction time, and solvent composition is necessary to prevent thermal degradation of heat-sensitive compounds and ensure reproducible results [14, 16].

4. Enzyme-Assisted Extraction (EAE)
Enzyme-assisted extraction (EAE) utilizes enzymes such as cellulase, pectinase, or protease to degrade the structural components of citrus waste and release entrapped bioactive compounds. Enzymes selectively target specific substrates within the citrus matrix, facilitating the extraction of target compounds while minimizing non-specific degradation. EAE offers advantages such as mild processing conditions, high specificity, and minimal solvent usage. However, the selection of appropriate enzymes, optimization of extraction conditions, and enzyme stability are critical factors influencing extraction efficiency [17].

5. Ultrasound-Assisted Extraction (UAE)
Ultrasound-assisted extraction (UAE) employs high-frequency ultrasound waves to disrupt the cell walls and enhance the extraction of bioactive compounds from citrus waste. Ultrasound energy generates cavitation bubbles within the solvent, leading to the formation of microjets and shockwaves that facilitate mass transfer and cell

disruption. UAE offers advantages such as rapid extraction, increased yields, and reduced solvent consumption compared to conventional methods. However, optimization of ultrasound parameters, including frequency, intensity, and extraction time, is essential to avoid excessive heating and degradation of heat-sensitive compounds [18].

12.5 Health Benefits and Functional Properties of Citrus Byproducts

Citrus byproducts, including peels, pulp residues, seeds, and membranes generated during fruit processing, are rich sources of bioactive compounds with various health-promoting properties and functional attributes. This review provides a comprehensive overview of the health benefits and functional properties of citrus byproducts, highlighting their potential applications in promoting human health and well-being.

1. Antioxidant Activity
Citrus byproducts are renowned for their high antioxidant content, attributed to the presence of flavonoids, polyphenols, vitamins, and other bioactive compounds. These antioxidants help neutralize free radicals and oxidative stress, protecting cells from damage and reducing the risk of chronic diseases such as cancer, cardiovascular diseases, and neurodegenerative disorders. The antioxidant activity of citrus byproducts contributes to their anti-aging, anti-inflammatory, and immune-boosting properties, supporting overall health and well-being [12, 13].

2. Anti-inflammatory Effects
In addition to their antioxidant properties, citrus byproducts exhibit anti-inflammatory effects attributed to bioactive compounds such as flavonoids and polyphenols. These compounds help modulate inflammatory pathways, reduce inflammation, and alleviate symptoms associated with inflammatory conditions such as arthritis, asthma, and inflammatory bowel disease. Consumption of citrus byproducts may help mitigate chronic inflammation and promote tissue repair and regeneration, supporting musculoskeletal health and immune function [11].

3. Cardiovascular Health
Citrus byproducts have been shown to benefit cardiovascular health through various mechanisms, including cholesterol-lowering, blood pressure regulation, and anti-platelet effects. Flavonoids such as hesperidin and naringin found in citrus byproducts help improve lipid profiles, reduce LDL cholesterol levels, and inhibit platelet aggregation, lowering the risk of atherosclerosis and cardiovascular events. Additionally, polyphenols and dietary fiber in citrus byproducts support endothelial function, blood vessel integrity, and overall cardiovascular function.

4. Weight Management

Dietary fiber, particularly soluble fiber, found in citrus byproducts plays a crucial role in weight management by promoting satiety, regulating appetite, and modulating energy metabolism. Consumption of citrus byproducts can help control food intake, prevent overeating, and support healthy weight loss or weight maintenance. Additionally, the low calorie and low-fat content of citrus byproducts make them suitable components of weight-loss diets and calorie-controlled meal plans, providing nutritional value without excess energy intake [19].

5. Digestive Health

Citrus byproducts are rich sources of dietary fiber, including soluble and insoluble fibers such as pectin, cellulose, and hemicellulose. Dietary fiber supports digestive health by promoting regular bowel movements, preventing constipation, and improving stool consistency. Fiber also acts as a prebiotic, nourishing beneficial gut bacteria and promoting a healthy gut microbiota composition. Consumption of citrus byproducts can help alleviate digestive discomfort, promote gastrointestinal motility, and support overall digestive function [20].

6. Metabolic Health

Bioactive compounds in citrus byproducts, including flavonoids, polyphenols, and vitamins, exert beneficial effects on metabolic health by regulating glucose metabolism, insulin sensitivity, and lipid metabolism. These compounds help modulate key metabolic pathways involved in energy homeostasis, lipid storage, and glucose uptake, reducing the risk of metabolic disorders such as type 2 diabetes and obesity. Additionally, citrus byproducts may have thermogenic effects, increasing energy expenditure and promoting fat oxidation, further supporting metabolic health and weight management [19–21].

7. Immune Function

Vitamins, minerals, and phytochemicals present in citrus byproducts play essential roles in supporting immune function and enhancing resistance to infections. Vitamin C, abundant in citrus fruits and their byproducts, is known for its immune-boosting properties, stimulating the production of white blood cells and antibodies, and enhancing the body's defense against pathogens. Additionally, flavonoids and polyphenols in citrus byproducts exhibit antimicrobial, antiviral, and immunomodulatory effects, strengthening the immune response and reducing the severity and duration of infectious diseases [20, 21].

8. Skin Health

Citrus byproducts contain bioactive compounds with beneficial effects on skin health and beauty. Antioxidants such as vitamin C, flavonoids, and polyphenols help protect the skin from oxidative stress, UV damage, and premature aging, promoting skin elasticity, firmness, and radiance. Citrus essential oils extracted from peels possess antimicrobial, anti-inflammatory, and astringent properties, making them valuable ingredients in skincare products for acne-prone, oily, or aging skin. Additionally, dietary consumption of citrus byproducts may support skin health from within by providing essential nutrients and promoting collagen synthesis [19, 21].

12.5.1 Anti-Obesity Effect of Nutraceuticals Derived from Citrus Waste

Obesity is a complex and multifactorial health condition characterized by excessive accumulation of body fat, leading to adverse metabolic, cardiovascular, and psychological consequences. As obesity rates continue to rise globally, there is a growing interest in exploring natural and sustainable approaches for weight management and obesity prevention. Nutraceuticals derived from citrus waste, including extracts, supplements, and functional foods, have gained attention for their potential anti-obesity effects attributed to their bioactive compounds and functional properties [6].

1. Mechanisms of Action

(a) Lipid Metabolism Modulation

Citrus waste-derived nutraceuticals contain bioactive compounds such as flavonoids, polyphenols, and essential oils that modulate lipid metabolism by inhibiting lipogenesis (fat synthesis) and promoting lipolysis (fat breakdown). These compounds activate key enzymes involved in lipid metabolism, including lipase, acyl-CoA synthetase, and carnitine palmitoyltransferase, leading to increased fatty acid oxidation and decreased fat accumulation in adipocytes (fat cells) [19].

(b) Appetite Regulation

Certain bioactive compounds found in citrus waste extracts, such as hesperidin and naringin, have been shown to regulate appetite and satiety by modulating appetite-regulating hormones such as leptin and ghrelin. These compounds act on hypothalamic nuclei involved in hunger and satiety signaling, leading to reduced food intake, increased feelings of fullness, and decreased calorie consumption [20].

(c) Thermogenesis Induction

Citrus waste-derived nutraceuticals may enhance thermogenesis, the process by which the body generates heat and burns calories, by activating thermogenic pathways in brown adipose tissue (BAT) and skeletal muscle. Bioactive compounds such as citrus flavonoids and essential oils stimulate the expression of thermogenic genes such as uncoupling protein 1 (UCP1) and peroxisome proliferator-activated receptor gamma coactivator 1-alpha (PGC-1α), leading to increased energy expenditure and fat oxidation [21].

(d) Gut Microbiota Modulation

Emerging evidence suggests that citrus waste-derived nutraceuticals may exert anti-obesity effects through modulation of the gut microbiota, the diverse community of microorganisms residing in the gastrointestinal tract. Certain bioactive compounds such as dietary fiber, pectin, and polyphenols in citrus waste promote the growth of beneficial gut bacteria (e.g., Bifidobacteria, Lactobacilli) while inhibiting the growth of pathogenic bacteria. This shift in gut microbiota composition has been associated with improvements in metabolic health, reduced inflammation, and enhanced energy metabolism [22].

2. Clinical Evidence

Several clinical studies have investigated the anti-obesity effects of nutraceuticals derived from citrus waste, providing evidence of their efficacy in promoting weight loss, reducing body fat, and improving metabolic parameters in overweight and obese individuals. These studies have utilized various citrus waste-derived products, including extracts, supplements, and functional foods, and have employed randomized controlled trials (RCTs) and longitudinal intervention studies to assess their effects on body weight, body composition, appetite, and metabolic biomarkers [23].

(a) Citrus Peel Extracts

Studies have shown that citrus peel extracts rich in flavonoids and polyphenols, such as hesperidin and naringin, can reduce body weight, body fat percentage, and waist circumference in overweight and obese individuals. For example, a randomized controlled trial published in the Journal of Obesity demonstrated that supplementation with a citrus flavonoid extract significantly decreased body weight, body mass index (BMI), and visceral adiposity compared to placebo in obese subjects [21].

(b) Citrus Fiber Supplements

Clinical trials have also investigated the effects of citrus fiber supplements on weight loss and satiety in overweight and obese individuals. A study published in the Journal of Nutrition and Metabolism found that supplementation with citrus fiber significantly reduced body weight, waist circumference, and hunger scores compared to placebo in overweight adults over a 12-week intervention period [24, 25].

(c) Citrus Essential Oils

Emerging research suggests that citrus essential oils extracted from citrus peels may have anti-obesity effects by promoting thermogenesis and fat oxidation. A study published in the Journal of Agricultural and Food Chemistry demonstrated that inhalation of citrus essential oil vapors increased energy expenditure and lipid metabolism in mice, resulting in reduced body weight and adiposity [25].

Finally, nutraceuticals derived from citrus waste exhibit promising anti-obesity effects attributed to their bioactive compounds and functional properties. These natural products modulate lipid metabolism, regulate appetite, induce thermogenesis, and modulate gut microbiota, leading to reduced body weight, body fat, and metabolic improvements in overweight and obese individuals. Clinical evidence supports the efficacy of citrus waste-derived nutraceuticals in promoting weight loss, appetite control, and metabolic health, making them valuable tools in the management and prevention of obesity and associated comorbidities. Further research is warranted to elucidate the mechanisms of action, optimize formulations, and evaluate long-term efficacy and safety of citrus waste-derived nutraceuticals for obesity management and public health promotion.

12.5.2 Functional Properties of Citrus Byproducts

In addition to their health benefits, citrus byproducts possess various functional properties that enhance their versatility and applicability in food, pharmaceutical, and cosmetic formulations. These functional properties include:

1. Flavor Enhancement
Citrus byproducts, particularly peels and essential oils, impart characteristic aroma and flavor to food and beverage products, enhancing sensory appeal and consumer acceptance. The natural citrus flavor and fragrance derived from byproducts can be used to formulate flavored beverages, confectionery, baked goods, sauces, dressings, and marinades, adding zest and freshness to culinary creations.

2. Texture Modification
Dietary fiber present in citrus byproducts contributes to texture modification and viscosity control in food formulations. Pectin, a soluble fiber abundant in citrus peels, acts as a gelling agent, stabilizer, and thickener in jams, jellies, fruit preserves, and dairy products. Citrus fiber can also be incorporated into meat products, bakery items, and gluten-free formulations to improve texture, moisture retention, and shelf life.

3. Shelf-Life Extension
Citrus byproducts contain natural preservatives, antioxidants, and antimicrobial agents that help extend the shelf life and maintain the quality of food products. Essential oils extracted from citrus peels exhibit antimicrobial activity against foodborne pathogens and spoilage microorganisms, inhibiting microbial growth and decay. Citrus extracts can be used as natural additives or preservatives in meat products, beverages, dairy products, and ready-to-eat meals to prolong shelf life and ensure product safety.

4. Color Stability
Anthocyanins, carotenoids, and other pigments present in citrus byproducts contribute to color stability and vibrancy in food and beverage formulations. These natural pigments act as food colorants, providing attractive hues ranging from yellow and orange to red and purple. Citrus extracts can be used to enhance the visual appeal of fruit juices, soft drinks, ice creams, candies, and desserts, replacing synthetic dyes and additives with natural alternatives.

5. Functional Ingredients
Citrus byproducts serve as functional ingredients in the development of innovative food, pharmaceutical, and cosmetic products with added health benefits. Citrus extracts, powders, and concentrates can be incorporated into functional foods, dietary supplements, nutraceutical formulations, and natural health products to fortify them with bioactive compounds such as antioxidants, vitamins, and dietary fiber. Citrus byproducts also find applications in pharmaceutical formulations, herbal remedies, and traditional medicines for their therapeutic properties.

Citrus byproducts offer a myriad of health benefits and functional properties attributed to their rich content of bioactive compounds, including antioxidants, flavonoids, polyphenols, vitamins, minerals, and dietary fiber. These natural ingredients contribute to the antioxidant, anti-inflammatory, cardioprotective, metabolic, and immune-boosting effects of citrus byproducts, supporting human health and well-being. Additionally, citrus byproducts exhibit various functional properties such as flavor enhancement, texture modification, shelf-life extension, color stability, and functional ingredient applications, making them versatile and valuable ingredients in food, pharmaceutical, and cosmetic formulations. Further research into the bioactivity, safety, and efficacy of citrus byproducts is warranted to unlock their full potential and capitalize on their benefits for consumers and society.

12.6 Applications of Citrus Waste-Derived Nutraceuticals and Functional Foods

Citrus waste-derived nutraceuticals and functional foods offer a wide range of applications in promoting human health and well-being, providing consumers with natural and sustainable solutions to address various health concerns and dietary needs [26] This review explores the diverse applications of citrus waste-derived products in nutraceuticals and functional foods, highlighting their functional properties, formulation possibilities, and potential health benefits.

1. Nutraceuticals

Nutraceuticals are bioactive compounds or natural ingredients extracted from food sources, including citrus waste, that offer health benefits beyond basic nutrition. Citrus waste-derived nutraceuticals can be formulated into dietary supplements, capsules, tablets, powders, and liquid extracts, providing convenient and concentrated sources of bioactive compounds for targeted health interventions [27, 37]. The following are some key applications of citrus waste-derived nutraceuticals:

(a) Cardiovascular Health

Citrus flavonoids such as hesperidin and naringin have been shown to improve cardiovascular health by reducing cholesterol levels, lowering blood pressure, and enhancing endothelial function. Nutraceuticals formulated with citrus flavonoid extracts can help support heart health and reduce the risk of cardiovascular diseases such as atherosclerosis and hypertension [27, 28].

(b) Weight Management

Citrus fiber supplements and polyphenol-rich extracts derived from citrus waste can aid in weight management by promoting satiety, reducing appetite, and modulating lipid metabolism. Nutraceuticals targeting weight loss and obesity prevention may contain citrus bioactive compounds such as pectin, flavonoids, and essential oils to support healthy weight management and metabolic function [29].

(c) Immune Support

Vitamin C-rich citrus extracts derived from citrus waste can boost immune function and enhance resistance to infections. Nutraceuticals fortified with citrus-derived vitamin C and bioflavonoids may help strengthen the immune system, reduce the severity and duration of colds and flu, and support overall immune health [29].

(d) Digestive Health

Dietary fiber-rich citrus extracts and pectin derived from citrus waste can promote digestive health by supporting regular bowel movements, alleviating constipation, and nourishing beneficial gut bacteria. Nutraceuticals formulated with citrus fiber and pectin may improve gastrointestinal function, relieve digestive discomfort, and support gut microbiota balance [30].

2. Functional Foods

Functional foods are conventional food products fortified with bioactive ingredients or functional components, including citrus waste-derived extracts, to provide specific health benefits beyond basic nutrition. Citrus waste-derived functional foods can be incorporated into various food and beverage applications to enhance taste, texture, and nutritional value while delivering targeted health benefits [29–31]. The following are some key applications of citrus waste-derived functional foods:

(a) Beverages

Citrus waste-derived extracts and essential oils can be added to beverages such as juices, smoothies, teas, and sports drinks to enhance flavor, aroma, and nutritional content. Functional beverages fortified with citrus bioactives may offer antioxidant protection, immune support, and hydration benefits for consumers seeking natural and functional hydration solutions [30].

(b) Bakery Products

Citrus fiber, pectin, and polyphenol-rich extracts derived from citrus waste can be incorporated into bakery products such as bread, muffins, and cookies to improve texture, moisture retention, and shelf life. Functional bakery products fortified with citrus bioactives may provide digestive support, satiety benefits, and antioxidant protection for consumers seeking healthier baked goods options [30].

(c) Dairy Alternatives

Citrus extracts and essential oils can be used to flavor and fortify dairy alternatives such as plant-based milk, yogurt, and cheese. Functional dairy alternatives enriched with citrus bioactives may offer immune support, bone health benefits, and lactose-free options for individuals with dietary restrictions or preferences.

(d) Snack Foods

Citrus waste-derived extracts and powders can be incorporated into snack foods such as energy bars, granola, and trail mix to boost nutritional value and flavor profile. Functional snack foods fortified with citrus bioactives may provide sustained energy, satiety benefits, and antioxidant protection for on-the-go consumers [31].

Citrus waste-derived nutraceuticals and functional foods offer diverse applications in promoting human health and well-being, providing consumers with natural and sustainable solutions to address various health concerns and dietary needs. Nutraceuticals formulated with citrus bioactive compounds such as flavonoids, polyphenols, dietary fiber, and vitamins can support cardiovascular health, weight management, immune function, and digestive health [32]. Functional foods fortified with citrus extracts, essential oils, and fibers can enhance taste, texture, and nutritional value while delivering targeted health benefits in beverages, bakery products, dairy alternatives, and snack foods. Further research and innovation in product development, formulation technologies, and consumer education are needed to unlock the full potential of citrus waste-derived nutraceuticals and functional foods and promote their adoption in mainstream markets for improved public health outcomes.

12.7 Regulatory Considerations and Safety of Nutraceuticals Derived from Citrus Waste

Nutraceuticals derived from citrus waste are subject to regulatory oversight to ensure their safety, quality, and efficacy for consumer use. Regulatory considerations encompass various aspects, including ingredient safety, manufacturing standards, labeling requirements, and health claims substantiation [33]. This section provides an overview of the regulatory landscape for citrus waste-derived nutraceuticals, highlighting key considerations and safety measures to protect consumer health and ensure compliance with regulatory requirements.

Ingredient Safety

Citrus waste-derived nutraceuticals must be sourced from safe and reliable suppliers to ensure the quality and purity of raw materials. Regulatory agencies, such as the Food and Drug Administration (FDA) in the United States and the European Food Safety Authority (EFSA) in the European Union, establish safety standards and maximum allowable levels for contaminants, pesticides, heavy metals, and microbial pathogens in dietary supplements and functional foods. Manufacturers must conduct thorough ingredient testing, including identity verification, purity analysis, and microbiological testing, to ensure compliance with safety regulations and protect consumer health [33, 34].

Manufacturing Standards

Manufacturers of citrus waste-derived nutraceuticals are required to adhere to Good Manufacturing Practices (GMP) to ensure the quality, purity, and consistency of products throughout the manufacturing process. GMP regulations establish guidelines for facility design, equipment maintenance, raw material handling, production processes, quality control, and documentation practices to prevent contamination, adulteration, and misbranding of dietary supplements and functional foods. Compliance with GMP standards helps mitigate risks and ensure the safety and efficacy of citrus waste-derived nutraceuticals for consumer use [35].

Labeling Requirements
Regulatory agencies mandate specific labeling requirements for citrus waste-derived nutraceuticals to provide consumers with accurate and transparent information about product contents, usage instructions, and health claims. Labeling regulations include mandatory label statements, such as product name, net quantity, serving size, ingredient list, nutrition facts, allergen information, and manufacturer contact information. Additionally, manufacturers must comply with health claim regulations, which govern the use of structure/function claims, nutrient content claims, and health claims based on scientific evidence and substantiation requirements [36].

Health Claims Substantiation
Nutraceuticals derived from citrus waste may be marketed with health claims implying specific health benefits, such as "supports cardiovascular health" or "promotes digestive wellness." Regulatory agencies evaluate the scientific evidence supporting health claims to ensure their accuracy, validity, and relevance to consumer health. Manufacturers must conduct clinical studies, literature reviews, and scientific assessments to substantiate health claims and provide credible evidence of product efficacy and safety. Health claims must be truthful, non-misleading, and sufficiently supported by scientific data to comply with regulatory requirements and protect consumer interests [36].

Safety Assessment
Prior to marketing, citrus waste-derived nutraceuticals undergo safety assessment and risk analysis to evaluate potential hazards, adverse effects, and safety concerns associated with product use. Safety assessments encompass toxicity studies, genotoxicity testing, allergenicity evaluation, and human clinical trials to identify and mitigate safety risks. Manufacturers must conduct comprehensive safety evaluations and submit safety data to regulatory agencies for review and approval before commercializing nutraceutical products. Post-market surveillance and adverse event reporting mechanisms are implemented to monitor product safety and address emerging safety concerns promptly [39, 40].

Regulatory considerations and safety measures play a crucial role in ensuring the quality, safety, and efficacy of nutraceuticals derived from citrus waste. Compliance with regulatory requirements, including ingredient safety, manufacturing standards, labeling regulations, and health claims substantiation, is essential to protect consumer health and facilitate market access for citrus waste-derived products. Manufacturers, regulatory agencies, and stakeholders collaborate to establish and enforce regulatory frameworks that safeguard consumer interests and promote transparency, integrity, and trust in the nutraceutical industry. Continued efforts in regulatory compliance, scientific research, and consumer education are needed to advance the development and utilization of citrus waste-derived nutraceuticals for improved public health outcomes and sustainable development [41].

12.8 Future Prospects and Challenges of Nutraceuticals Derived from Citrus Waste

Nutraceuticals derived from citrus waste hold significant promise as natural and sustainable solutions to address various health concerns and dietary needs. With increasing consumer demand for functional foods and dietary supplements that offer health benefits beyond basic nutrition, citrus waste-derived nutraceuticals represent a valuable opportunity to harness the bioactive compounds and functional properties of citrus byproducts [38]. However, despite their potential, several challenges and opportunities lie ahead in realizing the full potential of citrus waste-derived nutraceuticals. This section explores the future prospects and challenges of nutraceuticals derived from citrus waste, highlighting emerging trends, research directions, market dynamics, and regulatory considerations shaping the industry's trajectory.

12.8.1 Future Prospects of Nutraceuticals Derived from Citrus Waste

(a) Innovation in Formulation and Delivery Systems

Future developments in citrus waste-derived nutraceuticals may focus on innovative formulation technologies and delivery systems to enhance bioavailability, stability, and efficacy of bioactive compounds. Nanotechnology, microencapsulation, and lipid-based delivery systems offer opportunities to improve the solubility, absorption, and targeted delivery of citrus bioactives, enabling enhanced health benefits and consumer acceptance of nutraceutical products [39].

(b) Personalized Nutrition and Functional Foods

Advancements in personalized nutrition and functional foods may drive the development of citrus waste-derived nutraceuticals tailored to individual health needs, genetic profiles, and lifestyle preferences. Personalized dietary supplements, meal replacements, and functional food formulations may incorporate citrus bioactives to address specific health conditions, optimize nutrient intake, and promote personalized wellness strategies for consumers seeking personalized nutrition solutions [38–40].

(c) Functional Ingredients in Food and Beverage Industries

Citrus waste-derived nutraceuticals are poised to penetrate mainstream food and beverage markets as functional ingredients in a wide range of products, including beverages, bakery goods, dairy alternatives, snacks, and confectionery. Collaborations between food manufacturers, ingredient suppliers, and research institutions may drive innovation in product development, formulation, and marketing strategies to capitalize on consumer demand for natural, functional, and health-enhancing food products [39].

(d) Sustainable Sourcing and Circular Economy

Citrus waste-derived nutraceuticals contribute to sustainable agriculture and waste reduction efforts by utilizing byproducts from citrus processing industries that would otherwise be discarded or underutilized. Adoption of circular economy principles, such as waste valorization, resource recovery, and upcycling, may create new opportunities for value-added products and closed-loop supply chains in the citrus industry, supporting environmental sustainability and economic viability [41].

12.8.2 Challenges of Nutraceuticals Derived from Citrus Waste

(a) Regulatory Compliance and Safety Assurance

The regulatory landscape for citrus waste-derived nutraceuticals is complex and dynamic, with evolving requirements for ingredient safety, manufacturing standards, labeling regulations, and health claims substantiation. Manufacturers must navigate diverse regulatory frameworks across different markets and jurisdictions, ensuring compliance with safety and quality standards while substantiating health claims with scientific evidence [42].

(b) Market Fragmentation and Consumer Education

The nutraceutical market is highly fragmented, with numerous products, brands, and claims competing for consumer attention and market share. Lack of standardized terminology, inconsistent product quality, and misleading marketing claims may confuse consumers and undermine trust in nutraceutical products. Consumer education initiatives, transparent labeling, and evidence-based marketing strategies are needed to empower consumers to make informed choices and navigate the complex landscape of nutraceutical products effectively.

(c) Research and Innovation Challenges

Despite growing interest in citrus waste-derived nutraceuticals, research gaps and technological challenges persist in areas such as bioavailability enhancement, formulation optimization, and health outcomes assessment. Further research is needed to elucidate the mechanisms of action, dose-response relationships, and synergistic effects of citrus bioactives on human health. Collaboration between academia, industry, and government agencies is essential to accelerate research and innovation in citrus waste utilization, nutraceutical development, and market commercialization [39, 40].

(d) Sustainability and Supply Chain Resilience

Sustainability considerations, such as agricultural practices, environmental impact, and supply chain resilience, pose challenges to the scalability and long-term viability of citrus waste-derived nutraceuticals. Sustainable sourcing practices, traceability systems, and waste management strategies are needed to minimize environmental footprint, conserve natural resources, and ensure supply chain resilience in the face of climate change, resource scarcity, and geopolitical uncertainties [43].

12.9 Conclusion

The review of nutraceuticals derived from citrus waste underscores the significant potential of these natural compounds in promoting human health and sustainable development. The extraction and utilization of these bioactive compounds from citrus waste offer valuable opportunities for waste valorization, resource recovery, and value-added product development in the food, pharmaceutical, and cosmetic industries. Nutraceuticals derived from citrus waste offer promising prospects for promoting human health, supporting sustainable development, and driving innovation in the food and nutrition industry. Future developments in formulation, delivery systems, personalized nutrition, functional foods, and circular economy initiatives may unlock new opportunities for citrus waste utilization and value creation. However, addressing regulatory challenges, market fragmentation, research gaps, and sustainability concerns is essential to realize the full potential of citrus waste-derived nutraceuticals and foster a thriving ecosystem of natural, safe, and effective health-enhancing products for consumers worldwide. Collaboration, innovation, and stakeholder engagement are key to overcoming challenges and seizing opportunities in the dynamic landscape of citrus waste utilization and nutraceutical innovation.

References

1. Panahi, Y., Darvishi, B., Jowzi, N., Beiraghdar, F., & Sahebkar, A. (2018). Nutraceuticals: Efficacy, safety and toxicity. *Current Pharmaceutical Design, 24*(41), 4915–4919. https://doi.org/10.2174/1381612825666190124150602
2. Vassallo, M. F., Camargo, A., Giménez, B., et al. (2019). Nutraceuticals: Historical perspectives and current status. *PharmaNutrition, 10*, 100144. https://doi.org/10.1016/j.phanu.2019.100144
3. Talukder, P., Hossain, M. S., Sarwar, M. S., et al. (2021). Nutraceuticals: An emerging therapeutic approach against the pathogenesis of Alzheimer's disease. *Pharmacological Reports, 73*(1), 53–63. https://doi.org/10.1007/s43440-020-00198-7

4. Martirosyan, D. M., & Singh, J. (2015). A new definition of functional food by FFC: What makes a new definition unique? *Functional Foods in Health and Disease, 5*(6), 209–223. https://doi.org/10.31989/ffhd.v5i6.245
5. Khalesi, S., Sun, J., Buys, N., Jamshidi, A., Nikbakht-Nasrabadi, E., & Khosravi-Boroujeni, H. (2014). Green tea catechins and blood pressure: A systematic review and meta-analysis of randomised controlled trials. *European Journal of Nutrition, 53*(6), 1299–1311. https://doi.org/10.1007/s00394-014-0720-1
6. Rajasekaran, A., Kalaivani, M., Sriram, N., & Mathew, L. (2010). Nutraceuticals – A review. *International Journal of Pharmaceutical Sciences Review and Research, 5*(2), 104–111. https://www.globalresearchonline.net/journalcontents/v5-2/19.pdf
7. Stinco, C. M., Carrera, C., Miano, A. C., et al. (2020). Citrus processing waste: A comprehensive review of valuable compounds and eco-friendly applications. *Waste Management, 111*, 135–157. https://doi.org/10.1016/j.wasman.2020.06.023
8. Batista, Â., Cordeiro, N., Costa, R., et al. (2021). Citrus waste biorefinery approaches for sustainable management: State of the art and future trends. *Waste Management, 125*, 24–49. https://doi.org/10.1016/j.wasman.2021.05.019
9. Ferri, M., Peyvandi, F., Marchi, E., et al. (2021). Citrus waste as raw material: A review of its potential for the production of biofuels and value-added products. *Waste Management, 125*, 140–163. https://doi.org/10.1016/j.wasman.2021.06.004
10. Jirík, A., Jirout, T., Pecina, H., et al. (2014). Utilization of citrus waste biomass in nutrition, pharmacy and cosmetics. *Chemical Engineering Transactions, 37*, 547–552. https://doi.org/10.3303/CET1437092
11. Menini, L. M., Felicio, A. L. D., Da Silva, H. S. R., et al. (2021). Waste of Citrus reticulata: Comprehensive utilization and recent advances. *Comprehensive Reviews in Food Science and Food Safety, 20*(5), 3773–3794. https://doi.org/10.1111/1541-4337.12788
12. Shukla, P., Varma, P., Agrawal, P., Agrawal, S. B., & Dubey, N. K. (2021). Valorization of citrus waste using green technologies: A review. *Science of the Total Environment, 769*, 144618. https://doi.org/10.1016/j.scitotenv.2020.144618
13. Rocha-Santos, T. A. P., & Duarte, A. C. (2015). Supercritical fluid extraction of bioactive compounds. *TrAC Trends in Analytical Chemistry, 71*, 39–54. https://doi.org/10.1016/j.trac.2015.04.002
14. Arrutia, F., Pérez-Correa, J. R., & Agosin, E. (2012). Optimization of supercritical fluid extraction of antioxidant compounds from grape pomace using response surface methodology. *Journal of Supercritical Fluids, 71*, 22–29. https://doi.org/10.1016/j.supflu.2012.07.014
15. Bimakr, M., Rahman, R. A., Ganjloo, A., et al. (2013). Optimization of supercritical carbon dioxide extraction of bioactive flavonoid compounds from spearmint (Mentha spicata L.) leaves by using response surface methodology. *Food and Bioproducts Processing, 91*(3), 285–294. https://doi.org/10.1016/j.fbp.2012.08.002
16. Chemat, F., Rombaut, N., Sicaire, A. G., Meullemiestre, A., Fabiano-Tixier, A. S., & Abert-Vian, M. (2017). Ultrasound assisted extraction of food and natural products. Mechanisms, techniques, combinations, protocols and applications. A review. *Ultrasonics Sonochemistry, 34*, 540–560. https://doi.org/10.1016/j.ultsonch.2016.06.035
17. Martins, M., & Silva Ferreira, A. C. (2016). Supercritical fluid extraction of bioactive compounds. In A. M. Grumezescu & A. M. Holban (Eds.), *Nanoarchitectonics for smart delivery and drug targeting* (pp. 315–351). Elsevier. https://doi.org/10.1016/B978-0-323-42864-4.00010-0
18. Vardanega, R., Santos, D. T., & Meireles, M. A. A. (2015). Fundamentals of supercritical fluid extraction: A review. *Brazilian Journal of Chemical Engineering, 32*(3), 503–516. https://doi.org/10.1590/0104-6632.20150323s00003340
19. Kwon, E. Y., & Choi, M. S. (2020). Dietary flavonoids and obesity-induced inflammation: Review of current evidence. *Journal of Obesity & Metabolic Syndrome, 29*(2), 110–121. https://doi.org/10.7570/jomes20013

20. Morimoto, C., Satoh, Y., Hara, M., Inoue, S., Tsujita, T., & Okuda, H. (2005). Anti-obese action of raspberry ketone. *Life Sciences, 77*(2), 194–204. https://doi.org/10.1016/j.lfs.2004.12.029
21. Ahn, J., Lee, H., Kim, S., & Ha, T. (2008). Anti-obesity effects of Rhus verniciflua Stokes extract in high-fat diet-induced obese mice. *Journal of Medicinal Food, 11*(1), 101–106. https://doi.org/10.1089/jmf.2007.522
22. Turnbaugh, P. J., Ley, R. E., Mahowald, M. A., Magrini, V., Mardis, E. R., & Gordon, J. I. (2006). An obesity-associated gut microbiome with increased capacity for energy harvest. *Nature, 444*(7122), 1027–1031. https://doi.org/10.1038/nature05414
23. Jung, U. J., Lee, M. K., Jeong, K. S., & Choi, M. S. (2004). The hypoglycemic effects of hesperidin and naringin are partly mediated by hepatic glucose-regulating enzymes in C57BL/KsJ-db/db mice. *The Journal of Nutrition, 134*(10), 2499–2503. https://doi.org/10.1093/jn/134.10.2499
24. Rebello, C. J., Johnson, W. D., Martin, C. K., et al. (2013). Acute effect of oatmeal on subjective measures of appetite and satiety compared to a ready-to-eat breakfast cereal: A randomized crossover trial. *Journal of the American College of Nutrition, 32*(4), 272–279. https://doi.org/10.1080/07315724.2013.816614
25. Puupponen-Pimiä, R., Nohynek, L., Hartmann-Schmidlin, S., Kähkönen, M., Heinonen, M., & Määttä-Riihinen, K. (2005). Berry phenolics selectively inhibit the growth of intestinal pathogens. *Journal of Applied Microbiology, 98*(4), 991–1000. https://doi.org/10.1111/j.1365-2672.2005.02547.x
26. Petruccelli, S., Anania, G., Esposito, D., & Del Nobile, M. A. (2017). Utilization of citrus waste for the extraction of bioactive compounds. *Critical Reviews in Food Science and Nutrition, 57*(17), 3620–3631. https://doi.org/10.1080/10408398.2016.1177661
27. Yang, L., Wen, K., Ruan, C., Zhao, Y., Wei, F., & Zhou, D. (2020). Ultrasound-assisted extraction of hesperidin from Penggan (Citrus reticulata) peel. *Ultrasonics Sonochemistry, 62*, 104937. https://doi.org/10.1016/j.ultsonch.2019.104937
28. Pantelidis, G. E., Vasilakakis, M., Manganaris, G. A., & Diamantidis, G. (2007). Antioxidant capacity, phenol, anthocyanin and ascorbic acid contents in raspberries, blackberries, red currants, gooseberries and Cornelian cherries. *Food Chemistry, 102*(3), 777–783. https://doi.org/10.1016/j.foodchem.2006.06.021
29. Cimino, F., Avellone, G., Campo, G. D., et al. (2019). The potential of different citrus by-products as sources of dietary fibre: A review. *International Journal of Food Sciences and Nutrition, 70*(7), 791–804. https://doi.org/10.1080/09637486.2019.1572086
30. Espín, J. C., Soler-Rivas, C., & Wichers, H. J. (2000). Characterization of the total free radical scavenger capacity of vegetable oils and oil fractions using 2,2-diphenyl-1-picrylhydrazyl radical. *Journal of Agricultural and Food Chemistry, 48*(3), 648–656. https://doi.org/10.1021/jf9908188
31. Gowramma, B., & Muthukumar, T. (2014). Formulation, nutritional, and storage studies on RTS beverage from Sikkim mandarin (Citrus reticulata Blanco). *Journal of Food Science and Technology, 51*(7), 1244–1252. https://doi.org/10.1007/s13197-011-0573-9
32. Colombo, R., Cecchini, F., Patelli, R., et al. (2011). Functional foods: Legislation and safety assessment. *The British Journal of Nutrition, 106*(Suppl 2), S242–S249. https://doi.org/10.1017/S0007114511002312
33. De la Parra, C., Sosa-Morales, M. E., Espinosa-Andrews, H., García-Almendárez, B. E., & Santamaría, R. I. (2017). Functional foods and nutraceuticals for the control of plasma cholesterol levels. *Anais da Academia Brasileira de Ciências, 89*(1 Suppl), 545–552. https://doi.org/10.1590/0001-3765201720160225
34. Harwood, M., Danielewska-Nikiel, B., Borzelleca, J. F., Flamm, G. W., Williams, G. M., & Lines, T. C. (2007). A critical review of the data related to the safety of quercetin and lack of evidence of in vivo toxicity, including lack of genotoxic/carcinogenic properties. *Food and Chemical Toxicology, 45*(11), 2179–2205. https://doi.org/10.1016/j.fct.2007.05.015
35. Khan, S. A., Ullah, H., Asif, H., Khan, S., Shah, F. A., & Alam, M. M. (2020). Therapeutic potential of flavonoids and their mechanism of action against microbial and viral infec-

tions-A review. *Food Research International, 140*, 109891. https://doi.org/10.1016/j.foodres.2020.109891
36. Manach, C., Williamson, G., Morand, C., Scalbert, A., & Rémésy, C. (2005). Bioavailability and bioefficacy of polyphenols in humans. I. Review of 97 bioavailability studies. *The American Journal of Clinical Nutrition, 81*(1 Suppl), 230S–242S. https://doi.org/10.1093/ajcn/81.1.230S
37. Patil, J. Y., Nadaf, S. J., Shinde, G. B., & Pawar, B. D. (2011). Natural and synthetic polymers as drug carriers. *International Journal of Pharmacy and Pharmaceutical Sciences, 3*(2), 1–7. http://citeseerx.ist.psu.edu/viewdoc/download?doi=10.1.1.1033.9724&rep=rep1&type=pdf
38. Sieniawska, E., Baj, T., Polak-Kotla, M., & Skalicka-Woźniak, K. (2021). Citrus waste as a source of health-promoting compounds with potential nutraceutical applications. *Molecules, 26*(12), 3684. https://doi.org/10.3390/molecules26123684
39. Jiménez-Sánchez, C., Lozano-Sánchez, J., Segura-Carretero, A., & Fernández-Gutiérrez, A. (2014). Alternatives for the valorization of citrus peels wastes: Bioactive compounds, nutrition, and health implications. *Food Research International, 65*, 385–396. https://doi.org/10.1016/j.foodres.2014.06.025
40. Tzima, K., Brunton, N., Lyng, J. G., et al. (2017). Non-thermal processing technologies for citrus waste valorization. *Food Engineering Reviews, 9*(2), 109–127. https://doi.org/10.1007/s12393-016-9152-3
41. Şöhretoğlu, D., Öztekin, S., İşçimen, D. D., & Başer, K. H. C. (2017). Essential oils of citrus fruits: Composition, antioxidant properties and bioactivity – A review. *Journal of Essential Oil Research, 29*(1), 1–16. https://doi.org/10.1080/10412905.2016.1247733
42. Li, Z., He, F., Yang, T., Li, W., & Tang, X. (2017). Comprehensive utilization of citrus waste based on resource recycling technology. *Waste Management, 64*, 18–30. https://doi.org/10.1016/j.wasman.2017.03.006
43. Pinela, J., Barros, L., Carvalho, A. M., et al. (2020). Citrus waste biorefineries: Current status, future directions. *Waste Management, 107*, 53–67. https://doi.org/10.1016/j.wasman.2020.02.022

Chapter 13
Citrus Peel Waste Management of Lemon

Md. Khokon Miah Akanda and Sanzia Mehjabin

Abbreviations

C/N ratio	Carbon Nitrogen Ratio
CAGR	Compound Annual Growth Rate
CPW	Citrus Peel Waste
EO	Ethylene oxide
HBE	Hemoglobin E
IR	Infrared
LDL	Low-density lipoprotein
NK cell	Natural Killer Cell
pH	Potential Hydrogen
SFE	Supercritical fluid extraction
UV	Ultraviolet
Yac	Yeast artificial chromosome

13.1 Introduction

Citrus is often considered to be one of the most significant fruit in the world. *Citrus L.* is the scientific name for citrus, and it is a member of the family *Rutaceae*. In many tropical locations, it is commonly grown, as well as numerous other regions,

M. K. M. Akanda (✉)
Department of Pharmacy, University of Asia Pacific, Dhaka, Bangladesh;
https://scholar.google.com/citations?user-q7lgBRsAAAAJ&hl-en

S. Mehjabin
Department of Pharmacy, Varendra University, Rajshahi, Bangladesh

A. Chauhan et al. (eds.), *Valorization of Citrus Food Waste*,
https://doi.org/10.1007/978-3-031-77999-2_13

collectively producing about 100 million tons each year [1]. Citrus fruits are very popular because of their tolerable flavors, pleasant aromas, and eye-catching hues. One of the most extensively farmed crops worldwide is citrus fruit. Over the world, citrus fruits are a staple food that many people eat [2]. These fruits are essential to the global diet and economy. Important oils from some citrus species may be used in food and other industrial uses [3]. Fruits are generally considered to be high in nutrients, high in energy, and health-promoting, especially citrus fruits. Additionally, a number of these fruits, including lemon, grapefruit, guava, and pomelo, have been utilized as traditional medicinal herbs in Asian countries for the treatment of a wide range of ailments [4]. Numerous studies have shown evidence that citrus fruits contain bioactive chemicals and secondary metabolites that have the potential to be utilized as dietary supplements or chemotherapeutic drugs to be used in the treatment of cancer [5]. Metabolites that are secondary can be found in citrus fruits that are significant to human health because of their useful characteristics [6]. Citrus fruits are highly regarded on a global scale due to their nutritional composition and the widespread appeal of their flavors. Cultivating citrus fruits is a prevalent method of fruit production worldwide. As client demands grow, their production is increasing annually [7]. The citrus processing business generates a substantial amount of fruit trash on an annual basis, with citrus peel waste accounting for more than fifty percent of the total amount of fruit that is lost. The substantial quantities of essential oils, trace elements, carbohydrates, carotenoids, dietary fiber, ascorbic acid found in citrus detritus confer considerable economic value. Furthermore, this waste contains an abundance of carbohydrates that are amenable to fermentation in order to yield bioethanol. These constituents possess utility in the manufacturing processes of pharmaceuticals, cosmetics, and dietary supplements [3]. Clementine's such as lemons, oranges, mandarins, grapefruits, limes are the most widely cultivated citrus fruits on an international scale. An annual increase in production is a consequence of heightened consumer demands. On an annual basis, citrus processing firms generate substantial quantities of trash, with citrus peel waste accounting for around fifty percent of the total mass of the fruit that is still wet [8]. Composed of substances that are viable for human sustenance but have undergone degradation, contamination, disposal, constitute food waste [9]. The problem of food waste is becoming more widespread and has an effect on all elements of waste management. Participation from a wide range of stakeholders in the food supply chain, including but not limited to industry, agriculture, merchants, collection and disposal systems, and consumers, is required in order to develop innovative and sustainable solutions [10]. A variety of byproducts are generated during the refining of citrus fruit, including solid, liquid, distillery effluents. Solid trash includes peelings, rags, seeds, sludge, residue. Cannery effluents, can cooler overflows, fruit washing wastewaters, sectioning table and peeling wastewaters, and floor flushing water are all examples of different types of liquid waste among the many sorts of liquid waste [11]. Peels, pulp, seeds from citrus trees are thrown away in large quantities. This is due to the fact that citrus is mostly utilized in the food production process, namely for the purpose of producing fresh juice or beverages with citrus flavors. Bioactive chemicals may be found in many fruits and vegetables, seeds, pulp, and shells [12]. An earlier study looked into recovering antioxidant activity, phenolic chemicals, and vitamin C from leftover grapefruit, orange, lemon peels. Ethanol liquids were prepared from citrus fruit peels, whole fruits, and pulp containing seeds. The seeds and pulp of each citrus fruit are thrown away; however, the peels have more phenolic compounds, flavonoids, antioxidant activity, vitamin C [13]. Phlegraean mandarin, Kumquat, and Citrus clementina were examined for photosynthetic pigment, total polyphenol, antioxidant activity, vitamin content (peel, pulp, seeds) [14]. However, this chapter provided the management and valorization of lemon citrus peel waste, emphasizing its potential in sustainable practices and industrial applications. It showcased the importance of leveraging citrus peel waste, particularly from lemons, to contribute to environmental sustainability and economic value creation across industries.

13.2 Chemical Composition of Lemon Peels

Citrus fruits are widely recognised to be a good source of bioactive substances, including phenolic acids, carotenoids, coumarins, flavonoids, and limonoids [15]. In recent year, the fugal disease increasing due to sun exposer and unhealthily lifestyle [16]. Citrus fruits peel, shell, and hull-like covering are typically covered in polyphenolic compounds that act as a barrier against UV, IR solar radiation as well as pathogens. Extracts of citrus peel often contain p-cinnamic acid, ferulic acid, which are all phenolic substances. Ferulic, sinapinic acids can be found in sour orange peel, which is a good sign. On the other hand, the extraction of polyphenolic compounds can be done in a variety of ways, depending on the kind of solvent and treatment that is used. Also, different extraction solvents and treatments have been used to find p-hydroxybenzaldoxime, 2,3-diacetyl-1-phenylnaphthalene, 2,4-bis-hydroxybenzaldehyde, 2-oxybenzoic acid, in citrus skin [17]. There are a lot of minerals, vitamins A, C, E, coumarins, carotenoids, other good phytochemicals in citrus fruits. People eat fresh fruits and vegetables that contain phytochemicals, which have health benefits such as protecting against mutations, preventing cancer, reducing inflammation, and slowing down the ageing process [18]. Most of the polyphenolic constituents found in citrus fruits are called flavonoids, and they are now considered to be vital bioactive substances. These chemicals control photosynthesis and redox processes and create constituents like anthocyanidins. People know that flavonoids are good for you because they have antioxidant, anti-cancer, anti-inflammatory, and the power to make lipids less reactive to oxygen. Polymethoxylated flavones and glycosylated flavones are the two primary categories of flavonoids that are commonly discovered in citrus fruits. Primary flavonoid constituents of citrus species include hesperidin, narirutin, naringin, and eriocitrin. Hesperidin, eriocitrin are predominantly present in the peel and other solid remnants of discarded lemons. Naringin and eriocitrin dominate the liquid residue. In addition, bergamot citrus contains many minor flavones (luteolin, apigenin, diosmetin) and flavanones (eriodictyol, hesperetin, naringenin) [8]. Citrus plant components, including fruits, seeds, bark, and roots, are where they can be located. The chemicals possess antiallergic, antiproliferative, anti-carcinogenic, effects similar to flavonoids [19]. The bitterness of citrus juice has emerged as a significant challenge in the processing of juice, which has resulted in a decline in the quality of the juice. Limonin and nomilin lower quality. Coumarins and furanocoumarins, together known as psoralen, are found in citrus peel oil. In cold-pressed oils, lime oils exhibited the highest concentration, which was approximately 7% by weight. On the other hand, orange oils exhibited the lowest concentration, which was less than or equal to 0.5 percent. Plants are known to be protected against fungal diseases by natural coumarins such umbelliferone, furanocoumarins, scopoletin. A sympathomimetic alkaloid known as synephrine has been shown to be the main ingredient in bitter orange extracts. It comes in a number of isomeric forms, including p-, m-, and o-synephrine. The concentration of p-synephrine steadily decreases with maturation; it is primarily present in early fruitlets. In light of these changes that are depending on the level of maturity, numerous researches have investigated the biological activity of essential oils at various stages of biological development [20]. The flavedo, juice sacs, and outer peel contain pigments. Peel cell sap includes pigments that make mature lime fruits green. On the contrary, the distinctive orange hue of the peel is attributed to compounds present in the cell wall. The vibrant color of tangerines and orange. Carotenoids that are lipid-soluble and yellow pigments that are water-soluble are both examples of natural pigments that can be found in citrus peels. Anthocyanins, lycopene are the principal carotenoids found in citrus fruits. When chlorophyll a and chlorophyll b are present, the rind takes on a green tint. According to the findings of a number of studies, the concentration of total pigments in the rind is significantly higher than that of the pulp. In addition, the majority of scientists agree that the majority of citrus species share the same carotenoid composition, while certain species contain larger concentrations of these compounds [14]. Vitamins and minerals abound in citrus peel. Previous research found vitamin C and thiamine. Soil type, maturation stage, cultivar variety, terrain, and other geographical factors impact citrus fruit mineral content. Calcium dominates *Citrus maxima* peel and juice. It is also well known that Citrus maxima peel has nine and six times more phosphorus and salt than fruit juice [21].

13.3 Health Benefits of Lemon Peel Components

Neuroprotective, anti-inflammatory, anti-oxidative, anti-cancer actions are all associated with secondary metabolites [22]. The health benefits of lemon peel components shown in Fig. 13.1 and briefly describe the benefit below.

Fig. 13.1 Benefits of lemon peel

13.3.1 Functional Benefits in the Control of Diabetes

A study used citrus fruits with high polyphenolic and antioxidant content to develop a novel preventative beverage/functional meal. Because the lemon peel extracts contained bioactive components and nutraceuticals, the results showed promise for treating diabetes, cardiovascular disease, cancer [23]. Citrus flavonoids can increase peripheral glucose absorption and glucose tolerance, decrease insulin resistance and hepatic glucose, modify enzyme activity, increase insulin release and response, reduce inflammation [24].

13.3.2 Anti-cancer Properties

Cancer is the top cause of mortality that occurs unexpectedly around the globe. Researchers and scientists have become more interested in creating functional food products with superior anti-cancer qualities in recent decades. Certain functional food products are often associated with major negative outcomes that can negatively impact patient's overall quality of life, and they have multiple limitations regarding their applicability and usefulness [25]. The discovery of natural anti-cancer drugs has boosted interest in functional foods. A number of these drugs have shown promise in managing and treating cancer-causing processes, with minimal side effects. Because lemon peels and other similar extracts have a high flavonoid concentration, they have shown promising anti-cancer effects. In conclusion, in vivo study shows that citrus peel flavonoids, especially nobiletin and extracts of citrus peel, have powerful anti-tumor effects on prostate, liver, skin, and colon malignancies. Inflammation, angiogenesis, apoptosis, proliferation can be controlled as part of the therapeutic strategy [26]. Because lemon peels are high in vitamins, minerals, and fiber, they seem to be beneficial to human health. Citrus fruits have been proven to contain anti-inflammatory and antioxidant qualities that may be useful in preventing chronic degenerative diseases including cancer and atherosclerosis. Human atherosclerosis and cancer can be treated and prevented with the use of citrus flavonoids [27].

13.3.3 Mental Health and Metabolism

Age-related illnesses like Alzheimer's disease have become important global health issues as people live longer. Alzheimer's disease is a condition that is defined by the presence of amyloid accumulation, oxidative stress, hyperphosphorylation, and neuroinflammation [28]. Additionally, there is a higher prevalence of dementia in such lifestyle-connected illness as dyslipidemia, diabetes among others. Consequently, the emergence of new ways to protect brain activity and prevent dementia in adult and preclinical patients has become an interesting topic for scientists and researchers [29]. Flavanones are found in large quantities in citrus fruits and have been found to enhance brain function. According to this preclinical research it presents neuroprotective effects on dementia models [30]. In yet another clinical study done on humans' consumption of citrus peels and extract flavonoids showed increased cognitive performance while reducing dementia and depression. Citrus peels & extracts improved cognition with associated functions [31].

13.3.4 Prevention of Oxidative Damage and Cardiovascular Disease

Treatment resistance is becoming more widespread, and cardiovascular disorders are a major global health concern with several reported deaths each year [32]. In order to find new, natural ways to separate medicinal components from citrus fruit like lemons that is used to treat hypertension, scientists and researchers need to devote more time to this field of study. Extract from lemon peels can significantly lessen hemorrhages. Citrus flavonoids 7-hydroxy flavone, luteolin, and chrysin induce umbilical vein lipoproteins. LDL is oxidized by human endothelial cells, which results in an increase in reactive oxygen species within the cells. The body's bioactive substances that are more effective in treating hypercholesterolemia are altered as a result of what is known as LDL [33]. Citrus flavonoids improve insulin resistance, lipid metabolism, additionally free radical scavenging. Citrus fruits contain flavonoids, which have been shown to promote cardiovascular health [24].

13.3.5 Role in the Prevention of Obesity

Globally, obesity is a chronic and potentially fatal disease. Anti-obesity medications may induce severe adverse effects, and the long-term efficacy of these treatments cannot be guaranteed [34]. Complementary and alternative remedies have garnered significant interest in recent decades on account of their remarkable health benefits. Citrus phytochemicals have showed considerable potential in a range of approaches to treating obesity issues. Citrus fruits are recognized as the primary source of flavonoids and p-synephrine, which are active biological compounds. Numerous studies have demonstrated that citrus fruits possess anti-obesity characteristics via multiple pathways, such as regulating adipogenesis, modulating lipid metabolism, and controlling energy intake and expenditure [35].

13.3.6 Regulation of Lipoprotein Metabolism

It has long been known that lemon peels are a great resource of bioactive flavonoids, vitamins, carotenoids substances. Recent study has examined how citrus flavonoids alter metabolism of lipid, along with other metabolic values connected to metabolic syndrome, cardiac difficulties, atherosclerosis. Major citro flavonoids such as nobiletin, naringenin, tangeretin, hesperidin has shown promise as therapeutic and nutraceutical agents for the purpose of metabolic dysregulation treatment [36]. Consuming meals high in lemon flavonoids is linked to a decreased incidence of metabolic disorders, inflammation, and cardiovascular diseases, per epidemiological research. Citrus flavonoids are prospective antibacterial, antihypertensive medicines, according to another clinical study [37]. Citrus flavonoids, like those found in lemon peels, have been shown in animal models to decrease the risk of insulin resistance, dyslipidemia, and hepatic steatosis by enhancing fatty acid oxidation and blocking the synthesis of hepatic fatty acids. In metabolically essential organs like the liver, adipose tissue, kidneys, aorta, citrus flavonoids reduced inflammation. Researchers found that citrus flavonoids help with insulin resistance, obesity, hepatic steatosis [38].

13.4 Industrial Applications of Lemon Peel Compounds

Because citrus garbage has special functional components. With greater knowledge of trash valorization, researchers are finding many uses for citrus waste, reducing environmental impact from improper rubbish disposal. Citrus refuse including peels, pulp, pomace and seeds, contains pectin which is a type of dietary fiber [39]. Nutraceuticals and functional foods can be produced from chemicals found in citrus waste. These bioactive agents also have functional benefits besides protecting against bacteria, allergy or diseases [40].

13.4.1 Source of Bioactive Compounds

A number of investigators have labored persistently to find out a new, cost efficient and practicable way of reusing processed fruit waste for the production of clean, environmentally friendly goods. The scientific community is interested in extraction methods for preparation and processing of samples [41]. From the fruit waste, the first stages of bioactive components recovery are extraction, isolation and characterization. It is essential to make the proper moves which would protect against destruction, loss or alteration by extraction of possible bioactive compounds in the process. Moreover, consideration should be given to the kind of biologically active chemical being investigated when selecting a solvent. Methanol or ethanol are used as polar solvents for extracting water-soluble chemicals while dichloromethane or a mixture composed half with methanol and dichloromethane is used for fat-soluble compounds [42]. Verifying the appropriateness of the extraction technique utilized to recover certain bioactive compounds is also essential. Green extraction methods are becoming more popular because they operate better, use fewer or no organic solvents, utilize less time, energy [43].

13.4.1.1 Polyphenols

Various citrus fruit cultivars have various amounts of polyphenols. Citrus peels contain primarily flavonoids, which include flavonones, flavonols, isoflavones, anthocyanidins, flavones. The efficiency of green extraction techniques over conventional methods for phenol extraction from citrus peel has been documented in a number of researches. In a study that extracted phenols from citrus peel, the greatest yield of ultrasonic-assisted extraction was 26.52%, which was 1.77 times greater than the maximum yield of microwave-assisted extraction [44]. For the purpose of their experiment, Panesar and Bera, employed both ultrasonic and maceration aided extraction methods to extract polyphenols from the peels of citrus. As such, it is outlined that ultrasound-assisted extraction proved more effective than maceration. In her study, used combination of maceration and ultrasound extraction techniques for extracting polyphenols from peels of citrus fruits. Notably, the efficacy of ultrasound-assisted extraction surpassed that of maceration [45].

13.4.1.2 Carotenoids

In addition, a rise in energy input from 55 to 364 kJ/kg increased the bioavailability of carotenoids extracted from lemon peel by high voltage electric discharge by 62.5% [46].

13.4.1.3 D-Limonene

At a pressure of 300 bar, supercritical fluid extraction is performed that yielded 30.65% d-limonene from lemon peels [47].

13.4.1.4 Pectin

Food companies utilize citrus peel pectin to gel jelly, jams jellies [48]. Remaining peel from citrus peels is used to make around 85% of the commercial pectin produced worldwide. Fresh citrus fruit peels, such as those from lemons, grapefruits, and oranges, or washed and dried peels can be used as raw materials to make pectin. 1.5% to 3% of pectin can be found in fresh citrus peels, and 9% to 18% can be found in dried peels [49]. It was fascinating to note that researchers experimented with unconventional methods to extract pectin from orange peel. Citrus peels were microwaved to extract pectin at 160–480 W, 1: 10–1: 30 g/ml liquid-solid ratio, and 60–180 s irradiation period. With a power of 477 watts, a time of 128 seconds, and a solid-liquid ratio of 1:20, the pectin yield was 25.79% [49].

13.4.1.5 Essential Oils

Aromatic liquids containing volatile components are present in orange peel cuticles and oil sacs [50]. Citrus peel essential oil is used to create scents and flavors. It is important to remember, though, that the extraction technique employed affects the yield and chemical makeup of citrus peel essential oil. A solvent-free microwave extraction of citrus lime peel essential oils gave 1.26% limonene [51]. A related study recovered limonene from mandarin peel using ultra critical CO2 at 300 bar pressure, yielding 30.65% and 13.16% at 100 bar pressure [52].

13.4.2 Prebiotic

Even though citrus peel pectin is used to make jam, jellies, and marmalades, it has potent prebiotic properties. Oligosaccharides derived from citrus pectin seem to be an active prebiotic ingredient [48]. Recent research has suggested using lemon peel residue to make functional prebiotic foods with improved bio-functionality. Pectic oligosaccharides (POS) are a prebiotic generated from agro-waste, such as orange-peel waste, acquired through the partial hydrolysis of pectin [53]. Citrus peel pectic polysaccharides HBE-III were tested for anti-tumor activities by Lee et al. (2014). The study found that HBE-III suppressed lung metastases depending on dosage. Through the use of HBE-III, NK cytotoxicity towards Yac-1 tumor cells was significantly enhanced [54].

13.4.3 Single Cell Protein

Citrus waste can be used to make functional meals and single-cell proteins. Considered to be the majority of dried cells produced by microorganisms, yeast, bacteria etc., these single-cell proteins. These are biomass, microbial protein, or bio-protein [55]. Single-cell protein is a good alternative to vegetable protein since it requires less land and water [56].

13.4.4 Livestock Feed

Some research suggested feeding cattle citrus processing waste. According to the findings of Pourhossein et al., the incorporation of lemon zest into the diet of broiler chickens resulted in a notable elevation in the levels of immunoglobulin G, white blood cells, and immunoglobulin M [57]. Similarly, the study conducted by Kour, examined the effects of incorporating kinnow refuse into goat feed on both feed intake and nutrient utilization [58]. The study revealed that the calcium-to-phosphorus ratio increased significantly with the 40% addition of kinnow waste; this is something that should be generally considered when feeding animals. However, none of the goats' health was adversely affected by kinnow refuse, as determined by blood biochemical analysis. Quail exhibited an increase in body weight, a reduction in triglyceride levels, an enhancement in the feed conversion ratio subsequent to the incorporation of lemon peel extract into their diet [59]. A report indicated that the waste of orange peels is safe, cheap and hygienic for feeding chicken. However, it is necessary to use citrus remainders in feeds sparingly since their overabundance may cause gastrointestinal disturbances in livestock [60].

13.5 Nutraceutical and Pharmaceutical Applications

13.5.1 Encapsulating Agent

Over the years, there has been a lot of research in the field of nanotechnology that has led to citrus peel pectin being used as an encapsulant. Besides, other polymer mixtures are presently under investigation for their potential use as encapsulating agents to improve both efficiency and stability [15]. Take into consideration an oil emulsion produced from lemon peel waste. It was made through employing pectin and the Ca^{2+} ion is all hydrosoluble components of them, citric acid as well as citrus waste. The prepared emulsion was stable with smaller droplet sizes, based on this research [61]. Furthermore, nanoparticles made up of resveratrol were created using orange peel's zein in combination with pectin. The manufactured nanoparticles were at pH levels between 2 and 7, 80 °C, over 1 hour and salt concentrations below 50 mM. Besides, in-vitro studies demonstrated that encapsulated form of resveratrol had enhanced bioavailability compared to free resveratrol. Moreover, it also improved the efficacy of resveratrol's antioxidant properties [62].

13.5.2 Pharma Drug

In addition to food additives, colorants, bioactive compounds and encapsulating agents these citrus peels have a purpose in the pharmaceutical industry [63]. In relation to pharmaceutical formulations, it is used as an ingredient because of the presence of bioactive components. Nobiletin is one of such flavonoids which are found in the citrus peel especially the polymethoxylated flavones. Various pharmacological benefits have been provided by these flavonoids against cardiovascular diseases [64]. According to a number of researchers, there are a bunch of health benefits in citrus peels including; antimicrobial features, anti-aging effects, hepatoprotective effect, immunosuppression and cardioprotection [65]. Using the minimum bactericidal concentration of 8%, a zone of inhibition of about 25.48 ± 1.41 mm and an inhibitory concentration of about 4% v/v, Chang's research revealed that it had an inhibitory effect on planktonic Listeria monocytogenes cells. Furthermore, the anti-biofilm potential was demonstrated by the essential oils obtained from citrus peel (*Staphylococcus aureus*); this means that orange peel essential oil could be used as germicidal or preservatives in the drug industry [66]. Later research examined how Citrus sinensis peel extract polymethoxy flavonoid chemicals treated and controlled stomach ulcers in male albino mice. Peel extract raised stomach pH and decreased lesions, volume of gastric [67].

13.6 Food and Beverage Industry

13.6.1 Dietary Fiber

As a dietary supplement, as well as in both natural and engineered meals, dietary fiber is in high demand. Citrus peel fiber is high in both soluble and insoluble fiber and is well-known for its possible dietary also health advantages [49]. Citrus peels are utilized in the literature to extract dietary fiber by a range of technologies, such as enzymes, dry/wet treatment, acid-alkali treatment, ultra-sonication [68]. A study employed the steam explosion and dilute acid soak technique to extract soluble fiber (33.74%) from orange peels, indicating its potential application as a food additive [69].

13.6.2 Food Additive

Food additives are compounds that are included in food to enhance its flavor, color, and taste. Bioactive substances that have antioxidant properties are found in food preparation industry waste and could be used instead of artificial food additives [69]. Citrus peel, rind is utilized in confectionary and bakery products due to their unique taste, color, nutritional content. In 2020, the global market for bakery goods was worth about USD 507.38 billion. Over the next few years, the bakery industry is expected to rise 2.5%, reaching roughly USD 574.05 billion by 2026. Technology allowed citrus waste to be processed and baked into baked goods, increasing nutritional value and waste reduction. For example, muffins were made with 10% and 15% of orange bagasse, which is high in nutritional fiber. The produced muffins (15%) had a modest (15%) fat content and a high (15.3%) dietary fiber content. Adding dietary fiber-rich orange bagasse was found to reduce the expected glycemic index [70]. Moreover, wheat flour was combined with 2.50 percent fiber and 0.23% pectin from citrus peels to make bread. Citrus peel fiber made the dough 6.5% better at absorbing water in the farinograph, 6.4% better in the mixolab, 7% better in the mixograph, and 10% better in the baking. Fermented and unfermented processed citrus peels were employed as dough additives in the production of bread. Specifically, the incorporation of 4% and 6% unfermented peels, which were desiccated at 150 °C, 100 °C, resulted in enhanced sensory acceptability [71]. In order to create the sponge cake, different percentages of citrus peel powder were incorporated 40%. The research findings suggested that replacing 10% of the citrus peel powder in sponge cake resulted in an optimal compromise between taste and quality. Citrus peel is also widely used in food applications because peel pectin can be used to thicken, stabilize, emulsify, texturize, and gel fruit jam, jellies, and low-fat/low-sugar products [72]. As a hydrolysate, citrus rind pectin is also utilized. Dangi and Yadav (2020) examined the physicochemical, functional properties of acid-hydrolyzed lemon peel pectin hydrolysate [73]. The study suggested that improved-solubility pectin hydrolysates could be employed as dietary fiber in liquid foods, notably beverages [71].

13.7 Cosmetic and Personal Care Products

13.7.1 Antiaging Agents

All creatures age naturally, hormones, smoking, nutrition, direct, indirect solar disclosure are the main extrinsic factors affecting aging. This procedure derives unstable ROS, which impede cell matrix enzymes. Free radicals are neutralized by antioxidants. Natural phytochemical substances found in lime plants offer antiaging and antioxidant properties [74]. Numerous biological traits and physiological processes depend on these naturally occurring phytochemical substances. Lemon fruit waste and products made from it are thought to be inexpensive, sustainable sources of chemicals that could be used in cosmetic, pharmaceutical, nutraceutical industries [75].

13.7.2 Skin Care Products

Because agro-industrial waste is rich in chemicals, antioxidants, vitamins (particularly C and E), and polyphenolic compounds, it is being utilized more often as an active ingredient in skincare products [74]. One such agricultural waste that is widely used in the cosmetics business is citrus peel garbage. The antioxidants help prevent oxidative damage to the skin and postpone the aging process. They also help reduce skin-related concerns like dark marks on skin, acne, and so on. Nevertheless, research is being done on using citrus wastes to make skin-brightening cosmetic items [17]. An alcoholic extract of citrus peels, for example, was tested for its anti-aging qualities using in-vitro antioxidant, anti-enzyme assays; the results supported utilizing this extract in skin care formulations targeted at wrinkle prevention [76]. A recent study on the anti-tyrosinase activity of lemon peel extract led to the development of a whitening lotion that employs the peel to assist prevent the generation and build-up of melanin pigment. Significant anti-tyrosinase properties were demonstrated by lemon peel, with an IC50 value of 255.10 μg/ml. It is possible to diminish the melanin pigment in whitening cream by 17.33% when it is made using citrus peel waste (2%) weight by weight. As a result, citrus peel waste can be used to cure detrimental skin conditions such as hyperpigmentation. However, additional study on its usefulness as an ingredient for cosmetic items has to be conducted [77].

13.8 Environmental Impact of Citrus Peel Waste

13.8.1 Bio-adsorbents

Bio-adsorbents are chemical compounds that are employed to aid in the elimination of heavy metals. In contemporary times, agrowaste is extensively employed as an economical bio-adsorbent to eliminate heavy metals from solutions [78]. A lot of study has been done on using citrus peel as a green bioadsorbent to get rid of heavy metals [79]. Lemon peels were used to make a cheap and eco-friendly activated carbon that was activated by alkali chemicals and carbonization to get rid of ranitidine in pharmacy wastewater. At 39.85 °C, 140 rpm agitation speed at 6 pH, 200 mg of adsorbent, 10 ppm concentration, ranitidine was eliminated (85.63%) [80]. Furthermore, lemon peels were used to create an activated carbon adsorbent that was impregnated with FeCl3. Villen-Guzman et al., examined the bio-sorbent potential of alkali-modified lemon peel to remove nickel from industrial effluents

[81]. First five minutes of investigation showed 90% nickel and cadmium sorption. Another study looked at the impact of silicon sand and a bioabsorbent made from lemon peels on soil filtration utilizing seawater irrigation. It was possible to successfully retain salt from seawater by combining silicon sand with orange peel bioabsorbent. Consequently, lemon peel-based bio-absorbents are low-cost natural bio-absorbents [82].

13.8.2 Source of Biofuels

Biofuels, including biogas, bioethanol, and biodiesel, are also manufactured using lemon rind waste as a natural resource. Citrus peels become biofuel through fermentation and anaerobic digestion. Mandarin peel bioethanol (50–600 L/1000 kg) was produced via steam explosion pretreatment and microbiological fermentation [83]. Lemon peel waste was pretreated with steam explosion caused by acid, Saccharomyces cerevisiae F15 fermentation to produce bioethanol. The study reported an ethanol production of 0.495 g/g and productivity of 4.85 g/Lh [84]. Bioethanol synthesis from waste citrus peel generally starts with acid-catalyzed steam pretreatment and followed by enzymatic hydrolysis. The study showed an ethanol yield of 64% after 48 hours at pH 4 Trans-esterification of essential oils with alcohols forms biodiesel from orange peels. Using trash from lemon peels to generate sustainable energy [85].

13.8.3 Packaging Material

Recently, a new field has emerged; one that uses leftover citrus peels to make eco-friendly packaging materials. Lemon peel was used to create a wrapper for fresh strawberries [86]. In this case, edible packaging films using nanoparticles of titanium oxide (0.5–2.0% by weight) with 3:1 sweet potato starch to lemon peel pectin as were prepared. Lemon peel flour combined with tea polyphenols was used in making a bioactive edible packaging film through casting process. The results showed that composite packages based on pomelo peel and tea polyphenol containing 10% tea polyphenol exhibited better moisture barrier property along with mechanical, antibacterial and antimicrobial properties than other samples. Furthermore, they might also replace the storage of soybean oil for a month sustainably [87]. Recently, fish gelatin was amalgamated with blood lemon peel pectine to form the edible film. As a result, the film made from fish gelatin/Lemon peel pectin (50/50 ratio) had high tensile strength as well as antioxidant and antimicrobial activities. Use of such an edible film wrapping ricotta cheese has improved it physicochemical characteristics, texture behavior and microbiological stability [39].

13.9 Waste Management Strategies

The ways in which citrus peels can be disposed of differ depending on the specific variety, scale of production and local waste management systems. Beyond the citrus processing industry, citrus peel waste is a major byproduct which must be properly dealt with because the environment and economic reasons. Several waste management strategies, including composting, animal feed, traditional disposal method and anaerobic digestion. Each strategy is evaluated based on its environmental impact, economic feasibility, and technological requirements [88].

13.9.1 Traditional Disposal Methods

Traditional disposal methods for lemon peel waste management typically involve composting, landfill disposal. In this method lemon peels that cannot be reused or recycled are separated from other waste. They are dispersed in low-lying sections of a landfill in a thin layer. To cover the waste, a layer of dirt is placed after every layer of trash. Lemon peel waste can still yield valuable products. Through pyrolysis (thermal decomposition in the absence of oxygen), lemon peels can be converted into bio-oil, gas. The biochar obtained can be used for manufacturing carbon materials and as solid fuel. The bio-oil contains bioactive compounds like d-limonene, squalene, phenol, ß-Sitosterol. Both biochar and bio-oil from lemon peels exhibit bactericidal activity against various bacteria, including *Salmonella enterica, Staphylococcus aureus, Micrococcus luteus, Listeria monocytogenes* [89].

13.9.2 Composting Techniques

The leftover peel from lemons can be utilized to make high-quality compost, just like with other vegetal matrices. Lemon peel waste compost, like biomass made from other organic leftovers, is seen as a partial or complete replacement for nonrenewable resources [90]. The biochemical process involved in composting waste from lemon peels begins with a brief latency period, which is followed by a significant temperature increase that lasts for about 2 months, and a final phase that lasts for another month and involves slow cooling. Composting organic matter into stable compounds allows it to be handled, stored, transported, applied to land without harming the environment. The metabolic heat produced by microorganisms during composting kills illnesses and weed seeds [91]. Lemon peel compost has high lignocellulosic compounds, C/N ratio, and pH. Waste C/N ratio, pH, and water content should be 24:1, 6.3, and 60%, respectively. To change lemon peel waste's low pH (4.1), calcium hydroxide was used, which is used to compost acidic substrates because its exothermic reaction boosts microbial activity. Phytotoxicity and the carbon-to-nitrogen (C/N) ratio of lemon peel residue diminish during humification; this latter parameter does not typically induce mineral nitrogen immobilization in soil. Nevertheless, the high moisture content of moist CPW poses a challenge for composting using conventional methods employed for managing municipal solid waste. This is due to the fact that it necessitates a preliminary mixing with lignocellulosic support, such as straw or pruning residues, which incurs additional expenses [92].

13.9.3 Anaerobic Digestion Processes

The treatment of lemon peel residue via anaerobic digestion is a promising approach. Nevertheless, citrus residue anaerobic digestion is impeded by the presence of limonene. Antimicrobial limonene may inhibit the recovery of methane-producing bacteria, which require more time to restore their health compared to acid-producing bacteria that have been damaged. Thus, volatile fatty acid buildup reduces methane generation [92]. One potential solution to this issue is to divide the anaerobic digestion process into two distinct stages. Hydrolysis, acidogenesis are the objectives of the first stage; methanogenesis is the purpose of the second. Additionally, by separating the system, each stage could operate under its optimal conditions, resulting in a more stable process. Anaerobic digestion was conducted in this study in two phases using 120 ml glass bottle bioreactors in batch operations. In contrast to the free-cell bioreactor utilized in the initial phase, the membrane bioreactor was employed in both the free-cell and bioreactor stages of the second phase. Initially, the reactor was operated under "semi-aerobic" and "anaerobic" conditions to investigate the impact of oxygen on facultative anaerobic microbes during acid production. In the second phase, the membrane's limonene protection was assessed. First, 1.5 g VS of inoculum and 4.5 g VS of citrus waste were used to make the basal medium. The digesting process lasted 4 days at 55 °C [93].

13.9.4 Utilization in Animal Feed

Animal feed makes up as much as 70–80% of the production of lemon peel waste in several Mediterranean basin regions, including Southern Italy (though this number varies from processing plant to processing plant and year to year). Wet, dried lemon peel waste is used to substitute traditional animal diet feed, like grains. The type of cattle and their physiological state determine how much lemon peel waste should be progressively added to the diet [94]. Only during the short citrus processing season and for livestock near processing facilities can wet lemon peel garbage be used. Due to its high-water content, transportation costs reduce the financial benefits of using wet waste as animal feed. Lemon peel waste is cheap for cow breeders, but farm management activities like storage tanks, automated feed systems, and new animal diets may make it more expensive. More modern sheep and caprine farms can maximize the peel's nutritional value and pay more [95]. It has been extensively shown over the last 50 years that dried lemon peel waste offers a substantial nutritional capacity and is more appealing for animals than wet or ensiled peel [96]. Generally speaking, 5–10% of CPW production that is used to make animal feed is fuel-dried; nevertheless, this kind of drying is not very profitable, particularly for conventional low-efficiency drying plants. Research has shown that it is feasible to use solar energy in semi-arid rural areas to dry lemon peel waste, which will facilitate the reuse of CPW in livestock and agriculture. For example, a considerable portion (about 40–50%) of the waste produced from the solar-drying of lemon peels occurs in several regions of the Mediterranean basin, where the animals are mostly raised as livestock and kept in caprine zoos. Left in heaps or rows on agricultural land near the processing factory, the cleaner, upper portion of peel can be given to animals, while the dirtier, bottom portion, which may decay, is used for fuel or organic soil conditioner [96]. Infrequently, dried lemon peel waste that is added to animal feed is frozen to prevent microbial growth while the feed is being stored without interfering with rumen fermentation [20]. Lemon peel waste has potential applications, as summarized in Fig. 13.2 [114].

13.10 Challenges and Opportunities in Waste Management

Lemon peel waste, a byproduct of citrus processing industries, presents significant challenges to waste management due to its large-scale generation and environmental impacts. Traditional disposal methods such as landfilling and incineration are associated with environmental pollution and economic costs. The amount of trash produced by growing urban populations, their consequences on the environment, human health are too much for current technologies and systems to handle. The primary obstacle in employing agricultural residue for papermaking is that silica dissolves in pulping liquor when pulped with strong alkali and reacts with process equipment during liquor recovery operations. Silica is known to be capable of causing clogging in evaporator tubes leading to significant disturbance of lime mud set rate. In addition, global climate change is worsened by high input requirements for farming [97]. For effective management of lemon peel as waste, there should be inter-disciplinary work between researchers, industrialists, policy makers and community members. The use of interdisciplinary approaches helps in developing strategies for integrated waste management and also facilitates technology transfer and dissemination of knowledge. Moreover, through stakeholder engagement awareness campaigns are created that foster support for these initiatives thus improving sustainability for both implementation and long-term viability [98].

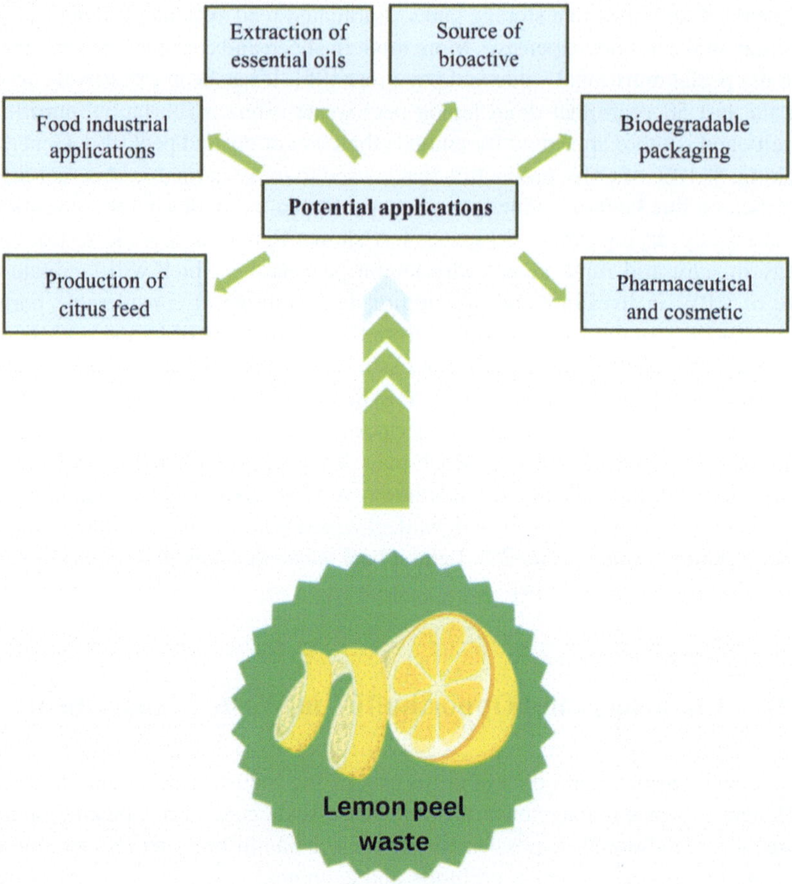

Fig. 13.2 Potential application of lemon peel waste

13.11 Emerging Technologies for Value Extraction

Lemon peel waste can be used in various emerging technologies with the aim of extracting valuable compounds. One such technology is microwave-assisted extraction (MAE), which uses microwave radiation instead of traditional heat to speed up the extraction, and requires less solvent [46]. There are essential oils leaking from ruptured oil glands present in citrus fruits when squeezed to make juice, mostly located at different levels of the peel and cuticles [72]. These chemicals are much needed in the food processing industries, pharmaceutical companies as well as the cosmetic manufacturing firms, or they could go directly for domestic distribution [74]. Many components that are taken out of citrus fruits during the juice-processing process are referred to as EO. Citrus essential oil (EO) is a mixture of oxygenated derivatives like alcohols, esters, and aldehydes, terpenic hydrocarbons. Lemon peel waste contains 5.4 kg of oil per gramme of raw biomass, 98–99% of which is dlimonene [99]. The majority of the essential oil (EO) is present in the oil sacs of the flavedo, while trace sums are present in other fruit sections such the juice cells. Although not identical, juice cell oil is usually better than fruit skin oil. The concentration of valencene, a sesquiterpene (0.3–2.0%) that is only present in trace amounts in other oils [98]. In addition to yielding a valuable product with a multitude of applications, extraction from lemon peel trash also contributes, as previously said, to the feasibility of using lemon peel garbage for other purposes like the production of bioethanol or biogas. Several studies state that in these situations, there are various ways to extract essential oil (EO) from lemon peel waste, either for the purpose of recovery or simple removal [100].

13.11.1 Biorefinery Concepts

A biorefinery is a facility that combines machinery and procedures for converting waste biomass into chemicals, fuels, and electricity. Because CPW contains a large number of high-value-added substances (like soluble sugars and cellulose), it can be valorized by using a biorefinery [101]. The value obtained from the biomass feedstock is maximized when numerous products are produced, and less trash needs to be processed as a result. Using a biorefinery strategy helps the bioeconomy expand while reducing the harmful environmental effects associated with the agro-industry [102]. According to Siles Làpez, the formation of ethanol and methane comes after the elimination of EO [94]. Prior to the leftover lignin being used as an energy source at the end, other high-value products are also removed, including pectin, industrial enzymes. To produce ethanol, biogas and limonene were the major aims of Pourbafrani. A byproduct of solvent extraction can yield up to 78% of total pectin content based on market requirements and cost implications. The solid or liquid part of hydrolyzed lemon peel waste gives rise to biogas through anaerobic digestion while sugars left over turn into ethanol during limonene retrieval [103].

13.11.2 Enzymatic Extraction Processes

Enzymatic extraction processes have given efficient and eco-friendly ways of getting out important agents from peeling waste. Which contributes to sustainable disposal of wastes. These methods use enzymes that break down cell walls and release bioactive substances such as pectin, flavonoids, essential oils present in lemon peels. Enzyme-assisted extraction normally encompasses a sequence of steps that includes pre-treatment to enhance the accessibility of enzymes lipid hydrolysis-breaking complicated molecules into simple ones along with subsequent compounds separation- all these are involved in enzymatically extracting extracts from lemon peel waste. Different types of enzymes are required for the recovery of specific compounds from wasted lemon peels. For instance, pectinases which pectinifies acts on pectin whereas cellulases disintegrate hemicellulose, with special reference to cellulose found in peels' cell wall structure. Lipases may also be employed in the removal of lipophilic chemicals plus essential oils [101]. Guo et al. examined commercially available pectinase enzymes for enzymatic extraction of pectin from lemon peel wastes. The authors found that enzymatic hydrolysis was the most effective way to liberate pectin from lemon peels at certain conditions which were best suited to obtain the highest possible yield of this substance. Guo et al.'s research demonstrates that rather than using chemical methods, enzymatic techniques can be applied in extracting pectin from lemon peels. In addition, it is possible to use this method to extract essential oils and flavonoids besides reclaiming a valuable substance called pectin out of lemons' skin. Therefore, after orange peeling process; enzymatic extractions are considered economical and ecologically viable means because they have several applications in food industry, pharmaceutical manufacturing as well as cosmetic sector where there are residues left after removal of the outer skin [104].

13.11.3 Supercritical Fluid Extraction

Superfluid extraction (SFE) is a technique used for extracting bioactive compounds from different sources of nature like debris of lemon peel or more. This process employs a supercritical fluid, typically carbon dioxide (CO2), which has both liquid and gaseous properties depending on the pressure and temperature employed. Many of these active compounds are found in the components that make up lemons. Citrus industry extracts essential oil such as those obtained by squeezing remaining peels after juice preparation and other derivatives [105]. For the last 20 years, quite a number of investigations have utilized supercritical extraction and fractionation processes for obtaining bioactive constituents from plant matrix volatile oils. The food industry, cosmetics, aromatherapy massage and medicine apply D-limonene due to its chemotherapeutic activity, anticancer effects against preclinical breast and liver cancers, antioxidant ability plus antidiabetic agents among others. It is a monocyclic terpene with no clinical toxicity whatsoever [106].

13.12 Future Perspective

The global community has recently shown increased interest in adopting sustainable methods to minimize garbage production across various sectors. The utilization of citrus peel for industrial food applications is now seen as a priority research area. Increased healthcare expenditure and the management of chronic diseases have led to a rise in medication [107]. Moreover, pharmaceuticals can also be derived from them by extraction of essential oils and antioxidants that are applicable for medicinal purposes [108]. Various novel green extraction processes have been employed for extracting bioactive components from orange peel [109]. There have been a great number of researches that have investigated ecologically friendly methods for extracting bioactive components from lemon peel. These methods include supercritical fluid extraction, ultrasound-assisted extraction, and microwave-facilitated extraction [110]. For commercial use, the food industry must overcome natural component stability. Thus, the encapsulation of bioactive substances has been the subject of various investigations. Future research on nano-encapsulation has emerged as a major goal. Pectin, which is extracted from the peels of citrus fruits like lemons, has been utilized in a variety of investigations as a wall material for the purpose of encapsulating bioactive compounds [111]. To efficiently use citrus-based components, however, the trend is to create emulsion-based delivery systems with hydro soluble and liposoluble (citrus oil) components [112]. Nevertheless, as technology advances, it is essential that research be directed towards the investigation of potential applications in the fourth industrial sector. These applications include the intelligent production of functional meals and personalized diets through the utilization of 3D food printing technology. Citrus peels contain bioactive compounds that can be utilized in an efficient manner to produce a variety of inks for use in three-dimensional printing of food. The significance of extracting bioactive compounds from citrus peels and the biological properties of these peels has already been proven by a number of studies. Again, research on how to use these chemicals in different foods is relatively scarce. Citrus peel extract has been studied to have possible uses for other kinds of food products such as a thickener, flavor or color enhancer, but only a few studies have concentrated on this. Immediate future need exists regarding findings from experiments made on citrus peel extract as the additive in variety of food matrixes as well as its storage stability [113].

13.13 Conclusion

Waste from citrus processing can serve as a potential source of bioactive compounds and other useful molecules for commercial applications including thickening agents, flavor enhancers and functional food ingredients. In addition, the industry is striving to develop environmentally friendly ways to extract bioactive substances and other chemicals from citrus peels on one hand, and more sustainable means of utilizing waste generated through the process on the other. It is important to properly manage lemon peel waste so as to achieve environmental sustainability and economic viability. Reusing lemon peel waste through various methods such as composting, extraction for essential oils, animal feed production or even generation of biofuel contributes towards mitigating climate change effects by reducing greenhouse gas emissions and land filling pressures. Nevertheless, scalability, logistics in transportation and technology have remained major challenges hence limiting mass adoption. Generally speaking, adopting innovative approaches to citrus peel wastage will not only reduce environmental impact but also unearth potential economic opportunities for redefining 'trash' into valuable items or products that can be put back into use.

References

1. Zou, Z., Xi, W., Hu, Y., Nie, C., & Zhou, Z. (2016). Antioxidant activity of citrus fruits. *Food Chemistry, 196*, 885–896. https://doi.org/10.1016/j.foodchem.2015.09.072
2. Liu, Y., Heying, E., & Tanumihardjo, S. A. (2012). History, global distribution, and nutritional importance of citrus fruits. *Comprehensive Reviews in Food Science and Food Safety, 11*(6), 530–545. https://doi.org/10.1111/j.1541-4337.2012.00201.x
3. Anwar, F., Naseer, R., Bhanger, M. I., Ashraf, S., Talpur, F. N., & Aladedunye, F. A. (2008). Physico-chemical characteristics of citrus seeds and seed oils from Pakistan. *Journal of the American Oil Chemists' Society, 85*(4), 321–330. https://doi.org/10.1007/s11746-008-1204-3
4. Uzzaman, S., Akanda, K. M., Mehjabin, S., & Parvez, G. M. M. (2018). A short review on a nutritional fruit: Guava. *Open Access: Toxicology & Research, 1*(1), 1–8.
5. Rawat, N., Kiran, S. P., Du, D., Gmitter, F. G., & Deng, Z. (2015). Comprehensive meta-analysis, co-expression, and miRNA nested network analysis identifies gene candidates in citrus against Huanglongbing disease. *BMC Plant Biology, 15*(1), 184. https://doi.org/10.1186/s12870-015-0568-4
6. Saini, R. K., Ranjit, A., Sharma, K., Prasad, P., Shang, X., Gowda, K. G. M., & Keum, Y.-S. (2022). Bioactive compounds of citrus fruits: A review of composition and health benefits of carotenoids, flavonoids, limonoids, and terpenes. *Antioxidants, 11*(2), Article 2. https://doi.org/10.3390/antiox11020239
7. Hussain, S. Z., Naseer, B., Qadri, T., Fatima, T., & Bhat, T. A. (2021). Citrus fruits—Morphology, taxonomy, composition and health benefits. In S. Z. Hussain, B. Naseer, T. Qadri, T. Fatima, & T. A. Bhat (Eds.), *Fruits grown in highland regions of the Himalayas: Nutritional and health benefits* (pp. 229–244). Springer International Publishing. https://doi.org/10.1007/978-3-030-75502-7_18
8. Sharma, K., Mahato, N., Cho, M. H., & Lee, Y. R. (2017). Converting citrus wastes into value-added products: Economic and environmently friendly approaches. *Nutrition, 34*, 29–46. https://doi.org/10.1016/j.nut.2016.09.006
9. Parvez, G. M. M., Akanda, K. M., Karim, R., Mou, S. M., Zahan, P., & Mosaddik, D. A. (2016). Comparrative phytochemical screening and antimicrobial evaluation of different varieties of banana (Musa sapientum). *International Journal of Innovative Pharmaceutical Sciences and Research, 4*(4), 372–383.
10. Arshad, M. S., Khalid, W., Khalid, M. Z., Maqbool, Z., Ali, A., Kousar, S., Afzal, M. F., Mukhtar, S., Rahim, M. A., Aziz, A., & Ansari, M. J. (2023). Sonication microwave synergistic extraction of bioactive compounds from plants, animals and others agro-industrial waste sources. In G. A. Nayik, M. Ranjha, X. A. Zeng, S. Irfan, & S. M. Zahra (Eds.), *Ultrasound and microwave for food processing* (pp. 345–361). Academic. https://doi.org/10.1016/B978-0-323-95991-9.00008-4
11. Maqbool, Z., Khalid, W., Atiq, H. T., Koraqi, H., Javaid, Z., Alhag, S. K., Al-Shuraym, L. A., Bader, D. M. D., Almarzuq, M., Afifi, M., & AL-Farga, A. (2023). Citrus waste as source of bioactive compounds: Extraction and utilization in health and food industry. *Molecules, 28*(4), Article 4. https://doi.org/10.3390/molecules28041636
12. Vattem, D. A., & Shetty, K. (2002). Solid-state production of phenolic antioxidants from Cranberry Pomace by Rhizopus Oligosporus. *Food Biotechnology, 16*(3), 189–210. https://doi.org/10.1081/FBT-120016667

13. Sir Elkhatim, K. A., Elagib, R. A. A., & Hassan, A. B. (2018). Content of phenolic compounds and vitamin C and antioxidant activity in wasted parts of Sudanese citrus fruits. *Food Science & Nutrition, 6*(5), 1214–1219. https://doi.org/10.1002/fsn3.660
14. Costanzo, G., Iesce, M. R., Naviglio, D., Ciaravolo, M., Vitale, E., & Arena, C. (2020). Comparative studies on different citrus cultivars: A revaluation of waste mandarin components. *Antioxidants, 9*(6), Article 6. https://doi.org/10.3390/antiox9060517
15. Panwar, D., Saini, A., Panesar, P. S., & Chopra, H. K. (2021). Unraveling the scientific perspectives of citrus by-products utilization: Progress towards circular economy. *Trends in Food Science & Technology, 111*, 549–562. https://doi.org/10.1016/j.tifs.2021.03.018
16. Akanda, M. K. M., Sultana, R., Rana, M. M., Hossain, M. A., Barik, S. M. A., Islam, M. R., Islam, M. N., & Parvez, G. M. M. (2024). Prevalence, causes, and management strategies of fungal diseases in northern regions of Bangladesh. *Sciences of Pharmacy, 3*(1), 24–33. https://doi.org/10.58920/sciphar0301191
17. Mahato, N., Sinha, M., Sharma, K., Koteswararao, R., & Cho, M. H. (2019). Modern extraction and purification techniques for obtaining high purity food-grade bioactive compounds and value-added co-products from citrus wastes. *Food, 8*(11), Article 11. https://doi.org/10.3390/foods8110523
18. Ke, Z., Pan, Y., Xu, X., Nie, C., & Zhou, Z. (2015). *Citrus* flavonoids and human cancers. *Journal of Food and Nutrition Research, 3*(5), Article 5. https://doi.org/10.12691/jfnr-3-5-9
19. Akanda, M. K. M., & Hasan, A. H. M. N. (2021). Characterization of pharmacological properties of methanolic seed and stem bark extracts of Ziziphus mauritiana (BAU Kul) using in-vitro and in-vivo animal (Swiss albino male mice) model. *Clinical Phytoscience, 7*(1), 8. https://doi.org/10.1186/s40816-020-00246-0
20. Rossi, R. C., da Rosa, S. R., Weimer, P., Lisbôa Moura, J. G., de Oliveira, V. R., & de Castilhos, J. (2020). Assessment of compounds and cytotoxicity of *Citrus deliciosa* Tenore essential oils: From an underexploited by-product to a rich source of high-value bioactive compounds. *Food Bioscience, 38*, 100779. https://doi.org/10.1016/j.fbio.2020.100779
21. Ani, P. N., & Abel, H. C. (2018). Nutrient, phytochemical, and antinutrient composition of Citrus maxima fruit juice and peel extract. *Food Science & Nutrition, 6*(3), 653–658. https://doi.org/10.1002/fsn3.604
22. Makni, M., Jemai, R., Kriaa, W., Chtourou, Y., & Fetoui, H. (2018). *Citrus limon* from Tunisia: Phytochemical and physicochemical properties and biological activities. *BioMed Research International, 2018*, e6251546. https://doi.org/10.1155/2018/6251546
23. Aruoma, O. I., Landes, B., Ramful-Baboolall, D., Bourdon, E., Neergheen-Bhujun, V., Wagner, K.-H., & Bahorun, T. (2012). Functional benefits of citrus fruits in the management of diabetes. *Preventive Medicine, 54*, S12–S16. https://doi.org/10.1016/j.ypmed.2012.02.012
24. Mahmoud, A. M., Hernández Bautista, R. J., Sandhu, M. A., & Hussein, O. E. (2019). Beneficial effects of citrus flavonoids on cardiovascular and metabolic health. *Oxidative Medicine and Cellular Longevity, 2019*, e5484138. https://doi.org/10.1155/2019/5484138
25. Hegde, P. S., & Chen, D. S. (2020). Top 10 challenges in cancer immunotherapy. *Immunity, 52*(1), 17–35. https://doi.org/10.1016/j.immuni.2019.12.011
26. Rawson, N. E., Ho, C.-T., & Li, S. (2014). Efficacious anti-cancer property of flavonoids from citrus peels. *Food Science and Human Wellness, 3*(3), 104–109. https://doi.org/10.1016/j.fshw.2014.11.001
27. Giammanco, M., Plescia, F., Giammanco, M. M., Leto, G., & Gentile, C. (2022). Bioactive effects of citrus flavonoids and role in the prevention of atherosclerosis and cancer. *Journal of Biological Research – Bollettino Della Società Italiana Di Biologia Lo Sperimentale, 95*(1), Article 1. https://doi.org/10.4081/jbr.2022.10313
28. Parvez, G. M. M., Sultana, R., Akanda, M. K. M., Mehjabin, S., & Mosaddik, A. (2018). Nature and cause of taking psychotropic drugs: A cross-sectional health survey in northern regions of Bangladesh. *International Journal of Pharmaceutical Sciences and Research, 9*(1), 361–368.

29. Sultana, R., Akanda, K. M., Mehjabin, S., & Parvez, G. M. M. (2017). Prescribing trends of psychotropic drugs in northern Bangladesh. *Specialty Journal of Medical Research and Health Science, 2*(2), 17–25.
30. Breijyeh, Z., & Karaman, R. (2020). Comprehensive review on Alzheimer's disease: Causes and treatment. *Molecules, 25*(24), Article 24. https://doi.org/10.3390/molecules25245789
31. Freedland, K. E., Skala, J. A., Steinmeyer, B. C., Carney, R. M., & Rich, M. W. (2021). Effects of depression on heart failure self-care. *Journal of Cardiac Failure, 27*(5), 522–532. https://doi.org/10.1016/j.cardfail.2020.12.015
32. Haque, M. U., Ahmmed, S. B., Akanda, M. K. M., Hasan, M. T., Mou, S. A., Sajon, S. R., & Islam, M. A. U. (2016). Prevalence and risk factors of obesity and hypertension among university students in Rajshahi City, Bangladesh. *Bangladesh Pharmaceutical Journal, 19*(2), 179–184. https://doi.org/10.3329/bpj.v19i2.29277
33. Kakutani, N., Yokota, T., Fukushima, A., Obata, Y., Ono, T., Sota, T., Kinugasa, Y., Takahashi, M., Matsuo, H., Matsukawa, R., Yoshida, I., Kakinoki, S., Yonezawa, K., Himura, Y., Yokota, I., Yamamoto, K., Tsuchihashi-Makaya, M., & Kinugawa, S. (2022). Impact of citrus fruit intake on the mental health of patients with chronic heart failure. *Journal of Cardiology, 79*(6), 719–726. https://doi.org/10.1016/j.jjcc.2021.12.004
34. Chooi, Y. C., Ding, C., & Magkos, F. (2019). The epidemiology of obesity. *Metabolism – Clinical and Experimental, 92*, 6–10. https://doi.org/10.1016/j.metabol.2018.09.005
35. Lavie, C. J., Laddu, D., Arena, R., Ortega, F. B., Alpert, M. A., & Kushner, R. F. (2018). Healthy weight and obesity prevention: JACC health promotion series. *Journal of the American College of Cardiology, 72*(13), 1506–1531. https://doi.org/10.1016/j.jacc.2018.08.1037
36. Fardoun, M. M., Maaliki, D., Halabi, N., Iratni, R., Bitto, A., Baydoun, E., & Eid, A. H. (2020). Flavonoids in adipose tissue inflammation and atherosclerosis: One arrow, two targets. *Clinical Science, 134*(12), 1403–1432. https://doi.org/10.1042/CS20200356
37. Maleki, S. J., Crespo, J. F., & Cabanillas, B. (2019). Anti-inflammatory effects of flavonoids. *Food Chemistry, 299*, 125124. https://doi.org/10.1016/j.foodchem.2019.125124
38. Mulvihill, E. E., Burke, A. C., & Huff, M. W. (2016). Citrus flavonoids as regulators of lipoprotein metabolism and atherosclerosis. *Annual Review of Nutrition, 36*(1), 275–299. https://doi.org/10.1146/annurev-nutr-071715-050718
39. Jridi, M., Abdelhedi, O., Salem, A., Kechaou, H., Nasri, M., & Menchari, Y. (2020). Physicochemical, antioxidant and antibacterial properties of fish gelatin-based edible films enriched with orange peel pectin: Wrapping application. *Food Hydrocolloids, 103*, 105688. https://doi.org/10.1016/j.foodhyd.2020.105688
40. Babbar, N., Oberoi, H. S., Uppal, D. S., & Patil, R. T. (2011). Total phenolic content and antioxidant capacity of extracts obtained from six important fruit residues. *Food Research International, 44*(1), 391–396. https://doi.org/10.1016/j.foodres.2010.10.001
41. Wu, F., Jin, Y., Xu, X., & Yang, N. (2017). Electrofluidic pretreatment for enhancing essential oil extraction from citrus fruit peel waste. *Journal of Cleaner Production, 159*, 85–94. https://doi.org/10.1016/j.jclepro.2017.05.010
42. Sasidharan, S., Chen, Y., Saravanan, D., Sundram, K. M., & Latha, L. Y. (2011). Extraction, isolation and characterization of bioactive compounds from plants' extracts. *African Journal of Traditional, Complementary and Alternative Medicines, 8*(1), Article 1. https://doi.org/10.4314/ajtcam.v8i1.60483
43. Anticona, M., Blesa, J., Lopez-Malo, D., Frigola, A., & Esteve, M. J. (2021). Effects of ultrasound-assisted extraction on physicochemical properties, bioactive compounds, and antioxidant capacity for the valorization of hybrid Mandarin peels. *Food Bioscience, 42*, 101185. https://doi.org/10.1016/j.fbio.2021.101185
44. Nipornram, S., Tochampa, W., Rattanatraiwong, P., & Singanusong, R. (2018). Optimization of low power ultrasound-assisted extraction of phenolic compounds from mandarin (*Citrus reticulata* Blanco cv. Sainampueng) peel. *Food Chemistry, 241*, 338–345. https://doi.org/10.1016/j.foodchem.2017.08.114

45. Saini, A., Panesar, P. S., & Bera, M. B. (2019). Comparative study on the extraction and quantification of polyphenols from citrus peels using maceration and ultrasonic technique. *Current Research in Nutrition and Food Science Journal, 7*(3), 678–685.
46. Gómez-Mejía, E., Rosales-Conrado, N., León-González, M. E., & Madrid, Y. (2019). Citrus peels waste as a source of value-added compounds: Extraction and quantification of bioactive polyphenols. *Food Chemistry, 295*, 289–299. https://doi.org/10.1016/j.foodchem.2019.05.136
47. Šafranko, S., Ćorković, I., Jerković, I., Jakovljević, M., Aladić, K., Šubarić, D., & Jokić, S. (2021). Green extraction techniques for obtaining bioactive compounds from mandarin peel (Citrus unshiu var. Kuno): Phytochemical analysis and process optimization. *Food, 10*(5), Article 5. https://doi.org/10.3390/foods10051043
48. Foti, P., Ballistreri, G., Timpanaro, N., Rapisarda, P., & Romeo, F. V. (2022). Prebiotic effects of citrus pectic oligosaccharides. *Natural Product Research, 36*(12), 3173–3176. https://doi.org/10.1080/14786419.2021.1948845
49. Berk, Z. (2016). By-products of the citrus processing industry. In Z. Berk (Ed.), *Citrus fruit processing* (pp. 219–233). Academic. https://doi.org/10.1016/B978-0-12-803133-9.00010-2
50. Prakash Maran, J., Sivakumar, V., Thirugnanasambandham, K., & Sridhar, R. (2014). Microwave assisted extraction of pectin from waste *Citrullus lanatus* fruit rinds. *Carbohydrate Polymers, 101*, 786–791. https://doi.org/10.1016/j.carbpol.2013.09.062
51. Golmakani, M.-T., & Moayyedi, M. (2016). Comparison of microwave-assisted hydrodistillation and solvent-less microwave extraction of essential oil from dry and fresh Citruslimon (Eureka variety) peel. *Journal of Essential Oil Research, 28*(4), 272–282. https://doi.org/10.1080/10412905.2016.1145606
52. Asbahani, A. E., Miladi, K., Badri, W., Sala, M., Addi, E. H. A., Casabianca, H., Mousadik, A. E., Hartmann, D., Jilale, A., Renaud, F. N. R., & Elaissari, A. (2015). Essential oils: From extraction to encapsulation. *International Journal of Pharmaceutics, 483*(1), 220–243. https://doi.org/10.1016/j.ijpharm.2014.12.069
53. Míguez, B., Gómez, B., Gullón, P., Gullón, B., & Alonso, J. L. (2016). Pectic oligosaccharides and other emerging prebiotics. In *Probiotics and prebiotics in human nutrition and health*. IntechOpen. https://doi.org/10.5772/62830
54. Lee, E. H., Park, H.-R., Shin, M.-S., Cho, S. Y., Choi, H.-J., & Shin, K.-S. (2014). Antitumor metastasis activity of pectic polysaccharide purified from the peels of Korean *Citrus* Hallabong. *Carbohydrate Polymers, 111*, 72–79. https://doi.org/10.1016/j.carbpol.2014.04.073
55. Sharif, M., Zafar, M. H., Aqib, A. I., Saeed, M., Farag, M. R., & Alagawany, M. (2021). Single cell protein: Sources, mechanism of production, nutritional value and its uses in aquaculture nutrition. *Aquaculture, 531*, 735885. https://doi.org/10.1016/j.aquaculture.2020.735885
56. Mekonnen, M. M., & Hoekstra, A. Y. (2014). Water footprint benchmarks for crop production: A first global assessment. *Ecological Indicators, 46*, 214–223. https://doi.org/10.1016/j.ecolind.2014.06.013
57. Pourhossein, Z., Qotbi, A. A. A., Seidavi, A., Laudadio, V., Centoducati, G., & Tufarelli, V. (2015). Effect of different levels of dietary sweet orange (Citrus sinensis) peel extract on humoral immune system responses in broiler chickens. *Animal Science Journal, 86*(1), 105–110. https://doi.org/10.1111/asj.12250
58. Kour, R., Rastogi, A., Sharma, R. K., & Singh, M. (2016). Evaluation of Kinnow mandarin fruit waste in rations of goats. *Indian Journal of Animal Nutrition, 33*(4), 416. https://doi.org/10.5958/2231-6744.2016.00074.8
59. Ciftci, M., Şimşek, Ü., Dalkiliç, B., Azman, M., Yilmaz, Ö., Mutlu, S., Özçelik, M., Baykalir, Y., Tonbak, F., & Bahşi, M. (2016). Effect of dietary orange peel extract on physiological, biochemical, and metabolic responses of Japanese quail reared under low ambient temperature. *Turkish Journal of Veterinary & Animal Sciences, 40*(3), 288–297. https://doi.org/10.3906/vet-1504-11
60. Seidavi, A., Zaker-Esteghamati, H., & Salem, A. Z. M. (2020). A review on practical applications of Citrus sinensis by-products and waste in poultry feeding. *Agroforestry Systems, 94*(4), 1581–1589. https://doi.org/10.1007/s10457-018-0319-2

61. Ren, W., Zhao, S., Lian, Y., Yang, Y., Tian, G., Zhao, C., Gao, W., & Zheng, J. (2020). Effects of hydrosoluble calcium ions and organic acids on citrus oil emulsions stabilized with citrus pectin. *Food Hydrocolloids, 100*, 105413. https://doi.org/10.1016/j.foodhyd.2019.105413
62. Huang, X., Liu, Y., Zou, Y., Liang, X., Peng, Y., McClements, D. J., & Hu, K. (2019). Encapsulation of resveratrol in zein/pectin core-shell nanoparticles: Stability, bioaccessibility, and antioxidant capacity after simulated gastrointestinal digestion. *Food Hydrocolloids, 93*, 261–269. https://doi.org/10.1016/j.foodhyd.2019.02.039
63. Parvez, G. M. M., & Akanda, K. M. (2019). Foods and arthritis: An overview. In *Bioactive food as dietary interventions for arthritis and related inflammatory diseases* (pp. 3–22). Elsevier. https://doi.org/10.1016/B978-0-12-813820-5.00001-5
64. Manthey, J. A., Guthrie, N., & Grohmann, K. (2001). Biological properties of citrus flavonoids pertaining to cancer and inflammation. *Current Medicinal Chemistry, 8*(2), 135–153.
65. Berk, Z. (2016). Introduction: History, production, trade, and utilization. In Z. Berk (Ed.), *Citrus fruit processing* (pp. 1–8). Academic. https://doi.org/10.1016/B978-0-12-803133-9.00001-1
66. Guo, J., Gao, Z., Li, G., Fu, F., Liang, Z., Zhu, H., & Shan, Y. (2019). Antimicrobial and antibiofilm efficacy and mechanism of essential oil from *citrus Changshan-huyou* Y. B. Chang against *Listeria monocytogenes*. *Food Control, 105*, 256–264. https://doi.org/10.1016/j.foodcont.2019.06.014
67. Aboul Naser, A., Younis, E., El-Feky, A., Elbatanony, M., & Hamed, M. (2020a). Management of Citrus sinensis peels for protection and treatment against gastric ulcer induced by ethanol in rats. *Biomarkers, 25*(4), 349–359. https://doi.org/10.1080/1354750X.2020.1759693
68. Yang, Y., Ma, S., Wang, X., & Zheng, X. (2017). Modification and application of dietary fiber in foods. *Journal of Chemistry, 2017*, e9340427. https://doi.org/10.1155/2017/9340427
69. Wang, L., Xu, H., Yuan, F., Fan, R., & Gao, Y. (2015). Preparation and physicochemical properties of soluble dietary fiber from orange peel assisted by steam explosion and dilute acid soaking. *Food Chemistry, 185*, 90–98. https://doi.org/10.1016/j.foodchem.2015.03.112
70. Romero-Lopez, M. R., Osorio-Diaz, P., Bello-Perez, L. A., Tovar, J., & Bernardino-Nicanor, A. (2011). Fiber concentrate from orange (Citrus sinensis L.) Bagase: Characterization and application as bakery product ingredient. *International Journal of Molecular Sciences, 12*(4), Article 4. https://doi.org/10.3390/ijms12042174
71. Shyu, Y.-S., Lu, T.-C., & Lin, C.-C. (2014). Functional analysis of unfermented and fermented citrus peels and physical properties of citrus peel-added doughs for bread making. *Journal of Food Science and Technology, 51*(12), 3803–3811. https://doi.org/10.1007/s13197-013-0938-6
72. Sharma, K., Mahato, N., & Lee, Y. R. (2019). Extraction, characterization and biological activity of citrus flavonoids. *Reviews in Chemical Engineering, 35*(2), 265–284. https://doi.org/10.1515/revce-2017-0027
73. Di Donna, L., Bartella, L., De Vero, L., Gullo, M., Giuffrè, A. M., Zappia, C., Capocasale, M., Poiana, M., D'Urso, S., & Caridi, A. (2020). Vinegar production from Citrus bergamia by-products and preservation of bioactive compounds. *European Food Research and Technology, 246*(10), 1981–1990. https://doi.org/10.1007/s00217-020-03549-1
74. Mahato, N., Sharma, K., Sinha, M., & Cho, M. H. (2018). Citrus waste derived nutra-/pharmaceuticals for health benefits: Current trends and future perspectives. *Journal of Functional Foods, 40*, 307–316. https://doi.org/10.1016/j.jff.2017.11.015
75. Afroja, S., Falgunee, F. N., Jahan, M. M., Akanda, K. M., Mehjabin, S., & Parvez, G. M. M. (2017). Antibacterial activity of different citrus fruits. *Specialty Journal of Medical Research and Health Science, 2*(1), 25–32.
76. Apraj, V., Pandita, N., Pandita, N., & Pandita, N. (2016). Evaluation of skin anti-aging potential of citrus reticulata Blanco Peel. *Pharmacognosy Research, 8*(3), 160–168. https://doi.org/10.4103/0974-8490.182913

77. Wuttisin, N., Boonmak, J., Thaipitak, V., Thitilertdecha, V., & Kittigowittana, K. (2017). Anti-tyrosinase activity of orange peel extract and cosmetic formulation. *International Food Research Journal, 24*(5), 2128–2132.
78. Abood, M. M., Rajendiran, J., & Azhari, N. N. (2015). Agricultural waste as low cost adsorbent for the removal of fe (ii) ions from aqueous solution. *Infrastructure University Kuala Lumpur Research Journal, 3*(1), 1–11.
79. Akkaya Sayğılı, G., Sayğılı, H., Yılmaz, C., & Güzel, F. (2020). Lead recovery from aqueous environment by using porous carbon of citrus fruits waste: Equilibrium, kinetics and thermodynamic studies. *Separation Science and Technology, 55*(15), 2699–2712. https://doi.org/10.1080/01496395.2019.1653917
80. Bhattacharyya, S., Das, P., & Datta, S. (2019). Removal of ranitidine from pharmaceutical waste water using activated carbon (AC) prepared from waste lemon Peel. In S. K. Ghosh (Ed.), *Waste water recycling and management* (pp. 123–141). Springer. https://doi.org/10.1007/978-981-13-2619-6_11
81. Villen-Guzman, M., Cerrillo-Gonzalez, M. M., Paz-Garcia, J. M., Rodriguez-Maroto, J. M., & Arhoun, B. (2021). Valorization of lemon peel waste as biosorbent for the simultaneous removal of nickel and cadmium from industrial effluents. *Environmental Technology & Innovation, 21*, 101380. https://doi.org/10.1016/j.eti.2021.101380
82. Falcó, M., García Raurich, J., & Arnal Madrid, O. (2021). Study of the influence of a bioabsorbent derived from orange peel on a filtering soil using seawater irrigation by capillarity. *International Journal of Environmental & Agriculture Research, 7*(3), 8–17.
83. Boluda-Aguilar, M., García-Vidal, L., del Pilar González-Castañeda, F., & López-Gómez, A. (2010). Mandarin peel wastes pretreatment with steam explosion for bioethanol production. *Bioresource Technology, 101*(10), 3506–3513. https://doi.org/10.1016/j.biortech.2009.12.063
84. Santi, G., Crognale, S., D'Annibale, A., Petruccioli, M., Ruzzi, M., Valentini, R., & Moresi, M. (2014). Orange peel pretreatment in a novel lab-scale direct steam-injection apparatus for ethanol production. *Biomass and Bioenergy, 61*, 146–156. https://doi.org/10.1016/j.biombioe.2013.12.007
85. John, I., Pola, J., Thanabalan, M., & Appusamy, A. (2020). Bioethanol production from Musambi Peel by acid catalyzed steam pretreatment and enzymatic saccharification: Optimization of delignification using Taguchi design. *Waste and Biomass Valorization, 11*(6), 2631–2643. https://doi.org/10.1007/s12649-018-00565-x
86. Muñoz-Labrador, A., Moreno, R., Villamiel, M., & Montilla, A. (2018). Preparation of citrus pectin gels by power ultrasound and its application as an edible coating in strawberries. *Journal of the Science of Food and Agriculture, 98*(13), 4866–4875. https://doi.org/10.1002/jsfa.9018
87. Wu, H., Lei, Y., Zhu, R., Zhao, M., Lu, J., Xiao, D., Jiao, C., Zhang, Z., Shen, G., & Li, S. (2019). Preparation and characterization of bioactive edible packaging films based on pomelo peel flours incorporating tea polyphenol. *Food Hydrocolloids, 90*, 41–49. https://doi.org/10.1016/j.foodhyd.2018.12.016
88. Zema, D. A., Calabrò, P. S., Folino, A., Tamburino, V., Zappia, G., & Zimbone, S. M. (2018). Valorisation of citrus processing waste: A review. *Waste Management, 80*, 252–273. https://doi.org/10.1016/j.wasman.2018.09.024
89. Abidi, S., Trabelsi, A., & Mihoubi, N. (2021). Pyrolysis of lemon peel waste in a fixed-bed reactor and characterization of innovative pyrolytic products. *Journal of Material Cycles and Waste Management, 25*, 235–248. https://doi.org/10.21203/rs.3.rs-656531/v1
90. Bernal-Vicente, A., Ros, M., Tittarelli, F., Intrigliolo, F., & Pascual, J. A. (2008). Citrus compost and its water extract for cultivation of melon plants in greenhouse nurseries. Evaluation of nutriactive and biocontrol effects. *Bioresource Technology, 99*(18), 8722–8728. https://doi.org/10.1016/j.biortech.2008.04.019
91. Van Dyk, J. S., Gama, R., Morrison, D., Swart, S., & Pletschke, B. I. (2013). Food processing waste: Problems, current management and prospects for utilisation of the lignocellulose

component through enzyme synergistic degradation. *Renewable and Sustainable Energy Reviews, 26*, 521–531. https://doi.org/10.1016/j.rser.2013.06.016
92. Millati, R., Lukitawesa, L., Permanasari, E. D., Sari, K. W., Cahyanto, M. N., Niklasson, C., & Taherzadeh, M. J. (2018). Anaerobic digestion of citrus waste using two-stage membrane bioreactor. *IOP Conference Series: Materials Science and Engineering, 316*(1), 012063. https://doi.org/10.1088/1757-899X/316/1/012063
93. Aggarwal, K. K., Khanuja, S. P. S., Ahmad, A., Santha Kumar, T. R., Gupta, V. K., & Kumar, S. (2002). Antimicrobial activity profiles of the two enantiomers of limonene and carvone isolated from the oils of Mentha spicata and Anethum sowa. *Flavour and Fragrance Journal, 17*(1), 59–63. https://doi.org/10.1002/ffj.1040
94. Siles, J. A., Vargas, F., Gutiérrez, M. C., Chica, A. F., & Martín, M. A. (2016). Integral valorisation of waste orange peel using combustion, biomethanisation and co-composting technologies. *Bioresource Technology, 211*, 173–182. https://doi.org/10.1016/j.biortech.2016.03.056
95. Chanakya, H. N., Sharma, I., & Ramachandra, T. V. (2009). Micro-scale anaerobic digestion of point source components of organic fraction of municipal solid waste. *Waste Management, 29*(4), 1306–1312. https://doi.org/10.1016/j.wasman.2008.09.014
96. Bampidis, V. A., & Robinson, P. H. (2006). Citrus by-products as ruminant feeds: A review. *Animal Feed Science and Technology, 128*(3), 175–217. https://doi.org/10.1016/j.anifeedsci.2005.12.002
97. Rafiq, S., Kaul, R., Sofi, S. A., Bashir, N., Nazir, F., & Ahmad Nayik, G. (2018). Citrus peel as a source of functional ingredient: A review. *Journal of the Saudi Society of Agricultural Sciences, 17*(4), 351–358. https://doi.org/10.1016/j.jssas.2016.07.006
98. Lauro, M. R., Crasci, L., Carbone, C., Aquino, R. P., Panico, A. M., & Puglisi, G. (2015). Encapsulation of a citrus by-product extract: Development, characterization and stability studies of a nutraceutical with antioxidant and metalloproteinases inhibitory activity. *LWT – Food Science and Technology, 62*(1, Part 1), 169–176. https://doi.org/10.1016/j.lwt.2015.01.017
99. Arce, A., Marchiaro, A., Martínez-Ageitos, J. M., & Soto, A. (2005). Citrus essential oil deterpenation by liquid-liquid extraction. *The Canadian Journal of Chemical Engineering, 83*(2), 366–370. https://doi.org/10.1002/cjce.5450830226
100. Espina, L., Somolinos, M., Lorán, S., Conchello, P., García, D., & Pagán, R. (2011). Chemical composition of commercial citrus fruit essential oils and evaluation of their antimicrobial activity acting alone or in combined processes. *Food Control, 22*(6), 896–902. https://doi.org/10.1016/j.foodcont.2010.11.021
101. Ruiz, B., de Benito, A., Rivera, J. D., & Flotats, X. (2016). Assessment of different pretreatment methods for the removal of limonene in citrus waste and their effect on methane potential and methane production rate. *Waste Management & Research, 34*(12), 1249–1257. https://doi.org/10.1177/0734242X16661053
102. Satari, B., & Karimi, K. (2018). Citrus processing wastes: Environmental impacts, recent advances, and future perspectives in total valorization. *Resources, Conservation and Recycling, 129*, 153–167. https://doi.org/10.1016/j.resconrec.2017.10.032
103. Pourbafrani, M., Forgács, G., Horváth, I. S., Niklasson, C., & Taherzadeh, M. J. (2010). Production of biofuels, limonene and pectin from citrus wastes. *Bioresource Technology, 101*(11), 4246–4250. https://doi.org/10.1016/j.biortech.2010.01.077
104. Guo, X., Han, D., Xi, H., Rao, L., Liao, X., Hu, X., & Wu, J. (2012). Extraction of pectin from navel orange peel assisted by ultra-high pressure, microwave or traditional heating: A comparison. *Carbohydrate Polymers, 88*(2), 441–448. https://doi.org/10.1016/j.carbpol.2011.12.026
105. Chemat, F., Vian, M. A., & Cravotto, G. (2012). Green extraction of natural products: Concept and principles. *International Journal of Molecular Sciences, 13*(7), Article 7. https://doi.org/10.3390/ijms13078615
106. Sovová, H. (2005). Mathematical model for supercritical fluid extraction of natural products and extraction curve evaluation. *The Journal of Supercritical Fluids, 33*(1), 35–52. https://doi.org/10.1016/j.supflu.2004.03.005

107. Akter, M. R., Akanda, M. K. M., Islam, M. S., & Rana, M. M. (2020). Assessment of drug utilization pattern among outpatients in the orthopedic department of several private clinics in Rajshahi city, Bangladesh. *International Journal of Pharmaceutical Science and Research, 5*(4), 13–16.
108. Briggs, P. K., Hogan, J. P., & Reid, R. L. (1957). Effect of volatile fatty acids, lactic acid and ammonia on rumen pH in sheep. *Australian Journal of Agricultural Research, 8*(6), 674–690. https://doi.org/10.1071/ar9570674
109. Bursać Kovačević, D., Gajdoš Kljusurić, J., Putnik, P., Vukušić, T., Herceg, Z., & Dragović-Uzelac, V. (2016). Stability of polyphenols in chokeberry juice treated with gas phase plasma. *Food Chemistry, 212*, 323–331. https://doi.org/10.1016/j.foodchem.2016.05.192
110. Lv, X., Zhao, S., Ning, Z., Zeng, H., Shu, Y., Tao, O., Xiao, C., Lu, C., & Liu, Y. (2015). Citrus fruits as a treasure trove of active natural metabolites that potentially provide benefits for human health. *Chemistry Central Journal, 9*(1), 68. https://doi.org/10.1186/s13065-015-0145-9
111. Panuccio, M. R., Attinà, E., Basile, C., Mallamaci, C., & Muscolo, A. (2016). Use of recalcitrant agriculture wastes to produce biogas and feasible biofertilizer. *Waste and Biomass Valorization, 7*(2), 267–280. https://doi.org/10.1007/s12649-015-9445-5
112. Duan, X., Zhu, Y., Shu, C., Gao, J., Liu, F., & Pan, S. (2022). Extraction of pectin from Satsuma mandarin Peel: A comparison of high hydrostatic pressure and conventional extractions in different acids. *Molecules, 27*(12), Article 12. https://doi.org/10.3390/molecules27123747
113. Terao, J. (1999). 17 – Dietary flavonoids as plasma antioxidants on lipid peroxidation: Significance of metabolic conversion. In L. Packer, M. Hiramatsu, & T. Yoshikawa (Eds.), *Antioxidant food supplements in human health* (pp. 255–268). Academic. https://doi.org/10.1016/B978-012543590-1/50018-4
114. Wedamulla, N. E., Fan, M., Choi, Y.-J., & Kim, E.-K. (2022). Citrus peel as a renewable bioresource: Transforming waste to food additives. Journal of Functional Foods, 95(105163), 105163. https://doi.org/10.1016/j.jff.2022.105163

Chapter 14
Valorization of Citrus Processing Waste Towards Sustainability

Maryam Batool, Zahra Batool, Manoj Swale, Qudsia Yousafi, Aleeha Ishtiaq, Hadia Fatima, and M. Abubaker Siddique

14.1 Introduction

With the increasing challenge of meeting consumer demand for processed and ready-to-eat foods, huge amounts of processing waste having artificial food additives waste are produced and released during processing risking the customers' health [1]. Globally, citrus fruits like oranges, lemons, limes, and grapefruits are cultivated. In this way, their cultivation helps the global economy [2]. However, the citrus fruit industry growth has caused questionable production of huge quantities of citrus processing waste (CPW) [3]. CPW includes many by-products like peels, seeds, pulp, and membranes that are produced after the extraction of juice. In the past, these by-products were taken as low-value waste and were disposed of through conventional methods e.g., landfilling or incineration. These methods caused environmental pollution and economic losses [4]. Recently, due to growing awareness of the environmental impact of citrus processing waste (CPW), there is a shift towards utilizing more sustainable by-products. This shift has been stimulated by many factors such as, increasing pressure to control environmental pollution, the growing concerns in circular economy principles, and the acknowledgment of the hidden ability of citrus by-products as key resources [5]. The improper management of the organic content present in CPW can result in water and soil pollution which is a main environmental concern. Besides, methane emissions are caused by the

M. Batool (✉) · Z. Batool · Q. Yousafi · A. Ishtiaq · H. Fatima · M. A. Siddique
Department of Biosciences, COMSATS University Islamabad, Sahiwal Campus, Sahiwal, Pakistan
e-mail: maryambatool@cuisahiwal.edu.pk; zahra@cuisahiwal.edu.pk; qudsia@cuisahiwal.edu.pk; abubaker.siddique@cuisahiwal.edu.pk

M. Swale
Department of Food Science, Purdue University, West Lafayette, IN, USA
e-mail: mswale@purdue.edu

A. Chauhan et al. (eds.), *Valorization of Citrus Food Waste*,
https://doi.org/10.1007/978-3-031-77999-2_14

disposal of citrus waste (CW) in landfills which is a strong greenhouse gas that affects climate [6]. These environmental challenges can be addressed if researchers and industry stakeholders try to discover different strategies and techniques to valorize and maximize CPW potential benefits. These strategies and techniques involve a variety of approaches, like bioconversion, extraction of compounds, and usage of waste as feedstock for bioenergy production. For example, bioconversion of citrus peels through anaerobic digestion or fermentation into biofuels like ethanol or biogas. Moreover, citrus essential oils (Eos) extracted from peels help in various economic opportunities because Eos are used in the food, fragrance and pharmaceutical industries [7]. CPW can also be reutilized for agricultural uses like soil improvements and animal feed supplements. Citrus by-products especially dried citrus pulp and citrus meal are full of nutrients like fiber, vitamins and minerals. So, they can be considered as excellent ingredients for livestock feed formulations. If citrus by-products are incorporated into animal feeds, feed costs can be lowered and animal nutrition can be enhanced thus, leading to both economic and environmental sustainability [8]. A sustainable management of CPW needs collaboration and innovation across the chain that involves citrus growers, processors, industry stakeholders, researchers, policymakers and consumers. Also, regulatory frameworks and initiatives are important to support and encourage investments in waste valorization techniques [9].

This chapter seeks to deliver a detailed review of sustainable CPW valorization. Various methodologies, techniques and technologies have been discussed for conversion of citrus by-products and waste into valuable products such as biofuels, biochemicals, dietary supplements, and functional additives. Moreover, environmental effects and economic benefits related to CPW valorization have also been examined in this chapter. The chapter aims to highlight the prospects and challenges in this field.

14.2 Citrus Fruit Waste Composition

Citrus fruit is generally composed of two parts outer one which is composed of peels and acts as a covering on the epidermis falvedo and the inner one is a covering which is albedo composed of a spongy layer of parenchymous cells [5] (Fig. 14.1).

The outer layer i.e. Flavedo contains Specific oil-producing glands having steroids, essential oils, waxes, enzymes, fatty acids, and phytochemicals (flavonoids, carotenoids, and phenolic compounds). As a result of ripening chlorophyll is converted into carotenoid which is a color pigment. This change leads to the color of the citrus fruit from green to orange [10]. The inner part of citrus fruit is also known as the endocarp which is composed of many segments partitioned by a thin epidermal layer of tissues containing different juice sacs called seeds and vesicles. White spongy tissues associated with albedo show the axis of the fruit. These core and segments are collectively called "rag" of the extracted juice [5]. Most of the citrus waste comprises of citrus peel containing bundle of nutrients and bioactive

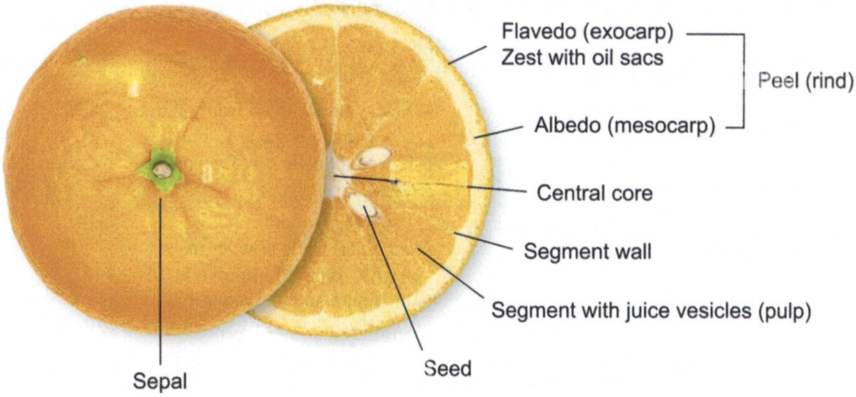

Fig. 14.1 Citrus Fruit Anatomy

compounds such as flavonoids, carotenoids, polyphenolic compounds, dietary fibers (DF) and Eos are found in peels. The unuseable part of the citrus fruit are seeds which encompass 20–40% by weight. Citrus fruit seeds usually contain 28.5% carbohydrate, 3.0% crude protein, 2.5% crude ash, 52% crude fat, and 5.5% crude fiber. Besides this they are a good source of PUFA. Citrus pomace is the richest nutritive waste product that is left after juice production. The higher level of moisture and sugar content make the fruit pomace a suitable commodity for fermentation and acute management is required to prevent spoilage and adverse effects on the environment [11]. Citrus processing is followed by the production of excessive wastewater produced during concerned unit operations such as cleaning, washing, canning, peeling etc. This wastewater contains sugars, essential oils and other bioactive components which can be utilized in a better way. Extraction of these compounds from CW could be done and utilized in functional food and nutraceutical manufacturing.

14.2.1 Citrus Fruit Waste Components

14.2.1.1 Citrus Peels (CP)

Citrus peel is the primary waste component that is attained as a result of juice preparation in about 30–34% [5]. CP is considered a rich and numerous source of bioactive components and phytochemicals like Phytochemicals like carbohydrates, carotenoids, D-limonene, polyphenols and flavouring compounds. Polyphenols include hesperidin, nobiletin tangerine, naringin. Carbohydrates include cellulose, hemicellulose, dietry fibre and pectin. [12]. Citrus peels are added to many foods, beverages, perfumery, drugs, and medicated cosmetics due to their high content of phytochemicals (Table 14.1).

Table 14.1 Applications of citrus peel waste in food and non-food sector

Sr#	Applications	role	Results	References
Citrus peels food applications				
1	CP as additive in food products	Dietary fiber-rich muffins Development	Dietary fiber (15%) from orange (C. sinensis) bagasse in muffins showed a decline in glycemic index.	[13]
		Manufacturing of nutrient biscuits	Using 6% mandarin (C. reticulata) peels powder in biscuits reported improved nutritional and AO property.	[14]
		Preparation of brioches a bakery product by using citrus peel oil as Fat replacer	Debittered orange fiber (50%) showed good fat replacement characters in making of bakery products.	[15]
2	Bioactive compounds	Polyphenols	MAE of C. inshiu resulted in phenol extraction with high hesperidin and narirutin content containing 5860 mg and 1310 mg per 100 g of extract respectively.	[16]
		Carotenoids	High voltage electric discharge resulted in carotenoid extraction from citrus peels that showed enhanced bioavailability of 82.5%.	[17]
		Pectin	Surfactant-based MAE of C. sinensis peels at 400 W power for 7 min and 1.2 pH resulted in 28% pectin content.	[8]
3	Prebiotics	Citrus pectic oligosaccharides activity as prebiotic	Significant prebiotic effects in the developed product showed antimicrobial and modulating-microbiota properties.	[18]
4	Encapsulating agent	Citrus pectin/zein nanoparticle containing Resveratrol as an oral delivery agent	Ctrus peel pectin and zein nanoparticles containing resveratrol showed good heat stability upto 80 °C for an hour and pH stability ranging from 2 to 7.	[19]
5	Animal feed	Feed for Animals (broiler, quail, cattle) etc.	Diet of broiler chickens supplemented with orange (C. sinensis) peel extract corrected their blood profile.	
Citrus peels non-food applications				
6	Bio-adsorbent	$FeCl_3$ impregnated activatedcarbon absorbant containg *Citrus limetta* peels	Eradication of heavy metals from waste water such as fluoride ions.	[20]

(continued)

Table 14.1 (continued)

Sr#	Applications	role	Results	References
7	biofuels	Bio-ethanol Production	50–60 L per 1000 kg of ethanol production from mandarin peel waste by using S. cerevisiae as a result of steam explosion, enzymatic hydrolysis and fermentation techniques	[21]
8	Edible packaging	Potent packaging material Biodegradable edible film	Edible coating can be made by using C. limon and C. aurantifolia peels for fresh strawberries.	[22]
9	Cosmetic products	Anti-aging agent skin whitening cream	Peel extracted from C. reticulata Blanco through Soxhlation method showed anti-elastase properties and anti-collagenase	[23]
10	Pharma Drug	Anti-ulcer effects	Diet of rats added with C. sinensis peel extract given resulted in their gastrointestinal health maintenance	[24]
		Anti-obesity effect	Aged sun-dried citrus peels (chenpi) given in the diet of rats showed weight loss, prevention of fat cell enlargement and accumulation of lipids in adipocytes.	[25]

14.2.1.2 Citrus Seeds

20–25% of Citrus fruits are comprised of seeds that are left behind after extracting juice. Essential oil and proteins are the major components of all kinds of seeds and they have high value from a commercial point of view [26]. Literature evidenced the nutritive enrichment of citrus seeds with tocopherols, polyphenols, limonoids, carotenoids and essential oils. These nutritive components could be utilized in many ways [27]. Their extracts have high value in the food, cosmetic and pharmaceutical industries [2]. They also have high nutritional value. It was reported that enhanced physicochemical and chemical attributes in orange seeds in unhulled and dehulled form. In unhulled citrus seed, high fat content 55%, 23% nitrogen-free extract, 3.12% protein, 2.9% ash, and 7% fiber were observed. However, the dehulled citrus seed showed high carbohydrate content 28.5% and energy value of 614 kcal/100 g. Among micronutrients, the unhulled seeds composed of Fe, K, Na and Ca is about (0.42 mg, 160 mg, 7.50 mg, 136 mg/100 g). Moreover, the oil extracted from citrus seeds has shown an iodine value of 100–110, (0.920–0.941 mg/mL) density at 24 °C, (1.4639–1.4670) refractive index at 40 °C, color (1-in. cell) (1.5–3.0R + 15.0–30.0Y), acid value (0.5–2.2 mg KOH/g of oil), (181.90–198.90) saponified matter, and (0.3–0.5%) un-saponifiable matter [28]. The citrus seed being a good source of nutrients and having functional properties could be part of our daily food. Various uses of citrus seeds (Table 14.2).

Table 14.2 Applications of citrus Seed waste in food and non-food sector

Sr#	Applications	role	Results	References
Citrus Seeds Foods Applications				
1	Essential oil	Limonoids	Pummelo (Citrus maxima) seeds were treated to extract 11.52 mg/g limonoid was	[29]
		(PUFAs)	n-Hexane solvent extraction of Lemon (C. limon) seed yielded oil containing 71.3% PUFAs having 8.7% linolenic acid, 30.3% oleic acid, 34% linoleic acid and 110.2 mg/kg α-tocopherol in the oil.	[30]
2	1. Food Additive	Preparation of food items	Citrus seeds could be used as food additive due to high polysaccharide level. Mandarin (C. reticulata) seed could be used as daily use oil for making food due to high EO percentage in it.	[31]
Citrus seed non-food applications				
3	Skin care product	Medicinalsoaps were prepared by using citrus seeds oil	Anti-bacterial, anti-fungal, and antioxidant characters were seen due to incorporation of citrus seed oil in soap.	[32]
4	1. Biofuels	Bio diesel production	Trans-esterification of C. sinensis seeds caused 76.93% yield of bio-diesel.	[33]

14.2.1.3 Citrus Pomace

The solid residue after juice extraction is termed citrus pomace. It is a combination of seeds, pulp, and rags. A high level of pectin, molasses, essential oils, and limonene are present in citrus pomace. Additionally, Pomace is a good source of polyphenols-flavonoids and antioxidants which have further demand in the industry for use in different product fields [34]. Its nutritional composition shows 10% water, 30–40% sugars, 14–25% pectin, 13–17% cellulose and hemicellulose, 0.4–1.6% fat, and 0.5% micronutrients [35]. When dried, it comprises of 5.9% protein, 1.52% total fat, 3% ash, and 89% carbohydrates [36]. Due to these functional and nutritional properties, it is widely used in food-based and non-food related industries (Table 14.3).

14.3 Biotechnological Approaches for Citrus Waste Management

Biotechnological treatments consist of anaerobic digestion that involves biogas production and solid-state fermentation that causes bioconversion into animal feed.

Table 14.3 Applications of citrus pomace waste in food and non-food sector

Sr#	Applications	Role	Results	References
Citrus Pomace food applications				
1	Pectin source	Pectin	At extraction temperature of 100 °C for 60 min, pre-mature (13.13%), mature (10.83%) and over ripped lemon (10.33%) (C. limon) yield respectively.	[37]
2	Bioactive compounds	Polyphenols	2.01 to 2.62% polyphenol were produced by CO2-SFE of orange (C. sinensis) pomace	[36]
3	1. Nano-particle	Nano-bacterial cellulose	At 30 °C temperature for 8 days, enzymolysis of citrus pomace resulted in 5.7 g/L yield of bacterial cellulose.	[38]
Citrus pomace non-food applications				
4	Bio- fertilizer	Improved fertility of soil	Removal of the heavy metals from the soil and anti microbial properties were observed in Citrus waste bio-fertilizers.	[39]

14.3.1 Anaerobic Digestion of Citrus Waste

The anaerobic digestion of citrus waste also produces methane that is used as a fuel. The CW biogas production includes two primary steps, acid generation and CH_4 formation. Limonene present in citrus oil inhibits the production of biogas from CW. 90% of citrus peel EO consists of D-limonene. Pretreatment of CW is done to remove the D-limonene to counter the inhibition of biogas production or by using the cell protection method [40].

14.3.2 Production of Feed Enzymes by Solid-State Fermentation

Many microorganisms are used that produce feed enzymes like cellulase, protease, and amylase in SSF utilizing waste material. CW processing due to hemicellulose, pectin, soluble sugar cellulose, lignin, minerals and low molecular weight hydrocarbons can be used for production of feed enzymes. Currently, citrus-based industry deals with fresh, processed juice and canned pieces. So, fruit peels as waste are produced in high quantities that can lead to any form of environmental pollution. This problem can be resolved by utilization of CW for feed enzymes. Aureobasidium pullulans and Aspergillus Niger species of fungi use citrus peel to produce pectinase. A. niger A-20 has reported to produce a multi- enzyme including pectinase, cellulase and xylanase [41].

14.3.3 Citrus Waste Composting

One of sources of poultry feed is the powdered citrus peel. CPW is rich in pectic PS, cellulose, and hemicelluloses that are hydrolysable and fermentable. CPW is recycled for compost production. The suitable parts for citrus peel compost are peels, cores and pulp. The composting efficiency can be increased by including a variety of vegetable and fruit waste in composite mixture. For compost, CPW with added Ca hydroxide and 60% moisture content are necessary. The duration of composting is completed in 90 days when the thermal process completes around 70 days [42].

14.4 Extracted Components and Sustainable Applications

14.4.1 Extracted Compounds

14.4.1.1 Bioactive Compounds

The most grown citrus fruits around the globe are lemons, grapefruits, oranges, limes, and mandarins. Citrus processing industries produce large amounts of CPW that alone represent about 50% of the wet fruit mass. High levels of phenolic compounds, antioxidants (Aos), ascorbic acid and flavonoids are present in CPW in each citrus species as compared to their seeds and pulp waste. If they are processed, many by-products can be obtained which are rich in biologically active substances like Eos, pectin, and water soluble and insoluble Aos [43].

14.4.1.2 Essential Oils (EO)

Citrus Essential Oils contain some major bioactive substances like sabinene, α-pinene, β-pinene, d-limonene, α-terpineol and β-myrcene. Eos are used as flavouring agents in beverages soaps, cosmetics and other household goods. They have natural insecticidal and antimicrobial activities. By using the hydro distillation technique, extraction of Essential oils was done using five different citrus species. Different phytochemicals in all five CEOs were discovered after gas chromatography/mass spectrometry (GC–MS) analysis. They included tannins, terpenoids, saponins, limonene and sterols. It was utilized to analyze and measure the chemical compounds present in EOs. D-Limonene is the main chemical compound found in cold-pressed peel/seed oils and it can be extracted from oranges, lemons, and limes [44] (Fig. 14.2).

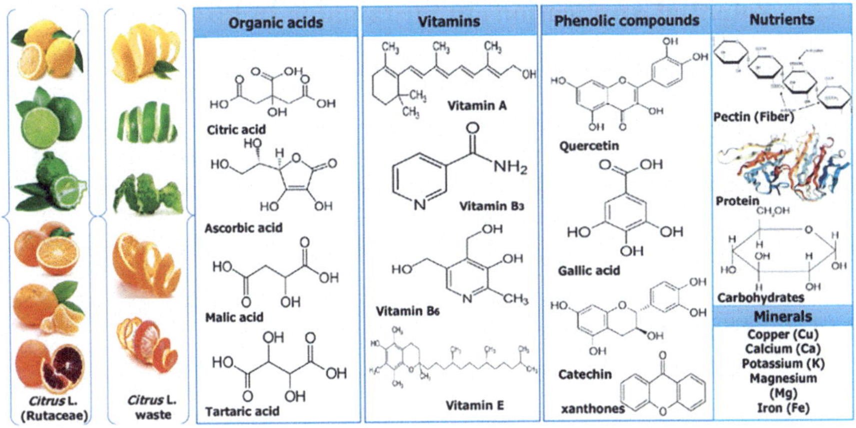

Fig. 14.2 Bioactive Compounds as Phytochemicals in Citrus L. Waste

14.4.1.3 Polyphenolic Compounds

Citrus fruit by-product extracts and EOs mainly has Limonene. It has wide applications in the food, cosmetic and medicine industries. They can be employed in various detergents, soaps, fragrances, cosmetics and domestic items. They exhibit anti-asthmatic anti-cancerous, bactericidal, insecticidal and fungicidal properties. Due to its property of dissolving cholesterol, D-limonene has been utilized to remove gallstones. It also provides relief from heartburn by lowering the action of gastric acid [44].

CF are rich in carotenoids, particularly lutein, alpha carotene, beta carotene, beta cryptoxanthin and zeaxanthin. Beta cryptoxanthin is found in pomace & peels of oranges, tangerines, and mandarins. They have a strong AO and provitamin A activity. Peel has high carotenoid content as compared to the pulp [45].

14.4.1.4 Dietary Fiber

Citrus fiber is a bioactive compound and act as a lipid oxidation inhibitor in meat products due to polyphenol components hence, increasing the stability of the meat. Peel, pulp, and seeds of oranges have been used as a substitute of fat in ice cream. Pectin, a soluble fiber is extracted from the peels, seeds, or pomace of orange, lime, lemon, and grapefruit. Pectin extraction can be done by UAE, MAE, and EAE. Pectin as a dietary fiber is a biologically active compound has anti-cancerous properties. It is also used as a jellying agent in food industry [43].

14.5 Potential Applications in Sustainable Industries

Citrus waste extracted Eos can be used in food preservation and nutraceutical manufacturing industries. Eos-based films, biodegradable polymers, sprays and nanoemulsion coatings have been produced using the bioactive chemicals present in Citrus Essential Oils. Due to their sweet fragrance and flavor, Eos can be used in many beverages, foods, cosmetic and pharmaceutical products.

14.5.1 Citrus Waste Derived Pectin Applications in Food Industry

CW is rich source of pectin which is used in jams and jellies as a solidifier and also used in dairy products, desserts, soft drinks, frozen foods. Pectin containing dietary fiber has high WHC and viscosity. That's why it can bind water more tightly to lower syneresis during food processing. When pectin rich orange fiber is added to yogurt, it stabilizes casein networks and acts as a filler [46].

14.5.2 Polysaccharide-Based Edible Films and Coatings Containing Citrus Waste Essential Oil

Excellent palatability, preservation, biodegradability properties of edible coatings and films of citrus waste have gained great interest. Recently, polysaccharides have been studied as an applicable substance for edible films. In Literature, Eos is incorporated with polysaccharide to increase the functional characteristics of edible coatings and films. Cold-stored strawberries coated with chitosan and lemon extracted Eo revealed no major effect on TSS content, acidity and pH of strawberries when stored at 5 °C. Numerous food-borne diseases can be overcome by Lemon peel extracted Eos as they have bactericidal properties [43].

14.5.3 Application of Citrus Peel Derived Dietary Fiber in Bakery Items

An orange bagasse dietry fibre increased the total fiber percentage of the experimental muffins by 40–63% as compared to control muffins. When wheat derived flour was partially replaced with this orange-derived DF, bakery items with high levels of total DF were produced [43].

14.5.4 Activated Carbon from Lemon Waste

It can be obtained by drying 140 g wet lemon peels that have 80% moisture content. They were chopped into less than 10 mm. Then carbonization was done through hydrothermal carbonizer at 180 °C for one hour. The pressure employed was between 300–400 psi. The slurry from the hydrothermal reactor is filtered and dried in an oven at 105 °C for 24 hrs, leaving behind charcoal that is activated carbon [47].

14.6 Economic Impact of Citrus Waste Valorization

The Italian word pastazzo is utilized for waste by-products from the industrial CPW like peels and squeezed pulp. The energy is produced from this biomass during anaerobic digestion that can be used in industries. This energy can also be used for civil heating and liquid fuel in biomass to generate electricity. These are good alternatives to fossil fuels. For Example, methane and diesel can be replaced by the burning of straw and scraps of wood. Gasoline can be replaced by bio-ethanol and diesel fuel can be replaced by bio-diesel. The transport industry uses 21% of the total energy consumption so these fuels can also be used as an alternative in the transport sector. The energy generated from biomass utilization is 10–12% of the total worldwide energy production. By mid-century, it is estimated that bio-fuels will share 38% of total fuel and 17% of electricity [48]. Generally, citrus seeds are considered useless and discarded as waste. Citrus seeds consist of about 30% oil and can be utilized as biodiesel. Biodiesel oils were made by using transesterification of citrus seeds oil in methanol. [2].

14.6.1 Environmental Impact Assessment of Citrus Waste

Citrus process waste (CPW) due to high carbohydrate content and fermentability can increase its degradation and can result in an environmental problem. Nowadays, CPW reduction methods, like incineration or landfilling, are not enough due to environmental impacts and energy inefficiency. Valorization of CPW can be done to obtain energy-generating derivatives such as by fermentation, and by physical treatment methods.

14.7 Current Challenges in Citrus Fruit Waste Valorization and Future Opportunities for Sustainable Research and Development

14.7.1 Quantity and Seasonality of Citrus Fruit Waste

Due to large market demand continuous production of citrus products generate large amount of waste. One of the aims of sustainable development goal such as responsible production and consumption can be achieved by waste valorization [49]. Citrus fruit processing industry produce generate huge amount of citrus waste, where the peels are the major representative with 50% and 60% share [50]. The global productivity of citrus fruit is about 158.49 million tons in 2020. The highest share in global citrus market is of Asia (47.7%) after that Africa, America, Europe and Oceania has second, third and fourth highest share respectively (43.7%, 8.1%, 0.4% and 0.1). The waste includes peels, seeds, pulp and waste water. Globally citrus fruit processing industry produce about ten million tons of waste annually. Dye to huge amount and perishable properties, citrus waste management causes serious challenges. Even peel contain about 80% water invite flies, fungus, bacteria and produce toxins. So, citrus waste management is necessary, not just an attempt to enhance income. Currently, the citrus fruit waste management industry is facing following challenges [8].

14.7.2 Ecological Challenges

Ecological effect of citrus waste starts from the citrus farms due to improper post-harvest handling practices. Large amount of citrus waste generates during citrus fruit production. Citrus waste has significant quantity of fibers, carbs, organic acids and phenolics. These composites may be source of environmental problems such as eutrophication, soil deprivation and reservoir. The large quantity of fermentable substrates accelerates its deprivation, ahead to quick and negative ecological effects [51]. In a case study life cycle assessment (LCA) technique is applied to assess ecological effect of citrus waste management. In study all citrus waste management techniques categorized into three categories, first is conventional techniques: feed, landfill (with and without energy recovery) and ignition, second is nutrient valorization: silage and production of single cell protein and third is biorefinery: extraction of pectin, D-limonene, flavonoids. According to LCA result feed system had the lowest ecological effect similarly multiple extraction unit had less effect. While the variation of ecological effect is high in the single cell protein (SCP) and silage production from citrus waste [51].

14.7.3 Economic Challenges

Citrus waste is generated from fruit washing, device and plants cleaning, cooling and essential oil extraction as well as peel and seed drying. It also contains effluent from the production of pectin, seed and peel oil as well as citrus molasses depending on treatment. Due to large amount and perishable as well as fermentable properties of solid, semi solid and water waste from citrus fruits need application of novel intensive treatment techniques. While simple and advance systems for waste management are apply in developed countries due to socio economic condition. Apart from large volume to be handled, extremely variable chemical properties required ecofriendly and simplified advance technology that result in high cost weighing on citrus processing industry budget [51]. Financial advantage is the use of integrated multi-feed biorefinery. This refinery system eliminates seasonality and support all year-round operation by coproducing bioactive compounds, bioenergy and biofuels [52].

14.7.4 Technological Limitations and Processing Constraints

To convert citrus biomass to value added products petroleum refineries are used which cause ecological and economical challenges. Biorefinery method of waste management is efficient and flexible way that convert biomass to generate high-quality value-added products, chemicals and fuel. Biorefineries are different from petroleum refineries which used to produce energy and value-added products [53]. To extract bioactive compounds, conventional methods are used in which many sensitive bioactive compounds are lost. In a case study techno-economic assessment of conventional and nonconventional extraction technologies of citrus waste was done. According to the result supercritical extraction technology is producing low yield of bioactive compound from citrus waste as compared to solvent extraction technology. So, supercritical extraction [50].

14.7.5 Future Opportunities for Sustainable Research and Development

Current study focus on specific treatment for the valorization of citrus waste meanwhile integrated method for sustainable citrus waste management is still absent [54].

Modernization of waste treatment machinery, automation of some segments in processing lines that efficient in energy consumption as well as development of durable and ecofriendly packaging material to prevent generation of citrus waste during transportation and sale of citrus products [55]. Biorefinery method of waste management is efficient and flexible way that convert biomass to generate high-quality value-added products, chemicals and fuel from biomass. Biorefineries are different from petroleum refineries which used to produce energy and value-added products [53]. Multi-feed biorefineries for citrus waste management can alleviate ecological and economical challenges as it is environmentally friendly as compare other techniques and promote all year round operation. This technique produces a wide range of products as compared to of single product, including full usage of the waste, minimum waste generation and diversification of the revenues by covering a large spectrum of markets. In result of that additional revenue in a circular.

According to assessment biorefinery system had less ecological impact than silage system and SCP similarly multiple extraction system has less ecological and economic impact. Future research in multiple feed biorefinery system can focus on increasing the energy efficiency of pretreatment or extraction chemicals reuse. Feed system for citrus waste valorization is suitable and by improving this technology, could achieved better performance in human toxicity, global warming and eutrophication. For silage and single cell protein system, the environmental impact can be minimized by enhancing energy utilization efficiency or usage of maximum renewable energy [53]. Financial advantage is the use of Integrated Multi-feed biorefinery. This refinery system eliminates seasonality and support all year-round operation by coproducing bioactive compounds, bioenergy and biofuels [52].

In future recommendations the application of the circular economy model in citrus waste management is vital. This economy model work on the principle in which the recovery of resources from material considered as waste, attaining of waste disposal at the same time and development of value-added products. Financial advantages occur by reuse citrus waste and waste water through correct processing techniques, supporting natural water sources and minimizing fertilizers costs, due to higher content of organic substances [51].

14.7.6 Existing Policies Relevant to Citrus Fruit Waste

Citrus Waste does not have any specific technical and managing policies in place. The implementation of good manufacturing practices in the citrus fruit processing industry is a way to reduce the amount of solid, and semi-solid waste and wastewater as well as composting waste. There is ongoing research and development on citrus waste management regulation and policies. While for citrus wastewater management there is a recently published new regulation (Regulation UE, 2020/741). This regulation provides guidelines to reuse of industrial waste water particularly in the food sector [51]. Indeed, in the transition from fossil-based linear economy to a balanced circular economy, food sector plays a vital role [56]. Some other general policies for food waste management include sustainable development goal (SDG). The target of SDG 12.3 is to halve per capita global food waste at the trade and consumer stages by 2030. In addition to that 'Zero Hunger challenge' of the UN,

SAVE FOOD is a worldwide initiative to reduce food waste. UN launched a program on the FAO platform for reduction of food waste [57].

14.7.7 Recommendations for Sustainable Policy Development

There should be measurable and sanctionable reduction targets for citrus fruit waste within the framework of circular economy and waste management laws and regulations. Furthermore, it is essential to develop laws and guidelines to remove challenges for citrus fruit waste management.

14.8 Conclusion

Sustainable valorization of citrus waste from secondary and primary processing through to consumer offers a wide range of benefits such as resource recovery, promotion of circular economy, and environmental sustainability. United Nations, sustainable development goal is also achieved through citrus waste valorization as it reduces pollution and helps conserve natural resources. The food application of compounds derived from citrus waste and citrus by products, the permitted limit, contact with the source material and in vitro and in vivo bioavailability analysis of these compounds as well as innovative extraction techniques present significant research gaps and future work.

References

1. Sharma, P., et al. (2022). Valorization of citrus peel waste for the sustainable production of value-added products. *Bioresource Technology, 351*, 127064.
2. Rashid, U., Ibrahim, M., Yasin, S., Yunus, R., Taufiq-Yap, Y. H., & Knothe, G. (2013). Biodiesel from Citrus reticulata (mandarin orange) seed oil, a potential non-food feedstock. *Industrial Crops and Products, 45*, 355–359.
3. Kundu, D., Das, M., Mahle, R., Biswas, P., Karmakar, S., & Banerjee, R. (2020). Citrus fruits. In *Valorization of fruit processing by-products* (pp. 145–166). Elsevier. Accessed March 13, 2024. [Online]. Available: https://www.sciencedirect.com/science/article/pii/B9780128171066000071
4. Russo, C., et al. (2021). The second life of citrus fruit waste: A valuable source of bioactive compounds. *Molecules, 26*(19), 5991.
5. Chavan, P., Singh, A. K., & Kaur, G. (2018). Recent progress in the utilization of industrial waste and by-products of citrus fruits: A review. *Journal of Food Process Engineering, 41*(8), e12895. https://doi.org/10.1111/jfpe.12895
6. Satari, B., & Karimi, K. (2018). Citrus processing wastes: Environmental impacts, recent advances, and future perspectives in total valorization. *Resources, Conservation and Recycling, 129*, 153–167.

7. Yadav, V., Sarker, A., Yadav, A., Miftah, A. O., Bilal, M., & Iqbal, H. M. (2022). Integrated biorefinery approach to valorize citrus waste: A sustainable solution for resource recovery and environmental management. *Chemosphere, 293*, 133459.
8. Suri, S., Singh, A., & Nema, P. K. (2022). Current applications of citrus fruit processing waste: A scientific outlook. *Applied Food Research, 2*(1), 100050.
9. Raimondo, M., Caracciolo, F., Cembalo, L., Chinnici, G., Pecorino, B., & D'Amico, M. (2018). Making virtue out of necessity: Managing the citrus waste supply chain for bioeconomy applications. *Sustainability, 10*(12), 4821.
10. Mamma, D., & Christakopoulos, P. (2014). Biotransformation of citrus by-products into value added products. *Waste and Biomass Valorization, 5*(4), 529–549. https://doi.org/10.1007/s12649-013-9250-y
11. Papoutsis, K., et al. (2016). Impact of different solvents on the recovery of bioactive compounds and antioxidant properties from lemon (Citrus limon L.) pomace waste. *Food Science and Biotechnology, 25*(4), 971–977. https://doi.org/10.1007/s10068-016-0158-8
12. Wang, X., Chen, Q., & Lü, X. (2014). Pectin extracted from apple pomace and citrus peel by subcritical water. *Food Hydrocolloids, 38*, 129–137.
13. Romero-Lopez, M. R., Osorio-Diaz, P., Bello-Perez, L. A., Tovar, J., & Bernardino-Nicanor, A. (2011). Fiber concentrate from orange (Citrus sinensis L.) bagase: Characterization and application as bakery product ingredient. *International Journal of Molecular Sciences, 12*(4), 2174–2186.
14. Ojha, P., & Thapa, S. (2017). Quality evaluation of biscuit incorporated with mandarin peel powder. *Scientific Study & Research. Chemistry & Chemical Engineering, Biotechnology, Food Industry, 18*(1), 19.
15. Caggia, C., et al. (2020). Employ of citrus by-product as fat replacer ingredient for bakery confectionery products. *Frontiers in Nutrition, 7*, 46.
16. Inoue, T., Tsubaki, S., Ogawa, K., Onishi, K., & Azuma, J. (2010). Isolation of hesperidin from peels of thinned Citrus unshiu fruits by microwave-assisted extraction. *Food Chemistry, 123*(2), 542–547.
17. Yousuf, A., et al. (2020). Pineapple juice preservation by pulsed electric field treatment. *Open Journal of Biological Sciences, 5*(1), 006–012.
18. Foti, P., Ballistreri, G., Timpanaro, N., Rapisarda, P., & Romeo, F. V. (2022). Prebiotic effects of citrus pectic oligosaccharides. *Natural Product Research, 36*(12), 3173–3176. https://doi.org/10.1080/14786419.2021.1948845
19. Huang, X., et al. (2019). Encapsulation of resveratrol in zein/pectin core-shell nanoparticles: Stability, bioaccessibility, and antioxidant capacity after simulated gastrointestinal digestion. *Food Hydrocolloids, 93*, 261–269.
20. Ayub, S., et al. (2020). Removal of heavy metals (Cr, Cu, and Zn) from electroplating wastewater by electrocoagulation and adsorption processes. *Desalination and Water Treatment, 179*, 263–271.
21. Boluda-Aguilar, M., García-Vidal, L., del Pilar González-Castañeda, F., & López-Gómez, A. (2010). Mandarin peel wastes pretreatment with steam explosion for bioethanol production. *Bioresource Technology, 101*(10), 3506–3513.
22. Muñoz-Labrador, A., Moreno, R., Villamiel, M., & Montilla, A. (2018). Preparation of citrus pectin gels by power ultrasound and its application as an edible coating in strawberries. *Journal of the Science of Food and Agriculture, 98*(13), 4866–4875. https://doi.org/10.1002/jsfa.9018
23. Apraj, V. D., & Pandita, N. S. (2016). Evaluation of skin anti-aging potential of Citrus reticulata Blanco peel. *Pharmacognosy Research, 8*(3), 160.
24. Aboul Naser, A., Younis, E., El-Feky, A., Elbatanony, M., & Hamed, M. (2020). Management of *Citrus sinensis* peels for protection and treatment against gastric ulcer induced by ethanol in rats. *Biomarkers, 25*(4), 349–359. https://doi.org/10.1080/1354750X.2020.1759693
25. Wang, Q., et al. (2023). Review of recent advances on health benefits, microbial transformations, and authenticity identification of *Citri reticulatae* Pericarpium bioactive compounds.

Critical Reviews in Food Science and Nutrition, 1–29. https://doi.org/10.1080/10408398.2023.2222834
26. Berk, Z. (2016). *Citrus fruit processing*. Academic.
27. Rosa, A., et al. (2019). Supercritical CO_2 extraction of waste citrus seeds: Chemical composition, nutritional and biological properties of edible fixed oils. *European Journal of Lipid Science and Technology, 121*(7), 1800502. https://doi.org/10.1002/ejlt.201800502
28. Rahman, M. M., Jahan, F. I., & Mim, S. A. (2019). A brief phytochemical investigation and pharmacological uses of citrus seed–A review. *Pharmacology Online, 1*, 94–103.
29. Qin, S., et al. (2018). Extraction, identification, and antioxidant property evaluation of limonin from pummelo seeds. *Animal Nutrition, 4*(3), 281–287.
30. Yilmaz, E., & Güneşer, B. A. (2017). Cold pressed versus solvent extracted lemon (Citrus limon L.) seed oils: Yield and properties. *Journal of Food Science and Technology, 54*(7), 1891–1900. https://doi.org/10.1007/s13197-017-2622-8
31. Zayed, A., Badawy, M. T., & Farag, M. A. (2021). Valorization and extraction optimization of citrus seeds for food and functional food applications. *Food Chemistry, 355*, 129609.
32. Atolani, O., et al. (2020). Chemical characterization, antioxidant, cytotoxicity, Anti-Toxoplasma gondii and antimicrobial potentials of the Citrus sinensis seed oil for sustainable cosmeceutical production. *Heliyon, 6*(2)., Accessed March 19, 2024. [Online]. Available: https://www.cell.com/heliyon/pdf/S2405-8440(20)30244-9.pdf
33. Ezekoye, V., Adinde, R., Ezekoye, D., & Ofomatah, A. (2019). Syntheses and characterization of biodiesel from citrus sinensis seed oil. *Scientific African, 6*, e00217.
34. Kim, J.-W., Lee, B. C., Lee, J.-H., Nam, K.-C., & Lee, S.-C. (2008). Effect of electron-beam irradiation on the antioxidant activity of extracts from Citrus unshiu pomaces. *Radiation Physics and Chemistry, 77*(1), 87–91.
35. Widmer, W., Zhou, W., & Grohmann, K. (2010). Pretreatment effects on orange processing waste for making ethanol by simultaneous saccharification and fermentation. *Bioresource Technology, 101*(14), 5242–5249.
36. Espinosa-Pardo, F. A., Nakajima, V. M., Macedo, G. A., Macedo, J. A., & Martínez, J. (2017). Extraction of phenolic compounds from dry and fermented orange pomace using supercritical CO2 and cosolvents. *Food and Bioproducts Processing, 101*, 1–10.
37. Azad, A. K. M., Ali, M. A., Akter, M. S., Rahman, M. J., & Ahmed, M. (2014). Isolation and characterization of pectin extracted from lemon pomace during ripening. *Journal of Food and Nutrition Sciences, 2*(2), 30–35.
38. Fan, X., et al. (2016). Production of nano bacterial cellulose from beverage industrial waste of citrus peel and pomace using Komagataeibacter xylinus. *Carbohydrate Polymers, 151*, 1068–1072.
39. Zema, D. A., Calabrò, P. S., Folino, A., Tamburino, V., Zappia, G., & Zimbone, S. M. (2018). Valorisation of citrus processing waste: A review. *Waste Management, 80*, 252–273.
40. Wikandari, R., Millati, R., Cahyanto, M. N., & Taherzadeh, M. J. (2014). Biogas production from citrus waste by membrane bioreactor. *Membranes, 4*(3), 596–607.
41. Tao, N., Shi, W., Liu, Y., & Huang, S. (2011). Production of feed enzymes from citrus processing waste by solid-state fermentation with *Eupenicillium javanicum*. *International Journal of Food Science Technology, 46*(5), 1073–1079. https://doi.org/10.1111/j.1365-2621.2011.02587.x
42. Ramasamy, M., Srirangarayan, R. S., & Paulin Thangam, R. *Composting of fruit peel from Rutaceae family*. Available at SSRN 3641904, 2020, Accessed 18 March 2024. [Online]. Available: https://papers.ssrn.com/sol3/papers.cfm?abstract_id=3641904
43. Maqbool, Z., et al. (2023). Citrus waste as source of bioactive compounds: Extraction and utilization in health and food industry. *Molecules, 28*(4), 1636.
44. Akhavan-Mahdavi, S., et al. (2022). Nanodelivery systems for d-limonene; techniques and applications. *Food Chemistry, 384*, 132479.
45. Andrade, M. A., et al. (2022). Citrus by-products: Valuable source of bioactive compounds for food applications. *Antioxidants, 12*(1), 38.

46. Huang, J., Liao, J., Qi, J., Jiang, W., & Yang, X. (2021). Structural and physicochemical properties of pectin-rich dietary fiber prepared from citrus peel. *Food Hydrocolloids, 110*, 106140.
47. Weldekidan, H., Patel, H., Mohanty, A., & Misra, M. (2024). Synthesis of porous and activated carbon from lemon peel waste for CO2 adsorption. *Carbon Capture Science & Technology, 10*, 100149.
48. Lanfranchi, M. (2012). Economic analysis on the enhancement of citrus waste for energy production. *Journal of Essential Oil Research, 24*(6), 583–591. https://doi.org/10.1080/10412905.2012.739788
49. Sarkar, R., Kundu, A., Dutta, A., Mandal, A., & Saha, S. (2022). Citrus peel as a source for waste valorization and its greener processing. In B. S. Patil, N. Dhowlaghar, J. Singh, & S. S. Pillai (Eds.), *ACS symposium series* (Vol. 1415, pp. 147–174). American Chemical Society. https://doi.org/10.1021/bk-2022-1415.ch011
50. Restrepo-Serna, D. L., & Cardona-Alzate, C. A. (2024). Pre-feasibility assessment to obtain an extract rich in hesperidin from orange peel: A comparison of extraction technologies conventional and non-conventional. *Sustainable Chemistry and Pharmacy, 38*, 101466.
51. Lucia, C., et al. (2022). Challenges and opportunities for citrus wastewater management and valorisation: A review. *Journal of Environmental Management, 321*, 115924.
52. Manhongo, T. T., Chimphango, A. F. A., Thornley, P., & Röder, M. (2022). Current status and opportunities for fruit processing waste biorefineries. *Renewable and Sustainable Energy Reviews, 155*, 111823.
53. Chen, W., Zhou, Y., & Chen, Y. (2021). The environmental impacts of citrus residue management in China: A case study in the three gorges reservoir region. *Waste Management, 133*, 80–88.
54. Corsino, S. F., Di Trapani, D., Capodici, M., Torregrossa, M., & Viviani, G. (2021). Optimization of acetate production from citrus wastewater fermentation. *Water Resources and Industry, 25*, 100140.
55. Bozzano, G., Raymo, M., Manenti, F., Rulli, M. C., Girotto, F., & Piazza, L. (2021). Prompting sustainability in the citrus derivates industry: A case study. *Cleaner Engineering and Technology, 4*, 100127.
56. Mak, T. M., Xiong, X., Tsang, D. C., Iris, K. M., & Poon, C. S. (2020). Sustainable food waste management towards circular bioeconomy: Policy review, limitations and opportunities. *Bioresource Technology, 297*, 122497.
57. U. Nations. Department of Economic and Social Affairs. *Population Division*. 2015. Accessed 19 March 2024. [Online]. Available: https://www.un.org/en/development/desa/policy/wess/wess_archive/1967wes_part2.pdf

Chapter 15
From Peel to Pulp: Maximizing Citrus Products and Value Addition

Jagbir Singh Aswal, Bipin Kumar Sati, and Anamika Chauhan

15.1 Introduction

Citrus is the most extensively cultivated fruit tree globally, boasting two distinct markets: fresh fruit and processed. Originating in Southeast Asia thousands of years ago, citrus fruits have spread worldwide. The Mediterranean, particularly Spain, is a major hub for fresh fruit production, although consumption patterns are shifting. While industrialized nations see a decline in fresh orange consumption, emerging economies like Mexico, India, Argentina, Brazil, and China show a rising interest. Mandarin production, including tangerines, clementines, and satsumas, has surged, surpassing fresh oranges in some regions. Sweet oranges dominate the citrus juice market, making up about 80% of production. Two regions stand out: Florida (USA) and Sao Paulo (Brazil). Brazil exports nearly all its orange juice (99%), while the US mainly consumes theirs domestically (90%). The European Union is the biggest importer, taking in over 80% of the world's orange juice (according to UNCTAD). Overall, citrus production and trade grew slightly (around 4%) in 2020–21, reaching 98 million metric tons. China is leading the production, followed by Brazil and collectively the European Union at number three. Orange makes up 50% of the citrus production, followed by Mandarins, lemons, and grapefruit. China is the top producer, with Brazil and the EU behind them. However, the US citrus industry has shrunk significantly. Citrus greening disease is a primary culprit, reducing production to less than a third of its level 20 years ago, and citrus acreage has shrunk by nearly 40%. The world exports about 11 million tons of citrus fruits, with oranges making up over 40% and tangerines/mandarins nearly 30%. China, South Africa,

J. S. Aswal · B. K. Sati
Council For Science & Technology, Jhajra Via Premnagar Dehradun, Uttarakhand, India

A. Chauhan (✉)
Department of Home Sciences, Chaman Lal Mahvidhyalya Landhaura Haridwar, Uttarakhand, India

A. Chauhan et al. (eds.), *Valorization of Citrus Food Waste*,
https://doi.org/10.1007/978-3-031-77999-2_15

and Turkey are major exporters of tangerines/mandarins, while Mexico, South Africa, and Turkey have increased lemon exports. South Africa is the current leader in citrus exports, followed by Turkey and Egypt. The US citrus export market has shrunk, mainly due to fewer oranges being exported and limited US participation in the growing global trade of tangerines and lemons. This is likely due to a decrease in exportable US citrus supplies. Despite the decline, tangerine/mandarin production is increasing in the US, making it the only citrus commodity with growing exports. While the US is now the seventh largest citrus exporter, their shipments have decreased for all products except tangerines/mandarins (data from USDA [1]).

Citrus bioactive compounds Citrus peels are much more than just waste products. They are packed with a surprising variety of health-promoting compounds. Beyond the fragrant essential oils, the peel contains fiber, pectin, protein, pigments, and a powerful arsenal of antioxidants. These antioxidants come in two primary forms: polyphenols and ferulic acid, including hesperidin and naringenin. Both types help protect your cells from damage. On top of that, citrus peels boast limonoids, natural compounds that contribute to antioxidant, anti-inflammatory, and even anti-cancer benefits. Finally, the essential oils themselves, particularly abundant in limes, are rich in coumarins and furanocoumarins, which help fight fungal infections. These constituents find applications across various industries. Synephrine, an alkaloid found in bitter orange extracts, undergoes alterations depending on fruit maturity. Pigments responsible for the vibrant hues of citrus fruits consist of carotenoids and yellow pigments. Notably, the peel contains higher pigmentation like beta-carotene, lycopene, anthocyanins, and chlorophyll a and b, contributing to this chromatic diversity. Furthermore, citrus peel is renowned for its rich vitamin content, particularly vitamin C and thiamine, alongside essential minerals like calcium. The mineral composition of citrus peel is subject to various factors, including soil type, maturity stage, cultivar variety, topography, and geographical conditions. With the escalating demand for citrus peel-based products, comprehensive research is imperative to ensure their safe utilization in the food industry. Despite traditionally being discarded, citrus peel represents a valuable resource, comprising sugars, fibers, organic acids, amino acids, essential oils, and vitamins in its substantial waste generated during processing. This waste finds utility in diverse sectors such as food, animal feed, biodegradable packaging, and cosmetics, offering molasses, pectin, and health-promoting compounds. In alignment with contemporary consumer preferences favoring health-beneficial products, citrus peel offers phytochemicals with various advantages, such as anti-diabetic, anti-bacterial, anti-cancer, antioxidant, and neuroprotective effects. Polymethoxylated flavonoids in citrus peels, including neohesperidin, naringin, hesperidin, and others, have demonstrated diverse healthiness benefits; its pivotal role in sensory and nutritional realms within the food, cosmetic, and pharmaceutical industries underscore citrus peel's multifaceted significance Fig. 15.1.

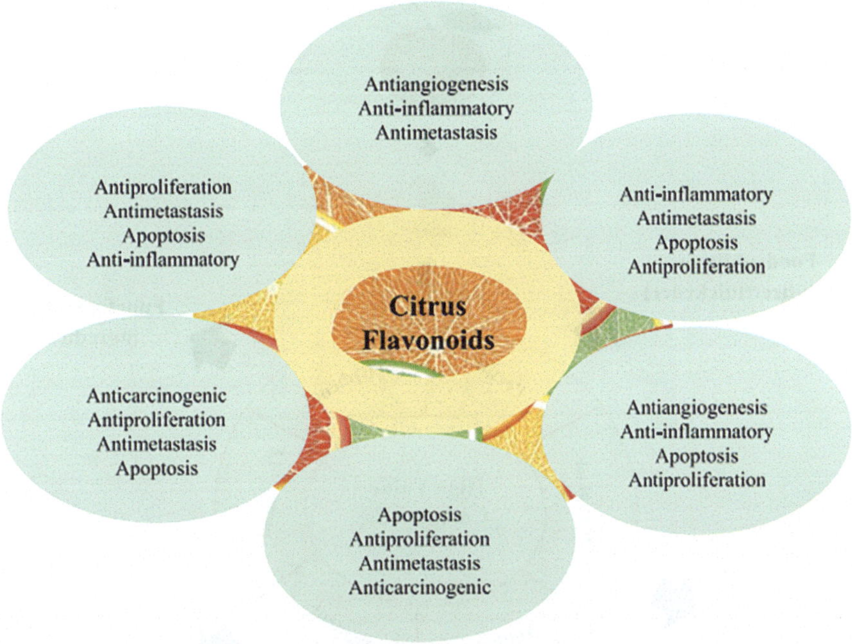

Fig. 15.1 Bioactive properties of citrus flavonoids. (Source: Nishala et al)

Fruits and vegetables are the most significant sources of fiber-based diets and are known for their benefits in fighting chronic health issues [2, 3]. This fiber helps with digestion, weight management, blood sugar control, and heart health [4, 5]. Interestingly, even the byproducts of fruits and vegetables offer valuable sources of fiber and other functional compounds [6, 7]. Phenolic compounds are crucial in regulating cholesterol, reducing the risk of heart disease, and fighting against cancer, and they are found abundantly in fruits and vegetables [8]. Citrus fruits, mainly, are a treasure trove of health-promoting elements. They are rich in vitamins, Minerals, essential oils, and carotenoids, offering significant therapeutic potential [9–15]. Additionally, citrus fruits contain prenyloxy coumarins known for their antioxidant properties [16–18]. The vitamin C in citrus fruits contributes to preventing degenerative diseases [19, 20]. As concerns rise around the toxicity of synthetic antioxidants, the focus is shifting towards natural sources like citrus fruits with identified health benefits. This highlights the exciting potential of utilizing byproducts from food processing industries. In light of concerns regarding the toxicity of synthetic antioxidants, there is a burgeoning preference for natural food sources containing identified bioactive compounds, thereby emphasizing the potential value addition from byproducts of food processing industries [21–23]. This review underscores the three primary constituents of citrus peel possessing significant functional properties (Fig. 15.2).

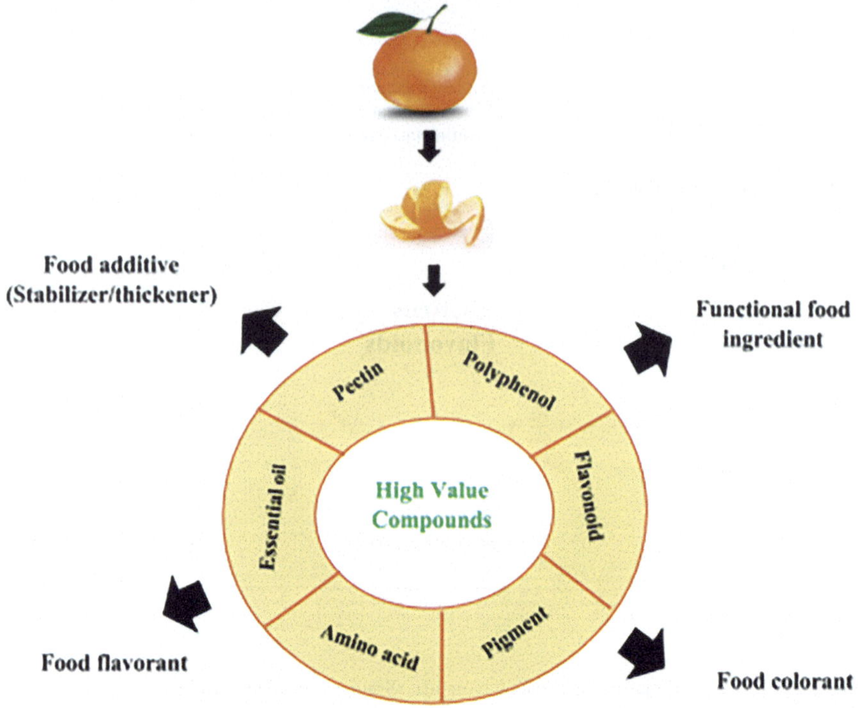

Fig. 15.2 Ingredients in citrus peel. (Source: Nishala et al)

15.2 Citrus Production and Trade Overview

Lemon byproducts stand out in the market due to their extensive use in various industries. As countries implement tariffs and import bans, international trade in lemons presents challenges and opportunities. Russia recently restricted lemon imports from the European Union, significantly impacting local producers. This opened doors for Argentina and Turkey to increase their lemon exports to Russia. Similarly, Argentina resumed trade with the US after a lengthy feud of 15 years. However, Spain faced high tariffs on its lemon exports to the US, limiting their market share. With competitive pricing, Argentina has the potential to capture a larger share of the US market, currently dominated by Mexico and Chile. India leads the world in lemon and lime production. Worldwide production is roughly split between Asia and the American continent, with 42.8% and 40.9% respectively. After peaking in 2008, global lemon and lime production and harvested area declined but rose again in 2011, with significant growth between 2017 and 2018. China experienced the most significant increase, with production jumping from 716,000 metric tons in 2006 to 4.14 million metric tons in 2008. Production has since stabilized around 2 million metric tons annually. Unlike oranges or tangerines, lemons are primarily used as ingredients or processed products rather than for direct consumption. This makes the lemon market susceptible to oversupply, as the processing industry plays a crucial role in maintaining equilibrium. Despite not producing lemons, the Netherlands is a significant re-exporter to European countries like Germany, France, and Poland. The global export value for lemons and limes has steadily increased from 2015 to 2019, with a slight dip between 2018 and 2019. The US, Germany, the Netherlands, France, and Russia import lemons and limes. Interestingly, lemon imports in Europe typically reach their lowest point in spring and rise rapidly throughout the summer, peaking in July and August before declining again in winter. Spain and Italy dominate the European lemon processing industry, which fuels a high demand for lemons. The top European producer, Spain, supplies lemons to most Northern European countries. Other producers like Turkey, Argentina, and South Africa compete with Spanish exports in the European market. The processing industry influences the fresh lemon market. Low prices for processed lemon products lead to decreased demand for fresh lemons, potentially resulting in a surplus. The Eureka variety of lemon is more prevalent in non-Mediterranean regions like California, Australia, South Africa, Argentina, and Israel. (2020 Industry Report: Lemon) Fig. 15.3.

15.3 Citrus By-Products: A Versatile Asset in the Food Industry

Various citrus by-products have found diverse applications in the food industry. For instance, citrus solid residues have been utilized as dietary fiber-rich flour in bakery products like biscuits and cereal bars [24]. Debittered citrus peel has been incorporated into bakery products to enhance flavor [25], while orange peel and lemon peel have been transformed into seasoning and powder forms for use in chocolate, cakes, and bakery items [26–29]. Additionally, citrus by-products have been explored for their colorant properties, such as β-carotene-loaded nanoemulsions from citrus reticulate peel for fruit juices [30]. Moreover, carotenoid lipid extracts from mandarin epicarp have been utilized in the bakery industry [31]. Polyethoxylated flavonoids from Citrus Limon have been incorporated into various food products, including sports drinks, dairy, bakery items, fats, ice creams, dried fruits, and cereals. Grapefruit peel, on the other hand, has been used as a pectin stabilizer/thickening agent in jams, jellies, marmalades, milk, and confectionery products [32]. Citrus peel, particularly bergamot oil, has flavored in sweets, jams, and flour-based confectionaries [33]. Furthermore, grapefruit peel extract has been employed as a food additive, specifically as an antioxidant, in meatballs. Source: Nishala et al. (Figs. 15.4 and 15.5).

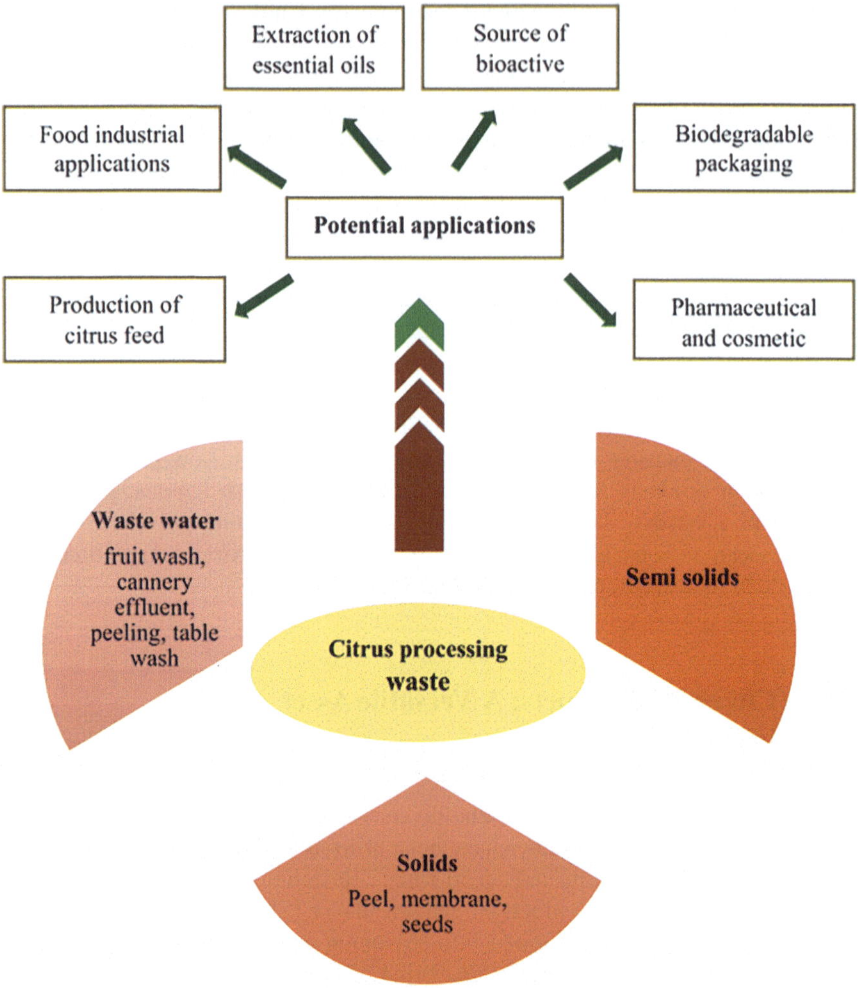

Fig. 15.3 Citrus Waste. (Source: Nishala et al)

15.4 Medicinal and Industrial Importance of Citrus and By-Product

The richness of components found in orange peel establishes it as a valuable source for economically significant substances and processes, both through direct and indirect applications. Direct uses involve utilizing the entire orange peel without separating its constituents, such as employing it as animal feed, organic fertilizer, and more. Citrus peel, a substantial waste product abundant in limonene, pectin, and molasses, is commonly dried and combined with dried pulps for cattle feed. Studies have shown that incorporating orange peel into cattle diets can significantly reduce

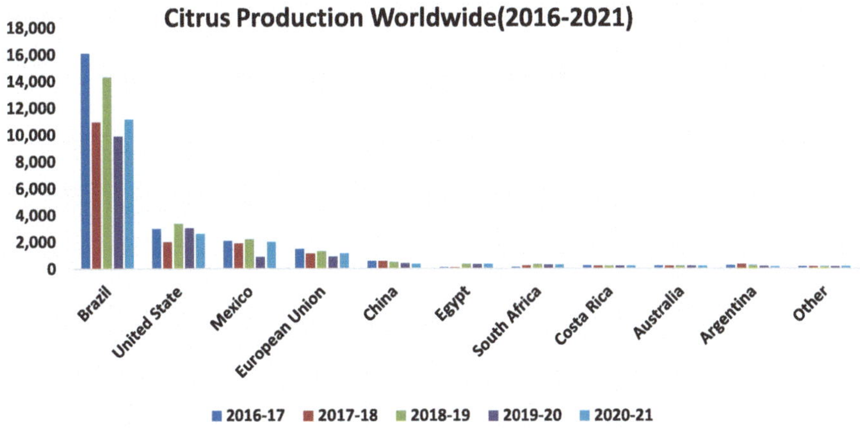

Fig. 15.4 Citrus Processing Worldwide (1000MT). (Source: US Agriculture Market)

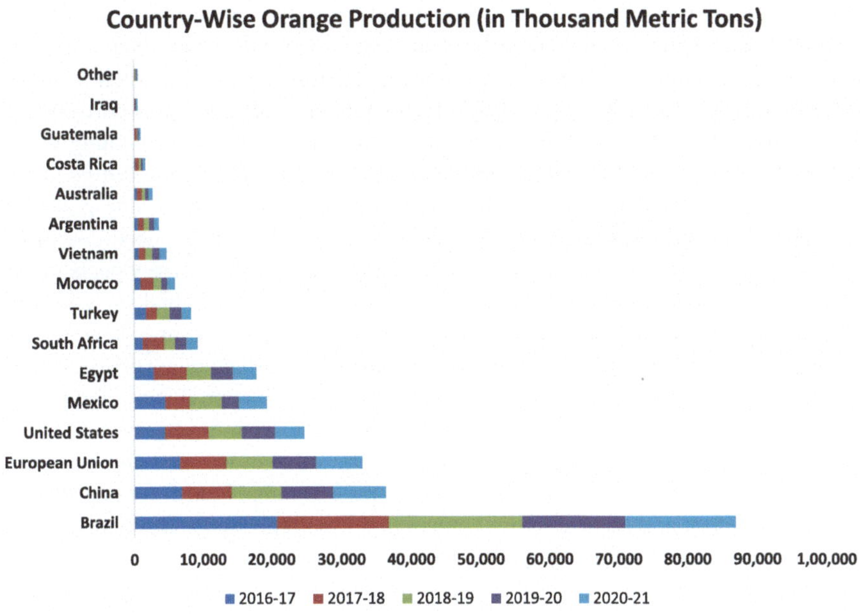

Fig. 15.5 Country-Wise Orange Production (in Thousand Metric Tons). (Source: US Agriculture Market)

harmful bacteria like *Salmonella* and *E. coli* in their intestines, consequently lowering the levels of hazardous microorganisms in the meat. Orange peel also serves as compost and organic fertilizer, possessing medicinal properties beneficial for preventing and treating flu, colds, scurvy, and vitamin deficiencies and combating bacterial and viral infections. Highly nutritious orange peels can be utilized as food supplements and drugs through this approach. Various bioactive compounds

Table 15.1 Specific components analyzed in the peels of different citrus fruits, along with the corresponding references

Citrus fruit	Analyzed components	References
Orange (*Citrus sinensis*)	Reducing sugars, Polyphenols	Patra et al.
Lemon	Polyphenols, Eriocitrin	Kaur et al. (2023)
Mandarin (*Citrus reticulata*)	Polyphenols, Eriocitrin	Kaur et al. (2023)
Grapefruit (*Citrus paradisi*)	Lycopene, Naringi	Kaur et al. (2023)
Lime (*Citrus aurantiifolia*)	Phenolic compounds, Antioxidants	Patra et al.

Source: Sakile Hannah et al.

extracted from orange peel find applications across different fields, as outlined in Table 15.1. Enzymes like invertase, laccases, pectinase, xylanases, and α-amylases can be produced using orange peel as a substrate, with applications ranging from vegetable oil extraction to wastewater treatment and food and beverage production. Techniques such as solid-state fermentation produce cellulase enzymes from orange peel. Citrus belongs to the Rutaceae family and is represented by the Citrus L. genus. It is recognized not only for its widespread popularity but also for its active photochemical contribution to overall health. Citrus peel, a primary byproduct, is a valuable source of molasses, pectin, and limonene. Historical practices, such as using oranges to combat scurvy [12], highlight the therapeutic applications of citrus fruits. Furthermore, the citrus peel has been harnessed to create health-promoting products, serving as a source of natural additives and sweeteners and contributing to the characteristic bitter flavor in some beverages (naringin) [34]. Regarding potential industrial applications, orange peel has been recognized as animal feed due to its nutritional properties. This property has been attributed to high-energy peels in the diets of microbial ruminants, which enhance lactation and microbial growth [35]. The orange peel waste has also been utilized as organic fertilizer through composting techniques. The bio-refinery approach for citrus peel utilization involves extracting valuable compounds through various processes for producing fuels and chemicals from waste biomass. Like petroleum refineries, a bio-refinery can extract multiple products from orange peel, enhancing the value of biomass and minimizing waste. Another application is using orange peel as a supplement for paper pulp. Its high cellulose and hemicellulose content and low lignin and ash concentrations make orange peel a promising option [36].

15.5 Post-harvest Industrial Application

Citrus peels, for instance, are rich in bioactive compounds such as flavonoids, limonoids, and pectin, which possess antioxidant, antimicrobial, and anti-inflammatory properties. These compounds have sparked interest in utilizing citrus peels as functional ingredients in various food products, including baked goods, beverages, and snacks, to enhance their nutritional profiles and impart health benefits. Moreover, the essential oils extracted from citrus peels are extensively used in the cosmetic industry for their aromatic properties and potential skincare benefits. Thus, the valorization of citrus byproducts not only mitigates waste but also opens avenues for the development of novel, value-added products across multiple sectors. As research in this field progresses, it holds promise for creating a more sustainable and resource-efficient citrus industry while simultaneously offering innovative solutions for improving human health and well-being (Table 15.1).

15.6 Conclusion

In conclusion, the exploration and utilization of bioactive compounds found in citrus peels hold significant promise for advancing health-promoting initiatives across various industries. Throughout this research chapter, we have delved into the rich diversity of bioactive components present in citrus peels, highlighting their potential to contribute to human well-being in multiple ways. Citrus peels, whether from oranges, lemons, mandarins, grapefruits, or limes, have been revealed to contain many beneficial compounds. From reducing sugars and polyphenols in orange peels to polyphenols and eriocitrin in lemon and mandarin peels, each citrus variety offers unique compositions that could have distinct health implications. Additionally, grapefruit peels are noted for their abundance of lycopene and naringi, while lime peels boast significant levels of phenolic compounds and antioxidants. This diversity underscores the versatility and potential of citrus peels as valuable resources for addressing various health needs. Furthermore, the findings presented in this research chapter highlight the vast array of potential applications for citrus peel extracts across different industries. In the food industry, citrus peel extracts can be incorporated into functional foods, beverages, and supplements to enhance nutritional value and promote health. Cosmetic formulations can benefit from the antioxidant and anti-inflammatory properties of citrus peels, offering natural alternatives for skincare and haircare products. Moreover, the pharmaceutical industry stands to gain from the bioactive compounds found in citrus peels, with potential applications in drug development and preventive medicine. The significance of citrus peels as a valuable reservoir of bioactive compounds cannot be overstated. Their potential to contribute to human health and well-being underscores the importance of continued exploration and research in this field. By harnessing the bioactive compounds present in citrus peels, we have the opportunity to develop innovative solutions for addressing various health challenges and improving quality of life. The findings presented in this research chapter affirm citrus peel as an invaluable resource with numerous potential health benefits. As we continue to uncover the therapeutic properties of citrus peels and explore their applications in different industries, it is clear that they hold immense promise for promoting human health and well-being.

Acknowledgements The study support to Uttarakhand State Council for Science and Technology UCOST, Government of Uttarakhand, Department of Information & Science Technology We express our gratitude to the Director General of UCOST for granting us access to the necessary facilities.

References

1. United States Department of Agriculture Foreign Agricultural Service. (2021). *Citrus: World markets and trade* (pp. 1–13).
2. Elleuch, M., Bedigian, D., Roiseux, O., Besbes, S., Blecker, C., & Attia, H. (2011). Dietary fibre and fibre rich by-products of food processing: Characterisation, technological functionality and commercial applications: A review. *Food Chemistry, 4*, 411–421.
3. Maeda-Yamamoto, M., Kawahara, H., Tahara, N., Tsuji, K., Hara, Y., & Isemura, M. (1999). Effect of tea polyphenols on the invasion and matrix metalloproteinases activities of human fibrosarcoma HT1080 cells. *Journal of Agricultural and Food Chemistry, 47*, 2350–2354.
4. Figuerola, F., Hurtado, M. L., Estevez, A. M., Chiffelle, I., & Asenjo, F. (2005). Fibre concentrates from apple pomace and citrus peel as potential fibre sources for food enrichment. *Food Chemistry, 91*, 395–401.
5. Shahidi, F. (1997). Natural antioxidants an over view. In F. Shahidi (Ed.), *Natural antioxidants* (pp. 1–11). AOCS Press.
6. Larrauri, J. A. (1999). New approaches in the preparation of high dietary fibre powders from fruit by-products. *Trends in Food Science and Technology, 10*, 3–8.
7. Meyer, A., Yi, O., Pearson, D., Waterhouse, A. L., & Frankel, E. (1997). Inhibition of human low-density lipoproteins oxidation in relation to phenolic antioxidants in grapes. *Journal of Agricultural and Food Chemistry, 43*, 1638–1643.
8. Carrol, K. K., Kurowska, E. M., Guthrie, N. (1999). *Use of citrus limonoids and flavonoids as well as tocotrienols for the treatment of cancer*. International Patent WO 9916167.
9. Alvarez-Gonzales, I., Madrigal-Bujaidar, E., & Sanchez-Garcia, V. Y. (2010). Inhibitory effect of grapefruit juice on the genotoxic damage induced by ifosfamide in mouse. *Plant Foods for Human Nutrition, 65*, 369–373.
10. Assis, C. R. C. L., Hermsdorff, H. H. M., & Bressan, J. (2013). Anti-inflammatory properties of orange juice: Possible favorable molecular and metabolic effects. *Plant Foods for Human Nutrition, 68*, 1–10.
11. Cha, J. Y., Cho, Y. S., Kim, I., Anno, T., Rahman, S. M., & Yanagita, T. (2001). Effect of hesperedin, a citrus flavonoid, on the liver triacylglycerol content and phosphatidatephosphohydrolase activity in oroticacidfed rats. *Plant Foods for Human Nutrition, 56*, 349–358.
12. Magiorkinis, E., Beloukas, A., Diamantis, A. (2011). Scurvy: past, present and future. *European Journal of Internal Medicine, 22*, 147–152. Manthey, J. A., Grohmann, K. (2001). Phenols in citrus peel byproducts. Concentrations of hydroxycinnamates and polymethoxylated flavones in citrus peel molasses. *Journal of Agricultural and Food Chemistry, 49*(7), 3268–3273.
13. Pak, C. Y. (2004). Medical management of urinary stone disease. *Nephron Clinical Practice, 98*, 49–53.
14. Sadek, E. S., Makris, D. P., & Kefalas, P. (2009). Polyphenolic composition and antioxidant characteristics of kumquat (Fortunella margarita) peel fractions. *Plant Foods for Human Nutrition, 64*, 297–302.
15. Sica, D. A. (2006). Interaction of grapefruit juice and calcium channel blockers. *American Journal of Hypertension, 19*, 768–773.
16. Epifano, F. (2014). *Analysis of biologically active oxyprenylated ferulic acid derivatives in citrus fruits*. Springer Science+Business Media.
17. Epifano, F., & Genovese, S. (2013). Recent acquisitions on naturally occurring oxyprenylated secondary plant metabolites. In G. Brahmachari (Ed.), *Chemistry and pharmacology of naturally occurring bioactive compounds* (1st ed., pp. 239–257). CRC Press.
18. Moure, A., Cruz, J. M., Franco, D., Domínguez, J. M., Sineiro, J., Domínguez, H., et al. (2001). Natural antioxidants from residual sources. *Food Chemistry, 72*(2), 145–171.
19. Farag, R. S., Badei, A. Z., Heweij, F. M., & El-Baroty, G. S. A. (1986). Antioxidant activity of some spices essential oil on linoleic acid in aqueous media. *Journal of the American Oil Chemists' Society, 66*, 792–799.

20. Wang, L., Wang, J., Fang, L., Zheng, Z., Dexian, Z., Wang, S., Li, S., Ho, C. T., & Zhao, H. (2014). Anticancer activities of citrus peel polymethoxyflavones related to angiogenesis and others. *BioMed Research International*. https://doi.org/10.1155/2014/453972
21. Burlow, S. M. (1990). Toxicological aspects of antioxidants used as food additives. In B. J. F. Hudson (Ed.), *Food antioxidants* (pp. 253–268). Elsevier.
22. Cozzi, R., Ricordy, R., Aglitti, T., Gatta, V., Petricone, P., & DeSalvia, R. (1997). Ascorbic acid and b-carotene as modulators of oxidative damage. *Carcinogenesis, 18*, 223–228.
23. Ito, N., Fukushima, S., Hasegawa, A., Shibata, M., & Ogiso, T. (1983). Carcinogenicity of butylated hydroxyanizole in F344 rats. *Journal of the National Cancer Institute, 70*, 343–344.
24. Ferreira, S. S., Silva, A. M., & Nunes, F. M. (2018). Citrus reticulata Blanco peels as a source of antioxidant and anti-proliferative phenolic compounds. *Industrial Crops and Products, 111*, 141–148. https://doi.org/10.1016/j.indcrop.2017.10.009
25. Caggia, C., Palmeri, R., Russo, N., Timpone, R., Randazzo, C. L., Todaro, A., & Barbagallo, S. (2020). Employ of citrus by-product as fat replacer ingredient for bakery confectionery products. *Frontiers in Nutrition, 7*, 1–9. https://doi.org/10.3389/fnut.2020.00046
26. Albak, F., & Tekin, A. R. (2014). Development of functional chocolate with spices and lemon peel powder by using response surface method: Development of functional chocolate. *Retrieved from Akademik Gıda, 12*(2), 19–25. http://www.academicfoodjournal.com
27. Espinosa-Garza, G., Antonyan, N., & Loera-Hernandez, I. (2018). Orange peel dehydration and creation of new edible products. *International Journal of Food Engineering, 4*(4), 327–331. https://doi.org/10.18178/ijfe.4.4.327-331
28. Iftikhar, M., Wahab, S., Haq, N. U., Malik, S. N., Amber, S., Taran, N. U., & Rehman, S. U. (2019). Utilization of citrus plant waste (peel) for the development of food product. *Pure and Applied Biology, 8*(3), 1991–1998. https://doi.org/10.19045/bspab.2019.80
29. Lu, S. Y., Chu, Y. L., Sridhar, K., & Tsai, P. J. (2021). Effect of ultrasound, high-pressure processing, and enzymatic hydrolysis on carbohydrate hydrolyzing enzymes and antioxidant activity of lemon (Citrus limon) flavedo. *LWT, 138*(2021). https://doi.org/10.1016/j.lwt.2020.110511
30. Barman, K., Chowdhury, D., & Baruah, P. K. (2020). Development of β-carotene loaded nanoemulsion using the industrial waste of orange (citrus reticulate) peel to improve in vitro bioaccessibility of carotenoids and use as natural food colorant. *Journal of Food Processing and Preservation, 44*(5), 1–10. https://doi.org/10.1111/jfpp.14429
31. Ordonez-Santos, L. E., Esparza-Estrada, J., & Vanegas-Mahecha, P. (2021). Ultrasoundassisted extraction of total carotenoids from mandarin epicarp and application as natural colorant in bakery products. *LWT, 139*, Article 110598. https://doi.org/10.1016/j.lwt.2020.110598
32. Mohamed, H. (2016). Extraction and characterization of pectin from grapefruit peels. *MOJ Food Processing & Technology, 2*(1), 31–38. https://doi.org/10.15406/mojfpt.2016.02.00029
33. Mustafa, N. E. M. (2015). Citrus essential oils: Current and prospective uses in the food industry. *Nutrition & Agriculture, 7*(2), 115–127.
34. Addi, M., Elbouzidi, A., Abid, M., Tungmunnithum, D., Elamrani, A., & Hano, C. (2022). An overview of bioactive flavonoids from citrus fruits. *Applied Sciences, 12*(1), 29.
35. Bampidis, V. A., & Robinson, P. H. (2006). Citrus by-products as ruminant feeds: A review. *Animal Feed Science and Technology, 128*(3–4), 175–217.
36. Kasaai, B., Caolo, V., Peacock, H. M., Lehoux, S., GomezPerdiguero, E., Luttun, A., et al. (2017). Erythro-myeloid progenitors can differentiate from endothelial cells and modulate embryonic vascular remodeling. *Scientific Reports, 7*(1), 43817.

Chapter 16
Citrus Waste: Potential Source of Beneficial Microbiota

Sohaib Afzaal, Ehsan-ul-Haque, Asad Manzoor, Shinawar Waseem Ali, Akbar Hayat, and Muhammad Asim

16.1 Introduction

The production of citrus fruits has seen a significant rise over the past few decades, due increased demand for fresh fruits and processed citrus products such as juices, marmalades, and essential oils. This increase in production has led to a significant amount of waste. Citrus waste sometimes reaches 60% of the production in developing countries due to poor value chain, export and policy problems. Oranges are main citrus cultivars in the world resulting in about 24.3 million tons of waste per year, mainly consisting of peels, seeds, and pulp [1]. The citrus processing industry creates a variety of by-products, including peels, seeds, and pulp, which account for almost half of the total fruit mass to manage the citrus waste. Citrus waste creates a significant environmental challenge upon ignoring its importance and using inappropriate waste management practices. Improper management of citrus waste can lead to soil and water pollution, increased greenhouse gases, and other ecological problems. Conventional disposal methods, such as landfilling and incineration, are not sustainable solutions due to their environmental impact and high cost [2].

On other hand, citrus waste is not just an environmental problem; it possesses a valuable resource that can be utilized in various beneficial applications. By transforming citrus waste into valuable products, we can reduce environmental hazards and create economic opportunities [3]. Citrus waste is a premium source of a lot beneficial microbiota including the bacterial and fungal communities. These microbial communities include beneficial bacteria, fungi, and yeasts, can be utilized for various applications such as probiotic formulations, bio-fertilizers, and the production of biofuels.

S. Afzaal (✉) · Ehsan-ul-Haque · A. Manzoor · A. Hayat · M. Asim
Citrus Research Institute, Sargodha, Pakistan

S. W. Ali
Department of Food Sciences, University of the Punjab, Lahore, Pakistan

A. Chauhan et al. (eds.), *Valorization of Citrus Food Waste*,
https://doi.org/10.1007/978-3-031-77999-2_16

The diverse microbiota found in citrus waste can play crucial roles in promoting gut health, enhancing plant growth, and facilitating sustainable agricultural processes [4]. Understanding the microbial diversity in citrus waste is crucial for the development of strategies to utilize the beneficial characteristics of microbiota.

The unique microbial communities found in citrus waste, including beneficial bacteria, fungi, and yeasts, offer significant potential in promoting gut health, enhancing plant growth, and supporting sustainable agricultural processes. One of the prominent microbial groups present in citrus waste is lactic acid producing bacteria (LAB). Species such as *Lactobacillus plantarum*, *Leuconostoc mesenteroides*, and *Pediococcus pentosaceus* are commonly found in citrus waste. These strains are having potential probiotic properties and can be utilized for formulation of probiotics based products.

Group of LAB also have significant contribution to bio-preservation and production of organic acids, hydrogen-peroxide, and bacteriocins responsible for altering the growth of undesirable spoilage organisms. This trait of LAB makes them effective to be used in food preservation where they can be added in food products to enhance the period of preservation and safety of the food. The numerous potential uses of these microbial communities from citrus waste include their use in probiotic supplements include *Lactobacillus plantarum* and *Leuconostoc mesenteroides* in the enhancement of the useful flora of the gut to aid in digestion and boost the immune system. These probiotics can also be added to foods for special purposes and in food supplements so that clients can take natural ways of improving their health.

Yeast, another group of microorganisms present in citrus waste is related to the process of fermentation. Among the yeasts isolated from citrus waste, include *Saccharomyces cerevisiae*, *Candida tropicalis*, and *Pichia kudriavzevii*. *Saccharomyces cerevisiae* is famously known for ethanol production, utilizes the converts the sugars in citrus waste to produce ethanol. This property plays important role in managing the citrus waste and to be used in the preparation of bio fuel, which is a better option than the fossil fuels. Yeast can also be used to prepare single cell protein which is nutritious and used for animal feed. Such species as *Candida tropicalis* and *Pichia kudriavzevii* are searched for this intention as it is the environmentally friendly method to solve the problems connected with animal feeding.

Various fungi in citrus waste i.e., *Aspergillus niger*, *Penicillium chrysogenum*, and *Trichoderma harzianum* are well-known due to their exportable enzyme profile that includes pectinases, cellulases and amylases that has extensive application in the food and textile industries. According to the findings, enzymes derived from these fungi using citrus waste are cheap, and reveal an added advantage of encouraging the utilization of wastes in the industrial processes. Moreover, mycorrhiza fungi (T*richoderma harzianum*) help the plants to take nutrients from the soil, help root growth, and protect the plants from pathogenic microorganisms. *Actinobacteria* present in citrus waste, are known for their ability to produce antibiotics and degrade complex organic compounds. The use of actinobacteria-enriched compost derived from citrus waste can significantly improve soil structure, enhance microbial diversity, and boost plant growth.

In the field of agriculture, microbial bio fertilizers isolated from citrus waste, for example fungi such *Trichoderma harsszianum* and bacteria including *Pseudomonas fluorescens* encourage sustainable farming by increasing the availability of nutrients and health of the soil as well as preventing diseases. These biofertilizers help to cut on the use of chemicals fertilizers, fight pollution and aid in organic farming.

Succus citratus, citrus waste, after being fermented by strains like Saccor recombes cerevisiae is a viable opportunity to develop bio fuels. The ethanol derived from the waste part of citrus fruits can be in a position to mitigate the use of fossil fuel and increase the use of renewable energy sources. This type of approach is useful in handling the problem of disposal of citrus wastes and at the same time offers a non-conventional source of fuel. Furthermore, the enzymes such as amylases, cellulase and protease produced by fungi like *Aspergillus niger* can be employed in variety of industrial applications out of which relatively few are value added products in citrus waste. For example, pectinases can be used in the production of pectin from citrus rinds which is a main ingredient in the manufacture of jams, jellies and gelling substances. This make adds economic value to citrus waste and contributed to its utilization in the food industry. The use of actinobacteria of citrus waste, i.e., *Streptomyces* spp., is possible for the development of bio plant protection agents. These products assist in managing plant diseases and decrease the occurrences of such diseases to enhance crops production.

This chapter aims to explore the potential of citrus waste as a source of beneficial microbiota. It will discuss the types of microorganisms present in citrus fruits and their roles, the techniques for isolating and characterizing these microorganisms, and the various applications of these microbial communities in different industries. Some of the opportunities and challenges have also been discussed associated with the valorization of citrus waste.

16.2 Microbial Diversity in Citrus Fruits

Citrus fruit contains a complex microbial diversity possessing on various parts. These microorganisms are found on the surface of the fruit as well as within the fruit. Surface microflora is comparatively difficult to isolate rather than from inner parts of the fruit. The microbial community inside the fruit collectively called as endophytes; plays important roles in the fruit's life cycle, impacting the process of ripening, spoilage, and flavor development.

16.2.1 Bacteria

Bacteria comprise the major portion among the microorganisms present in citrus waste. Various major genera of bacteria, which can be isolated from citrus waste, include the groups of Lactic acid producing bacteria (LAB), *Bacillus*, *Pseudomonas*, and *Bifidobacteria*. *Lactobacillus* is the major genus among the group of LAB which can be isolated from the citrus waste. *Lactobacillus* species are known for their fermentation abilities and probiotic properties. They produce lactic acid, which helps in preserving food by lowering the pH and inhibiting the growth of spoilage organisms and pathogens. *Lactobacillus* species can improve gut health, improve the immune system, and enhance the nutritional value of fermented products. A brief about the bacteria is given in Table 16.1, with their various uses in industrial biotechnology [5–10]. Bacteria isolated from citrus waste are vigorous and can tolerate low pH levels, making them suitable for various fermentation processes. These bacteria contribute to the overall health benefits associated with consuming citrus fruits and their derivatives [11]. They can enhance gut health by improving the intestinal microflora balance, inhibiting harmful bacteria, and strengthening the immune responses. These bacteria are anaerobic and thrive in the low-oxygen environment of the gut. Citrus waste is a rich source of *Acetobacter* species, which are well known for their ability to convert ethanol into acetic acid. They are essential in the production of vinegar and can also play a role in the initial stages of citrus waste degradation [12].

16.2.2 Fungi

Wasted citrus contains various infectious fungi causing various diseases to citrus crop, but some of the beneficial fungi are also inhabitant of citrus waste [13]. Beneficial mycobiota in citrus waste primarily contains *Penicillium* and other *Phytophthora* species which may be isolated and employed in industrial biotechnology. *Penicillium chrysogenum*, is a beneficial fungi which can be easily isolated from citrus waste and can be used for producing antibiotics (e.g., penicillin). However, pathogenic fungi like *Colletotrichum* species can cause diseases such as root rot and fruit decay, causing challenges for citrus production [14–16]. A brief description of fungi which can be isolated from citrus waste and can be employed in human wellbeing is given in Table 16.2.

16.2.3 Yeasts

Yeast is a minor group found in citrus waste. *Saccharomyces cerevisiae* is primarily inhabitant to citrus waste, which in involved in fermentation processes and can be used to produce products like biofuels and functional foods. *Saccharomyces cerevisiae* isolated from citrus fruits can be utilized in the production of alcoholic beverages and has applications in biotechnology for the production of bioethanol from citrus waste. Some of the yeasts of citrus resources along-with their industrial applications are given in Table 16.3 [17–19]. Other yeast species from citrus waste include *Pichia* species, well-known for their ability to assimilate a wide range of

Table 16.1 Table listing various beneficial bacteria that can be obtained from citrus waste

Species	Source	Uses in Industrial Biotechnology
Lactobacillus plantarum	Waste Citrus pulp	Probiotic production, lactic acid fermentation, and food preservation
Bacillus subtilis	Citrus pulp and rag	Enzyme production (amylase, protease), bio-pesticide production, and plant growth promotion
Pseudomonas fluorescens	Citrus peel and roots	Bioremediation, plant growth promotion, and biocontrol agent
Clostridium butyricum	Citrus waste	Butyric acid production, probiotic production, and biofuel production
Enterococcus faecium	Citrus juice and shoots	Probiotic production, antimicrobial peptide production, and dairy fermentation
Escherichia coli	Citrus waste and leafs	Genetic engineering applications, bioremediation, and production of recombinant proteins
Streptococcus thermophilus	Citrus juice and rag	Dairy fermentation, probiotic production, and starter culture in yogurt production
Lactobacillus casei	Citrus juice vesicles	Probiotic production, dairy fermentation, and food preservation
Pseudomonas putida	Citrus peel and waste	Bioremediation, bioplastic production, and plant growth promotion
Bacillus cereus	Citrus waste	Enzyme production (protease, amylase), bio-pesticide production, and soil health improvement
Lactobacillus acidophilus	Citrus juice, rag and fiber	Probiotic production, lactic acid fermentation, and food preservation
Lactobacillus brevis	Citrus waste	Probiotic production, fermentation, and bio-preservation
Lactobacillus rhamnosus	Citrus peel	Probiotic production, dairy fermentation, and gut health improvement
Lactobacillus fermentum	Citrus waste	Probiotic production, immune system support, and cholesterol reduction
Lactobacillus reuteri	Citrus peel and juice	Probiotic production, antimicrobial production (reuterin), and gut health improvement
Lactococcus lactis	Wasted Citrus rag	Dairy fermentation, cheese production, and nisin production (natural preservative)
Leuconostoc mesenteroides	Citrus waste	Fermentation, sauerkraut and kimchi production, and dextran production
Leuconostoc lactis	Citrus peel	Fermentation, dairy products, and probiotic production
Leuconostoc citreum	Citrus waste	Fermentation, production of exopolysaccharides, and bio-preservation
Leuconostoc fallax	Citrus peel	Fermentation, production of mannitol, and bio-preservation
Leuconostoc gelidum	Citrus waste	Fermentation, meat preservation, and bio-preservation
Leuconostoc pseudomesenteroides	Citrus peel	Fermentation, probiotic production, and bio-preservation
Pediococcus pentosaceus	Citrus peel	Fermentation, probiotic production, and bio-preservation

Table 16.2 Table listing various beneficial mycobiota that can be obtained from citrus waste

Species	Source	Uses in Industrial Biotechnology
Aspergillus niger	Citrus peel	Enzyme production (pectinase, cellulase), citric acid production, bio-remediation, and organic acid production
Aspergillus oryzae	Citrus waste	Enzyme production (amylase, protease), fermentation, and traditional Asian food production
Penicillium spp.	Citrus waste	Antibiotic production, enzyme production (lipase), and mycotoxin degradation
Rhizopus stolonifer	Citrus peel	Enzyme production (amylase, lipase), organic acid production, and fermentation
Trichoderma viride	Citrus waste	Enzyme production (cellulase, xylanase), biocontrol agent, and plant growth promotion
Geotrichum candidum	Citrus peel	Enzyme production (lipase, protease), cheese ripening, and bioremediation

Table 16.3 Table listing various beneficial yeasts that can be obtained from citrus waste

Species	Uses in Industrial Biotechnology
Brettanomyces bruxellensis	Wine and beer fermentation, bioethanol production, and flavor compound production
Saccharomyces cerevisiae	Bioethanol production, baking, and brewing
Candida tropicalis	Xylitol production, bioethanol production, and bioremediation
Candida utilis	Single-cell protein production, enzyme production, and animal feed

carbon sources, making them versatile in biotechnological applications. They are used in the production of biofuels, bio-chemicals, and recombinant proteins. *Pichia pastoris* is widely used for the production of heterologous proteins [20].

16.3 Techniques for Isolation and Characterization of Microorganisms from Citrus Fruits

Isolation and characterization of microbiota from citrus waste depends upon the target microorganism. It involves several steps and methods to ensure the right identification and application of beneficial microbes. These techniques can be categorized into two types i.e., (i) traditional culture-based methods and (ii) advanced molecular techniques.

16.3.1 Traditional Culture-Based Methods

16.3.1.1 Selective Culturing on Specific Media

Selective culturing involves using specific growth media to isolate particular types of microorganisms. Selective growth media helps for initial screening of the microbes from the complex communities. A lot of selective media targeting various specific groups are available commercially, or may be formulated using various combinations of reagents i.e., various lactic acid bacteria (LAB) especially *Lactobacilli* can be screened and isolated using MRS (de Man, Rogosa, and Sharpe) agar, which supports their growth while inhibiting other bacteria. Some of the coliforms are selective to MacConkey media. Using selective media allows for the enrichment and identification of target microorganisms by promoting their growth in a controlled environment [21].

16.3.1.2 Enrichment Cultures

Enrichment cultures are used to increase the relative abundance of a specific microorganism in a sample, which makes it easy to isolate. This technique involves providing favorable conditions for the growth of the target microorganisms while suppressing others. These favorable conditions may include the varied composition of media i.e., using an acidified medium, using various suppressing agents like antibiotics to inactivate the undesired microbes etc. This method is useful for isolating the microorganisms present in minute quantities in the original samples [22].

16.3.1.3 Streak Plate Method

The streak plate method is a basic technique in microbiology for isolation of the purified colonies of any microbe. Sample preparation for streak plate method is crucial step to avoid any contamination. Sample is streaked with a pre-sterilized metallic loop on an agar plate having selective media. Culture is diluted using three ways streaking and individual colonies grow on the plate after incubation period. Each colony represents a pure strain of a microorganism that can be further studied and identified [23].

16.3.2 Advanced Molecular Techniques

16.3.2.1 Polymerase Chain Reaction (PCR)

Purified individual colonies are subjected to polymerase chain reaction (PCR) for further identification and confirmation. PCR is considered as an effective molecular tool used to identify the microbes. It involves the amplification of target segments from the DNA sequence for identification. Various types of PCR reactions are employed in microbiology for identification purposes including nested PCR, multiplex PCR, RCA PCR, and real time PCR. Basic requirement for PCR reaction for identification include the total genomic DNA isolated from the purified culture. This DNA is amplified with the specific primers designed according the requirements. DNA polymerase enzyme serves basic role in PCR and PCR steps consists of the repeated cycles of denaturation, annealing and extension at specified temperatures for specific times. Annealing is a critical step in PCR reaction and conditions for annealing step should be optimized prior final identification. PCR is a highly sensitive reaction and can detect even small amounts of microbial DNA segments [24].

16.3.2.2 Metagenomics

Metagenomics involves sequencing of genetic material directly from the sample, providing the comprehensive information of the microflora present in the sample. This technique bypasses the need for culturing of the microbes and can identify a wide range of microorganisms, including those that are difficult to culture. Metagenomics is useful for studying complex microbial communities and understanding their functions within the ecosystem [23].

16.3.2.3 Next-Generation Sequencing (NGS)

NGS technologies enable high-throughput sequencing of DNA, allowing for detailed analysis of microbial diversity. NGS provides understanding about the genetic makeup of microorganisms in a sample and aids to identify novel strains. NGS technique is used along-with strong bioinformatics skills to analyze the large volumes of the data [25].

For gaining the effective results, it is recommended to use the combination of the traditional and molecular techniques i.e., initial isolation might be performed using selective culturing, followed by molecular identification using PCR or sequencing [26]. This integrated approach ensures the accurate identification and characterization of microorganisms, facilitating their potential applications in various industries.

16.4 Applications of Microorganisms Isolated from Citrus Waste

The microorganisms isolated from the citrus waste have a wide range of industrial applications including the production of biofuels and bioactive compounds, formulation of probiotics and bio-fertilizers etc. Some of the key applications are given below:

16.4.1 Biofuel Production

Biofuel production from the microbes is an efficient approach for citrus waste management. Microorganisms such as yeasts and bacteria can ferment citrus waste to produce biofuels like bioethanol and biogas. Citrus waste works as efficient substrate for the fermentation due to its nutritional composition. *Saccharomyces cerevisiae* is being commercially used for production of ethanol which is ultimately used as a renewable energy source [27]. Various parts of citrus waste i.e., peels, pulp, rag etc., contains much nutrients and organic matter that can be utilized by specific microorganisms to produce biofuels such as ethanol and biogas [28].

Following is a brief insight in the processes involved in the biofuel production:

16.4.1.1 Fermentation

Fermentation is a biological process where microorganisms, primarily yeasts and some bacteria, convert sugars and other carbohydrates into ethanol, carbon dioxide, and other byproducts [28]. The key microorganisms involved in the fermentation of citrus waste for production of biofuels include *Saccharomyces cerevisiae* and *Zymomonas mobilis*. *Saccharomyces cerevisiae* bears high efficiency in fermenting sugars into ethanol and carbon dioxide. It can utilize the sugars present in citrus waste, such as sucrose, glucose, and fructose [29]. *Zymomonas mobilis* is a bacterium which produces ethanol from sugars with higher yields and productivity compared to traditional yeast fermentation. It can be particularly useful in industrial biofuel production processes [30].

16.4.1.2 Anaerobic Digestion

Anaerobic digestion is a process where microorganisms break down organic matter in the absence of oxygen, producing biogas, consisting of methane and carbon dioxide. Anaerobic digestion involves complex microbial communities and reaction i.e., hydrolytic bacteria, Acidogenic bacteria, Acitogenic bacteria and Methanogenic Archaea. In this process, hydrolytic bacteria initiate the breakdown of complex organic compounds like carbohydrates, proteins, and lipids into simpler molecules. Acidogenic bacteria further convert the simple molecules into volatile fatty acids, alcohols, and gases such as hydrogen and carbon dioxide. Acetogenic Bacteria convert the volatile fatty acids into acetic acid, hydrogen, and carbon dioxide. Methanogenic Archaea These microorganisms convert acetic acid, hydrogen, and carbon dioxide into methane and water, producing biogas.

16.4.2 Types of Biofuels from Citrus Waste

16.4.2.1 Ethanol

Ethanol is a renewable biofuel that can be used as a substitute for gasoline in internal combustion engines. The fermentation process converts the sugars in citrus waste into ethanol, which can be purified and mixed with gasoline to produce bioethanol [31]. Ethanol production from citrus waste involves several steps. Citrus waste is pretreated to release fermentable sugars by mechanical, chemical, or enzymatic processes to break down the cell walls and release the sugars. The pretreated waste is fermented by yeasts or bacteria, converting the sugars into ethanol and carbon dioxide. The fermented mixture is distilled to separate and purify the ethanol [32].

16.4.2.2 Biogas

Biogas, mainly consist of methane and carbon dioxide, is produced through anaerobic digestion. It can be used as a renewable energy source for heating, electricity generation, and as a vehicle fuel [33]. The production of biogas from citrus waste involves several stages e.g., (i) Formation of homogenous slurry from citrus waste, (ii) induction of the slurry into an anaerobic digester, where the microbial community breaks down the organic matter, (iii) production of raw biogas, (iv) purification of raw biogas.

The pretreatment of citrus waste to release fermentable sugars can be costly and energy-intensive. Research into more efficient and cost-effective pretreatment methods is need of time to improve the overall feasibility of biofuel production. It is crucial to optimize the efficiency of the microbial fermentations and anaerobic digestion processes. Genetic engineering and microbial pool development can enhance the performance and yield of biofuels. Scaling up biofuel production from laboratory to industrial scale presents technical and logistical challenges. Integrating biofuel production with existing waste management and energy systems ultimately improving the efficiency and reduces the costs.

16.5 Probiotics from Citrus Waste

Probiotics are live microorganisms that, when delivered in adequate amounts, provide health benefits to the host. The development of novel probiotic formulations from citrus waste involves isolating beneficial bacteria, particularly lactic acid bacteria (LAB) and *Bifidobacterium*, and utilizing them to create products that can improve human and animal health [10]. Citrus waste, including peels, pulp, and juice residues, is a rich source of LAB and *Bifidobacterium* species.

Lactic acid bacteria (LAB) isolated from citrus waste exhibit potential probiotic properties. These bacteria can be formulated into probiotic products that improve gut health and boost the immune system. Probiotic strains like *Lactobacillus* and *Bifidobacterium* are particularly known for their health benefits and are used in dietary supplements and functional foods [34]. Here, we explore the steps involved in developing these formulations, the specific health benefits they offer, and their applications in various industries [35].

16.5.1 Characterization of Probiotic Properties

Probiotic strains must survive and remain active during the harsh conditions in the gastrointestinal tract, including low pH in the stomach and bile salts in the intestine. In vitro assays simulate these conditions to test the survival rate of the isolated strains. The ability of probiotics to adhere to intestinal epithelial cells is crucial for colonization and efficacy. Adhesion assays using human epithelial cell lines help determine the adhesion potential of the strains. Probiotic strains should exhibit antimicrobial activity against pathogenic bacteria. This is assessed through co-culture experiments and testing for the production of bacteriocins (antimicrobial peptides).

16.5.2 Formulation Development

Probiotics can be delivered through various carriers, including dairy products (yogurt, cheese), non-dairy products (juices, cereals), and dietary supplements (capsules, tablets). To enhance the stability and viability of probiotics during storage and gastrointestinal track, microencapsulation techniques are utilized. Encapsulation materials such as alginate, chitosan, and whey proteins protect the bacteria. Probiotic formulations undergo rigorous stability testing under different conditions (temperature, humidity) to ensure shelf-life and efficacy.

16.5.3 Health Benefits of Probiotic Formulations

Probiotics enhance gut health by maintaining a balanced gut microbiota, preventing the overgrowth of harmful bacteria, and improving intestinal barrier function. They can alleviate symptoms of gastrointestinal disorders such as irritable bowel syndrome (IBS), inflammatory bowel disease (IBD), and diarrhea [36]. They are involved to boost the immune system by enhancing the activity of immune cells, producing antimicrobial substances, and modulating inflammatory responses. This can lead to a reduced incidence of infections and improved overall immune health. Probiotics may positively influence metabolic health by improving lipid profiles, reducing inflammation, and helping in the weight management. They can help in managing conditions like obesity, type II diabetes, and various metabolic syndromes [37]. Emerging researches suggest a linkage between gut health and mental health, known as the gut-brain axis. Probiotics may improve mental health by modulating gut microbiota, reducing inflammation, and producing neuroactive compounds that affect brain function [38]. Some of microbes are given in Table 16.4 having indicating their health benefits and therapeutic properties [5, 10, 39–41].

Table 16.4 Potential probiotic bacterial strains that can be obtained from citrus waste, their modes of applications and their therapeutic effects

Probiotic Strain	Mode of Applications	Health Benefits and Therapeutic Effects
Lactobacillus plantarum	Probiotic supplements, gut health, immune support, and food preservation	Improves gut health, enhances immune function, reduces irritable bowel syndrome symptoms, and lowers cholesterol
Lactobacillus casei	Probiotic supplements, dairy products, digestive health, and immune support	Enhances digestion, supports immune function, alleviates diarrhea, and may improve mood
Lactobacillus rhamnosus	Probiotic supplements, dairy products, gut health, and immune support	Reduces the risk of gastrointestinal infections, improves gut barrier function, and supports oral health
Lactobacillus acidophilus	Probiotic supplements, yogurt, digestive health, and immune support	Aids in digestion, prevents and reduces diarrhea, supports immune health, and may help reduce cholesterol
Lactobacillus fermentum	Probiotic supplements, immune support, cholesterol reduction, and gut health	Boosts immune system, reduces cholesterol levels, and has anti-inflammatory properties
Lactobacillus reuteri	Probiotic supplements, antimicrobial production (reuterin), gut health, and oral health	Produces antimicrobial substances, improves oral health, reduces colic in infants, and enhances gut health
Leuconostoc mesenteroides	Probiotic supplements, fermentation, sauerkraut, and kimchi production	Supports digestive health, enhances immune response, and has antioxidant properties
Leuconostoc lactis	Probiotic supplements, dairy products, and fermentation	Improves gut microbiota balance, supports immune health, and has anti-inflammatory effects
Leuconostoc citreum	Probiotic supplements, production of exopolysaccharides, and biopreservation	Enhances gut health, produces beneficial exopolysaccharides, and supports immune function
Pediococcus pentosaceus	Probiotic supplements, fermentation, food preservation, and biopreservation	Improves gut health, produces bacteriocins (natural antimicrobials), and supports overall well-being
Enterococcus faecium	Probiotic supplements, antimicrobial peptide production, and digestive health	Enhances gut health, supports immune system, and helps prevent gastrointestinal infections

(continued)

Table 16.4 (continued)

Probiotic Strain	Mode of Applications	Health Benefits and Therapeutic Effects
Lactococcus lactis	Probiotic supplements, dairy fermentation, cheese production, and nisin production (natural preservative)	Improves gut health, supports immune function, and has antimicrobial properties
Bifidobacterium bifidum	Probiotic supplements, gut health, immune support, and digestive health	Enhances gut microbiota, improves digestion, boosts immune system, and helps in the prevention of infections
Bifidobacterium longum	Probiotic supplements, gut health, and immune support	Alleviates gastrointestinal disorders, supports immune health, and has anti-inflammatory effects
Bifidobacterium breve	Probiotic supplements, gut health, and respiratory health	Improves gut health, enhances immune response, and may help reduce respiratory infections
Streptococcus thermophilus	Probiotic supplements, dairy fermentation, and yogurt production	Enhances lactose digestion, supports gut health, and boosts immune function

16.5.4 Applications of Probiotic Formulations

Citrus-derived probiotics can add a unique value proposition to functional food products i.e., yogurt, cheese, milk etc. Probiotic supplements in the form of capsules, tablets, and powders may also be administered and offer a concentrated dose of beneficial bacteria. These supplements can be targeted for specific health conditions or general wellness. Along-with human consumption, probiotics are also valuable in animal nutrition, promoting gut health, improving nutrient absorption, and enhancing growth and immunity in livestock. This can lead to healthier animals and improved productivity in the agriculture industry. Probiotics are increasingly being incorporated into personal care products such as skin creams, lotions, and oral care products. They can help maintain healthy skin microbiota, reduce inflammation, and improve skin barrier function [42].

16.6 Bio-fertilizers from Citrus Waste

The global citrus industry generates vast quantities of waste, primarily peels and pulp, creating environmental challenges. However, this waste is a potential source of biofertilizers. Biofertilizers are eco-friendly alternatives to chemical fertilizers, offering numerous benefits to soil health, plant growth, and environmental sustainability. Citrus waste microbiota, comprising diverse bacteria and fungi, possess unique capabilities that can be explored and utilized to enhance crop productivity and reduce reliance on synthetic inputs [43].

Plant growth-promoting bacteria (PGPB) isolated from citrus waste can be used to produce bio-fertilizers. These bio-fertilizers enhance soil fertility, promote plant growth, and improve crop yield by fixing nitrogen, solubilizing phosphorus, and producing plant hormones. This reduces the reliance on chemical fertilizers and promotes sustainable [44]. Table 16.5 indicates some beneficial microbes from citrus waste, their mode of application to crops and their effects on crop health [44–47]. These bacteria colonize the rhizosphere and enhance plant growth through various mechanisms. PGPR in citrus waste include species like *Pseudomonas*, *Bacillus*, and

Table 16.5 Biofertilizers derived from citrus waste, detailing their applications and benefits for crop health

Microbial Agents	Mode of Applications	Effect on Crop's Health
1. Bacterial Agents		
Bacillus subtilis	Soil inoculant, foliar spray	Promotes plant growth, enhances disease resistance, and improves nutrient uptake
Pseudomonas fluorescens	Soil inoculant, seed treatment	Suppresses soil-borne pathogens, promotes root growth, and enhances nutrient availability
Azotobacter chroococcum	Nitrogen fixation, soil amendment	Fixes atmospheric nitrogen, enhances soil fertility, and improves crop yield
Rhizobium leguminosarum	Legume inoculant, nitrogen fixation	Forms symbiotic relationships with legumes, fixes nitrogen, and improves soil health
Azospirillum brasilense	Soil inoculant, seed treatment	Fixes atmospheric nitrogen, promotes root development, and enhances plant growth
Bacillus thuringiensis	Bio-pesticide, soil amendment	Produces natural insecticides, controls pests, and enhances plant growth
Pseudomonas putida	Soil inoculant, bio-remediation	Degrades organic pollutants, promotes plant growth, and improves soil health
Actinomycetes (Streptomyces spp.)	Soil inoculant, composting	Produces antibiotics, controls soil-borne pathogens, and enhances nutrient cycling
Bacillus amyloliquefaciens	Soil inoculant, foliar spray	Produces antibiotics, promotes plant growth, and enhances disease resistance
Bacillus megaterium	Phosphate solublization, soil amendment	Solubilizes phosphate, enhances nutrient availability, and promotes plant growth
Enterobacter spp.	Soil inoculant, nitrogen fixation	Fixes atmospheric nitrogen, enhances soil fertility, and promotes plant growth
2. Fungal Agents		
Aspergillus niger	Soil inoculant, enzyme production	Produces organic acids, solubilizes phosphate, and improves soil health
Penicillium spp.	Soil inoculant, composting	Solubilizes phosphate, promotes plant growth, and improves soil health
Trichoderma harzianum	Soil inoculant, composting	Controls soil-borne diseases, promotes root growth, and enhances nutrient uptake
Mycorrhizae (Glomus spp.)	Soil inoculant, root dip	Enhances nutrient and water absorption, improves soil structure, and increases plant stress tolerance

Azospirillum. They can fix atmospheric nitrogen, solubilize phosphorus, produce phytohormones (e.g., auxins, gibberellins), and enhance nutrient uptake by plants [48]. Arbuscular Mycorrhizal Fungi (AMF) form symbiotic associations with plant roots, extending their reach into the soil and improving nutrient acquisition, particularly phosphorus. They also enhance plant stress tolerance and disease resistance. Citrus waste contains AMF spores that can be incorporated into biofertilizers [49]. *Penicillium* and *Aspergillus* species secrete organic acids that solubilize insoluble phosphorus compounds in the soil, making them available for plant uptake.

16.7 Conclusion

The valorization of citrus waste presents a promising area of beneficial microbiota for sustainable development. The bacteria, yeasts, and fungi found in citrus waste offer numerous benefits, from promoting gut health and enhancing plant growth to contributing to biofuel production and sustainable agricultural practices. Using probiotic and bio-fertilizers from the organic sources can act as natural remedies for human and crop health. By exploiting the diverse microbial communities from citrus waste, we can address environmental challenges, create economic opportunities, and develop health-promoting products.

References

1. Aboagye, D., Banadda, N., Kiggundu, N., & Kabenge, I. (2017). Assessment of orange peel waste availability in Ghana and potential bio-oil yield using fast pyrolysis. *Renewable and Sustainable Energy Reviews, 1*(70), 814–821.
2. Sharma, K., Mahato, N., Cho, M. H., & Lee, Y. R. (2017). Converting citrus wastes into value-added products: Economic and environmently friendly approaches. *Nutrition, 1*(34), 29–46.
3. Calabrò, P. S., Pontoni, L., Porqueddu, I., Greco, R., Pirozzi, F., & Malpei, F. (2016). Effect of the concentration of essential oil on orange peel waste biomethanization: Preliminary batch results. *Waste Management., 1*(48), 440–447.
4. Consoli, S., Caggia, C., Russo, N., Randazzo, C. L., Continella, A., Modica, G., Cacciola, S. O., Faino, L., Reverberi, M., Baglieri, A., & Puglisi, I. (2023). Sustainable use of citrus waste as organic amendment in orange orchards. *Sustainability., 15*(3), 2482.
5. Ortinero, C. V., Mariano, A. P., Kalaw, S. P., & Rafael, R. R. (2017). Bioconversion of Citrofortunella microcarpa fruit waste into lactic acid by Lactobacillus plantarum. *Journal of Ecological Engineering, 18*(4), 35–41.
6. Agrawal, R., Satlewal, A., & Verma, A. K. (2017). Utilization of Citrus sinensis waste for the production of β-glucosidase by solid-state fermentation using a Bacillus subtilis mutant. *Environmental Engineering & Management Journal (EEMJ), 16*(7), 1465.
7. Saadoun, J. H., Ricci, A., Cirlini, M., Bancalari, E., Bernini, V., Galaverna, G., Neviani, E., & Lazzi, C. (2021). Production and recovery of volatile compounds from fermented fruit by-products with Lacticaseibacillus rhamnosus. *Food and Bioproducts Processing, 1*(128), 215–226.

8. López, M., Madrid, J., Hernández, F., Ros, M. A., Segura, J. C., López, M. J., Pallarés, F. J., Sánchez, C. J., & Martínez-Miró, S. (2021). Effect of feed supplementation with clostridium butyricum, alone or in combination with carob meal or citrus pulp, on digestive and metabolic status of piglets. *Animals, 11*(10), 2924.
9. Mavani, H. A., Tew, I. M., Wong, L., Yew, H. Z., Mahyuddin, A., Ahmad Ghazali, R., & Pow, E. H. (2020). Antimicrobial efficacy of fruit peels eco-enzyme against enterococcus faecalis: An in vitro study. *International Journal of Environmental Research and Public Health, 17*(14), 5107.
10. Fathy, H. M., Abd El-Maksoud, A. A., Cheng, W., & Elshaghabee, F. M. (2022). Value-added utilization of citrus peels in improving functional properties and probiotic viability of acidophilus-bifidus-thermophilus (ABT)-type synbiotic yoghurt during cold storage. *Food, 11*(17), 2677.
11. Ma, J., Hong, Y., Deng, L., Yi, L., & Zeng, K. (2019). Screening and characterization of lactic acid bacteria with antifungal activity against Penicillium digitatum on citrus. *Biological Control, 1*(138), 104044.
12. Van Heerden, I., Cronjé, C., Swart, S. H., & Kotzé, J. M. (2002). Microbial, chemical and physical aspects of citrus waste composting. *Bioresource Technology, 81*(1), 71–76.
13. Gade, R. M. (2012). Biological and chemical management of phytophthora root rot/collar rot in citrus nursery. *The Bioscan, 7*(4), 631–635.
14. Suri, S., Singh, A., & Nema, P. K. (2022). Current applications of citrus fruit processing waste: A scientific outlook. *Applied Food Research, 2*(1), 100050.
15. Schinner, F., & Burgstaller, W. (1989). Extraction of zinc from industrial waste by a Penicillium sp. *Applied and Environmental Microbiology, 55*(5), 1153–1156.
16. Ahmed, I., Zia, M. A., Hussain, M. A., Akram, Z., Naveed, M. T., & Nowrouzi, A. (2016). Bioprocessing of citrus waste peel for induced pectinase production by aspergillus Niger; its purification and characterization. *Journal of Radiation Research and Applied Sciences., 9*(2), 148–154.
17. Kyriakou, M., Patsalou, M., Xiaris, N., Tsevis, A., Koutsokeras, L., Constantinides, G., & Koutinas, M. (2020). Enhancing bioproduction and thermotolerance in Saccharomyces cerevisiae via cell immobilization on biochar: Application in a citrus peel waste biorefinery. *Renewable Energy, 1*(155), 53–64.
18. Sorour, N., Herzallah, S., Alkhalaileh, N., Mamkagh, A., El-Baz, A., Shalaby, E., Dmoor, H., & Abbas, R. (2023). Biofuel production by Candida tropicalis from orange peels waste using response surface methodology. *Slovak Journal of Food Sciences, 17*, 862–885.
19. Multari, S., Guzzon, R., Caruso, M., Licciardello, C., & Martens, S. (2021). Alcoholic fermentation of citrus flavedo and albedo with pure and mixed yeast strains: Physicochemical characteristics and phytochemical profiles. *LWT, 1*(144), 111133.
20. Goodrich, R. M., Braddock, R. J., Parish, M. E., & Sims, C. A. (1998). Bioconversion of citrus aroma compounds by Pichia pastoris. *Journal of Food Science, 63*(3), 445–449.
21. Blacutt, A., Ginnan, N., Dang, T., Bodaghi, S., Vidalakis, G., Ruegger, P., Peacock, B., Viravathana, P., Vieira, F. C., Drozd, C., & Jablonska, B. (2020). An in vitro pipeline for screening and selection of citrus-associated microbiota with potential anti-"Candidatus Liberibacter asiaticus" properties. *Applied and Environmental Microbiology, 86*(8), e02883–e02819.
22. Yu, S., Li, L., Zhao, H., Tu, Y., Liu, M., Jiang, L., & Zhao, Y. (2023). Characterization of the dynamic changes of ruminal microbiota colonizing citrus pomace waste during rumen incubation for volatile fatty acid production. *Microbiology Spectrum, 11*(2), e03517–e03522.
23. Mushtaq, S., Shafiq, M., Ashraf, T., Haider, M. S., Atta, S., Almaary, K. S., & Elshikh, M. S. (2022). Enumeration of citrus endophytic bacterial communities based on illumine metagenomics technique. *PLoS One, 17*(4), e0263144.
24. Peroni, L. A., Dos Reis, J. R., Della Coletta-Filho, H., de Souza, A. A., Machado, M. A., & Stach-Machado, D. R. (2008). Assessment of the diagnostic potential of Immmunocapture-PCR and Immuno-PCR for citrus variegated chlorosis. *Journal of Microbiological Methods, 75*(2), 302–307.

25. Ginnan, N. A., Dang, T., Bodaghi, S., Ruegger, P. M., Peacock, B. B., McCollum, G., England, G., Vidalakis, G., Roper, C., Rolshausen, P., & Borneman, J. (2018). Bacterial and fungal next generation sequencing datasets and metadata from citrus infected with 'Candidatus Liberibacter asiaticus'. *Phytobiomes, 2*(2), 64–70.
26. Mahato, N., Sharma, K., Sinha, M., Dhyani, A., Pathak, B., Jang, H., Park, S., Pashikanti, S., & Cho, S. (2021). Biotransformation of citrus waste-I: Production of biofuel and valuable compounds by fermentation. *PRO, 9*(2), 220.
27. Volpe, M., Panno, D., Volpe, R., & Messineo, A. (2015). Upgrade of citrus waste as a biofuel via slow pyrolysis. *Journal of Analytical and Applied Pyrolysis, 115*, 66–76.
28. Wilkins, M. R., Widmer, W. W., & Grohmann, K. (2007). Simultaneous saccharification and fermentation of citrus peel waste by Saccharomyces cerevisiae to produce ethanol. *Process Biochemistry, 42*(12), 1614–1619.
29. Joshi, S. M., Waghmare, J. S., Sonawane, K. D., & Waghmare, S. R. (2015). Bio-ethanol and bio-butanol production from orange peel waste. *Biofuels, 6*(1–2), 55–61.
30. Su, H., Tan, F., & Xu, Y. (2016). Enhancement of biogas and methanization of citrus waste via biodegradation pretreatment and subsequent optimized fermentation. *Fuel, 181*, 843–851.
31. Hu, X., Zeng, J., Shen, F., Xia, X., Tian, X., & Wu, Z. (2022). Citrus pomace fermentation with autochthonous probiotics improves its nutrient composition and antioxidant activities. *LWT, 157*, 113076.
32. Kimoto-Nira, H., Moriya, N., Nogata, Y., Sekiyama, Y., & Toguchi, Y. (2019). Fermentation of Shiikuwasha (Citrus depressa Hayata) pomace by lactic acid bacteria to generate new functional materials. *International Journal of Food Science & Technology, 54*(3), 688–695.
33. Monteiro, S. S., de Oliveira, V. M., & Pasquali, M. A. (2022). Probiotics in citrus fruits products: Health benefits and future trends for the production of functional foods—A bibliometric review. *Foods, 11*(9), 1299.
34. Sen, A., Manuel, S. G., & Kale, R. D. (2014). Fruit waste pectin in enhancing the establishment of probiotic bacteria. *Journal of Nutritional Health & Food Engineering, 1*, 124–126.
35. Satari, B., & Karimi, K. (2018). Citrus processing wastes: Environmental impacts, recent advances, and future perspectives in total valorization. *Resources, Conservation and Recycling, 129*, 153–167.
36. Kumar, A., Naroju, S. P., Langthasa, M., Behera, I., Tyagi, S., Sharma, N., & Gautam, R. K. (2023). Bioremediation potential of green wastes and plant growth promoting rhizobacteria and its enhancement by their combination: A review. *Environmental Advances, 12*, 100379.
37. Medina, A., Jakobsen, I., Vassilev, N., Azcón, R., & Larsen, J. (2007). Fermentation of sugar beet waste by aspergillus Niger facilitates growth and P uptake of external mycelium of mixed populations of arbuscular mycorrhizal fungi. *Soil Biology and Biochemistry., 39*(2), 485–492.
38. Benny, N., Shams, R., Dash, K. K., Pandey, V. K., & Bashir, O. (2023). Recent trends in utilization of citrus fruits in production of eco-enzyme. *Journal of Agriculture and Food Research, 13*, 100657.
39. Ricci, A., Díaz, A. B., Lazzi, C., & Blandino Garrido, A. M. (2023). Valorization of orange peels exploiting fungal solid-state and lacto-fermentation. *Journal of the Science of Food and Agriculture, 103*(9), 4614–4624.
40. Ras El Gherab, F. Z., Hassaine, O., Zadi-Karam, H., & Karam, N. E. (2019). Statistical optimization for the development of a culture medium based on the juice of waste-dates for growth of Lactococcus lactis LCL strain by using the Plackett–Burman and response surface methodology. *Waste and Biomass Valorization, 10*, 2943–2957.
41. Alamoudi, S. A., Saad, A. M., Alsubhi, N. H., Alrefaei, G. I., Al-Quwaie, D. A., Binothman, N., Aljadani, M., Alharbi, M., Alanazi, H., Babalghith, A. O., & Almuhayawi, M. S. (2022). Upgrading the physiochemical and sensory quality of yogurt by incorporating polyphenol-enriched citrus pomaces with antioxidant, antimicrobial, and antitumor activities. *Frontiers in Nutrition, 9*, 999581.

42. Lucia, C., Laudicina, V. A., Badalucco, L., Galati, A., Palazzolo, E., Torregrossa, M., Viviani, G., & Corsino, S. F. (2022). Challenges and opportunities for citrus wastewater management and valorisation: A review. *Journal of Environmental Management, 321*, 115924.
43. Panwar, D., Panesar, P. S., & Chopra, H. K. (2021). Recent trends on the valorization strategies for the management of citrus by-products. *Food Reviews International, 37*(1), 91–120.
44. Dorta, S. D., Balbinotte, J., Monnerat, R., Lopes, J. R., da Cunha, T., Zanardi, O. Z., de Miranda, M. P., Machado, M. A., & de Freitas-Astúa, J. (2020). Selection of Bacillus thuringiensis strains in citrus and their pathogenicity to Diaphorina citri (Hemiptera: Liviidae) nymphs. *Insect Science, 27*(3), 519–530.
45. Manasi, J., Madhura, N., & Ravindra, A. (2013). Use of citrus limetta peels for pectinase production by marine Bacillus subtilis. *Innovative Romanian Food Biotechnology, 7*(12), 75–83.
46. Steddom, K., Becker, O., & Menge, J. A. (2002). Repetitive applications of the biocontrol agent Pseudomonas putida 06909-rif/nal and effects on populations of Phytophthora parasitica in citrus orchards. *Phytopathology, 92*(8), 850–856.
47. Abd-Alla, M. H., Gabra, F. A., Danial, A. W., & Abdel-Wahab, A. M. (2019). Enhancement of biohydrogen production from sustainable orange peel wastes using Enterobacter species isolated from domestic wastewater. *International Journal of Energy Research., 43*(1), 391–404.
48. Leong, Y. K., & Chang, J. S. (2022). Valorization of fruit wastes for circular bioeconomy: Current advances, challenges, and opportunities. *Bioresource Technology, 359*, 127459.
49. Du, C., Munir, S., Abad, R., & Lu, D. (2020). Valorization of organic waste into biofertilizer and its field application. *Waste Biorefinery, 1*, 179–2020.

Chapter 17
Citrus Waste in Agriculture: Fertilizers and Soil Amendments

Zunaira Basharat, Bushra Basharat, Maria Noor, and Zahra Batool

17.1 Introduction

Citrus fruits belong to *Rutaceae* family and considered as one of the extensively cultivated fruits around the globe. Grapefruit (*Citrus paradisi*), lemon (*Citrus limon*), lime (*Citrus aurantiifolia*), mandarin (*Citrus reticulate*), and orange (*Citrus sinensis*) are of significant importance from commercial perspectives [1]. Orange production accounts for 50–60% of total citrus fruit production whereas 22, 11, and 6% are mandarin, lime and lemon, and grape fruit respectively [2, 3]. Tropical and sub-tropical regions are highly suitable for citrus production due to their promising climatic and soil conditions. More than two-third of global citrus production is contributed by the USA, China, India, Brazil, Spain, and Mexico [2, 4].

Citrus fruits are enriched with several phytochemicals and bioactive components (carotenoids, polyphenols, and flavonoids) having health-promoting attributes. Furthermore, they are excellent source of dietary fibers, folic acid, vitamin C, potassium, and pectin [5, 6]. Bioactive compounds present in citrus fruits reduce the occurrence of different chronic conditions, including cardiovascular diseases and age-related degenerative disorders. Additionally, these compounds exhibit certain medicinal characteristics such as antioxidants, anticancer and anti-tumor properties.

Z. Basharat (✉)
School of Food & Biological Engineering, Jiangsu University, Zhenjiang, China

B. Basharat
Department of Physiology, Government College University Faisalabad, Faisalabad, Pakistan

M. Noor
Department of Food Science and Technology, Government College University Faisalabad, Faisalabad, Pakistan

Z. Batool
Institute of Quality and Technology Management, University of the Punjab, Lahore, Pakistan

A. Chauhan et al. (eds.), *Valorization of Citrus Food Waste*,
https://doi.org/10.1007/978-3-031-77999-2_17

Moreover, they demonstrate antidiabetic, anti-inflammatory, and antimicrobial activities [7, 8].

Citrus fruits are widely consumed due to their enticing aroma, taste, and nutritional benefits. They equally important for industrial processing and fresh-consumption. Approximately 18% of total global citrus production is exploited for industrial purpose particularly for juice manufacturing [9]. Whereas other processing area include marmalade, jams, and jellies production, recovery of bioactive components and extraction of essential oils [10]. Processing of citrus fruits generates enormous amount of waste approximately 10 million tons annually, ranging from 50–70% of processed fruits. This citrus processing waste (CPW) entails solid/semisolid residues (seeds and peels) and waste water. The peels accounts for 50–55% of the total fruit mass whereas 20–40% are seeds. On the other hand, waste water is blend of deteriorated fruit, pulp, peels, and seeds [11, 12].

Citrus waste is reservoir of numerous compounds of profound economic impact such as essential oils, carotenoids, dietary fiber, flavonoids, polyphenols, and sugars [7]. Increased consumption of citrus fruits is associated with surge in citrus waste posing environmental burden [13]. According to Berk [14] and Zema et al. [11], citrus waste management isn't essential from economic perspectives but also required to ensure environmental safety as inappropriate disposal can lead to soil toxicity and destruction of aquatic eco-system. As stated by Calabrò et al. [12], sustainable amount of citrus processing waste and its peculiar nature poses significant challenges for proper management influenced by economic and environmental constraints.

CPW is highly perishable and extremely suitable for the growth of microbes and insects followed by mycotoxins production. Furthermore, CPW exhibit antimicrobial properties attributed to its low pH and presence of organic acids and essential oils, making its biological management problematic [15]. The conventional approaches adopted for CPW management was landfilling and incineration, now considered inadequate in terms of environment sustainability [16]. Currently various valorization approaches have been evaluated to avoid environmental damage and reduce management. However, these approaches influenced by multiple factors eventually hampering economic viability [15].

Citrus fruit waste is a bio-resource of significant amounts of bioactive components that confers many functional and therapeutic properties. Based on these properties, it can be used to prepare designer foods and food additives such as pectin, colorants, dietary fiber, and flavorings [10]. It can be used for extraction of essential oils and bioactive ingredients that can be utilized as raw materials in cosmetics [17] and pharmaceutical industry [18]. The CPW finds various non-food implication such as bio-adsorbent [19, 20], bio-fuel [3, 21], bio-fertilizer [22], soil-amendments [23], bio-degradable packaging material [24, 25], and substrate for mushroom production [26].

Furthermore, CPW can be used as soil conditioner or animal feed after simplified processing or can be subjected to bio-refineries for further processing although hindered by economic constraints [15]. This chapter is focused on sustainable use of citrus processing waste in agriculture sector peculiarly as bio-fertilizer and organic soil-amendments. Additionally, chapter will provide a brief insight to various categories of citrus processing waste and their composition. The sustainable management approaches for citrus waste and potential challenges are also discussed in this chapter.

17.2 Characterization of Citrus Waste

The major by-products citrus processing waste includes peels, pomace, seeds, and waste water that are an increasing concern for environmental sustainability [27]. As compositional insights suggested that citrus fruits and waste both are rich source of valuable functional constituent such as vitamins, organic acids, dietary fiber, pectin, limonoids, carotenoids, polyphenols, and essential oils [5, 28]. Owing to this extraordinary nutritional profile researches are focused to citrus waste valorization to reduce its environmental impact in a sustainable way.

17.2.1 Citrus Peel

Citrus peel is regarded as the primary waste product during the extraction of juice that accounts for 30–34% approximately [5]. The literature suggested the presence of various phytochemicals in citrus peels, including carbohydrates (cellulose, dietary fiber, hemicellulose, and pectin), carotenoids, essential oils (D-limonene), flavoring compounds, polyphenolic-flavonoids (hesperidin, nobiletin, naringin, poly-ethoxylate flavones, tangerine), and pigments [29].

17.2.2 Citrus Seeds

The processing of citrus fruit yields around 20–25% of the seeds as waste that are natural and economic reservoirs of certain commercially important compounds such as essential oils and proteins [14]. The literature of Rosa et al. [30], confirmed the potential applications of citrus seeds in diverse domains interlinked to its phytochemicals such as carotenoids, tocopherols, limonoids, and polyphenols.

Beyond functional properties citrus seeds are of significant nutritional value, the whole seeds constitute 58.4% fat, 23% NFE, 7% fiber, 3.15% protein, and 2.85% ash. Additionally, whole seeds provide some important minerals like calcium, potassium, and iron with 136 mg/100 g, 160 mg/100 g and 0.42 mg/100 g respectively. On the other hand, the dehulled seed have high carbohydrates (28.5%) and contribute to 614 kcal/100 g [31].

17.2.3 Citrus Pomace

Citrus pomace is defined as the leftover of the citrus extraction that includes rags, seeds, and pulp residues. According to Kim et al. [32], citrus pomace is a good source of essential oils, limonene, molasses, and pectin along with natural antioxidants such as flavonoids and polyphenols that make it of great industrial significance. Nutritional analysis conducted by Widmer et al. [33], showed that citrus pomace retains 10% moisture, 30–40% sugars, 14–25% pectin, 13–17% cellulose and hemicellulose, 0.4–1.6% fats, and 0.5% micronutrients. Among citrus fruits, orange is extensively used, the study of Espinosa-Pardo et al. [8], depicted that dried orange pomace comprise carbohydrates, protein, fat, and ash as 89.5, 5.9, 1.52, and 3% respectively.

17.3 Applications of Citrus Waste

Citrus processing waste found extensive implications in agricultural settings and biorefineries.

17.3.1 Non-Food Applications

The rich mineral profile of citrus waste enables it to be utilized as a soil amendment [34] and bio-fertilizer [35]. Furthermore, it is employed as bio-absorbent for waste water treatments [36]and heavy metal extraction from industrial waste [37]. Enormous amounts of essential oils are responsible for its extensive implication in skin care products and fragrances [30, 38]. Additionally, it has potential to be used as feed stock material for the production of animal feed [39] and biofuels [21, 40] (Fig. 17.1).

17.4 Citrus Waste as Bio-Fertilizer

The agricultural sector contributes to more than 21% of greenhouse gases emissions with a global production of about 23.7 million tons food per day [41]. The intensification in agricultural production is adversely affecting the climate by depleting natural resources. In current era, the global imperative is to mitigate the environmental degradation by implementing the measures allied with the goals established by the United Nations 2019 [42]. The European Green Deal is focused on 3R (reducing, reusing and recycling) to enhance the effective resource utilization. Hence, transition to circular economy by mitigating pollution, combating climate change, and reverting biodiversity [43].

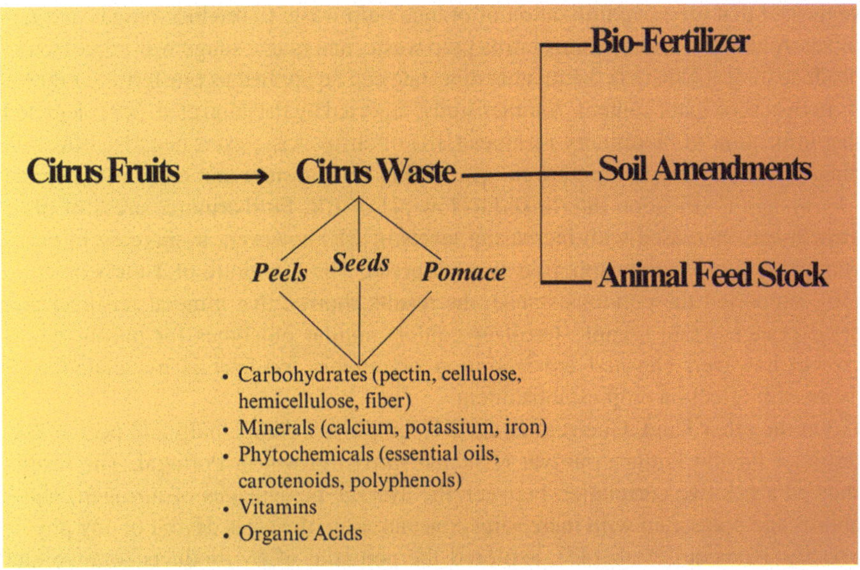

Fig. 17.1 Implications of Citrus Processing By-Products in Agriculture

The required measures have to contemplate that waste management and resource depletion should be resolved in tandem by employing green and sustainable technology for efficient use of waste as a resource with lower carbon foot prints [23, 44]. Being devoid of any toxic elements, still citrus processing waste could impart detrimental impact on environment because of its low pH and preeminent levels of polyphenols and salts [45, 46]. Subsequently, the high nutritional value of citrus waste present numerous opportunities for its utilization. The citrus processing industry faces a significant financial burden linked to its residue management, rendering it economically and environmentally non-viable, that is only possible by adopting low-cost valorization approaches and technologies.

One of the cost-effective strategies is to employ it as bio-fertilizer through composting [47]. According to Barros-Rodríguez et al. [48], fertilizer is any substance, microorganism or mixture intended to be applied on plants with the intention of improving the growth and nutritional efficacy of the plants. On the other hand, biofertilizers are defined as organic, eco-friendly substances constituting various microorganisms that are capable of improving nutrient uptake by plants through rhizosphere colonization consequently, making the nutrients available for plant root hairs. These are regarded as efficient substitutes for chemical fertilizers due to their economic viability, chemical composition, and sustainability [23].

The bio-fertilizer manufactured from citrus waste can be implicated to enhance soil fertility due to its micronutrients including carbon, nitrogen, and potassium. Furthermore, low pH and high lignocellulose content of citrus based bio-fertilizers has potent antibacterial qualities and effectively removes harmful heavy metals from soil begin a bio-absorbent [11]. The study reported by Panuccio et al. [22], suggested that effective utilization of orange pulp waste to produce biogas and fertilizer. A blend of animal dung, citrus pulp waste, and maize silage was processed to produce biogas, whereas the residue digestate can be applied as bio-fertilizer due to its high anti-oxidant content. Another study, reported by Patsalou et al. [49], depicted that utilization of biorefinery residues (BR) of citrus waste exert negative effect on plant growth by hindering nutrient availability (magnesium and calcium ions) and root system toxification interlinked to low pH of BR. Furthermore, stress of plant growth was increased with increasing levels of BR, however, an increase in nitrogen, potassium, and phosphorous was observed. The Literature of Tuttobene et al. [50], supported the previous stance, the results showed that mineral fertilizer and dried orange waste organic fertilizer confers similar outcomes for durum wheat growth however, elevated concentration of organic fertilizer is associated with depressive effect on crop establishment.

On the other hand, Guerrero et al. [51], employed orange pulp and peel as soil fertilizer for the Lettuce cultivar (*Lactuca sativa*) grown in Portugal. The results showed a positive correlation between the average percentages of nitrogen, phosphorus and potassium with increasing concentration of wastes devoid of any phytotoxicity. Panuccio et al. [47], explored the potential of by-products (orange and olive) obtained after aerobic and an-aerobic digestion as a fertilizer. They studied the impact of organic fertilizer on the soil characteristics, anti-oxidant properties and growth of garlic (*Allium sativum*). Improved soil properties were observed for all by-products however orange waste was considered more effective. Whereas, increased garlic growth and elevated anti-oxidant activity was evidenced when soil was enriched with all by-products. In another study, orange and sweet lime peels were processed into extracts and powder ($F_1 = 1$ g, $F_2 = 3$ g, $F_3 = 6$ g) to utilize as natural growth enhancer for fenugreek seeds. Durning 45 days study, best results were obtained for F1, suggesting that citrus waste can only impart beneficial impact to plant as a fertilizer, if applied in controlled concentrations [52].

Qader et al. [53], evaluated the influence of orange and banana peels on different growth indicators of chickpea plants. The organic fertilizer was applied to plants at various concentration levels as powder, foliar spray, and extract. The results elucidated significant improvement in growth indicators such as plant height, branch number, pods number, dry weight, total chlorophyll content and carotenoids. Another study was conducted by El Barnossi et al. [54], to evaluate the potential implication of fruit peels (tangerine, banana, pomegranate) as an eco-friendly bio-organic fertilizer. They studied the effect of individual fruit peel extracts and their mixtures on germination and growth parameters of *Pisum sativum* at various stages of decomposition. The obtained results confirmed the positive impact of extract on growth and germination of *P.sativum*. Furthermore, the microbiological and physiochemical characterization established the stance that fruit peel extracts can be exploited as natural substitute of chemical fertilizers. Khairnar et al. [55], also proposed sweet lime peel powders a sustainable and cost-effective fertilizer based on their mineral content (Table 17.1).

Table 17.1 Application of citrus processing waste as fertilizer

Composition of Organic Fertilizer	Observations	Outcomes	References
Biorefinery residues of citrus waste	Plant growth Phytotoxicity	Restricted plant growth due to impaired nutrient availability Root system toxification associated with low pH Plant growth stress increased at higher concentrations N, K, P increased	[49]
Dried orange waste fertilizer	Growth indicators of durum wheat	Higher administration levels exert negative effect on crop establishment	[50]
Orange by-products of aerobic and an-aerobic digestion	Soil characteristics Anti-oxidant capacity of garlic	Improved soil properties Elevated anti-oxidant activity Augmented growth	[47]
Orange and lime peel (powders & extracts)	Growth enhancement in fenugreek seeds	Positive effect only at low concentrations	[52]
Orange and banana peels (powder, extract & foliar spray)	Growth indicators for chickpea plants	Significant improvements in height, pods number, dry weight, total chlorophyll content and carotenoids.	[53]
Tangerine, banana, pomegranate peels (sole extracts/ combined extracts)	Germination and growth parameters of *Pisum sativum*	Positive results on germination and growth rate	[54]

17.5 Citrus Waste as Soil Amendments

The ability of the soil to sustain and facilitate crop production while protecting the environment is referred to as soil health when referring to agricultural land [56]. Healthy soil is capable of yielding good results, responding favorably to agricultural inputs, and withstanding biotic or environmental stresses [57]. Maintaining organic matter and microbial load in soil is crucial for nutrient recycling. A decrease in soil fertility is the primary cause of soil erosion, compaction, salinity, and nitrogen loss. Therefore, the decrease in organic matter levels is the primary driver of all of the aforementioned soil degradation processes [58].

Maintaining and ensuring the proper amount of soil organic matter and the effective cycling of nutrients are essential for the success of agricultural productivity [59, 60]. Hence, the sustainable agriculture is designated as fusion of biological, chemical, and physical processes to design novel approaches that are eco-friendly and economically viable [61]. Among them, implication of soil amendments is adopted to optimize crop yield and efficient nutrients utilization. Soil amendments are defined as substances that increase soil fertility and crop yield begin direct nutrient source and/or modify soil cation exchange capacity, soil pH, soil porosity, surface interactions, and water retention to enhance nutrient availability [62, 63].

Anaerobic digestate, biochar, biofertilizer, compost, engineered-biochar, vermicompost are categorized as soil amendments [64].

Citrus industry produces significant amount of citrus sewage sludges (CSSs), if properly treated is considered an excellent source of nitrogen, phosphorous and certain macronutrients that makes it a potential material to use as organic fertilizer. Lucia et al. [65], studied the effect of (CSSs) on soil fertility and lettuce performance. They concluded that irrespective of applied dosage, CSSs based amendment improvise soil fertility by enhancing total organic C. However, elevated microbial biomass and P availability was evident only at maximum dose. Ultimately, enhancing the biomass production without compromising its quality. Belligno et al. [66], accessed the efficacy of orange waste as soil conditioners for durum wheat production and growth in Mediterranean region. The results showed an overall positive impact of conditioning on various soil parameters however, higher dosage administration induced plant stress (vegetation vigour) limiting the crop productivity.

Another study, the utilization of citrus waste was investigated to augment phosphorus availability and facilitate pH adjustments in soils when used alone or combined with cattle manure. It was observed that citrus incorporation caused a short-lived acidification followed by alkalization with immediate elevation of water extractable phosphorous. The study confirmed that citrus waste enrichment in cattle manure amendment will increase the P availability in soil as well as exert the alkalization effect on acidic soils [67]. Similar outcomes were reported by Gharaie [68], he stated that increasing concentrations of lemon waste enhances minerals (P, Fe, Zn, Cu, Mn) availability in treated soils. Navarro et al. [69], designed biochar materials using grape fruit peels to evaluate their efficacy in immobilization of copper and lead in germination columns. Additionally, the effect of biochar on germination and biomass production of radish seed was also investigated. The findings suggested that GP biochar can be a potential soil additive that allows heavy metal retention while enhancing the vegetation establishment.

Karimi et al. [70], reported the utilization of dried grapefruit waste to mitigate the root knot-nematode *Meloidogyne javanica* in tomato roots. They also evaluated the effect of amendments on chlorophyll, total phenolics, C:N ratio, and growth indicators of infected plants. The outcomes elucidated that there is a positive correlation between increasing concentration of waste and growth indices. Furthermore, infected plants showed an increase in leaf chlorophyll and root phenols on implication of amendments. Digested Assam lemon waste was processed into bio-manure using *F. muscicola* (cyanobacterial culture). The efficacy of enriched bio-manure for mung bean (*Vigna radiata*) production was evaluated alone or combined with inorganic fertilizer. The combined fertilizer augmented the seed yield while regulating the physio-chemical properties of soil [71]. Moon et al. [72], studied the impact of carbonized tangerines prepared via plasma technology on the chemical variations in soil and growth attributes of lettuce. They found that plasma technology could be a promising alternate for by-product processing with minimal effect on soil composition and elevated crop growth characteristics (Table 17.2).

Table 17.2 Application of citrus processing waste as soil amendment

Composition of Organic Soil amendment	Observations	Outcomes	References
Citrus sewage sludges-based soil amendments	Soil fertility Lettuce growth	Improved total organic carbon content Only at higher dosage, microbial biomass and P availability increased. Increased biomass production	[65]
Orange waste soil conditioner	Soil parameters for durum wheat	Higher concentration induces plant stress, restricted crop productivity	[66]
Citrus waste CW combined with cattle manure	P-availability pH adjustments	Enhanced P-availability indicated by water extractable phosphorous Alkalization of acidic soils	[67]
Grapefruit peel bio-char	Cu and Pb immobilization Reddish seeds germination and yield	99% copper and lead retention Improved vegetation conditions	[69]
Dried grape fruit waste	Root-knot nematode mitigation in tomato roots Growth indicators of infected plants	Growth indicators improved with increasing dosage Augmented leaf chlorophyll and root phenolics	[70]
Lemon waste bio-manure with cyanobacteria (*F. muscicola*) + inorganic fertilizer	Mung bean production	Enhanced seed yield Improved soil characteristics	[71]
Plasma treated carbonized tangerine leftovers	Chemical modifications in soil Lettuce growth performance	Non-significant variations in soil composition Elevated crop growth	[72]

17.6 Challenges and Considerations

17.6.1 Environmental Impact

The biggest challenge offered by citrus processing industry is improper disposal of citrus processing waste that is associated with serious environmental concerns. Organic citrus waste contains elevated levels of carbohydrates that make it super fermentable and hence cause of microbial spoilage and consequently promotes infection [73]. Furthermore, traditional strategies such as landfilling and incineration are not considered energy efficient and eco-friendly [16]. According to EU Report (2008), citrus waste cannot be dumped into landfills without prior processing [74].

17.6.2 Mitigation Strategies

Adoption of suitable valorization approach is challenging because of various constraints, most importantly the economic viability and environmental sustainability of process. Previously adopted methods such as landfilling and incineration were highly associated with adverse environmental foot prints and also incurs high processing cost [15]. However, current valorisation approaches such as production of development of value-added products [7], nano-particles fabrication [75], inference as natural fertilizer, soil-amendment and bio-adsorbent can be suggested as green approaches that supports the narrative of bio-circular economy [34].

17.7 Future Prospects

Currently, citrus waste has been tremendously used in various industries such as cosmetics, food, and pharmaceuticals due to its bioactive components that imparts health endorsing attributes as well as functional characteristic to products. It is emerged as a sustainable source of biofertilizers, biosorbents, biopolymers, biofuels, bioethanol, biogas, and animal feed [10, 76]. In future, the implication of citrus processing waste as bio-fertilizer or soil amendments entails various promising directions. The efficient and targeted applications require detailed research on nutrient composition of citrus waste and its influence on soil strength. Furthermore, innovative processing methods should be developed to design specialized citrus fertilizers with improved nutritional value. Citrus waste as soil amendments should be integrated with precision agricultural practices to encounter the specific requirements of crops and soil conditions subsequently, ensuring the optimized inference.

The sustainable waste management of citrus waste requires adoption of circular economy approach, where it is recycled with in agricultural systems with out hampering environmental safety. Additionally, commercial processing of citrus waste into soil conditioners and bio-fertilizers will enhance its economic viability and farmers accessibility. The future insinuation of citrus processing waste in agriculture whether as soil amendments or fertilizers requires collaborative efforts across several sectors including a combination of technological advancements, sustainable practices, and regulatory support. Through considering economic, environmental, and social aspects the possible benefits of citrus waste can be augmented while diminishing its environmental footprint.

17.8 Conclusion

Citrus processing industry generates enormous amounts of citrus waste, if disposed without proper waste management could be a serious environmental hazard. The low pH of citrus waste and high fermentability are key factors that renders citrus waste management a challenging task. Traditional approaches include landfilling and incineration both are not considered cost-effective and eco-friendly management strategies. On the other hand, citrus waste is enriched with variety of bioactive and functional components like cellulose, polyphenols, essential oils, pectin, flavonoids, vitamins, and minerals that makes it a valuable raw material for various processing sectors such as agriculture, food industry, cosmetic industry, and pharmaceuticals. The adoption of close loop management or bio-circular economy allows the utilization of citrus processing waste in a sustainable manner. In agricultural setting citrus processing waste can be exploited as organic fertilizer or soil amendments after simple processing. Citrus based organic fertilizers or soil amendments improve soil characteristic, nutrient availability, and plant growth. Furthermore, elevated crop yields are obtained, without any quality defects due to increased biomass production. However, inference of organic fertilizer or soil amendments should be regulated as higher concentration of citrus waste biofertilizers can cause depressive effect on plant growth. It can be concluded, that citrus processing waste can be used as eco-friendly and cost-effective substitute of chemical fertilizers. Furthermore, this approach will reduce the carbon foot prints of conventional citrus waste management strategies with high energy and economic efficiency.

References

1. Rashid, U., Ibrahim, M., Yasin, S., Yunus, R., Taufiq-Yap, Y. H., & Knothe, G. (2013). Biodiesel from Citrus reticulata (mandarin orange) seed oil, a potential non-food feedstock. *Industrial Crops and Products, 45*, 355–359.
2. Satari, B., & Karimi, K. (2018). Citrus processing wastes: Environmental impacts, recent advances, and future perspectives in total valorization. *Resources, Conservation and Recycling, 129*, 153–167.
3. Taghizadeh-Alisaraei, A., Hosseini, S. H., Ghobadian, B., & Motevali, A. (2017). Biofuel production from citrus wastes: A feasibility study in Iran. *Renewable and Sustainable Energy Reviews, 69*, 1100–1112.
4. Paggiola, G., Stempvoort, S. V., Bustamante, J., Barbero, J. M. V., Hunt, A. J., & Clark, J. H. (2016). Can bio-based chemicals meet demand? Global and regional case-study around citrus waste-derived limonene as a solvent for cleaning applications. *Biofuels, Bioproducts and Biorefining, 10*(6), 686–698.
5. Rafiq, S., Kaul, R., Sofi, S. A., Bashir, N., Nazir, F., & Nayik, G. A. (2018). Citrus peel as a source of functional ingredient: A review. *Journal of the Saudi Society of Agricultural Sciences, 17*(4), 351–358.

6. Mahato, N., Sharma, K., Sinha, M., & Cho, M. H. (2018). Citrus waste derived nutra–/pharmaceuticals for health benefits: Current trends and future perspectives. *Journal of Functional Foods, 40*, 307–316.
7. Khan, U. M., Sameen, A., Aadil, R. M., Shahid, M., Sezen, S., Zarrabi, A., et al. (2021). Citrus genus and its waste utilization: A review on health-promoting activities and industrial application. *Evidence-based Complementary and Alternative Medicine, 2021*, 1–17.
8. Espinosa-Pardo, F. A., Nakajima, V. M., Macedo, G. A., Macedo, J. A., & Martinez, J. (2017). Extraction of phenolic compounds from dry and fermented orange pomace using supercritical CO2 and cosolvents. *Food and Bioproducts Processing, 101*, 1–10. https://doi.org/10.1016/j.fbp.2016.10.002
9. Fresh, F. C. F. (2017). *Processed statistical bulletin 2016*. Rome, Italy.
10. Suri, S., Singh, A., & Nema, P. K. (2022). Current applications of citrus fruit processing waste: A scientific outlook. *Applied Food Research, 2*(1), 100050.
11. Zema, D. A., Calabrò, P. S., Folino, A., Tamburino, V. I. N. C. E. N. Z. O., Zappia, G., & Zimbone, S. M. (2018b). Valorisation of citrus processing waste: A review. Waste management, 80, 252–273
12. Calabrò, P. S., Pontoni, L., Porqueddu, I., Greco, R., Pirozzi, F., & Malpei, F. (2016). Effect of the concentration of essential oil on orange peel waste biomethanization: Preliminary batch results. *Waste Management, 48*, 440–447.
13. Mahawar, M. K., Jalgaonkar, K., Bibwe, B., Bhushan, B., Meena, V. S., & Sonkar, R. K. (2020). Post-harvest processing and valorization of Kinnow mandarin (Citrus reticulate L.): A review. *Journal of Food Science and Technology, 57*, 799–815.
14. Berk, Z. (2016b). By-products of the citrus processing industry. In Z. Berk (Ed.), *Citrus fruit processing* (pp. 219–233). Academic. https://doi.org/10.1016/b978-0-12-803133-9.00010
15. Zema, D. A., Calabrò, P. S., Folino, A., Tamburino, V. I. N. C. E. N. Z. O., Zappia, G., & Zimbone, S. M. (2018b). Valorisation of citrus processing waste: A review. *Waste Management, 80*, 252–273.
16. Wei, Y., Li, J., Shi, D., Liu, G., Zhao, Y., & Shimaoka, T. (2017). Environmental challenges impeding the composting of biodegradable municipal solid waste: A critical review. *Resources, Conservation and Recycling, 122*, 51–65.
17. Pinto, D., de la Luz Cádiz-Gurrea, M., Silva, A. M., Delerue-Matos, C., & Rodrigues, F. (2021). Cosmetics—Food waste recovery. In *Food waste recovery* (pp. 503–528). Academic.
18. Sharma, K., Mahato, N., Cho, M. H., & Lee, Y. R. (2017). Converting citrus wastes into value-added products: Economic and environmently friendly approaches. *Nutrition, 34*, 29–46.
19. Villen-Guzman, M., Cerrillo-Gonzalez, M. D. M., Paz-Garcia, J. M., Rodriguez-Maroto, J. M., & Arhoun, B. (2021). Valorization of lemon peel waste as biosorbent for the simultaneous removal of nickel and cadmium from industrial effluents. *Environmental Technology & Innovation, 21*, 101380.
20. Akkaya Sayğılı, G., Sayğılı, H., Yılmaz, C., & Güzel, F. (2020). Lead recovery from aqueous environment by using porous carbon of citrus fruits waste: Equilibrium, kinetics and thermodynamic studies. *Separation Science and Technology, 55*(15), 2699–2712.
21. John, I., Pola, J., Thanabalan, M., & Appusamy, A. (2020). Bioethanol production from musambi peel by acid catalyzed steam pretreatment and enzymatic saccharification: Optimization of delignification using taguchi design. *Waste and Biomass Valorization, 11*(6), 2631–2643. https://doi.org/10.1007/s12649-018-00565-x
22. Panuccio, M. R., Attinà, E., Basile, C., Mallamaci, C., & Muscolo, A. (2016). Use of recalcitrant agriculture wastes to produce biogas and feasible biofertilizer. *Waste and Biomass Valorization, 7*, 267–280.
23. Consoli, S., Caggia, C., Russo, N., Randazzo, C. L., Continella, A., Modica, G., Cacciola, S. O., Faino, L., Reverberi, M., & Baglieri, A. (2023). Sustainable use of citrus waste as organic amendment in orange orchards. *Sustainability, 15*(3), 2482.
24. Jridi, M., Abdelhedi, O., Salem, A., Kechaou, H., Nasri, M., & Menchari, Y. (2020). Physicochemical, antioxidant and antibacterial properties of fish gelatin-based edible films enriched with orange peel pectin: Wrapping application. *Food Hydrocolloids, 103*, 105688.

25. Dash, K. K., Ali, N. A., Das, D., & Mohanta, D. (2019). Thorough evaluation of sweet potato starch and lemon-waste pectin based-edible films with nano-titania inclusions for food packaging applications. *International Journal of Biological Macromolecules, 139*, 449–458.
26. Kumla, J., Suwannarach, N., Sujarit, K., Penkhrue, W., Kakumyan, P., Jatuwong, K., et al. (2020). Cultivation of mushrooms and their lignocellulolytic enzyme production through the utilization of agro-industrial waste. *Molecules, 25*(12), 2811.
27. Zheng, X., Zhu, K., Sun, Q., Zhang, W., Wang, X., Cao, H., Tan, M., Xie, Z., Zeng, Y., Ye, J., Chai, L., Xu, Q., Pan, Z., Xiao, S., Fraser, P. D., & Deng, X. (2019). Natural variation in CCD4 promoter underpins species-specific evolution of red coloration in citrus peel. *Molecular Plant, 12*, 1294–1307.
28. Di Donna, L., Bartella, L., De Vero, L., Gullo, M., Giuffre, A. M., Zappia, C., et al. (2020). Vinegar production from citrus bergamia by-products and preservation of bioactive compounds. *European Food Research and Technology, 246*(10), 1981–1990. https://doi.org/10.1007/s00217-020-03549-1
29. Wang, X., Chen, Q., & Lu, X. (2014). Pectin extracted from apple pomace and citrus peel by subcritical water. *Food Hydrocolloids, 38*, 129–137. Berk, Z. (2016a). Introduction: History, production, trade, and utilization. In Z. Berk (Ed.), *Citrus fruit processing* (pp. 1–8). Academic Press.
30. Rosa, A., Era, B., Masala, C., Nieddu, M., Scano, P., Fais, A., et al. (2019). Supercritical CO2 extraction of waste citrus seeds: Chemical composition, nutritional and biological properties of edible fixed oils. *European Journal of Lipid Science and Technology, 121*(7), Article 1800502. https://doi.org/10.1002/ejlt.201800502
31. Akpata, M. I., & Akubor, P. I. (1999). Chemical composition and selected functional properties of sweet orange (Citrus sinensis) seed flour. *Plant Foods for Human Nutrition, 54*(4), 353–362. https://doi.org/10.1023/A:1008153228280
32. Kim, J. W., Lee, B. C., Lee, J. H., Nam, K. C., & Lee, S. C. (2008). Effect of electronbeam irradiation on the antioxidant activity of extracts from Citrus unshiu pomaces. *Radiation Physics and Chemistry, 77*(1), 87–91. https://doi.org/10.1016/j.radphyschem.2007.02.082
33. Widmer, W., Zhou, W., & Grohmann, K. (2010). Pretreatment effects on orange processing waste for making ethanol by simultaneous saccharification and fermentation. *Bioresource Technology, 101*(14), 5242–5249. https://doi.org/10.1016/j.biortech.2009.12.038
34. Suri, S., Singh, A., & Nema, P. K. (2021). Recent advances in valorization of citrus fruits processing waste: A way forward towards environmental sustainability. *Food Science and Biotechnology, 30*(13), 1601–1626. https://doi.org/10.1007/s10068-021-00984-y
35. Zema, D. A., Andiloro, S., Bombino, G., Caridi, A., Sidari, R., & Tamburino, V. (2016). Comparing different schemes of agricultural wastewater lagooning: Depuration performance and microbiological characteristics. *Water, Air, & Soil Pollution, 227*, 1–9.
36. Panwar, D., Saini, A., Panesar, P. S., & Chopra, H. K. (2021). Unraveling the scientific perspectives of citrus by-products utilization: Progress towards circular economy. *Trends in Food Science & Technology, 111*, 549–562. https://doi.org/10.1016/j.tifs.2021.03.018
37. Li, M., & Buschle-Diller, G. (2017). Pectin-blended anionic polysaccharide films for cationic contaminant sorption from water. *International Journal of Biological Macromolecules, 101*, 481–489. https://doi.org/10.1016/j.ijbiomac.2017.03.091
38. Atolani, O., Adamu, N., Oguntoye, O. S., Zubair, M. F., Fabiyi, O. A., Oyegoke, R. A., et al. (2020). Chemical characterization, antioxidant, cytotoxicity, anti-toxoplasma gondii and antimicrobial potentials of the Citrus sinensis seed oil for sustainable cosmeceutical production. *Heliyon, 6*(2), e03399 Article. https://doi.org/10.1016/j.heliyon.2020.e03399
39. Seidavi, A., Zaker-Esteghamati, H., & Salem, A. Z. (2020). A review on practical applications of Citrus sinensis by-products and waste in poultry feeding. *Agroforestry Systems, 94*(4), 1581–1589. https://doi.org/10.1007/s10457-018-0319-2
40. Ezekoye, V., Adinde, R., Ezekoye, D., & Ofomatah, A. (2019). Syntheses and characterization of biodiesel from citrus sinensis seed oil. *Scientific African, 6*, e00217 Article. https://doi.org/10.1016/j.sciaf.2019.e00217

41. Gerber, P. J., Steinfeld, H., Henderson, B., Mottet, A., Opio, C., Dijkman, J., et al. (2013). *Tackling climate change through livestock: A global assessment of emissions and mitigation opportunities*. Food and Agriculture Organization of the United Nations (FAO).
42. Global Sustainable Development Report. (2019). *The future is now—Science for achieving sustainable development*. United Nations.
43. European Commission. (2020). *Circular economy action plan. For a cleaner and more competitive Europe* (Vol. 28). Publications Office of the European Union.
44. O'Connor, G. (2021). Shifting the value of food and organic waste management in the food services sector in Brisbane, Australia. *Resources, Conservation & Recycling Advances, 12*, 200052.
45. Khdair, A. I., Abu-Rumman, G., & Khdair, S. I. (2019). Pollution estimation from olive mills wastewater in Jordan. *Heliyon, 5*, e02386.
46. Ashraf, M. A., Maah, M. J., & Yusoff, I. (2014). Soil contamination, risk assessment and remediation. In *Environmental risk assessment of soil contamination* (Vol. 1, pp. 3–56). Intech Open.
47. Panuccio, M. R., Marra, F., Maffia, A., Mallamaci, C., & Muscolo, A. (2022). Recycling of agricultural (orange and olive) bio-wastes into ecofriendly fertilizers for improving soil and garlic quality. *Resources, Conservation & Recycling Advances, 15*, 200083.
48. Barros-Rodríguez, A., Rangseekaew, P., Lasudee, K., Pathom-Aree, W., & Manzanera, M. (2020). Regulatory risks associated with bacteria as biostimulants and biofertilizers in the frame of the European Regulation (EU) 2019/1009. *Science of the Total Environment, 740*, 140239.
49. Patsalou, M., Chrysargyris, A., Tzortzakis, N., & Koutinas, M. (2020). A biorefinery for conversion of citrus peel waste into essential oils, pectin, fertilizer and succinic acid via different fermentation strategies. *Waste Management, 113*, 469–477.
50. Tuttobene, R., Avola, G., Gresta, F., & Abbate, V. (2009). Industrial orange waste as organic fertilizer in durum wheat. *Agronomy for Sustainable Development, 29*, 557–563.
51. Guerrero, C. C., de Brito, J. C., Lapa, N., & Oliveira, J. S. (1995). Re-use of industrial orange wastes as organic fertilizers. *Bioresource Technology, 53*(1), 43–51.
52. Mercy, S., Mubsira, B. S., & Jenifer, I. (2014). Application of different fruit peels formulations as a natural fertilizer for plant growth. *International Journal of Scientific & Technology Research, 3*(1), 300–307.
53. Qader, H. (2019). Influence combination of fruits Peel and fertilizer methods on growth and yield of chickpea (Cicer areitinum) L. Plants. *ZANCO Journal of Pure and Applied Sciences, 31*(3), 45–51.
54. El Barnossi, A., Moussaid, F. Z., Saghrouchni, H., Zoubi, B., & Iraqi Housseini, A. I. (2022). Tangerine, pomegranate, and banana peels: A promising environmentally friendly bioorganic fertilizers for seed germination and cultivation of Pisum sativum L. *Waste and Biomass Valorization, 13*(8), 3611–3627.
55. Khairnar, M. D., & Nair, S. S. (2019, September). Study on eggshell and fruit peels as a fertilizer. In *Proceedings of international conference on sustainable development* (Vol. 978, pp. 25–27). Springer.
56. Doran, J. W., & Zeiss, M. R. (2000). Soil health and sustainability: Managing the biotic component of soil quality. *Applied Soil Ecology, 15*(1), 3–11.
57. Srivastava, A. K., & Ngullie, E. (2009). Integrated nutrient management: Theory and practice. *Dynamic Soil, Dynamic Plant, 3*(1), 1–30.
58. Lal, R. (2015). Sequestering carbon and increasing productivity by conservation agriculture. *Journal of Soil and Water Conservation, 70*(3), 55A–62A.
59. Vanlauwe, B., Bationo, A., Chianu, J., Giller, K. E., Merckx, R., Mokwunye, U., Ohiokpehai, O., Pypers, P., Tabo, R., & Shepherd, K. D. (2010). Integrated soil fertility management: Operational definition and consequences for implementation and dissemination. *Outlook on Agriculture, 39*(1), 17–24.

60. Bationo, A., Kihara, J., Vanlauwe, B., Waswa, B., & Kimetu, J. (2007). Soil organic carbon dynamics, functions and management in West African agro-ecosystems. *Agricultural Systems, 94*(1), 13–25.
61. Lichtfouse, E., Navarrete, M., Debaeke, P., Souchère, V., Alberola, C., & Ménassieu, J. (2009). Agronomy for sustainable agriculture: A review. *Agronomy for Sustainable Agriculture, 29*, 1–7.
62. El Barnossi, A., Moussaid, F., & Housseini, A. I. (2021). Tangerine, banana and pomegranate peels valorisation for sustainable environment: A review. *Biotechnology Reports, 29*, e00574.
63. Lu, J., & Xu, S. (2021). Post-treatment of food waste digestate towards land application: A review. *Journal of Cleaner Production, 303*, 127033.
64. Palansooriya, K. N., Dissanayake, P. D., Igalavithana, A. D., Tang, R., Cai, Y., & Chang, S. X. (2023). Converting food waste into soil amendments for improving soil sustainability and crop productivity: A review. *Science of the Total Environment, 881*, 163311.
65. Lucia, C., Pampinella, D., Palazzolo, E., Badalucco, L., & Laudicina, V. A. (2023). From waste to resources: Sewage sludges from the citrus processing industry to improve soil fertility and performance of lettuce (Lactuca sativa L.). *Agriculture, 13*(4), 913.
66. Belligno, A., Di Leo, M. G., Marchese, M., & Tuttobene, R. (2005). Effects of industrial orange wastes on soil characteristics and on growth and production of durum wheat. *Agronomy for Sustainable Development, 25*(1), 129–135.
67. Paredes, C., Staunton, S., Simeone, G. D. R., Rao, M. A., Violante, A., & de la Luz Mora, M. (2021). Citrus residue enhances the effectiveness of beef cattle manure improving the phosphorus availability in acidic andisol. *Communications in Soil Science and Plant Analysis, 52*(21), 2767–2781.
68. Gharaie, H. (2008). Effect of lemon waste on soil pH and availability of micronutrient in calcareous soils of Fars province, southern Iran.
69. Navarro, M. A. (2022). *Grapefruit Peel Biochar applications for immobilizing copper and lead in soil*. [Doctoral dissertation, The University of Texas Rio Grande Valley].
70. Karimi, S., Azami-Sardooei, Z., & Fekraf, F. (2022). Effects of the grapefruit wastes and sesame oil cake to control of the root-knot nematode Meloidogyne javanica on tomato plants.
71. Kundu, D., Banerjee, S., Karmakar, S., & Banerjee, R. (2021). Enrichment of N and bioavailability of P and K of lemon wastes through biotechnological intervention with special reference to Mung bean production. *Bioresource Technology Reports, 15*, 100794.
72. Moon, S. Y., Kim, H. S., Jang, D., & Lee, J. Y. (2021). Effect of carbonized tangerines prepared using plasma as fertilizer: Case study—Lettuce farming. *Sustainability, 14*(1), 243.
73. Lin, C. S. K., Pfaltzgraff, L. A., Herrero-Davila, L., Mubofu, E. B., Abderrahim, S., Clark, J. H., et al. (2013). Food waste as a valuable resource for the production of chemicals, materials and fuels. Current situation and global perspective. *Energy & Environmental Science, 6*(2), 426–464.
74. Negro, V., Ruggeri, B., Fino, D., & Tonini, D. (2017). Life cycle assessment of orange peel waste management. *Resources, Conservation and Recycling, 127*, 148–158.
75. Marino, M., Lopes da Silva, L., Duran, N., & Tasic, L. (2015). Enhanced materials from nature: Nanocellulose from citrus waste. *Molecules (Basel, Switzerland), 20*(4), 5908–5923. https://doi.org/10.3390/molecules20045908
76. Varmie, E. B., & Thakur, M. (2021). Utilization of citrus processing waste: A review. *The Pharma Innovation Journal, 10*, 682–697.

Chapter 18
The Flavor of Sustainability: Citrus Waste in Distillation

Aliza Batool, Umar Farooq, Nida Firdous, Afshan Shafi, Hamna Saqib, Ramsha Akram, and Hunaina Saleem

18.1 Introduction

In recent decades, improvements in living standards have resulted in an unabated increase in the global population. This has led to an increase intake of expedients and, as a result, a significant amount of trash being produced. Anything that is not utilized to its full potential or is unused is referred to as "waste." Thus, ineffective trash disposal management may lead to excessive waste generation. A substantial amount nearly about 120 million tons of citrus waste is produced during processing by the industrial processing of nearly 5% of all citrus cultivars, mostly for the manufacturing of their juices, which are the most popular fruit liquids consumed globally [1]. The remainder, which includes the peel (flavedo), which makes up 27% of the fruit weight, the pulp (albedo and endocarp), which makes up 26% and the seeds (2%), which make up the remainder, are disposed of. These numbers are further increased by whole fruits that are thrown because they do not meet quality standards. Simultaneously, pruning is a common technique used to eliminate tree branchlet or enhance the fruit quality, suggesting the formation of many leaves a residue that ascending the vast amount of Citrus waste [2].

This is frequently tossed into landfills or rivers, especially in underdeveloped nations, where it contaminates the water and lowers the amounts of dissolved oxygen. Due to the simple fermentation method, citrus waste becomes the high in pollution; it is copious, biodegradable, and chemically complex, requiring a lot of oxygen to break down. It typically has a pH of 3–4, is heavily concentrated in

A. Batool · U. Farooq · N. Firdous (✉) · H. Saqib · R. Akram · H. Saleem
Faculty of Food and Home Sciences, MNS-University of Agriculture, Multan, Pakistan
e-mail: Nida.firdous@mnsuam.edu.pk

A. Shafi
Department of Food Sciences, Cholistan University of Veterinary and Animal Sciences, Bahawalpur, Pakistan

A. Chauhan et al. (eds.), *Valorization of Citrus Food Waste*,
https://doi.org/10.1007/978-3-031-77999-2_18

80–90% water, and contains organic matter is 95% (total residue). Moreover, alternative for getting rid of CF waste is to burn it to produce thermal energy. This is a risky process that produces a lot of carbon, nitrogen, and sulfur oxides into the atmosphere, so it is no longer a good way to get rid of CF waste. Therefore, a number of alternatives of Citrus Fruit waste were put forth in an effort to lessen environmental pollution and generate revenue for the food for better management, cosmetic industries, and pharmaceutical. These included the production organic fertilizers, enzymes, biofuel and ethanol, macro and micronutrients extraction, in confectionary products fiber-rich ingredients is used and as well as the animal fortified feed is use [3].

The establishment of waste biorefineries in developed nations has the potential to yield substantial economic and environmental advantages. These advantages include the recovery of value-added product and energy, reduction of greenhouse gas emissions, creation of new jobs, savings on land, soil, and groundwater, energy, and so on. However, biorefineries frequently have to pay for certain expenses while processing CF waste. For instance, for handling and transportation to the bio-refinery the assuming cost is 10 USD/ton, the projected total cost to ethanol manufacturing is utilize 100,000 tons of garbage per year would be 0.91 USD/L. However, Citrus Fruit waste is regarded as a significant, sustainable, and economically advantageous source for the health, cosmetic industries (i.e., for the production of functional foods, Nutraceuticals, and skin care products like essential oils, body lotions, soaps, and body sprays), and pharmaceutical because of it heterogeneity. Furthermore, by implementing these re-use tactics, the food industry may be able to lower waste generation and the associated expenses of disposing of it as organic matter, resulting in the creation of new commercial goods such as food packaging, antimicrobials, encapsulating agents, additives, and prebiotics [4].

18.2 Deviation of Citrus Waste

Oranges, lemons, grapefruits, mandarins, and limes are widely grown citrus fruits globally. Production increases annually as a result of rising consumer needs. The citrus processing industries generate a lot of garbage every year; just the half of the wet mass of fruit is peel waste. Food waste is made up of components that may be consumed by humans but are spoiled, contaminated, and thrown away. A growing problem that impacts every aspect of waste management is food waste. All parties included involved such as agriculture, retailer, consumers, business, and the collection to disposal processes, in the food supply chain, must collaborate in order to provide long-term solutions [5].

The processing of Citrus fruit produces the variety of wastes, such as solid, liquid, and distillery effluents. Solid trash includes things like residue, rags, peels, and sludge. Different kinds of liquid waste do exist, nevertheless, such as can cooler overflows, floor flushing water, fruit washing and peeling wastewaters, cannery effluents and sectioning table. In the food manufacturing procedure, the juices or drinks from the citrus is basically considered the main use of citrus. As a result, peels, pulp, and seeds from citrus are wasted annually in large quantities. On the other hand, a large number of bioactive chemicals may be present in the pulp, seeds, and skins of many vegetables and fruits. The restoration of antioxidants, phenolic compounds and vitamin C activity from citrus fruit waste (orange, grapefruit, and lemon) was studied previously. Ethanol extracts were made from the peel, entire fruit, and pulp containing seeds of every citrus fruit. For each citrus species, the peels were shown to have larger quantities of vitamin C, phenolic compounds, flavonoids, and antioxidant activity than the internal, discarded components (pulp and seeds). Citrus japonica (kumquat), Citrus reticulata (Phlegraean mandarin), and Citrus clementina (peel, pulp, and seeds) were compared and their vitamin C, total polyphenol content, photosynthetic pigment content, and antioxidant activity, were all measured [6].

18.2.1 Citrus Peel

Citrus peel waste (CPW) is especially has become a viable and sustainable choice for biorefinery without competing with human diets and animal feeds in the effort to reduce global food losses and waste. Recent research indicates that CPW is produced in large quantities and has the potential to biologically transform the chemicals into fuels at industry level [7].

The most popular usage of CPW is for the extraction of essential oils (limonene) as a ingredients in food and cosmetic additives and pectin. All across the world, agricultural activities generate a lot of fruit waste. This trash is often just dumped into the ocean or landfills. Bioethanol is make from the fruit waste because that contain a wide range of sugar that can undergoes in the process of fermentation, including fructose, sucrose, and glucose. Certain fruit wastes, like citrus peel waste (CPW), can having ingredients that are essential producing bioethanol efficiently, that might halt fermentation. An innovative technique to convert CPW from one of the following sources (Orange, Mandarin, Grapefruit, Lemon or Lime) or CPW coupled with CPW from one of the following sources (Orange, Mandarin, Grapefruit, Lemon or Lime) or CPW combined with orange [8].

18.2.2 Citrus Seeds

To extract phytoconstituents from the seeds, a number of downstream processes must be optimized once the seeds are collected, washed, dried, and ground. In downstream processes some of the steps including solvents, particle size, time, pressure, and temperature. Citrus seeds are extracted using a variety of methods, including as ultrasonic-assisted extraction, cold pressing, solvent extraction and supercritical CO2 extraction [9].

Because they are considered to be useless, citrus seeds are usually thrown away as waste. State that 4.8% of dried citrus pulp typically contains seeds. Due to their high protein content, these wasted seeds can be profitably supplemented with protein for cattle. Citrus seeds also contain 30% (by weight) oil, which makes them highly promising for use as biodiesel. The flour made from dehulled and unhulled citrus seeds has the following approximate composition: 28.5% carbs (dry basis), 5.5% crude fiber, 5.2% fat, 2.5% ash, and 3.1% crude protein used sodium methoxide to catalyze the trans-esterification of citrus seed oil with methanol [10].

18.2.3 Citrus Pomace

The surplus material left over after processing of juice or other products from the citrus fruit is called citrus pomace. Approximately half of the volume of citrus fruits disposed during the processing of citrus fruits into industrial citrus juice is citrus pomace, which poses major ecological issues. The pomace produced by the agro-fruit business when insufficient citrus is used has a detrimental effect occur on the environment and will cause large loss of financially. Pomegranate fruits are abundant in a range of health-promoting bioactive elements, such as phenolic compounds, polysaccharides, dietary fiber, carbs, phytochemicals, and natural antioxidants [11].

In a prior investigation, the impact of diverse solvents was assessed concerning the retrieval of extractable solids (ES), vitamin C, flavonoids, and phenolic compounds from discarded lemon pomace. The outcomes demonstrated how several solvents, including acetone, methanol, ethanol, and water, affected the results. A variety of solvents was used to assess the antioxidant activity, total flavonoids, vitamin C, and total phenolic components of lemon pomace. For dried citrus pomace, the main components are 10% moisture, 0.4–1.6% fat, 30–40% sugars, 14–25% pectin, 0.5% micronutrients, and 13–17% hemicellulose and cellulose [12].

18.3 Distillation

The most traditional extraction/separation techniques have been employed for many years in the desalination of water and the creation of alcoholic spirits. Distillation is now used more widely for drying, purifying, and extracting more thermolabile chemicals thanks to the development of vacuum pumps in the last century. The boiling point of distilled compounds is lowered by around 25 degrees for every tenfold drop in pressure when distillation is carried out under reduced pressure. Thus, in high vacuum distillation, compounds with high boiling points and those that are thermally degradable can be recovered as a distillate. In most laboratories, using rotatory evaporators to concentrate and dry extracts during distillation has become standard procedure, with vacuum applied during the process [13].

It is showed in many studies that the researchers are discussed about the utilization of citrus peel as a food application is either value addition, recovery of bioactive compounds or fortification of citrus peel in functional foods are discussed in following.

18.3.1 Distillation Techniques

Distillation is a process that is used for the conversion of any compound from liquid to vapour phase. Distillation is done by many different methods or techniques to get the suitable distillation of our research work. Citrus waste is generated from the processing of citrus fruits like oranges, lemons, and limes, gives an opportunity for value addition through distillation techniques. These techniques aid to extract valuable compounds such as essential oils, pectin, and bioactive compounds from citrus waste while minimizing environmental effect. Some methods are employed for citrus waste distillation offering its advantages and applications Fig. 18.1 [14].

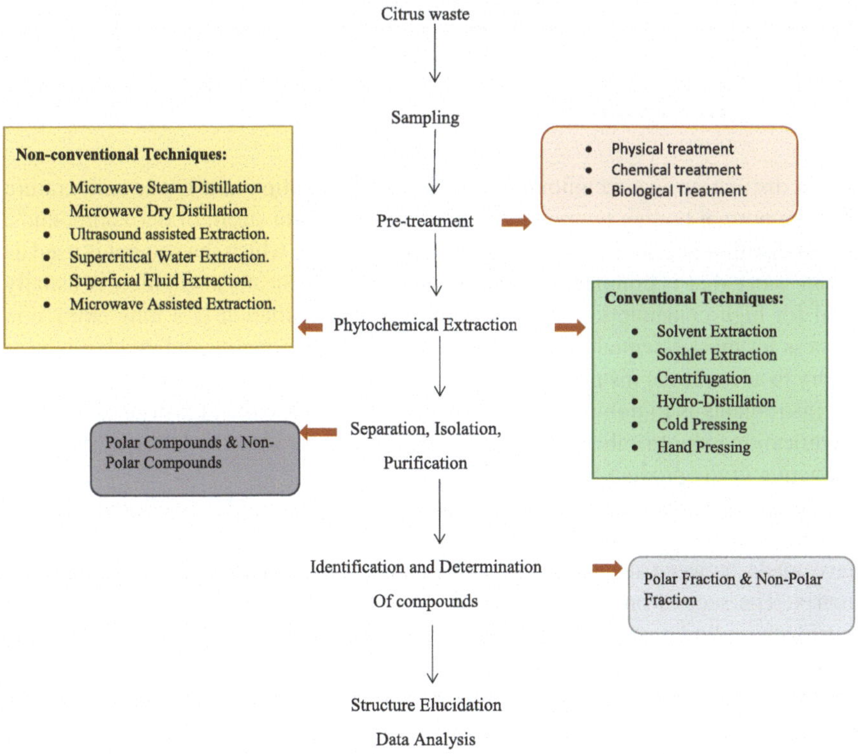

Fig. 18.1 Extraction of different compound from citrus by product

18.3.1.1 Solvent Extraction

Organic solvents such as ethanol and hexane are extracted from this technique. Moreover, this technique uses the organic solvents for the extraction of the bioactive compounds and essential oils through the citrus by-product waste. The waste is typically macerated or subjected to cold pressing to release the oils, which are then dissolved in the solvent. After extraction, the solvent is evaporated to obtain concentrated extracts. Solvent extraction offers high extraction efficiency and allows for the recovery of a maximum range of citrus waste compounds [14].

18.3.1.2 Supercritical Fluid Extraction (SFE)

Supercritical fluid extraction applies carbon dioxide (CO2) as a solvent under supercritical conditions to extract essential oils from citrus waste. In this method, CO2 is compressed above its critical pressure and temperature; keep in view the properties of both liquid and gas. CO2 acts as a selective solvent, extracting main compounds while leaving undesirable components behind. Supercritical fluid extraction offers high selectivity, low solvent residues, and the possibility of fractionation [14].

18.3.2 Steam Distillation

Steam distillation is a technique in which water is boiling at a lower temperature than the normal boiling temperature to get the maximum steam of our compounds. Steam distillation is an environment-friendly method in which the solvent is used as steam water that is compared to the organic solvent. Steam distillation is usually used for high consumption and low yield. Steam distillation is a long extraction process for most essential oil industries. To get a higher quality and yield, it is necessary to advance the instruments for steam distillation. The main reason to advance the instruments is to minimize energy usage and increase the industrial yield. This modification helps to enhance the steam distillation operations and the quality of the extraction of the product Fig. 18.2 [15].

Two types of steam distillation are operated in industries. First is dry distillation in which steam is created outside the cylinder at medium pressure so no contact can have made between the steam and water and steam can easily pass through the matrix. The second one is known as direct steam distillation, steam is generated outside the cylinder and in this type, there is no thermal degradation [16]. One of the common techniques that are used for extraction of the essential oils from the by-product waste of citrus is known as steam distillation. In this process citrus waste is used. Steam is passed through the citrus waste by-product which helps to break the oil glands and also release the aromatic compounds. The steam carrying the volatile oils is then condensed, allowing the separation of essential oils from the aqueous phase. Steam distillation is preferred for its simplicity, efficiency, and ability to produce high-quality essential oils [16].

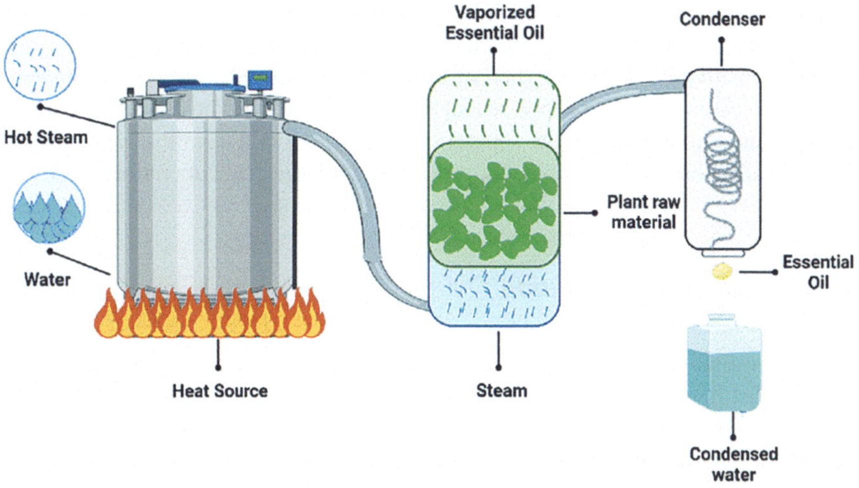

Fig. 18.2 Steps Involved in the steam distillation

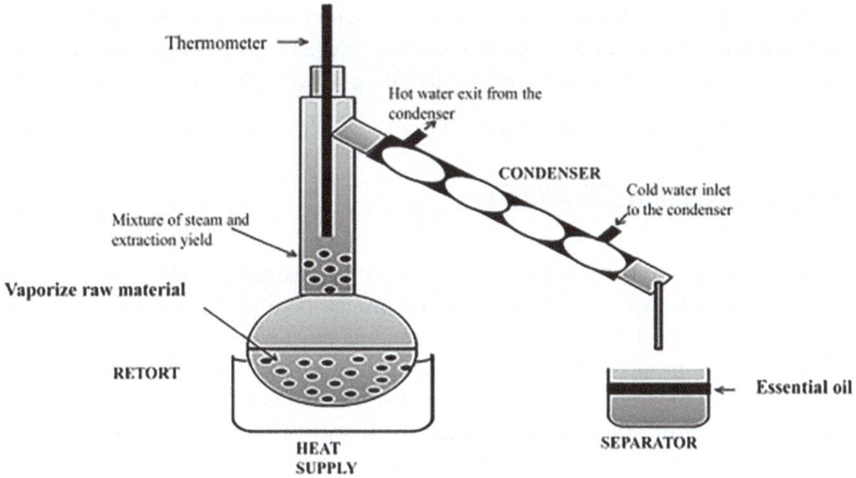

Fig. 18.3 Overview of Hydro-distillation

18.3.3 Hydro Distillation

In the hydro-distillation technique, water directly contacts the matrix. It is more like a liquid-solid interaction. Due to the chances of hydrolysis in some compounds during hydro-distillation, it is performed in a controlled environment [17]. It is simply used to get the volatile compounds. In the hydro distillation technique, Clevenger apparatus is used to extract the volatile compounds. In the Clevenger apparatus, the result that is obtained is the two layers, one layer is volatile compounds and the other is an aqueous layer. The procedure of the Clevenger apparatus is that the water is boiled in the cylinder to evaporate the volatile compounds. To separate the volatile compounds (polar) is a very difficult procedure Fig. 18.3 [18].

Hydro distillation involves the direct boiling of citrus waste in water to extract essential oils. Unlike steam distillation, hydro distillation uses water as the solvent for heat-sensitive compounds. The citrus waste is dissolved in water and heated, causing the essential oils to vaporize and rise with the steam. The oil-laden steam is then condensed, resulting in the separation of essential oils from the water phase. Hydro-distillation is a cheap method, especially for large-scale operations [19].

18.3.4 Vacuum Distillation

In vacuum distillation, a high vacuum is used the decrease the normal pressure. Vacuum distillation is utilized for compounds that are very sensitive and easily degrade in the normal atmosphere or boil at very high temperatures. There exist substances with rather high boiling points. Even at temperatures below the boiling point, they are nevertheless susceptible to distortion. The process of extracting these chemicals under atmospheric pressure is challenging and insufficient. Therefore, in order to prevent deterioration, it is more beneficial to lower the pressure rather than raise the temperature; this process is known as vacuum distillation. It is well known that when the boiling point drops in response to a drop in external pressure, the compound's vapor pressure is dropped and boiling with distillation commences [20].

18.3.5 Molecular Distillation

A physical separation technique called molecular distillation (MD) was created with a number of uses in the food, pharmaceutical, petrochemical, and cosmetics sectors in mind. Vegetable oil refinement found its most widespread early commercial applications in the food and petrochemical industries. Over the last 20 years, MD has attracted increasing interest and its uses in the manufacturing of nutraceuticals and the purification and enrichment of bioactive components from natural product extracts have grown. In order to prevent thermal damage, special instruments with a tiny distance between the evaporator and condenser were created during the 1930s for high vacuum distillation (<1 mbar). This allowed the target compounds to be heated for only a few seconds. In order to ensure consistent heating and boost evaporation efficiency, the mixture to be distilled (the feed) is spread out as a thin layer across a heated surface in order to achieve this short time. Short path distillation gets its name from the process of allowing evaporated components (the distillate, light fraction) to condense on a cooled surface (the condenser) positioned close together to prevent the distillate from returning to the evaporation surface. Molecular distillation, or MD, is the term used when the distillation gap, or the distance between the evaporating surface and the condenser, is near to the mean free route of evaporated molecules. This prevents other molecules from clashing with the vapor molecules as they approach the condenser surface. Because molecules do not collide with one another or return to the evaporating surface after reaching the condensation surface, they are eliminated as distillate, making molecular distillation a non-equilibrium process [21].

Since its creation, MD has been used extensively in the food, cosmetics, and petrochemical sectors. When used on an industrial scale, MD separations and purifications of oils and extracts offer a number of advantages over other methods of separation. These advantages include the use of no organic solvent, which lowers environmental hazards, the possibility of applying the process to substances with a wide range of molecular weights and vapor pressures, including molecules with high boiling points, suitability for thermolabile substances, and the ability to apply the process at different scales due to continuous feed application. These days, one of the two primary evaporators is employed in MD for the recovery, purification, and concentration of various chemicals. The first is the falling film method, which uses rolling wipers and gravity to apply and distribute feed on the heated walls of a cylindrical evaporator (still). The distillate is condensed and pumped out of the system through the cooling condenser, which is housed inside the still. In the second technique, known as the centrifugal short path evaporator, feed is positioned in the center of a spinning, heated plate and distributed uniformly across its surface using centrifugal force. Where distillate vapors condense and gather is a short distance from a cooled surface [22].

18.3.6 Membrane Distillation

In the thermally induced separation process known as membrane distillation (MD), only vapour molecules pass through a microporous hydrophobic membrane. The vapour pressure differential caused by the temperature differential across the hydrophobic membrane is what propels the MD process. There are several uses for this procedure, including the food sector, wastewater treatment, and desalination. When using membrane distillation, MD A microporous hydrophobic membrane's two sides have different vapour pressures, which encourages the transfer of water vapor. Surface tensions prevent aqueous solutions from penetrating the membrane's pores unless a transmembrane pressure is applied that is higher than the liquid entry pressure of the membrane [23]. Nevertheless, in MD, a temperature gradient rather than a concentration gradient is the physical source of the vapour pressure difference since cold water is employed as a stripping permeate and the feed is kept at a high temperature. Moreover, the membrane's mechanical resistance is significantly decreased. Another benefit of MD over conventional pressure-driven membrane processes is its high rejection rate (theoretically 100%) against ions, macromolecules, colloids, cells, and other non-volatile substances because it functions on the principles of vapour-liquid equilibrium. The primary disadvantage of the MD process is the same as that of the OD process: there is a chance that the hydrophobic membrane will get wet, which could reduce permeate flux if the pore is partially soaked, or negatively impact permeate quality if the pore is fully soaked. The process solutions must therefore be watery and appropriately diluted. This restricts the use of MD to processes like desalination, concentrating non-volatile aqueous solutions, and extracting trace volatile components from effluent [24].

Direct contact membrane distillation (DCMD), in which the feed and the permeate liquid are in direct contact with the membrane in their respective compartments, is the most popular configuration of MD utilized for the concentration of liquid meals. Within the membrane module, volatile molecules condense on the cold liquid/vapour interface and evaporate at the hot liquid/vapor interface. They then traverse the membrane pores in the vapour phase. Vacuum is provided to the permeate side of the membrane in vacuum membrane distillation (VMD) using a vacuum pump, and condensation occurs outside the membrane module. An air gap separates the condensing surface from the membrane in the air gap membrane distillation (AGMD) process. Both the air and the membrane pores are crossed by volatile molecules [25].

18.3.7 Product Obtained from Distillation

The by-product of the citrus is very nutritious compounds that contain pectin, dietary fibers and flavonoids which are used in many food industries as an additive, functional foods ingredients and food thickeners and emulsifier. So, it is proof that oil extraction from the by-products of citrus can easily volarized [26]. An inventive, economical, and successful way to use processed fruit waste for clean and green production has been the focus of much research. The scientific community is interested in the extraction techniques used for sample preparation and processing. The first steps in recovering bioactive chemicals are the extraction, separation, and characterization of bioactive components from fruit waste. To guarantee that the possible bioactive components are not lost, altered, or destroyed during the extraction process, the proper steps should be taken (Table 18.1). The type of bioactive substance being studied should also be taken into consideration when choosing a

Table 18.1 Application of waste of citrus in different food industry

Source of Citrus	Part	Composition	Food Application	References
Citron	Peel, Pulp	Vitamins C, Flavonoids, Antioxidants, Phenols	Carbonated drinks, Syrups, Jams, Beverages, Alcoholic.	[49]
Lemon	Peel	Vitamin C, Phenols, Carotenoids, Pectin	Use as film matrixes, Edible coatings.	[50]
Sweet Orange	Peel	Quinic acid, Sakuranetin, Narirutin, Naringin	Desserts, candied orange peel, Salad dressings, oxidative stability, Enhance vegetable oil.	[51, 52]
Grapefruit	Peel	Naringin, Phenols, Flavonoids, Naringenin	Sauces, Jams making, Flavouring of beverages, Dessert recipes.	[53]

solvent. Polar solvents like methanol, ethanol, or ethyl acetate are used to extract compounds that are soluble in water, while dichloromethane or a 1:1 mixture of dichloromethane and methanol is used to extract compounds that are soluble in fat. Additionally, it is necessary to verify the suitability of the extraction method used to recover particular bioactive chemicals. The adoption of green extraction procedures has gained significant relevance due to their increased performance, limited or zero use of organic solvents, and lower costs associated with time, energy, and other resources [27].

18.3.7.1 Polyphenols

Citrus fruit varieties vary in the amount of polyphenols they contain. In citrus peels, the primary phenolic chemicals found are flavonoids, which include flavonols, flavanones, flavones, isoflavones, and anthocyanidins. Green extraction techniques are more effective than traditional methods for extracting phenols from citrus peel, according to a number of studies. When phenols were extracted from citrus peel using ultrasonic assistance, the maximum yield was 26.52%, 1.77 times greater than that of microwave assistance. Maceration and ultrasound-assisted extraction used to extract polyphenols from the peels of kinnow mandarin (*Citrus reticulata*) and mousambi (*Citrus limetta*), with the latter approach proving to be more effective than the former. The extract of citrus peels showed high levels of hesperidin, ranging from 280 to 673 mg/g. Furthermore, ultra critical CO 2 extraction was used to extract polyphenols from citrus peel, and the waste citrus peel included a high concentration of hesperidin. However, at high temperatures, 5-hydroxymethylfurfural (5-HMF) and chlorogenic acid also found [28].

Using an ultrasound-assisted extraction technique, the extraction of polyphenols from three hybrid cultivars of mandarin (Clemenvilla, Ortanique, and Nadorcott) was studied at a power of 400 W, temperature of 40 °C, and duty cycle of 80%. Shorter ultrasonic extraction times resulted in better mandarin peel extract physicochemical profile, bioactive component content, and antioxidant activity. This proves that ultrasonic technology is a practical and sustainable extraction technique [29].

18.3.7.2 Pectin

Pectin is a polysachharide that is used as a thickening agent or stabilizer in different food industries. Pectin is most commonly used in acidified milk beverages as a stabilizer. They are very popular because of their good nutritional profile and taste. They are also used as stabilizer in jam or jallies. Pectin is used as a coating on the strawberries and it increases the shelf life of strawberries [30].

In the food business, citrus peel pectin is utilised as a gelling agent when making jams and jellies. Remaining peel from citrus peels is used to make around 85% of the commercial pectin produced worldwide. Pectin can be made using fresh citrus fruit peels, such as those from lemon, grapefruit, and oranges, or dried and cleaned peels. 1.5% to 3% of pectin can be found in fresh citrus peels, and 9% to 18% can be found in dried peels. It was fascinating to note that researchers experimented with unconventional methods to extract pectin from orange peel. Pectin was extracted using a microwave at different power levels (160–480 W), liquid-solid ratios (1: 10–1: 30 g/ml), and irradiation times (60–180 s). The results showed that a pectin yield of 25.79% was achieved at power levels (477 W), times (128 s), and solid-liquid ratios (1:20). Furthermore, a pectin yield of 28% observed at power (400 W), time (7 min), and pH (1.2) in Citrus sinensis peels pectin recovery using surfactant-based microwave-assisted extraction [31].

The pectin yield from the sequential microwave-assisted solvent extraction of tangerine (Kinnow) peel 27.58%. Furthermore, jams can be made with the pectin that is extracted from citrus peels. Pectin is used in food as emulsifier or thickening agent in different food industries sector such as in supplements of dietary fiber and confectionery. For the extraction of pectin from citrus waste, citrus peels are heated at 100 degree centigrade then acidic conditions are given. Citrus dried peel is added in beaker which contains water and acid and heated for 45 minutes. With the help of filter paper, they are separated [32].

18.3.7.3 Essential Oils

Aromatic liquids composed of a variety of volatile compounds are called essential oils. Usually, citrus peels' cuticles and oil sacs contain them. Flavors and perfumes are made with citrus peel essential oil. The production and chemical makeup of citrus peel essential oil, however, differ according on the extraction technique employed. Limonene (1.26%) recovered from the peel of a citrus lime using a solvent-free microwave extraction approach. Using the supercritical fluid method for oil extraction from citrus limon peels produced a higher yield of d-limonene (4.5%). Mandarin (Citrus unshiu Kuno) peels were subjected to a super critical CO2 extraction process, which yielded a high yield of 30.65% at 300 bar pressure and 13.16% at 100 bar pressure. The extremely crucial method was effective in extracting oil. The main drawback of super critical fluid extraction, however, was the high cost and maintenance of the equipment [33].

18.4 Dietary Fiber

As a nutritional supplement and an element in natural and modified meals, dietary fiber has become very popular recently. Rich in both soluble and insoluble fiber, orange peel fiber is well-known for its possible dietary and health advantages. Citrus peels are utilized in the literature to extract dietary fibre by a range of technologies, such as enzymes, dry/wet treatment, acid-alkali treatment, ultra-sonication, and combinations of these techniques [34]. Citrus sinensis peels were successfully used in a study to extract dietary fibre using the steam explosion and diluted acid soak method. The yield of soluble fibre (33.74%) used as a potential food ingredient. Therefore, it is evident from the literature reviewed above that citrus peels have the potential to be used to recover bioactive chemicals using energy-efficient extraction techniques. However, there is still a lot of study to be done in this area to investigate the physiologically active substances derived from citrus peels and how they work to prevent disease [35].

18.4.1 Food Additive

Food additives are compounds added to food products to give them flavour, colour, taste, and nutritional value. The use of natural additives derived from agro-food processing waste is becoming more and more important as the demand for organic food products rises. Bioactive components with antioxidant activity found in food processing industry waste may provide a natural alternative to artificial food additives [36].

Citrus peel and rind are commonly used as food additives in the confectionary and bakery sectors to prepare candied goods because of their distinct flavour, colour, and nutritional value. In 2020, the global market for bakery products was estimated to be valued at approximately USD 507.38 billion. In addition, the bakery industry is predicted to grow at a compound annual growth rate (CAGR) of 2.5% in the upcoming years, reaching a valuation of approximately USD 574.05 billion by 2026. In the bakery industry worldwide, biscuits hold the most market share, followed by bread & rolls, cakes & pastries, and rusk. Technology advancements allowed for the processing and incorporation of various citrus waste pieces into baked goods, increasing their nutritional content and making efficient use of waste. For example, muffins were made with 10% and 15% of orange bagasse, which is high in nutritional fiber. The produced muffins (15%) had a low (15%) fat content and a high (15.3%) dietary fiber content. The addition of dietary fiber-rich orange bagasse was found to reduce the anticipated glycemic index. Additionally, wheat flour was combined with 2.50 percent fiber and 0.23% pectin from citrus peels in order to prepare bread. Citrus peels that had been processed, both fermented and unfermented, were added to the dough to make bread. The addition of 4% and 6% of unfermented peels that had been separately dried at 150 °C and 100 °C showed higher levels of oral acceptance [37].

Similarly, different amounts of debittered orange waste fibre (30, 50, and 70%) utilised to substitute fat in baked goods (brioches, a type of bread). Comparing the 50% debittered orange fibre inclusion to the 10% fat control brioche without any debittered orange fibre, the fat content increased by 4.5%. The desired fiber content, technological, and microbiological characteristics were noted from the first day of manufacturing to the fifth day of storage [38].

Citrus peel is also widely used in food applications, where peel pectin is used to prepare fruit jams, jellies, and low-fat/low-sugar products as a thickener, stabiliser, emulsifier, texturizer, and gelling agent. In addition, citrus peels are also used as a component in the creation of probiotic beverages, gelato, and other foods. Made ice cream with 3% unbleached and 5% frozen blanched kinnow peel. The writers noted improvements to the citrus peel ice cream's flavour, look, and general appeal. Perez-created a probiotic dietary fiber-rich salami using probiotic *Lactobacillus rhamnosus* strain, citrus fibre (1.5%), and herbal extracts of coffee, tea, and olives (0.3%), The addition of citrus fibre significantly increased the salami's short-chain fatty acids and antioxidant activity. Pectin from citrus peels is also used as a hydrolysate. As a dietary fibre source for the manufacturing of liquid food products, particularly beverages, the study suggested that pectin hydrolysates with improved solubility have a lot of potential. Moreover, citrus fruit waste is rich in carotenoid, which the food and beverage industry might employ as a natural pigment [39].

18.4.2 Prebiotic

Citrus peel pectin is used to make jam, jellies, and marmalades, but it also has potent prebiotic properties. Oligosaccharides derived from citrus pectin seem to be a dynamic prebiotic component. Prebiotics are becoming more and more important in daily life. They are usually referred to as "a non-digestible food material that provides benefits to the host by specifically stimulating the growth and/or activity of one or more number of microorganisms in the colon, and further improves host health" [40].

Citrus peel waste has been shown in a number of recent studies to have potential applications in the production of prebiotic foods with enhanced bio functionality. Pectic oligosaccharides (POS), originating from the partial hydrolysis of pectin, are a newly discovered prebiotic found in agricultural waste, such as peels from citrus fruits (Citrus bergamia). The strain of Acetobacter aceti was used to prepare vinegar at a pH of 2.9. The study demonstrated how bergamot might be used to make citrus vinegar with a high concentration of bioactive substances including brutieridin and melitidin and utilizing suitable micro- bial starters. Orange peel fibre may be used to create high-value oligosaccharides. When compared to pectin from the commercial market, pectin derived from the enzymatic hydrolysis of orange peel-based fibre was found to have higher prebiotic activity. Therefore, pectic oligosaccharides derived from citrus trash can be used as emergent prebiotics with increased action [41].

18.4.3 Mushroom Production

Mushrooms are spore-forming saprophytic fungi that can originate from either an epigeous or hypogenous source and have a characteristic body. These mushrooms significantly contribute to human nutrition because of their high protein content, vitamin and mineral content, and therapeutic properties. In a regulated context, mushrooms are cultivated using the agro-industrial or lingo-cellulosic waste leftover as a substrate. A notable biotechnology method for efficiently valuing agricultural and food processing waste is the creation of mushrooms. By converting agricultural leftovers employing a range of microorganisms, it has additionally proven its strength in ecological and economic aspects [42].

The world's waste management issues are lessened when mushrooms are grown from agricultural and industrial trash. Furthermore, it represents the biological process by which this waste is transformed into palatable protein. For example, lingo-cellulosic agricultural waste has been extensively studied for its potential use as an inexpensive substrate for mushroom production. The method of producing mushrooms (Pleurotus ostreatus) by solid-phase fermentation with leftover ingredients from the food processing sector, such as rice straw, papaya fruit waste, and lemon pulp. According to the study's findings, fruiting bodies had dry weight amounts of fat (2.0–5.9%), digestible protein (26.0–31.5%), and total sugars (20.9–33.0%). For every 100 grammes of dried mushrooms, the fruiting bodies had an energy value ranging from 189 to 253 Kcal. Dried fruit peels from oranges, pineapples, avocados, mangos, bananas, and watermelons were used to cultivate the mushrooms, with wheat straw acting as a control substrate. The study found that the fungus derived from fruit waste has a high content of total tannins, total flavonoids, and total phenols, which are antioxidants. Therefore, using fruit waste to grow mushrooms is a cost-effective waste management technique [43].

18.4.4 Single Cell Protein

Citrus waste can be utilised to make single-cell proteins in addition to being included into a variety of functional foods. The majority of dried cells produced by microorganisms, yeast, bacteria, fungi, etc. are thought to be these single-cell proteins. These can also be referred to as biomass, microbial protein, or bio-protein [44]. Since single-cell proteins don't require as much land or water to produce as vegetable protein sources, they can be a great substitute. The most crucial step in lowering the cost of producing single-cell proteins is choosing inexpensive, appropriate substrates or biodegradable food and agricultural waste as a source of nutrients for microorganisms to proliferate and produce huge quantities of protein. For this, a variety of substrates have been employed. Different fruit waste use in the synthesis of single-cell protein. They used Saccharomyces cerevisiae and submerged fermentation to produce single-cell protein utilising orange and cucumber peels as a substrate. The findings indicated that by using the right bacteria, this fruit waste could be transformed into SCP. The commercially and nutritionally valuable products derived from the biotransformation of agricultural waste are high in protein [45].

18.4.5 Encapsulating Agent

Since more research has been done in the area of nanotechnology in recent years, citrus peel pectin has been used as an encapsulant. Other polymer blends are also being investigated as encapsulating agents in an effort to improve the stability and efficacy of encapsulation [46]. For example, an oil emulsion based on citrus peel waste was created using ascorbic acid, citric acid, Ca2+ ion, and pectin, which is a hydro-soluble component of citrus waste. According to the investigation, the produced emulsion had a smaller droplet size and was stable. Additionally, zein and pectin from citrus peels were utilised to create resveratrol-enriched nanoparticles. Furthermore, compared to free resveratrol, the in-vitro studies indicated that encapsulated resveratrol had a high bio-accessibility. Additionally, an improvement in resveratrol's antioxidant qualities. Consequently, it is possible that the citrus by-products could be used as an encapsulant [47].

18.4.6 Livestock Feed

Citrus processing waste has been used in several experiments as a livestock feed source. Feeding grill chickens orange peel increased the concentration of lymphocytes, white blood cells, immunoglobulin G, and immunoglobulin M. Similarly, the effects of adding kinnow waste (Citrus reticulate) in goat feed and nutrition utilization. According to the study, adding kinnow waste almost 40% resulted in an increase in the ratio of calcium to phosphorus, which is something that should always be considered when feeding animals. However, a blood biochemical profile revealed that there was no detrimental impact of kinnow waste on the goats' health. Orange peel extract to quail diets increased body weight, decreased triglyceride levels, and enhanced the ratio of feed conversion. Orange peel waste has as a sanitary, healthful, and modest feed for chicken diets. However, because too much consumption of the citrus waste could upset the livestock's digestive system, it is advised to use it sparingly in animal feed [48].

18.5 Recent Advances in Valorisation of Citrus

18.5.1 Waste Management of Citrus to Develop Eco-Friendly Environment

Citrus is one of the popular fruit which is commonly used in most of the countries. Citrus contain very high amount of bioactive components which help in different fields such as medical and other allied sciences like therapeutic field. Many processes are take place in industries on citrus but the main focus that industries are doing on the production of citrus juice production but as a by-product many waste are produce such as seeds, peeling of citrus, wastewater and pomace. But the waste management in industries are poor so there is high risk of environmental pollution. So, maintain the waste management in industries of citrus and provide the sustainable and polluted free environment scientist used the waste materials of citrus to produce the biofuels, Nutraceuticals, macro and micronutrients, extract the essential oils and ethanol etc. Many industries now use the "Zero waste theory method", where they use the by-products of citrus and used to make the other useful products so less environmental pollution is occurring due to industrialization waste. This vaporization of citrus is reduced the pollution and develop a sustainable and eco-friendly environment. This method also added the value compounds in citrus industry (Figs. 18.4 and 18.5) [48].

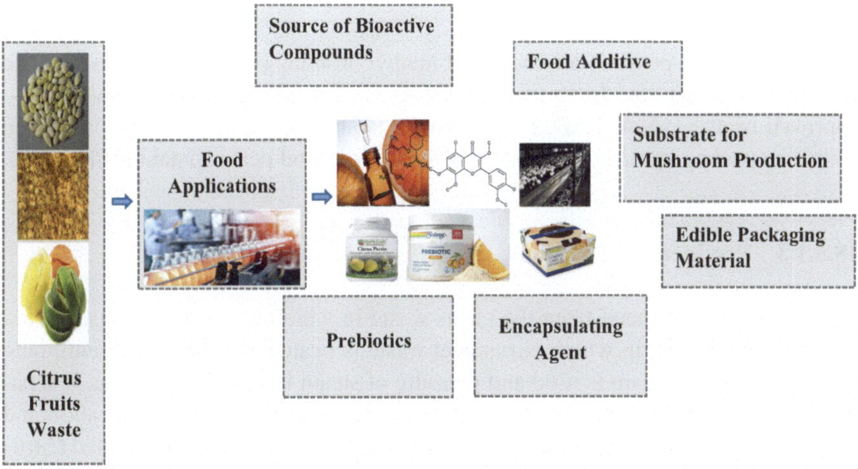

Fig. 18.4 Food Applications of By-Product of Citrus Waste that obtained during Processing

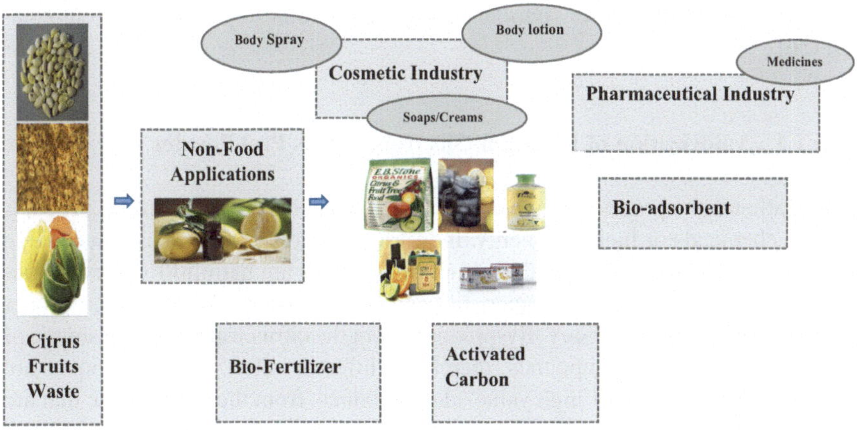

Fig. 18.5 Non- Food Applications of By-Product of Citrus Waste that obtained during Processing

18.5.1.1 Production of Lime Essential Oil with the Help of Lauric Acid with the Support of CO2

As we know lime oil is considered as a volatile oil which easily mortify when expose in oxidative-thermal condition and insoluble in water media. So, in recent advances scientists introduced PGSS technology which used to amplify the non-destructive production of volatile oils with moderate temperature and high pressure. In this technique, they used PEG or Lauric acid micro particles which to increase the extraction rate of volatile oils [54].

18.5.1.2 Improving the Import Packaging of Citrus by Using the Ethyl Formate

Ethyl formate is a possible alternative of methyl bromide (MB) which is used in the fumigation of imported fruits and vegetables such as citrus. Ethyl formate is also approved by the U.S.A. FAD. Ethyl formate is also maintained the flavour of the fruits in a packaging. Ethyl formate is also an insect and pest resistant [55].

18.5.1.3 Production of Ethanol

Ethanol is also produced from the citrus waste in which four steps are used. First step is pretreatment, in which citrus peel waste is heated at 150 degree centigrade and high pressure steam is used and pressure of steam is 70 psi. Limonene is separated by pre-treatment process. Second step is simultaneously saccharification and fermentation. In this step, agitation is continuing for 2–4 hours of pre-treated Citrus peel waste by the addition of yeast and enzymes (pectinase, cellulose, glucosidase and calcium carbonate). Then, it is allowed to ferment and hydrolyze until all cellulose digested and ethanol is produced by the complete conversion of sugars. Distillation is installed after simultaneously scarification step to separate out ethanol [56].

18.5.1.4 Valorization of Essential Oils from Citrus Peel Powder

Valorization is the process of conversion of waste and biomass into useful products such as chemicals or fuel and energy. Industries of citrus processing produce large quantities of by-products from the citrus waste, which are harmful for the environment due to their different characteristics and organic nature, therefore their proper management is very necessary. By-products from the citrus waste are rich source of different value added compounds, therefore, different valorization techniques are used for the production of high value added products from the citrus waste that are used in food, pharmaceutical and cosmetic sectors. Hence, valorization techniques provide double benefits: one is providing high value added products and other is environment management. Biofuels are also produced from the citrus waste through valorization approach. Products that are made from valorization of citrus waste provide economic gain that is very helpful for the country. By-products from citrus waste also reduce the hunger [57]. Citrus waste also used as fertilizer by combining with solid municipal waste on agriculture lands that decrease the soil erosion. Citrus wastes utilization reduces the environmental pollution. The main use of the valorization of citrus peel is its utilization and extraction of different type of volatile and aromatic compounds extraction that in present in the citrus peel. These compounds are used as industrial aspects. These extraction methods are very eco-friendly and less polluted and also very cheap. Essential oils are aromatic compounds that are volatile and used in different sectors such as in food, cosmetics and pharmaceutical. Essential oils from citrus waste were extracted through the process of hydro-distillation [58].

For extraction of essential oils, citrus peels are collected and outer part of peel that is coloured is separated from pith because most of the oil is present in them. For 2 hours, they are heated. After pre-treatment, distillation process is used which consist of different parts such as conical flask, basket type heater, distillation flask and condenser. Citrus peels (50 g) that are pre-treated are added in distillation flask and water (100 ml) is also added in distillation flask. Heat is also given with the help of basket heater. Conical flask is used for the distillate and it is collected in the form of layers that is two. For the separation of layers, funnel is used. In the funnel, less dense layer is collected that is oil from the citrus peels. To identify the composition of oils Gas chromatography –Mass Spectroscopy is used. The major constituent in essential oils is d-limonene and thirty-three different compounds are also found in essential oils [59].

18.5.2 Role of Citrus Waste in Food Security and Sustainability

Waste management and utilization is a major challenge for the food industries. Food waste is used in food industries as coating materials to protect the food from environmental factors. Citrus peel is used as coating in the food industry because they have antibacterial and antioxidant properties [60]. Citrus waste is very important in the food industries because of their properties that are very useful for the health of the customers. Vitamin C is present in citrus wastes, therefore, it is used in different drinks and beverages. Essential oils are also present in citrus wastes that are added in cosmetics, pharmaceuticals or beverages because of their good flavour and fragrance. Citrus waste contains different polyphenols that are used as bio fertilizers. Citrus wastes contain essential oils that are very complex because of bioactive compounds and they are used as substitutes in different food industries. Essential oils have different properties such as antifungal, antibacterial or antioxidant. Essential oils play role in food safety because of different compounds such as terpens and aromatic volatile compounds. Essential oils are used as preservatives because of their properties that increase the freshness of different food products [61].

The major compound that is present in essential oils is Limonene and it is less toxic. Therefore, under federal regulations they are safe (GRAS) and used as flavour. Citrus peel is used as a source of dietary fibre in muffins preparation of good nutritional profile biscuits and also used in the replacement of fat. After extraction of different components from citrus peel and pomace, it is transformed into packaging material that's biodegradable such as trays of eggs or fruit trays. In this way, waste can be again reutilized [62].

18.6 Conclusion

These days, managing CF waste properly is essential to improving the cultivation's sustainability. Reassessing CF by-products might, in fact, limit their environmental buildup and enable the full potential of these resources to be utilized. This is because CF by-products are a cheap and renewable source of compounds with high added value; therefore, researchers should take a comprehensive approach to understand the entire CF, including its waste, and utilize better techniques for its recovery and value-adding. Different environmentally friendly and sustainable techniques for the efficient reuse of garbage have been developed recently in this regard. As an alternative, thermochemical and biochemical processes like pyrolysis, thermolysis, gasification, and combustion can be used to turn CF waste into biofuels. This gives the use of CF by-products as nutraceuticals scientific justification. Future research and creative techniques for recovering CF waste are constantly expanding, but industry and academia working together will create the framework for a more productive circular economy that will make money out of waste because one man's trash is another man's treasure.

References

1. Chavan, P., Singh, A. K., & Kaur, G. (2018). Recent progress in the utilization of industrial waste and by-products of citrus fruits: A review. *Journal of Food Process Engineering, 41*(8), e12895.
2. Leporini, M., Tundis, R., Sicari, V., & Loizzo, M. R. (2021). Citrus species: Modern functional food and nutraceutical-based product ingredient. *Italian Journal of Food Science, 33*(2), 63–107.
3. Muscat, A., de Olde, E. M., Ripoll-Bosch, R., Van Zanten, H. H., Metze, T. A., Termeer, C. J., et al. (2021). Principles, drivers and opportunities of a circular bioeconomy. *Nature Food, 2*(8), 561–566.
4. Russo, C., Maugeri, A., Lombardo, G. E., Musumeci, L., Barreca, D., Rapisarda, A., et al. (2021). The second life of citrus fruit waste: A valuable source of bioactive compounds. *Molecules, 26*(19), 5991.
5. Sharma, K., Mahato, N., Cho, M. H., & Lee, Y. R. (2017). Converting citrus wastes into value-added products: Economic and environmently friendly approaches. *Nutrition, 34*, 29–46.

6. Sir Elkhatim, K. A., Elagib, R. A., & Hassan, A. B. (2018). Content of phenolic compounds and vitamin C and antioxidant activity in wasted parts of Sudanese citrus fruits. *Food Science & Nutrition, 6*(5), 1214–1219.
7. Jeong, D., Park, H., Jang, B. K., Ju, Y., Shin, M. H., Oh, E. J., et al. (2021). Recent advances in the biological valorization of citrus peel waste into fuels and chemicals. *Bioresource Technology, 323*, 124603.
8. John, I., Yaragarla, P., Muthaiah, P., Ponnusamy, K., & Appusamy, A. (2017). Statistical optimization of acid catalyzed steam pretreatment of citrus peel waste for bioethanol production. *Resource-Efficient Technologies, 3*(4), 429–433.
9. Zayed, A., Badawy, M. T., & Farag, M. A. (2021). Valorization and extraction optimization of citrus seeds for food and functional food applications. *Food Chemistry, 355*, 129609.
10. Rashid, U., Ibrahim, M., Yasin, S., Yunus, R., Taufiq-Yap, Y. H., & Knothe, G. (2013). Biodiesel from Citrus reticulata (mandarin orange) seed oil, a potential non-food feedstock. *Industrial Crops and Products, 45*, 355–359.
11. Iqbal, A., Schulz, P., & Rizvi, S. S. (2021). Valorization of bioactive compounds in fruit pomace from agro-fruit industries: Present insights and future challenges. *Food Bioscience, 44*, 101384.
12. Maqbool, Z., Khalid, W., Atiq, H. T., Koraqi, H., Javaid, Z., Alhag, S. K., et al. (2023). Citrus waste as source of bioactive compounds: Extraction and utilization in health and food industry. *Molecules, 28*(4), 1636.
13. Mahrous, E. A., & Farag, M. A. (2022). Trends and applications of molecular distillation in pharmaceutical and food industries. *Separation & Purification Reviews, 51*(3), 300–317.
14. Chandra, M. S., Srinivasulu, M., Yadav, P. S., Ramesh, B., Narasimha, G., & Chandrasekhar, T. (2021). Value added products from agriculture, paper and food waste: A source of bioenergy production. In *Bioenergy research: Commercial opportunities & challenges* (pp. 91–126). Springer.
15. Machado, C. A., Oliveira, F. O., de Andrade, M. A., Hodel, K. V. S., Lepikson, H., & Machado, B. A. S. (2022). Steam distillation for essential oil extraction: An evaluation of technological advances based on an analysis of patent documents. *Sustainability, 14*(12), 7119.
16. Prado, J. M., Vardanega, R., Debien, I. C. N., de Meireles, M. A. A., Gerschenson, L. N., Sowbhagya, H. B., & Chemat, S. (2015). Conventional extraction. In *Food waste recovery* (pp. 127–148). Elsevier. ISBN 9780128003510.
17. Pateiro, M., Barba, F. J., Domínguez, R., Sant'Ana, A. S., Khaneghah, A. M., Gavahian, M., Gómez, B., & Lorenzo, J. M. (2018). Essential oils as natural additives to prevent oxidation reactions in meat and meat products: A review. *Food Research International, 113*, 156–166.
18. Ashraf, R., Ghufran, S., Akram, S., Mushtaq, M., & Sultana, B. (2020). Cold pressed coriander (Coriandrum sativum L.) seed oil. In *Cold pressed oils* (pp. 345–356). Academic.
19. Sovová, H., & Aleksovski, S. A. (2006). Mathematical model for hydrodistillation of essential oils. *Flavour and Fragrance Journal, 21*(6), 881–889.
20. Zhang, Y., Deng, J., Jiang, W., Mei, Q., & Liu, D. (2018). Application of vacuum distillation in refining crude lead. *Vacuum, 148*, 140–148.
21. Bethge, D. (2014). Short path and molecular distillation. In *Vacuum technology in the chemical industry* (pp. 281–294). Wiley.
22. Ketenoglu, O., & Tekin, A. (2018). Computer simulation and experimental molecular distillation of olive pomace oil deodorizer distillate–A comparative study. *LWT, 96*, 636–641.
23. Alkhudhiri, A., Darwish, N., & Hilal, N. (2012). Membrane distillation: A comprehensive review. *Desalination, 287*, 2–18.
24. Pei, J., Gao, S., Sarp, S., Wang, H., Chen, X., Yu, J., et al. (2021). Emerging forward osmosis and membrane distillation for liquid food concentration: A review. *Comprehensive Reviews in Food Science and Food Safety, 20*(2), 1910–1936.
25. Cassano, A., Conidi, C., & Drioli, E. (2020). A comprehensive review of membrane distillation and osmotic distillation in agro-food applications. *Journal of Membrane Science and Research, 6*(3), 304–318.

26. Zhou, W., Widmer, W., & Grohmann, K. (2008, December). Developments in ethanol production from citrus peel waste. In *Proceedings of the Florida State Horticultural Society* (Vol. 121, pp. 307–310).
27. Suri, S., Singh, A., & Nema, P. K. (2021). Recent advances in valorization of citrus fruits processing waste: A way forward towards environmental sustainability. *Food Science and Biotechnology, 30*(13), 1601–1626.
28. Naseri, A. T., Thibault, J. F., & Ralet-Renard, M. C. (2008). Citrus pectin: Structure and application in acid dairy drinks. In *Tree and forestry science and biotechnology* (Vol. 2 Special Issue 1, np). Global Science Books.
29. Anticona, M., Blesa, J., Lopez-Malo, D., Frigola, A., & Esteve, M. J. (2021). Effects of ultrasound-assisted extraction on physicochemical properties, bioactive compounds, and antioxidant capacity for the valorization of hybrid Mandarin peels. *Food Bioscience, 42*, 101185.
30. Muñoz-Labrador, A., Moreno, R., Villamiel, M., & Montilla, A. (2018). Preparation of citrus pectin gels by power ultrasound and its application as an edible coating in strawberries. *Journal of the Science of Food and Agriculture, 98*(13), 4866–4875.
31. Su, D. L., Li, P. J., Quek, S. Y., Huang, Z. Q., Yuan, Y. J., Li, G. Y., & Shan, Y. (2019). Efficient extraction and characterization of pectin from orange peel by a combined surfactant and microwave assisted process. *Food Chemistry, 286*, 1–7.
32. Ngouémazong, E. D., Christiaens, S., Shpigelman, A., Van Loey, A., & Hendrickx, M. (2015). The emulsifying and emulsion-stabilizing properties of pectin: A review. *Comprehensive Reviews in Food Science and Food Safety, 14*(6), 705–718.
33. Šafranko, S., Ćorković, I., Jerković, I., Jakovljević, M., Aladić, K., Šubarić, D., & Jokić, S. (2021). Green extraction techniques for obtaining bioactive compounds from mandarin peel (Citrus unshiu var. Kuno): Phytochemical analysis and process optimization. *Foods, 10*(5), 1043.
34. Yang, Y. Y., Ma, S., Wang, X. X., & Zheng, X. L. (2017). Modification and application of dietary fiber in foods. *Journal of Chemistry, 2017*, 9340427.
35. Wang, L., Xu, H., Yuan, F., Fan, R., & Gao, Y. (2015). Preparation and physicochemical properties of soluble dietary fiber from orange peel assisted by steam explosion and dilute acid soaking. *Food Chemistry, 185*, 90–98.
36. Zhu, Z., Gavahian, M., Barba, F. J., Roselló-Soto, E., Kovačević, D. B., Putnik, P., & Denoya, G. I. (2020). Valorization of waste and by-products from food industries through the use of innovative technologies. In *Agri-food industry strategies for healthy diets and sustainability* (pp. 249–266). Academic.
37. Shin, G. M. (2015). Quality characteristics of sponge cake added with citrus peel powder. *Culinary Science and Hospitality Research, 21*(5), 88–97.
38. Caggia, C., Palmeri, R., Russo, N., Timpone, R., Randazzo, C. L., Todaro, A., & Barbagallo, S. (2020). Employ of citrus by-product as fat replacer ingredient for bakery confectionery products. *Frontiers in Nutrition, 7*, 46.
39. Ordóñez-Santos, L. E., Esparza-Estrada, J., & Vanegas-Mahecha, P. (2021). Ultrasound-assisted extraction of total carotenoids from mandarin epicarp and application as natural colorant in bakery products. *LWT, 139*, 110598.
40. Davani-Davari, D., Negahdaripour, M., Karimzadeh, I., Seifan, M., Mohkam, M., Masoumi, S. J., & Ghasemi, Y. (2019). Prebiotics: Definition, types, sources, mechanisms, and clinical applications. *Food, 8*(3), 92.
41. Foti, P., Ballistreri, G., Timpanaro, N., Rapisarda, P., & Romeo, F. V. (2022). Prebiotic effects of citrus pectic oligosaccharides. *Natural Product Research, 36*(12), 3173–3176.
42. Randive, S. D. (2012). Cultivation and study of growth of oyster mushroom on different agricultural waste substrate and its nutrient analysis. *Advances in Applied Science Research, 3*(4), 1938–1949.
43. Otieno, O. D., Mulaa, F. J., Obiero, G., & Midiwo, J. (2022). Utilization of fruit waste substrates in mushroom production and manipulation of chemical composition. *Biocatalysis and Agricultural Biotechnology, 39*, 102250.

44. Sharif, M., Zafar, M. H., Aqib, A. I., Saeed, M., Farag, M. R., & Alagawany, M. (2021). Single cell protein: Sources, mechanism of production, nutritional value and its uses in aquaculture nutrition. *Aquaculture, 531*, 735885.
45. Mekonnen, M. M., & Hoekstra, A. Y. (2014). Water footprint benchmarks for crop production: A first global assessment. *Ecological Indicators, 46*, 214–223.
46. Panwar, D., Saini, A., Panesar, P. S., & Chopra, H. K. (2021). Unraveling the scientific perspectives of citrus by-products utilization: Progress towards circular economy. *Trends in Food Science & Technology, 111*, 549–562.
47. Ren, W., Zhao, S., Lian, Y., Yang, Y., Tian, G., Zhao, C., et al. (2020). Effects of hydrosoluble calcium ions and organic acids on citrus oil emulsions stabilized with citrus pectin. *Food Hydrocolloids, 100*, 105413.
48. Suri, S., Singh, A., & Nema, P. K. (2022). Current applications of citrus fruit processing waste: A scientific outlook. *Applied Food Research, 2*(1), 100050.
49. Chhikara, N., Kour, R., Jaglan, S., Gupta, P., Gat, Y., & Panghal, A. (2018). Citrus medica: Nutritional, phytochemical composition and health benefits–a review. *Food & Function, 9*(4), 1978–1992.
50. Jiang, H., Zhang, W., Xu, Y., Chen, L., Cao, J., & Jiang, W. (2022). An advance on nutritional profile, phytochemical profile, nutraceutical properties, and potential industrial applications of lemon peels: A comprehensive review. *Trends in Food Science & Technology, 124*, 219–236.
51. Shehata, M. G., Awad, T. S., Asker, D., El Sohaimy, S. A., Abd El-Aziz, N. M., & Youssef, M. M. (2021). Antioxidant and antimicrobial activities and UPLC-ESI-MS/MS polyphenolic profile of sweet orange peel extracts. *Current Research in Food Science, 4*, 326–335.
52. Rezzadori, K., Benedetti, S., & Amante, E. R. (2012). Proposals for the residues recovery: Orange waste as raw material for new products. *Food and Bioproducts Processing, 90*(4), 606–614.
53. Khan, U. M., Sameen, A., Aadil, R. M., Shahid, M., Sezen, S., Zarrabi, A., et al. (2021). Citrus genus and its waste utilization: A review on health-promoting activities and industrial application. *Evidence-based Complementary and Alternative Medicine, 2021*, 1–17.
54. Ruiz, B., & Flotats, X. J. W. M. (2014). Citrus essential oils and their influence on the anaerobic digestion process: An overview. *Waste Management, 34*(11), 2063–2079.
55. Park, M. G., Lee, B. H., Yang, J. O., Kim, B. S., Roh, G. H., Kendra, P. E., & Cha, D. H. (2021). Ethyl formate as a methyl bromide alternative for fumigation of citrus: Efficacy, fruit quality, and workplace safety. *Journal of Economic Entomology, 114*(6), 2290–2296.
56. Panwar, D., Panesar, P. S., & Chopra, H. K. (2021). Recent trends on the valorization strategies for the management of citrus by-products. *Food Reviews International, 37*(1), 91–120.
57. Varmie, E. B., & Thakur, M. (2021). Utilization of citrus processing waste: A review. *The Pharma Innovation Journal, 10*, 682–697.
58. Teixeira, B., Marques, A., Ramos, C., Neng, N. R., Nogueira, J. M., Saraiva, J. A., & Nunes, M. L. (2013). Chemical composition and antibacterial and antioxidant properties of commercial essential oils. *Industrial Crops and Products, 43*, 587–595.
59. Nidhi, P., Rolta, R., Kumar, V., Dev, K., & Sourirajan, A. (2020). Synergistic potential of Citrus aurantium L. essential oil with antibiotics against Candida albicans. *Journal of Ethnopharmacology, 262*, 113135.
60. Anis, A., Pal, K., & Al-Zahrani, S. M. (2021). Essential oil-containing polysaccharide-based edible films and coatings for food security applications. *Polymers, 13*(4), 575.
61. Khalid, W., Arshad, M. S., Aziz, A., Rahim, M. A., Qaisrani, T. B., Afzal, F., et al. (2023). Chia seeds (Salvia hispanica L.): A therapeutic weapon in metabolic disorders. *Food Science & Nutrition, 11*(1), 3–16.
62. Shan, Y. (2016). Production of biodegradable packages using citrus peel. In *Comprehensive utilization of citrus by-products* (pp. 91–92). Academic Press.

Chapter 19
Citrus Waste Valorization: From Color to Wellness

Ehsan-ul-Haque, Sohaib Afzaal, Shinawar Waseem Ali, Akbar Hayat, Muhammad Shakeel Hanif, Muhammad Asim, Shakeel Ahmad, and Sajjad Hussain

19.1 Introduction

Citrus is a popular fruit type worldwide, and citrus fruits are one of the most consumed varieties globally for their sweet-sour delicious flavors laden with health benefits. But eating them creates a lot of trash in the form of peels, seeds and pith. What used to be traditionally treated as non-performing residues, these citrus wastes are now being valorized into dozens of industry applications including textiles and wellness products. In this chapter, the different valorizations of citrus waste are investigated into details through natural dyes (Sect. 19.2), essential oils (Sect. 19.3) and citrus cloudy concentrated juices (Sect. 19.4).

First part of this chapter discusses the application to produce beneficial natural dyes from citrus waste, related to sustainable fashion industry. These findings further support the potential use of nutrient-rich citrus peels as an alternative source to synthetic dyes, with a significant reduction in their environmental impact. The extraction processes of citrus peel color pigments are discussed, including solvent extraction, enzymatic methods and supercritical fluid extraction. In addition to

Ehsan-ul-Haque · S. Afzaal (✉) · A. Hayat · M. Asim
Citrus Research Institute, Sargodha, Pakistan

S. W. Ali
Department of Food Sciences, University of the Punjab, Lahore, Pakistan

M. S. Hanif
Fodder Research Institute, Sargodha, Pakistan

S. Ahmad
Department of Agronomy, Bahauddin Zakariya University, Multan, Punjab, Pakistan

S. Hussain
Citrus Center, Texas A&M University, Kingsville, Weslaco, TX, USA

A. Chauhan et al. (eds.), *Valorization of Citrus Food Waste*,
https://doi.org/10.1007/978-3-031-77999-2_19

extracting the pigment more efficiently, these methods are also upholding a sustainable approach for eco-friendly fashion.

This is interrelated with the process of value chain from extraction right to application including integrating circular dyeing processes to the regular textile chain. Different methods of exhaust dyeing are discussed and how the citrus-based dyes gives fabulous hues to fabrics with lesser harm to the environment are presented here. Moreover, we also look at the prospects of the citrus-based dyes its applicability and acceptance in today's creation of sustainable fashion.

Citrus waste, on the other hand, offers something more than just color: Essential oils - mostly obtained through citrus peels [1]. In this part, we will discuss the techniques used for extracting these oils like steam distillation and cold-press expression. This section is focused regarding the maintenance of bioactive compounds: which possess the medicinal properties of essential oils. The focus in many uses of citrus-based essential oils is on the antiseptic benefits they offer, and how this makes them desirable for skin care uses as well as cleaning product recipes and aromatherapy. In this Chapter, we will examine the individual components within these oils responsible for their antibacterial activity and describe how they can be used to improve personal hygiene and infection control. Additionally, the chapter explores all regulatory and safety issues when employing citrus oils in a range of consumer products.

As the finishing touch, this chapter covers a rather not well known but equally important front of citrus waste valorization: cloudy concentrated citrus juices. They are not clear like the popular drinks but they have extra nutrients and antioxidants from the part of citrus; we all tend to throw out as a waste. Future technologies of juice extraction techniques, combining mechanical pressing and enzymatic methods, are explored to produce Cloudy Concentrated Citrus Juices (CCCJ) that maximizes fruit utilization and minimizes waste.

The nutritional virtues of the Cloudy Concentrated Citrus Juices (CCCJ) and economic potential for producers offer additional value. Based on the higher nutritional value and rich antioxidant Cloudy Concentrated Citrus Juices is considered not only nutritious food but additionally endorsed as economical alternative in beverage industry. Natural nutrient-rich products that are in-line with sustainable consumption trends have increasingly seen demand, as reflected by evolving market tastes and consumer choices [2].

This chapter demonstrates the high potential for using citrus waste as a transforming agent rather than just discarded residues. You can reclaim a vast majority of the wastes from producing fruit juice, in some cases creating highly profitable new products like natural dyes or essential oils. Unfortunately these materials are being allowed to spoil due to improper handling and ignorance. By using waste citrus peels for natural dyes, industries could avoid up-to 15% of their overall environmental foot print.

19.2 Citrus Waste-Based Natural Dyes

19.2.1 Introduction to Citrus Waste as a Source of Natural Dyes

It is trendy to raise concerns that fashion holds one of the leading roles in polluting our society and thus, its social and global impact. The production and consumption of synthetic textiles and dyes have led to significant environmental changes: from polluting water with waste residues, epilating clothing made with versatile natural resources such as petroleum. In the recent past, these have been addressed under sustainability in fashion especially on behalf of the environment and injustices in their supply systems [3].

Eco-dyeing is one of the most essential steps in sustainable fashion. Artificial dyes have dominated the textile sector for several years where numerous among them arises from petrochemicals and have dangerous compounds; they are mostly non-degradable and ultra-violet sensitive. These chemicals have been known to discharge harmful effluents into the water sources, a factor, which had severely impacted aqueous systems and the human society in general.Over the last decades there has been an increase in concern in the use of citrus waste for natural dye production because of the environmental and economic benefits. Oranges, lemon, grape fruits, and many other citrus fruits are grown in many parts of the world; therefore, large amounts of wastes are produced in the processing of these fruits. This waste mainly includes peel, membrane, and seed which are often thrown away, utilized for animal feed or compost purpose. The citrus industry disposes a colossal sum of waste annually and therefore is a good candidate for sustainable practices especially in the fashion industry whose dyeing process is becoming worrying in terms of the impacts on the environment. However, natural dyes are way better than synthetic dyes as they are more environmental friendly. They are obtained from natural resources including plant, animal and minerals and are usually believed to have less harm to the surroundings as compared to man-made dyes. Citrus peel wastes consisting of rich pigments like flavonoids and carotenoids- bioactive compounds that could be profited from to create natural dyes. These compounds not only add bright shades to the final products but also contain antioxidant and antimicrobial effects and make dyed textile products to be functional products [4].

The fashion industry on the whole is very much aggressive and considered to be causing more harm to the environment but with the awareness arising for sustainability practices, fashion industries have started using natural dyes that are obtained from agricultural wastes such as citrus wastes. Thus, such waste can actually be recycled to produce valuable products i.e., natural dyes to reduce on the use of synthetic chemicals and deploying an economy where wastes are not dumped but transformed into useful products and services.

In conclusion, the study of citrus waste as natural dyes showed the potential of this promising line, which will be beneficial in the production of citrus products and designing the sustainable fashion industry. Thus, this chapter is dedicated to shedding the light on the possible use of citrus waste natural dyes describing methods of extraction and dyeing, color fastness, and the potential application for textile industries [5, 6]. The properties and opportunities of citrus waste outlined above give one an idea about the possibilities that can lead to the creation of a favorable environment for the sustainable utilization of agricultural by-products, thus contributing to the improvement of economic efficiency and the preservation of the environment [7–10].

19.2.2 Dyeing Process Using Citrus-Based Natural Dyes

Through this technique of dyeing, the natural dyes from citrus fruits are used and prove to be sustainable in the processing of textile fabrics as they also help to deal with wastes from citrus processing sector and give out brilliant colors. This section looks at different facets of dyeing, beginning with pre-treatment of textile and extending to the methods of application of dyes and aims at giving tips towards sustainable fashion application [11].

Preparation for the treatment of textiles before dyeing is very important to enhance the dyeing efficiency and the fastness of natural dyes gotten from citrus peel waste. Fabrics are also treated before they are dyed through processes such as scouring, bleaching, and mordanting. Scouring helps to remove natural oils, waxes and other materials from the fibers so that getting an even dye penetration can be achieved. Bleaching aims at increasing the general light colors of the textiles specifically by producing a smooth surface of the companion items. Mordanting is the process where fibers are subjected to metal salts so that maximum impact can be given to the dye, absorptiveness and the fastness of the colors [12]. In the case of natural dyes based on citrus fruits, the type of pre-mordanting and the further outcome considerably depends on the type of the used fiber material, such as cotton, wool, or silk and the like. Techniques that can be used when applying the orange natural dyes include the exhaust dyeing process together with pad-dry cure process. Exhaust dyeing consists in putting the textiles in a dye bath containing orange extract based dyes and allowing the penetration of the dye till it is at its optimum. This method is applicable to a limited scale production and the concentration of the dye and intensity of the colors can be easily regulated. Pad-dry-cure means applying the dye solution on the textile materials by the help of padding equipment, drying and then curing to set the dye. This method is best suitable when making large products and the dye penetration and color fastness can be easily controlled [13].

Some specific stories of successful application in the field of Sustainable fashion are available which illustrates the real life application of citrus natural dyes. Many of these dyes are gradually being used in companies and design collections for releasing clothing in the market in order to fulfill the demand of consumers with the perspective of sustainable and non-harmful products. Successful examples show that citrus-based natural dye appeals to a diverse and vast market due to its applicability to different kinds of textiles and different designs of the garments. These examples demonstrate application of sustainable strategies in fashion industry which involves the utilization of citrus waste products in the fashion industry to help minimize its effects on the environment and embrace the circular economy [14, 15].

Hence, it is revealed that the process of dyeing through the natural dyes extracted from citrus fruit is much eco-friendly than the synthetic dyes used in the textile industry. Consequently, it is possible to attain highly vibrant color and excellent color fastness of the dyed textile goods through successfully adopting proper pretreatment procedures and application techniques within the textile manufactures and product designers. The examples described in the cases show that it is possible to use citrus waste for creating natural dyes and promote sustainable processes in the fashion industry with the progress of the sustainable fashion market around the world.

19.2.3 Utilization of Citrus-Based Dyes in the Sustainable Fashion Industry

Thus, the use of products made from citrus-based dyes in sustainable fashion contributes positively to the protection of the environment considerably. This section looks into the advantages of natural Citrus dyes over synthetic dyes based on environmental analysis, and current market trends, consumers' perception of natural dyes, challenges facing this natural dye and the prospects thereof.

Advantages that come with using citrus based dyes over synthetic dyes are many and cut across the environmental sphere. Citrus-based dyes can be synthesized from waste; mostly the peels of citrus fruits which if not processed would make their way to the dump site or be incinerated. Hence, when using these waste materials they minimize the effects of the regular dyeing techniques on the environment. They usually involve the use of fewer chemicals and less water while embodying the extraction and dyeing of the fabric leading to less pollution of water bodies and energy use in general. Also, the use of natural dyes derived from citrus fruits shows compliance with ecological disposal at the end of a product's useful life [15, 16].

Newly, the fashion industry looks forward to sustainability of dyes, and latest market trends together with the acceptance of these citrus-based dyes. The consumer becomes more aware of the factors that influence the environment and the society hence going for products that are environmentally friendly. Fashion brands and designers are no longer turning a blind eye to the problems of environmental pollution and are now using citrus in dyes presented in their collections. The surveys revealed that now people are more conscious for wearing clothes which are environment friendly and the use of citrus dyes will be a distinct selling proposition in this respect.

Some of the current and potential future uses of citrus based dyes include; matters of application like productivity and intensity of the product, escalation of the color intensity and cost comparison that characterize the mass production (Fig. 19.1). A few challenges are: the variation in composition of the pigments within the citrus wastes, how best to extract the maximum quantities of dye from the wastes, how best to obtain adequate color fastness that is the durability of the dye and this is similar to synthetic dyes. Awareness in extraction techniques, formulation of dye and dyeing process and procedures play the decisive factor in eliminating the aforesaid challenges. The further prospective is concentrated on the investigation of the further enhancement of biotechnological methods for the efficient production of pigments and creation of new shades from citrus by-products. Therefore, it is very crucial that researchers, industries, and policymaker's coordinate in the promotion as well as the extension of the use of citrus based dyes in the fashion industries all across the globe.

In conclusion, the introduction of the sustainable fashion Industry, with the use of natural dyes extracted from citrus also points to the possible strategy to avoid further harm on nature and use of circular economy. Hence, application of the natural dyes derived from citrus placed it in the map of supporting sustainable and responsible fashion that is more preferred by consumers. From the above observations, Citrus based dyes' future development will depend on innovation, education,

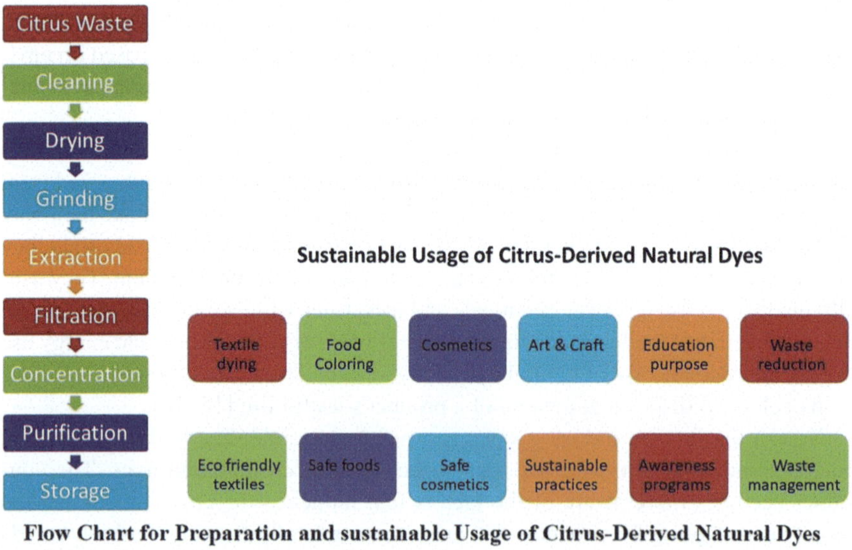

Fig. 19.1 Demonstration of flow chart for preparation of dyes from Citrus waste

and the consumers than any other element, and this has to be read as the way of using the citrus based dyes for the eventual change of the nature of the textile industry on the sustainable way.

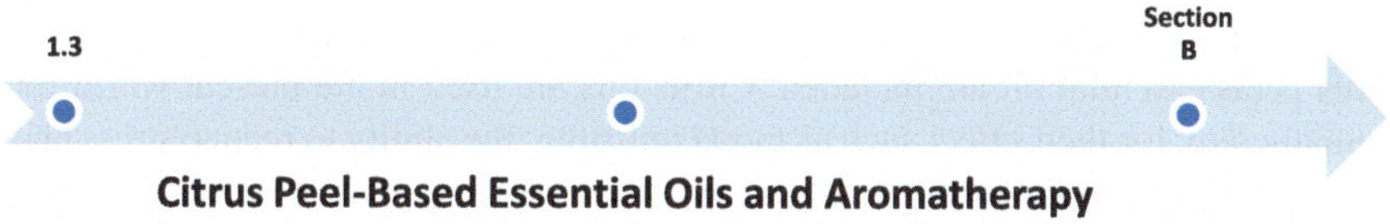

19.3 Citrus Peel-Based Essential Oils and Aromatherapy

19.3.1 Introduction to Citrus Peel Essential Oils

Citrus peel oils are obtained from the outer layer or skin of citrus fruits including, orange, lemon, lime, grape fruit, and bergamot. The major features of these oils are the freshness, the citrus and vitalizing notes. The primary components of citrus peel essential oils include following components [17, 18]:

- Limonene: The major constituent, even in the oils of oranges and lemons recognized for their high citrus character and antioxidant effects [19].
- Linalool: Shared in bergamot and lime oil which gives floral and spicy character and possess calming tendencies [20].
- Bergamotene: Found in bergamot oil with a concept of warm woody notes [21].
- Citral: This happens in Lemon and Lime oils; the one with the lemony scent and the ability to combat bacterial growth [22].
- Alpha-Pinene: Present in a lesser concentration, contributing to the freshness and that of the pine tree [23].
- Myrcene: It is also sweet and this one adds an earthy scent to the mix and can be found in grapefruit oil [24].
- Neral and Geranial: These are both citral components; it becomes: sweet, lemony note, and antimicrobial properties [22].

Citrus oils have a rich history in the practice of aromatherapy, tracing back to ancient civilizations. Citrus oils have a rich history in the practice of aromatherapy, tracing back to ancient civilizations:

In traditional Chinese remedies, citrus fruits and their juices were utilized as the treatment for digestive tract disorders, and to stabilize the mood. In ancient Egypt, Citrus oil in cosmetics was as a fragrance, perfumes, and as a disinfectant in the process of mummification. In the Medieval Europe; the citrus fruits were adopted in Europe through trades, and their oils were applied in herbalist to cure diseases such as Scurvy and other conditions that gave a general body health boost. During seventeenth Century; originally, citrus oils were used during the middle ages and especially during the Renaissance in aromatherapy, and they can be nowadays estimated in perfumery and medicine due to their invigorating effect [25, 26].

Modern Aromatherapy was originally named and defined by the French chemist René-Maurice Gattefossé at the beginning of the first half of the twentieth century. Aromatherapy refers to the curative properties of various essential oils among which citrus peels essential oil are included. Citrus oils are used in the present world for aromatherapy for their effect such as mood uplifting, the ability to reduce stress and their immune boost properties [27].

Citrus peels oil play the same importance they have always since they can be used in aromatherapy for diffusers, for body massages, and other uses as well as in products like soaps, cleaners and lotions. With pleasant smell, many qualities that are helpful to human health, they are fit to be used either as perfumes or aromatherapy [28, 29].

19.3.2 Antibacterial Benefits of Citrus-Based Essential Oils

Among the citrus-based essential oils, the antibacterial efficiency is remarkably high owing to several active compounds. Limonene a monoterpene that is present in abundance in oils of citrus fruits such as orange and lemon contains antimicrobial properties. This one interferes with the bacterial cell membranes, and hence, it kills the bacteria [30, 31]. Another substance that can be found in summer fruits' oils is citral which also possesses antimicrobial activity regarding a wide range of bacteria because it suppresses bacterial proliferation and the development of biofilms. The compound derived from bergamot and lime oils, linalool gives the oils a floral note and successfully hampers bacterial cell walls' envelope integrity contributing antibacterial properties. Pinene present in lesser amount also contributes to the antimicrobial effects of the citrus oils because of its antibacterial as well as antifungal characteristics [32, 33].

The antibacterial properties of citrus-based essential oils make them highly valuable in a variety of applications, particularly in skincare, cleaning products, and personal hygiene [34]. Citrus oils are very useful in skincare products; their main uses being the cure for acne and other skin infections [35]. They are beneficial for the acne treating process because they reduce the amount of bacteria that leads to formation of acne, they also are anti-inflammatory in nature hence they work wonders in such sensitive skin. When applied on the skin, facial cleansers, toners and serums, for instance, have the potential of benefiting from these oils and hence lead to healthy skin [36].

There is commonly evidence that citrus essential oils are widely added to natural cleaning products. Surface disinfection, getting rid of obnoxious smell and widespreading germs and bacteria is quite pleasant task completed through essential oils due to their antimicrobial properties. Moreover, smell fresh and very revitalizing make the cleaning procedure stimulating and the smelling of clean fruits like oranges improve not only the cleaning results but also the quality of space that seem fresh. Citrus Oils are common in liquid hand wash soaps, hand sanitizers, and perfumes in that they offer a natural method of controlling bacteria. Their presence assists to wash off bacteria on the skin, which reduces instances of infections as well as fouling smells. Also, the invigorating fragrance of citrus oils creates an additional value for these products as customers are willing to buy products that not only will produce the desired effect but also give a pleasant scent to the room [37–39].

Considering that a number of essential oils containing citrus bases are known to have antibacterial properties as a result of their active ingredients, they are useful for boosting health and hygiene in numerous contexts. These characteristics also coincide well with the recent trends for green and organic products which are beneficial for peoples' health, which confirms the place of natural minerals in the modern cosmetic, cleaning, and hygienic uses.

19.3.3 Additional Uses for the Citrus Essential Oils

Citrus-based essential oils are considered to be especially useful in aromatherapy due to their stress relieving and mood uplifting properties [36]. Cold-pressed oils include foods such as orange, lemon, bergamot, grape fruit which are remarkable in dispensing anxiety, relieving tension, and creating a relaxing mood and raising the emotional level [40]. The respiratory function of these oils helps recognize the limbic area of the brain which controls both emotions and reflexes and the after effects are rather soothing and invigorating (Table 19.1).

Thus, the blends prepared by combination of one or more citrus oils results not only in the strengthening but also in the adjustment of the therapeutic properties of the finished product. The composition of blends commonly involves blending of citrus oils with other types of oils such as lavender, peppermint or eucalyptus in specific ratios that enhance the intended performances. For instance, the chemical compounds of lemon and lavender are known to produce both stimulating and relaxing reactions; hence, it can be used for the relaxation purposes. These formulations are applied in the diffusers; massage oils, and bath products thus boosting their utility in wellbeing solution services [41].

The incorporation of citrus based essential oils in the products should meet the legal requirements and safety measures to be observed in order to protect the consumer. Such things as the permitted concentrations, the labels to be used, and quality control measures are normally set by regulations. Some of the citrus oils are phototoxic, and as such, care must be taken especially when using bergamot because it may cause skin rash when one is in contact with the sun. It is for this reason that the dilution and usage instructions must be closely followed in order to avoid negative reactions. Adherence to the rules and regulations put down by different bodies like the International Fragrance Association, and the local health departments make it certain that products in circulation are safe for use diligently by the general public [42, 43].

Table 19.1 Some proven health benefits of citrus essential oils

Benefit Category	Specific Benefits
Enhancing Immune system	Antiviral activity
	Antioxidant Activity
	Antimicrobial Effects
Mental and Emotional Well-being	Stress Management
	Mood Enhancement
	Cognitive Function
Digestive Health	Digestive aid
	Anti-inflammatory Properties
	Detoxification
Skin Care	Antimicrobial/Antifungal
	Astringent Effects
	Antioxidant Benefits
Respiratory Health	Bronchodilator/Decongestant Agents
	Expectorant Effects
	Anti-inflammatory Benefits
Weight Management	Appetite Suppressant
	Metabolism Boost
	Lymphatic Support
Anticancer Properties	Cellular Protection
	Immune Support
Pain Relief	Anti- Inflammatory
	Analgesic Properties
	Muscle Relaxation

The utilization of citrus essential oils in aromatherapy and wide range of other products depict their multifaceted benefits and versatility. However utilization should be ensured to adhere with prevailing regional laws and international regulations to guarantee their effective and safe incorporation enhancing consumer confidence.

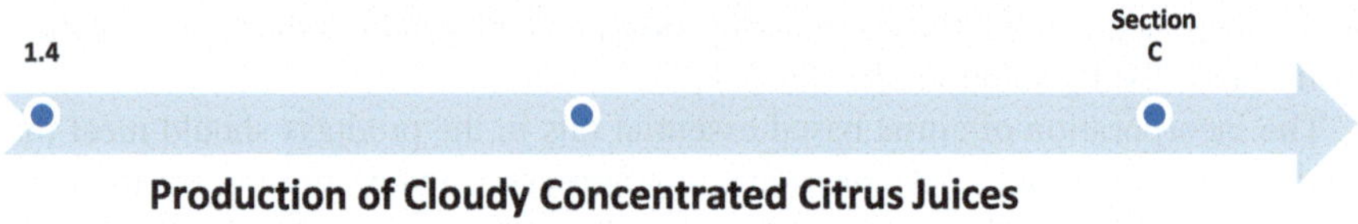

19.4 Production of Cloudy Concentrated Citrus Juices

19.4.1 Introduction to Cloudy Concentrated Citrus Juices

Citrus cloudy beverages contain the pulps, fibers, and oil of the citrus fruits, hence their clouded appearance and strong citrus taste; while these components are filtered out during the production of clear juice beverages. Cloudy juices retain a shallower degree of processing that seems to be closer to the natural, non-concentrated fruit juice experience.

Foods that are cloudy are categorized by their natural opaque appearance, due to suspended particles. Cloudy juices are slightly hazed and have an intense taste, thick, and full-bodied with the pulp, pectin, and other fruit fibers staying suspended in the liquid. These elements do not only improve the haptic and olfactory quality of the juice, beauty but also have an impact on the mouth feel of the juice. There is cloudiness because small particles of fruit pulp and essential oils are deliberately left behind during processing. Due to their ability to retain some of the natural elements, cloudy juices are even richer in color and flavor than clear juices. Moreover, these suspended particles can also contribute to a slight degree of the thickness that is desirable in juices, when consumed [44].

Typically, clear concentrated juices available in the market are considered to be of low nutritional standard than cloudy concentrated juices [45]. They are a source of higher amounts of dietary fiber that is beneficial to digestion and is also auspicious in managing hunger. The holding of the pulp and fiber also contributes to the fact that cloudy juices contain higher amount of Vitamin C, minerals as well as antioxidants. They offer a higher content of vitamin C, flavonoids, and carotenoids which play significant roles in immunity and health improvement. The main ingredient that is usually retained in cloudy juices is dietary fiber and this assists in the regulation of the digestion system, cholesterol and blood sugar level. The fiber content also supports the aspect of satiety; therefore, these juices are more likely to leave you fuller [46].

Some of the pulp and peel particles that are found in cloudy juices include the vitamins and minerals like vitamin c which helps in strengthening of the immune system and skin, folate which is used in DNA synthesis and repair. Some of the solid contents of cloudy juices such as flavonoids and carotenoids have high antioxidant activity due to their high concentration in cloudy juices. These compounds assist to counter free radicals, hence acting as an antioxidant to decrease oxidative diseases and chronic illnesses. This attribute makes cloudy juices even more healthful than the clear ones as they have other qualities, which include improved digestion, better cardiovascular health and increased bioavailability of bioactive compounds [47, 48]. Also it is worthy to note that the various natural constituents that are found in cloudy juices are anti-inflammatory and have immuno-enhancing properties; a role in wellness.

In general, the use of citrus peel in preparing cloudy concentrated juices enhances the quality and nutritional value of the fruits while satisfying the consumers' evolving palates for natural and tastier products. The action of retaining such components proves the ever-growing consumption of minimally processed foods that are not only good for one's health but also possessing enhanced sensory qualities.

19.4.2 Production Process for Cloudy Concentrated Citrus Juices

Cloudy concentrated citrus juices are specialized products targeted for health enthusiast population segments produced through innovative technologies. The processes are designed to retain natural fiber, pulp, essential oil and vital nutrients using characteristic technologies, which contribute to production of healthy and tasty cloudy citrus juices (Fig. 19.2). Extraction, enzymatic treatment to concentration and packing process is started from collection of clean citrus material and their crushing. Mechanical Pressing technique entails squeezing the fruits physically to get the juice with the pulp and oil used in making the food. Mechanical presses can actually start off as simple hand tool and advanced as Industrial device set to achieve the utmost degree of yield, while preserving the natural ingredients intact [49, 50]. The action creates a product with a desirable texture that associated with the juiciness, mass haze and pleasant taste. Enzymatic Extraction process uses certain enzymes to digest the cell walls of the fruit thereby insisting more juice as well as vital oil and allowing the pulp and the fiber. Nevertheless, it can raise yield and additionally control the steadiness of the juice. Additives such as pectinase will help in enhancing the yield and the stability of the pulp particles in the juice and thus keeping it cloudy [51–53].

After extraction, the juice undergoes clarification and concentration processes to achieve the desired quality and concentration. Clarification stands to note that while clear juices must undergo highly processing to be clarified, cloudy juices only need a little processing to help maintain the pulp and fibers. Settling, decanting and centrifugation are techniques used here to remove gross haze and pulp from the juice without causing a large change in the cloudiness of the juice. Debittering is an optional process adopted after clarification to remove bitterness and acids to retain them up to the desired levels targeted for the final product to be produced from cloudy concentrated juices [52, 54].

The juice is then concentrated through process such as vacuum evaporation or freeze concentration. These methods reduce the quantity of water in the juice and thus make the concentrate of the flavor, nutrients and particles in suspension. Vacuum evaporation is regarded as preferred due to minimizes heat input to the juice where the concentration is made at relatively low temperature that helps to retain the nutritional and sensory features of the juice [55].

Some of the key steps that are followed while concentrating the citrus cloudy juices include their quality check and packing. The concentrated juices are normally contained in steel barrels or plastic barrels lined with food grade liners which shields

Fig. 19.2 A brief process summary of cloudy concentrate process

the juice from light and air hence taking its nutrient value as well as its sensory attributes. Prior to packing, juice is heat processed, a process known as pasteurization in order to eradicate microbial loads which pose health hazards as well as to increase its shelf life. Correct packing is vital to avoid contact between the food and oxygen hence limiting the effects of enzymatic browning. It was noted that storage conditions should be cool and more preferably constant not exposed to extreme sun since the vitamins and antioxidants in the store may deteriorate. Furthermore, there is a necessity to provide the consumers with specific information on the composition of a product, its energy value, and storing conditions [55–57].

Maintaining high quality throughout the production process is crucial for producing consistent and safe cloudy concentrated citrus juices. Through monitoring pulp and fiber content one can be certain that the juice contains the right amount of pulp and fiber to be choice according to the finished products to be made. Pulp content is responsible for the juice's texture and nutritional profile. Microbial safety measures that are observed include washing and pasteurization to ensure that microorganisms that are likely to infest the product are dealt with. For sensory evaluation; taste, aroma, and texture are often tested since the appearance of the end product is an important aspect from the client's perspective. For chemical and physical testing; periodic checks of pH, Brix (sugar level), and essential oils ensure that the quality of the juice is up to the required standard by consumers [58, 59]. Altogether, the production technology of cloudy concentrated citrus juices is designed mainly to preserve the natural ingredients of the product. Stringent measures in checking quality are used hence to ensure the production of well-quality nutritious and palatable product that the consumers require.

19.4.3 Nutritional and Economic Benefits

The cloudy concentrated citrus juices contain natural pulp, fiber and essential oils and that is why these are rich in essential vitamins and antioxidants. These juices normally contain more amounts of vitamin C, a nutrient that is vital in immunity boosting, better skin and body health. Also, they have high antioxidant; flavonoids and carotenoids which assist in reducing oxidative stress and prevent chronic diseases [47, 60]. In this way, cloudy juices are much more nutritional in comparison to clear juices due to their distinguished taste.

Another likely factor that affects the economic feasibility of cloudy concentrated citrus juices is processing techniques and market trend. While making cloudy juices the process might be slightly more elaborate and might entail the use of a mechanical press or enzymes the process gives higher yield and less spoil. Also, the minimal clarification process leads to cutting of expenses that might have been incurred in filtration. Although there are increased costs in equipment and technology, the rate of processing fruits and producing cloudy juices can be more effective and costs can be contained in the long run due to increased efficiency and the potential for obtaining a better price in the market [61].

A new trend is the need for cloudy concentrated citrus juices in the market mainly due to twenty-first-century consumers' pull towards natural and processed food that is free from chemicals and additives that enhances the taste and nutritional value of the product. As consumers knowledge about the health benefits of natural pulp and fiber on juices increases those willing to pay for those sort of juices, as they are perceived to be healthier are also increasing [62]. This is coupled by the fact that cloudy juices bear a natural and fresh feel and mouth-feel respectively as, compared to bright juices. The market trends reveal that people are moving towards healthier lifestyle which will help drive the sales of these high-end juice brands.

In general, the nutritional values and cost effectiveness of cloudy concentrated citrus juices that are attributed to this product imply a highly effective product in the beverage market areas. The vitamins and antioxidants which these organisms are rich in together with the low production costs and high popularity make them quite ideal for the market.

19.5 Conclusion and Future Innovations

Modern studies are more concerned with re-profiling of waste citrus; especially the peels in order to add value. Researchers are looking for possibilities of developing methods for the extraction of valuable compounds like essential oils, flavonoids, and dietary fibers from citrus by products that might be used in the production of pharmaceuticals, functional foods, and cosmetics. Such techniques as, supercritical fluid extraction, and green chemistry are in the process of being developed to make the various processes more efficient as well as environmentally friendly. It's about achieving an enhanced economic and environmental worth of waste citrus in order to create a better market returns.

Other possible uses are proposing that pulp can be used in feed, milk, juice, and other products. Citrus waste besides juice production can be used in other industries as pointed out in the stated three sections. In the food industry citrus peel extracts are used for preservation and flavoring of foods owing to their natural antiseptic properties. In the agricultural field they are used as organic manure and feed supplements to animals. Orange extracts can be used in cosmetics as they contain bioactives that are effective antioxidants; they are also anti-inflammatory. Also, the use of materials from citrus waste for biodegradable packaging is still in development as a contender to conventional plastics.

Applying sustainable solution to the management and disposal of citrus waste is very important to reduce negative effects on the environment and support the principles of circular economy. Some of the practices include composting, which is a process of recycling of organic residuals and use of the resulting materials as fertilizers and soil conditioners, anaerobic digestion which is the process of producing biogas for renewable energy. To achieve zero waste utilization of citrus waste the various stakeholders in the chain ranging from the farmers and the manufacturers must consider the best way of utilizing the waste that is produced in the course of processing the fruits. Besides, promoting sustainable practices apprehend the environmental rendering that is vital to the citrus industry, such as the costs of waste disposal as well as providing an extra source of income from by-products.

Finally, future work and development on citrus waste management is focused on improving on the prospects of a sustainable and practical citrus industry. The industry could improve the environmental track and the solidity by engaging more innovations of researches, applying the technology in outlets other than juice production, and managing the untoward wastes adequately.

References

1. Ruiz, B., & Flotats, X. J. (2014). Citrus essential oils and their influence on the anaerobic digestion process: An overview. *Waste Management, 34*(11), 2063–2079.
2. González-Miquel, M., & Díaz, I. (2020). Valorization of citrus waste through sustainable extraction processes. In *Food industry wastes* (pp. 113–133). Academic Press.
3. Romdhani, Z., Sakji, N., & Hamdaoui, M. (2022). Eco-friendly dyeing of wool fabrics with natural dye extracted from citrus sinensis L peels. *Fibers and Polymers, 23*(6), 1621–1630.
4. Wedamulla, N. E., Fan, M., Choi, Y. J., & Kim, E. K. (2022). Citrus peel as a renewable bioresource: Transforming waste to food additives. *Journal of Functional Foods, 95*, 105163.
5. Werede, E., Jabasingh, S. A., Demsash, H. D., Jaya, N., & Gebrehiwot, G. (2023). Eco-friendly cotton fabric dyeing using a green, sustainable natural dye from Gunda Gundo (Citrus sinensis) orange peels. *Biomass Conversion and Biorefinery, 13*(6), 5219–5234.
6. Martínez-Abad, A., Ramos, M., Hamzaoui, M., Kohnen, S., Jiménez, A., & Garrigós, M. C. (2020). Optimisation of sequential microwave-assisted extraction of essential oil and pigment from lemon peels waste. *Food, 9*(10), 1493.
7. Dar, N. G., Hussain, A., Paracha, G. M., & Akhter, S. (2015). Evaluation of different techniques for extraction of antioxidants as bioactive compounds from citrus peels (industrial by products). *American-Eurasian Journal of Agricultural and Environmental Sciences, 15*(4), 676–682.
8. Rajput, S., Kaur, S., Panesar, P. S., & Thakur, A. (2023). Supercritical fluid extraction of essential oils from Citrus reticulata peels: Optimization and characterization studies. *Biomass Conversion and Biorefinery, 13*(16), 14605–14614.
9. de Andrade, L. M., Kestekoglou, I., Charalampopoulos, D., & Chatzifragkou, A. (2019). Supercritical fluid extraction of carotenoids from vegetable waste matrices. *Molecules, 24*(3), 466.
10. Abou Elmaaty, T., Sayed-Ahmed, K., Elsisi, H., & Magdi, M. (2022). Optimization of extraction of natural antimicrobial pigments using supercritical fluids: A review. *Processes, 10*(10), 2111.
11. Haque, E. U., Hayat, A., Asim, M., Afzaal, S., Hanif, M. S., Murtaza, M. A., Din, A., Sabir, S., & Ahmad, S. (2022). The supply and value chain of citrus fruit: Producer to consumer. In *Citrus production* (pp. 405–420). CRC Press.
12. Granone, L. I., Hegel, P. E., & Pereda, S. (2022). Citrus fruit processing by pressure intensified technologies: A review. *The Journal of Supercritical Fluids, 188*, 105646.
13. Vinha, A. F., Rodrigues, F., Nunes, M. A., & Oliveira, M. B. (2018). Natural pigments and colorants in foods and beverages. In *Polyphenols: Properties, recovery, and applications* (pp. 363–391). Woodhead Publishing.
14. Benucci, I., Lombardelli, C., Mazzocchi, C., & Esti, M. (2022). Natural colorants from vegetable food waste: Recovery, regulatory aspects, and stability—A review. *Comprehensive Reviews in Food Science and Food Safety, 21*(3), 2715–2737.

15. Gupta, R., & Barros, R. (2024). Bioconversion: Citrus fruit peel waste from hospitality sector to fashion sector. *Journal of Environmental Protection and Ecology, 25*(3), 813–824.
16. Mahmood, A., Bilal, B., Naeem, Z., & Iram, S. (2021). Physical, chemical, and biological remediation techniques for textile effluents in context with developed and developing countries. In *Rhizobiont in bioremediation of hazardous waste* (pp. 409–441). Springer.
17. Hardin, A., Crandall, P. G., & Stankus, T. (2010). Essential oils and antioxidants derived from citrus by-products in food protection and medicine: An introduction and review of recent literature. *Journal of Agricultural & Food Information, 11*(2), 99–122.
18. Chatzimitakos, T., Athanasiadis, V., Kotsou, K., Bozinou, E., & Lalas, S. I. (2023). Response surface optimization for the enhancement of the extraction of bioactive compounds from citrus limon peel. *Antioxidants, 12*(8), 1605.
19. Negro, V., Mancini, G., Ruggeri, B., & Fino, D. (2016). Citrus waste as feedstock for bio-based products recovery: Review on limonene case study and energy valorization. *Bioresource Technology, 214*, 806–815.
20. Bazemore, R., Rouseff, R., & Naim, M. (2003). Linalool in orange juice: Origin and thermal stability. *Journal of Agricultural and Food Chemistry, 51*(1), 196–199.
21. Marzocchi, S., Baldi, E., Crucitti, M. C., Toselli, M., & Caboni, M. F. (2019). Effect of harvesting time on volatile compounds composition of bergamot (Citrus × Bergamia) essential oil. *Flavour and Fragrance Journal, 34*(6), 426–435.
22. Rodov, V., Ben-Yehoshua, S., Fang, D. Q., Kim, J. J., & Ashkenazi, R. (1995). Preformed antifungal compounds of lemon fruit: Citral and its relation to disease resistance. *Journal of Agricultural and Food Chemistry, 43*(4), 1057–1061.
23. Mitiku, S. B., Ukeda, H., & Sawamura, M. (2001). Enantiomeric distribution of α-pinene, b-pinene, sabinene and limonene in various citrus essential oils. In *Food flavors and chemistry: Advances of the new millennium* (Vol. 274, p. 216). RSC.
24. Viuda-Martos, M., Ruiz-Navajas, Y., Fernández-López, J., & Pérez-Álvarez, J. A. (2009). Chemical composition of mandarin (C. reticulata L.), grapefruit (C. paradisi L.), lemon (C. limon L.) and orange (C. sinensis L.) essential oils. *Journal of Essential Oil Bearing Plants, 12*(2), 236–243.
25. Boelens, M. H. (1991). A critical review on the chemical composition of citrus oils. *Perfumer & Flavorist, 16*, 17–34.
26. Vekiari, S. A., Protopapadakis, E. E., Papadopoulou, P., Papanicolaou, D., Panou, C., & Vamvakias, M. (2002). Composition and seasonal variation of the essential oil from leaves and peel of a Cretan lemon variety. *Journal of Agricultural and Food Chemistry, 50*(1), 147–153.
27. Machado, C. A., Oliveira, F. O., de Andrade, M. A., Hodel, K. V., Lepikson, H., & Machado, B. A. (2022). Steam distillation for essential oil extraction: An evaluation of technological advances based on an analysis of patent documents. *Sustainability, 14*(12), 7119.
28. Masango, P. (2005). Cleaner production of essential oils by steam distillation. *Journal of Cleaner Production, 13*(8), 833–839.
29. Cakaloglu, B., Ozyurt, V. H., & Otles, S. (2018). Cold press in oil extraction. A review. *Ukrainian Food Journal, 7*(4), 640–654.
30. Darjazi, B. B. (2014). Comparison of peel components of pummelo (Citrus grandis) obtained using cold-press and hydrodistillation method. *Journal of Life Science and Biomedicine, 4*(2), 71–77.
31. Aydeniz, B., Güneşer, O., & Yılmaz, E. (2014). Physico-chemical, sensory and aromatic properties of cold press produced safflower oil. *Journal of the American Oil Chemists' Society, 91*(1), 99–110.
32. Afroja, S., Falgunee, F. N., Jahan, M. M., Akanda, K. M., Mehjabin, S., & Parvez, G. M. (2017). Antibacterial activity of different citrus fruits. *Specialty Journal of Medical Research and Health Science, 2*(1–2017), 25–32.
33. Ambrosio, C. M., Diaz-Arenas, G. L., Agudelo, L. P., Stashenko, E., Contreras-Castillo, C. J., & Da Gloria, E. M. (2021). Chemical composition and antibacterial and antioxidant activity of a citrus essential oil and its fractions. *Molecules, 26*(10), 2888.

34. Mahato, N., Sharma, K., Koteswararao, R., Sinha, M., Baral, E., & Cho, M. H. (2019). Citrus essential oils: Extraction, authentication and application in food preservation. *Critical Reviews in Food Science and Nutrition, 59*(4), 611–625.
35. Arena, M. E., Alberto, M. R., & Cartagena, E. (2021). Potential use of citrus essential oils against acute respiratory syndrome caused by coronavirus. *Journal of Essential Oil Research, 33*(4), 330–341.
36. Agarwal, P., Sebghatollahi, Z., Kamal, M., Dhyani, A., Shrivastava, A., Singh, K. K., Sinha, M., Mahato, N., Mishra, A. K., & Baek, K. H. (2022). Citrus essential oils in aromatherapy: Therapeutic effects and mechanisms. *Antioxidants, 11*(12), 2374.
37. Lertsatitthanakorn, P., Manwiwattanakun, K., Paengnakorn, N., & Khunkitti, W. (2014). Antibacterial activity of an effective essential oil formulated in liquid soap against skin bacteria. *Warasan Khana Witthayasat Maha Witthayalai Chiang Mai, 41*, 71–83.
38. Sandeep, D. S., Charyulu, R. N., Nayak, P., Maharjan, A., & Ghalan, I. (2016). Formulations of antimicrobial polyherbal hand wash. *Research Journal of Pharmacy and Technology, 9*(7), 864–866.
39. Erfando, T., Cahyani, S. R., & Rita, N. (2019). The utilization of citrus hystrix and citrus limon as an organic demulsifier formulation. *IOP Conference Series: Materials Science and Engineering, 509*(1), 012145. IOP Publishing.
40. Dosoky, N. S., & Setzer, W. N. (2018). Biological activities and safety of citrus spp. essential oils. *International Journal of Molecular Sciences, 19*(7), 1966.
41. Thangaleela, S., Sivamaruthi, B. S., Kesika, P., Bharathi, M., Kunaviktikul, W., Klunklin, A., Chanthapoon, C., & Chaiyasut, C. (2022). Essential oils, phytoncides, aromachology, and aromatherapy—A review. *Applied Sciences, 12*(9), 4495.
42. Irshad, M., Subhani, M. A., Ali, S., & Hussain, A. (2020). Biological importance of essential oils. *Essential Oils-Oils of Nature, 1*, 37–40.
43. Palazzolo, E., Laudicina, V. A., & Germanà, M. A. (2013). Current and potential use of citrus essential oils. *Current Organic Chemistry, 17*(24), 3042–3049.
44. Beveridge, T. (2002). Opalescent and cloudy fruit juices: Formation and particle stability. *Critical Reviews in Food Science and Nutrition, 42*(4), 317–337.
45. Akyildiz, A., & Ağçam, E. (2014). Citrus juices technology. In *Food processing: Strategies for quality assessment* (pp. 37–103). Springer New York.
46. Sreenath, H. K., Crandall, P. G., & Baker, R. A. (1995). Utilization of citrus by-products and wastes as beverage clouding agents. *Journal of Fermentation and Bioengineering, 80*(2), 190–194.
47. Uçan, F., Ağçam, E., & Akyildiz, A. (2016). Bioactive compounds and quality parameters of natural cloudy lemon juices. *Journal of Food Science and Technology, 53*, 1465–1474.
48. Sandhu, K. S., & Minhas, K. S. (2006). Oranges and citrus juices. In *Handbook of fruits and fruit processing* (pp. 309–357). Wiley.
49. Chandler, B. V., & Robertson, G. L. (1983). Effect of pectic enzymes on cloud stability and soluble limonin concentration in stored orange juice. *Journal of the Science of Food and Agriculture, 34*(6), 599–611.
50. Baldwin, E. A., Bai, J., Plotto, A., Cameron, R., Luzio, G., Narciso, J., Manthey, J., Widmer, W., & Ford, B. L. (2012). Effect of extraction method on quality of orange juice: Hand-squeezed, commercial-fresh squeezed and processed. *Journal of the Science of Food and Agriculture, 92*(10), 2029–2042.
51. Shearon, W. H., & Burdick, E. M. (1948). Citrus fruit processing. *Industrial & Engineering Chemistry., 40*(3), 370–378.
52. Chemat, F., Fabiano-Tixier, A. S., Vian, M. A., Allaf, T., & Vorobiev, E. (2017). Solvent-free extraction. In *Comprehensive analytical chemistry* (Vol. 76, pp. 225–254). Elsevier.
53. Lozano, J. E. (2006). Processing of fruits: Ambient and low temperature processing. In *Fruit manufacturing: scientific basis, engineering properties, and deteriorative reactions of technological importance* (pp. 21–54). Springer.

54. Mushtaq, M. (2018). Extraction of fruit juice: An overview. In *Fruit juices* (pp. 131–159). Academic Press.
55. Cassano, A., Drioli, E., Galaverna, G., Marchelli, R., Di Silvestro, G., & Cagnasso, P. (2003). Clarification and concentration of citrus and carrot juices by integrated membrane processes. *Journal of Food Engineering, 57*(2), 153–163.
56. Oikeh, E. I., Omoregie, E. S., Oviasogie, F. E., & Oriakhi, K. (2016). Phytochemical, antimicrobial, and antioxidant activities of different citrus juice concentrates. *Food Science & Nutrition, 4*(1), 103–109.
57. Jesus, D. F., Leite, M. F., Silva, L. F., Modesta, R. D., Matta, V. M., & Cabral, L. M. (2007). Orange (Citrus sinensis) juice concentration by reverse osmosis. *Journal of Food Engineering, 81*(2), 287–291.
58. Winniczuk, P. P., & Parish, M. E. (1997). Minimum inhibitory concentrations of antimicrobials against micro-organisms related to citrus juice. *Food Microbiology, 14*(4), 373–381.
59. Lima Tribst, A. A., de Souza Sant'Ana, A., & de Massaguer, P. R. (2009). Microbiological quality and safety of fruit juices—Past, present and future perspectives. *Critical reviews in microbiology., 35*(4), 310–339.
60. Romeo, R., Bruno, A. D., Piscopo, A., Medina, E., Ramírez, E., Brenes, M., & Poiana, M. (2020). Effects of phenolic enrichment on vitamin C and antioxidant activity of commercial orange juice. *Brazilian Journal of Food Technology, 23*, e2019130.
61. Zema, D. A., Calabrò, P. S., Folino, A., Tamburino, V. I., Zappia, G., & Zimbone, S. M. (2018). Valorisation of citrus processing waste: A review. *Waste Management, 80*, 252–273.
62. Satari, B., & Karimi, K. (2018). Citrus processing wastes: Environmental impacts, recent advances, and future perspectives in total valorization. *Resources, Conservation and Recycling, 129*, 153–167.

Chapter 20
Utilization of Citrus Waste to Develop Delivery Systems for Drugs and Nutraceuticals

Iahtisham-Ul-Haq, Saeeda Fatima, Kanza Saleem, Urwa Tariq, Basma Hatim, Eesha Ahmad, and Wajeeha Siraj

20.1 Introduction

Fruits and vegetables have long been essential to humans for nourishment and well-being, providing versatile nutrients and health benefits. Agricultural products not just fulfill the demand of a diverse range of food items but also meets other necessities including textile, medicinal and ornamental purposes. The use of agricultural products has been established for centuries as preservatives, additives, flavoring agents, aromatics, dyes and drugs *etc.*, playing significant role in human nutrition due to their substantial dietary composition [1, 2]. Epidemiological investigations have featured the impressive medical advantages of consuming leafy vegetables such as decreased risk of certain types of cancers and coronary heart anomalies [3]. These positive health benefits of food commodities are attributed to dietary fibers and secondary metabolites e.g., carotenoids and flavonoids have been known for their anticarcinogenic properties [4]. Remarkably, flavanols like catechin, epicatechin, and anthocyanidin found in apples and grapes, as well as flavanones prevalent in citrus fruits, are predominating flavonoids with therapeutic efficiency [5, 6].

Fruits, vegetables, and their extracted juices serve as the fundamental sources of micronutrients vital to human health. Amongst, citrus fruits are eminent which are cultivated in over 100 countries across the globe. Varieties including mandarin, lime, lemon, orange, and grapefruit hold significant prominence in citrus fruits being consumed worldwide. These organic products are not only consumed as fresh

Iahtisham-Ul-Haq (✉) · S. Fatima · K. Saleem · B. Hatim · E. Ahmad · W. Siraj
Kauser Abdulla Malik School of Life Sciences, Forman Christian College (A Chartered University), Lahore, Punjab, Pakistan
e-mail: iahtisham@fccollege.edu.pk

U. Tariq
Department of Human Nutrition and Dietetics, School of Food and Agricultural Sciences, University of Management and Technology, Lahore, Pakistan

A. Chauhan et al. (eds.), *Valorization of Citrus Food Waste*,
https://doi.org/10.1007/978-3-031-77999-2_20

fruits but also hold great influence in processing industries. However, during industrial processing, a huge amount of citrus food undergoes to waste bins leading to disposal of crucial components of citrus fruits such as peels, seeds and pulp residues [7, 8].

Citrus fruits including lemons, oranges, grapefruits, tangerines, limes, and mandarins stand out as the most extensively cultivated fruits over the globe, with consistent increase in production to satisfy the consumer's demand. Despite the notable consumption, a significant portion of citrus fruit is lost during industrial processing, with citrus peel constituting around half of the total fruit mass. However, citrus waste is rich in valuable components including flavonoids, sugars, dietary fiber, essential oils, polyphenols, carotenoids, ascorbic acid, and trace elements, offering significant economic potential. Moreover, the significant concentration of sugar present in citrus waste makes it more suitable for the production of bioethanol through fermentation [9].

Citrus family encompasses a variety of fruit cultivars that have been widely accepted as a prominent source of dietary bioactive compounds [10]. Among these, sweet orange (*Citrus sinensis*) is the most produced and consumed citrus food, followed by tangerine or mandarin (*Citrus reticulata*), lime (*Citrus aurantifolia*), grapefruit (*Citrus vitis*), and lemon (*Citrus lemonum*). Citrus fruits are well known for their heightened content of minerals, dietary fiber, vitamins, and many other health promoting bioactive components i.e., polyphenols, flavonoids, and carotenoids etc., which play vital role in human growth and development. Moreover, these bioactive mixtures have been extensively studied for their numerous therapeutic potential i.e., antioxidant, anti-inflammatory and anticancer properties [11]. Citrus fruits are typically cultivated in tropical and subtropical regions with an overall annual production of approximately 88 million tons. As the consumer demand is increasing for products with unique qualities as low-fat carbohydrates, low-sodium minerals, vitamins particularly B-complex and C and dietary fiber, the citrus fruits are becoming an integral component of human diet [12].

Approximately 80% of citrus fruits are utilized in the production of different products such as juice, jams, jellies, and preservatives, resulting in generation of roughly 40 million tons of waste, per annum [13]. Citrus by-products constitute soluble sugars, starch, soluble and insoluble fiber, pectin, gelatin, lignin, fat, protein, and various bioactive mixtures, providing potential opportunities for their applications across pharmaceuticals, nutraceuticals, beverages, cosmetics, and biofuel production. These renewable resources offer economic benefits by reducing the production cost, limiting the use of synthetic ingredients and minimizing the risk of environmental pollution [9, 14].

Overall, citrus fruits and byproducts play a significant role in human nutrition and wellbeing, offering a hub of bioactive components with diverse applications. Incorporating citrus waste provides an excellent opportunity for value addition and sustainability across different industries, underscoring the ways to enhance the potential benefits of these resources. In this context, citrus waste is gaining attention for its employment in the development of drug delivery systems. This chapter delves in the understanding of incorporating citrus-waste bioactive mixture in pharmaceutical industry to maximize drug efficiency.

20.2 Citrus Waste Byproducts and Bioactive Compounds

20.2.1 Citrus Waste Byproducts

Globally the production and cultivation of citrus fruit is increasing each year with an increase in consumer demand. However, a major part of these fruits is wasted as a result of direct consumption or industrial processing. These waste byproducts, consisting of peel, seeds, and pulp residue, are rich source of various bioactive compounds that exhibit therapeutic properties such as antioxidant, neuroprotective, anticancer, anti-inflammatory, and cardioprotective effects [11, 15]. Flavonoids, dietary fiber, ascorbic acid, essential oils, pectin, carotenoids, and polyphenols are among the significant compounds identified in citrus waste [16]. Citrus peels constitute a substantial portion of fruit weight and are discarded as waste after processing of fruits (canning and juicing). This waste contains elevated concentration of the aforementioned bioactive compounds as compared to other parts of the fruit such as seeds and pulp [15, 17]. Therefore, among all the waste products of citrus fruits such as seeds, pomace and peels, the citrus peel is most commonly used for the formulation of commercial products. Citrus peels are rich in sugars, which leads to production of bioethanol during fermentation and also contain greater amount of D-limonene, the compound with powerful antimicrobial properties [12, 18]. The yield of bioactive compounds extracted from citrus waste depends on the fruit processing techniques and extraction methodologies employed. The inherent structural and chemical properties of these compounds make them valuable in the development of delivery systems for drugs and nutraceuticals [15].

20.2.2 Potential Role of Citrus Bioactive Compounds in Drugs and Nutraceuticals Delivery Systems

20.2.2.1 Pectin

Pectin, a polysaccharide, is a component of the plant cell wall, which plays a vital role in their growth [19]. It is particularly abundant in citrus peels and most of the pectin commercially used around the world is extracted from apple and citrus waste [17, 20]. Pectin has become an important compound in delivery systems for drugs and nutraceuticals due to its abundance in nature, non-toxicity, bio-compatibility, biodegradability, and flexible structural properties Moreover, pectin and its modified nanocomposites are being used in pharmacological industries for their antioxidant, antidiabetic, anticancer, antibacterial, anti-inflammatory, and immunoregulatory properties [21]. Furthermore, it is mainly used as an encapsulate in drug delivery systems in a variety of formulations such as films, hydrogels, nanoparticles, and microspheres, which are tailored to target organ-specific medications [17, 22, 23]. Similarly, pectin-based emulsions and hydrogel technology have also been utilized in the controlled delivery of exogenous nutraceuticals [24]. Due to its hydrophilic nature, pectin holds gel-forming properties, making it a suitable stabilizing agent for controlled drug release [19, 21].

20.2.2.2 Essential Oils (EOs)

EOs are naturally occurring oily liquid with a mixture of volatile & aromatic compounds produced in different parts of plants resultant to the secondary metabolism [25]. EOs are primarily composed of aldehydes, ketones, carbohydrates, ether, alcohol, sesquiterpenes, and monoterpenes [26]. Terpenes, like d-limonene and linalool, are the main components present in citrus oils [27]. The bioactive components found in Citrus EOs exert various therapeutic properties such as antimicrobial, anti-inflammatory, and antioxidant effects that can be incorporated in the development of drug delivery systems [11]. The major barrier for permeation of drugs across the skin is its outermost layer, the stratum corneum. EOs with high content of terpene have been found as promising candidate to increase the permeation of drugs through the skin [26]. Therefore, citrus EOs have potential applications in the development of transdermal drug delivery systems. Moreover, they have also been used in nano emulsions, showing increased solubility in a water-based system, making them ideal for the effective transportation and absorption of active compounds [28] (Table 20.1).

20.2.2.3 Polyphenols

Polyphenols are organic compounds produced as secondary metabolites exclusively in plants. Polyphenols are naturally present in vast dietary sources like nuts, fruits, vegetables, seeds, barks, flowers and in beverages such as tea, wine, and beer etc. The primarily occurring polyphenols with their known health benefits include epigallocatechin, epigallocatechin gallate, catechin, quercetin, procyanidin, gossypol,

Table 20.1 Citrus-waste based bioactive compounds and their therapeutic effects

Bioactive Compound	Therapeutic Effects	References
Pectin	Anticancer, antidiabetic, antioxidant, anti-inflammatory, immunoregulatory, antibacterial	[17, 21]
Essential Oils	Anti-inflammatory, antioxidant, antimicrobial	[11, 26]
Polyphenols (flavonoids & phenolic acids)	Antioxidant, anti-inflammatory, anti-allergic, anticarcinogenic, neuroprotective, antimicrobial	[15]
Carotenoids	Antioxidant, anti-inflammatory, anticancer, cardioprotective	[11, 29]

gallic acid, curcumin, caffeic acid, and many others [30]. Considering the citrus fruits, polyphenols are found throughout the whole fruit. However, recent studies have reported that inedible portions of citrus fruits i.e., albedo and flavedo parts constitute a higher concentration of polyphenols compared to the edible fruit. In an attempt to explore the comparative content of polyphenols in pomace and juice of lemon, orange, mandarin, and bergamot, it was observed that polyphenol content was 3–4 times higher in pomace in contrast to the respective juices [31]. Citrus peels and seeds are also rich in phenolic acids and flavonoids. Among flavonoids, six distinct types have been found in citrus peels depending upon their chemical structures such as flavanones, flavones, flavanols, flavanols, isoflavones and anthocyanidins. These compounds offer many health benefits due to their antioxidant, anti-inflammatory, anti-allergic, anticarcinogenic, neuroprotective, vasodilation, and antimicrobial properties [15].Polyphenols exhibit potential to transport small drug molecules by serving as constituent in creation of biomaterials through non-covalent interactions or dynamic covalent bonds. Moreover, polyphenols from various natural sources have been employed in the production of coatings, capsules, films, hydrogels, and nanoparticles. Nanoparticles derived from polyphenols offer various advantages in drug delivery systems by improving solubility in biological media, stability, increased cellular uptake, controlled release etc. Moreover, the inherent therapeutic properties of polyphenols could enhance the efficacy of medications [30, 32].

20.2.2.3.1 Carotenoids

Carotenoids are widely distributed organic pigments that impart bright colors i.e., yellow, orange, and red to plants. Carotenoids contribute to various vital processes in plants such as photosynthesis, pollination, hormone biosynthesis and quenching free radicle species [33]. Citrus fruits contain carotenoids like lutein, lycopene, zeaxanthin,β-cryptoxanthin, α-carotene and β-carotene which are present in considerable amount in flavedo (the outer layer of peel) compared to the fruit pulp and other components [11, 34]. These compounds have antioxidant properties, promote immune function, and reduce oxidative stress, subsequently reducing the risk of developing cancer [35], arteriosclerosis, metabolic syndrome [36], macular degeneration, and arthritis. Additionally, some carotenoids like β-carotene acts as precursor of vitamin A and hence, prevents from the deficiency of vital vitamin [11]. Furthermore, carotenoids such as β-carotene, and zeaxanthin have been used as active ingredients in the production of drugs and supplements as oral and topical formulations [37]. Moreover, lycopene found in pink grapefruit exhibits synergistic effects with other bioactive molecules and has demonstrated the ability to combat several types of cancer through apoptosis. Co-administration of β carotene with antidiabetic medication i.e., Metformin, has been found to enhance the prognosis of diabetes and hyperlipidemia due to its antioxidant properties [29] (Table 20.2).

Table 20.2 Functions of citrus-waste based bioactive compounds in drug delivery systems

Bioactive Compound	Function in Drug Delivery Systems	References
Pectin	Encapsulant, stabilizing agent, gelling agent, controlled release of drugs and nutraceuticals	[17, 21, 22, 24]
Essential Oils	Enhances permeability of transdermal drugs, Potential drug and nutraceutical carrier	[26, 28]
Polyphenols (flavonoids & phenolic acids)	Potential role in developing encapsulants	[30, 32]
Carotenoids	Synergistic effects with other bioactive molecules	[29]

20.2.3 Extraction of Bioactive Compounds from Citrus Waste

20.2.3.1 Conventional Methods

Conventional methods to extract bioactive compounds include cold pressing, steam distillation, hydro distillation, solvent extraction, maceration and Soxhlet extraction. Cold pressing is a process in which EOs are mechanically extracted without heat so the EO retains its volatile compounds [38]. Steam distillation and hydro distillation are among the widely employed techniques to extract EOs and polyphenols. In these processes, the bioactive compounds are vaporized by the application of steam, subsequently condensed and separated [38, 39]. In solvent extraction citrus peels undergo soaking in a solvent, typically water, ethanol, or methanol for extraction of pectin, polyphenols and EOs [40]. Maceration involves immersion of peels in a solvent, followed by agitation, facilitating the dissolution of bioactive compounds, and is used to extract polyphenols and carotenoids. Soxhlet extraction involves continuous extraction from a solvent, typically in a reflux setup and is utilized to extract EOs, carotenoids and polyphenols from citrus [41].

20.2.3.2 Non-conventional Methods

Non-conventional methods of extraction include ultrasound-assisted extraction (UAE), microwave-assisted extraction (MAE), supercritical fluid extraction (SFE) and enzyme-assisted extraction (EAE). UAE involves using ultrasound waves to induce cellular disruption which expedites the extraction of bioactive compounds [40–42]. In MAE, the cell walls in citrus peels undergo breakdown. Both methods, MAE and SFE, are utilized to extract pectin, polyphenols, carotenoids and EOs. SFE encompasses utilization of fluids heated and pressurized to a state where they exhibit characteristics of both gases and liquids. Carbon dioxide is commonly employed as a fluid for extracting EOs, carotenoids and polyphenols [40–42]. EAE has been utilized for extracting pectin, carotenoids, and polyphenols. It uses enzymes that modifies the structure of cell wall, thereby increasing its permeability and ultimately leads to increased extraction yields [40, 43].

20.3 Utilization of Citrus Waste in Drug Delivery Systems

As listed earlier in the chapter, different citrus waste-based byproducts hold the potential for use as active drugs and nutraceutical delivery systems. The following sections highlight the strategies to utilize these by-products for the production of delivery systems (Fig. 20.1).

20.3.1 Formulation and Types of Delivery Systems of Drugs and Nutraceuticals

The formulation of novel drug delivery systems is based upon the principles of nanotechnology. In the field of medicine science, nanoparticles are of immense consideration because of their promising implementations. Nanoparticles based drug delivery systems enhance bioavailability and biocompatibility of active substances, improve on site targets, and reduce off site targets [44]. The geometrical shape, size,

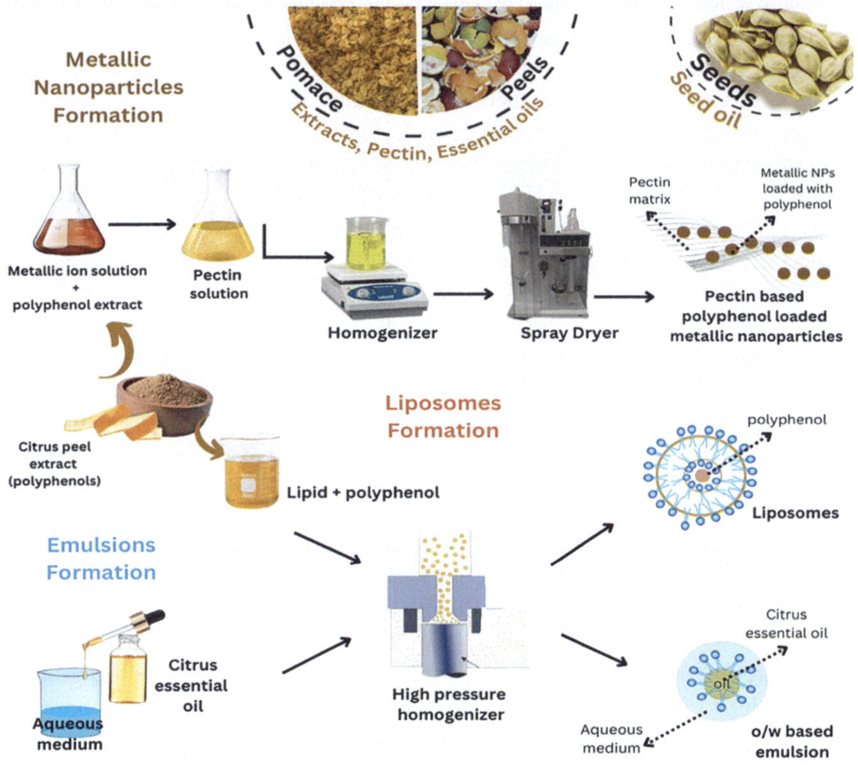

Fig. 20.1 Formulation of drug delivery systems from citrus waste

raw material and the techniques employed for the synthesis of nanoparticles are the determinant factors which play the crucial role in the formation of drug delivery systems [45, 46]. The consumers demand for biomaterial and green technology-based drug delivery systems is increasing [47]. Keeping this in view, medical researchers are paying attention to the development of advanced delivery systems originated from natural sources by valorization of low graded foods and food wastes. On the basis of materials, the drug delivery systems are majorly categorized into three groups namely polymeric (dendrimers, micelle, nanospheres), lipid based (liposomes, emulsions) and inorganic (metallic nanoparticles, silica nanoparticles, carbon nanotubes) [45, 46, 48]. Metallic nanoparticles, emulsions and liposomes are widely synthesized from citrus waste by-products [49, 50].

Metallic nanoparticles are considered one of the potent drug delivery systems because of their distinct physicochemical attributes. Silver [51, 52], gold [53], cadmium oxide [44], copper [54–56] and iron [56] based nanoparticles are prepared and characterized from citrus peel extracts. Such metallic nanoparticles are known for their antioxidant, antimicrobial anti-inflammatory and anti-carcinogenic properties [44]. Metallic nanoparticles enhance active and passive targeting as well as half-life of the therapeutic substances, however they pose toxic effects on the human cells if exceeds its GRAS (generally regarded as safe) limit [57]. The metals conjugated with nanocarriers generate reactive oxygen species (ROS) which interact with the cell wall and cell membrane leading to the cell death [53].

Emulsions serve as another drug delivery system developed by utilizing citrus essential oils. Emulsions are thermodynamically stable mixture oil, aqueous medium and surfactant in the form of small droplets. *Citrus aurantium L.* (bitter orange) essential oil was used in formulation of self-nanoemulsifying drug delivery system against PANC1 and MCF7 cancer cell lines. Such drug delivery system enhanced the cytotoxic effects against the targeted cell lines [58]. Likewise, mandarin peel pectin was used to develop an emulsion-based drug delivery system for β-carotene incorporating corn, olive, and soybean oils. Pectin based emulsions paved a pathway for the efficient encapsulation of β-carotene, thereby enhancing its bioavailability [59]. Oil in water nanoemulgels were prepared by using lemon essential oil and pectin obtained from citrus peels [60]. Emulsions have a promising role in the development of novel drug delivery systems as they are considered safe for human cells, their preparation is relatively easy and cost effective. However, emulsions have low drug entrapment and encapsulation efficiency resulting in compromised long term stability [58].

Liposomes are also utilized as delivery vehicles for drugs and nutraceuticals. Such a delivery system is compatible for both hydrophilic and hydrophobic therapeutic agents. Liposomes are artificially created vesicles composed of at least one layer of phospholipids or cholesterol [61]. Bioactive extracts derived from citrus waste represent the potential sources for the preparation of liposomes. Liposomes were developed for the encapsulation of orange peel extract in order to increase its antibacterial activity against different bacterial strains [62]. Citrus peels constitute polymethoxyflavonoids which exhibit anti-lipase activity. Polymethoxyflavonoids extract was encapsulated in liposomes as the potential dietary supplement against obesity. The loading of polymethoxyflavonoids in liposomes elevated its encapsulation efficiency and sustained release capabilities [63]. Though the liposomes are generally nontoxic on human cells and demonstrate compatibility with hydrophilic and hydrophobic active substances, their production cost is relatively high with lower stability and shorter half-life [64].

Pharmacists and medical researchers are exploring the potentials of citrus waste in order to synthesize further novel drug delivery system such as microbeads [65] hydrogels [66] and carbon-based quantum dots [67]. However, there is limited data available on these emerging drug delivery systems and needs further investigations for safe and timed-release of drugs and nutraceuticals.

20.3.2 Methods of Encapsulation and Loading of Therapeutic Substances

Encapsulation refers to a process wherein the therapeutic and bioactive agents (core materials) are entrapped and protected into a coating material, typically polymers to promote targeted drug delivery. Microencapsulation represent a novel drug delivery system in which tiny particles of active ingredients are entrapped within a coating ranging from 1–100 μm [68]. Microencapsulation can be executed by a variety of techniques to develop targeted drug and nutraceutical delivery systems. Amongst, spray drying, freeze drying *in situ* polymerization, ionotropic gelation and high-pressure homogenization stand out as widely applied green encapsulation methods.

Among the encapsulation techniques, Spray drying is faster and more feasible and is utilized vastly for encapsulating the food components due to its low cost and flexibility. It is a thermal process which is performed above 80 °C for heat stable bioactive substances. In contrast for the encapsulation of thermosensitive substances, freeze drying is employed [69, 70]. In an attempt to explore the comparative stability of spray dried and freeze dried encapsulates, hesperidin obtained from citrus peel extract was encapsulated in maltodextrin by using both spray drying and freeze drying. Spray drying technique enhanced the structural stability and functional characteristics of hesperidin in comparison to freeze drying [71].

In situ polymerization is also applied to encapsulate the drug molecules within layers of different polymers. The polymeric structures are stabilized by the addition of cross linkers such as calcium chloride solution. Limonin and nomilin, extracted from *Citrus maxima* (pomelo) seeds were encapsulated in zein-alginate system showcasing promising oral delivery system to enhance chemo preventive activity [72].

Ionotropic gelation is another technique used for encapsulation of active substances. This technique works on the principle of electrostatic interactions between two ionic species; namely polymer and therapeutic agent. The bioactive substance is entrapped within the polymeric matrix and allow sustained release under favorable conditions [73]. This techniques has been used widely to formulate drug delivery systems i.e., the phenolics extracted from *Citrus medica L* peels were loaded in Ca-alginate microbeads using ionotropic gelation technique [65]. Additionally, pectin extracted from *Citrus junos* peel was used to encapsulate quercetin by ionic gelation within hydrogel beads [66]. Furthermore, a liposomal drug delivery system was developed by using high pressure homogenization technique to encapsulate polymethoxyflavonoids extracted from citrus peel [63].

20.3.3 Controlled Release Mechanisms of Drug Delivery Systems

Oral administration is the most common route of medications due to its advantage i.e., patient compliance and non-invasiveness. Whenever a drug is ingested, it enters into the acidic environment of stomach where it disintegrates and dissolves with stomach fluids (dissolution). Subsequently, the small molecules are transported to the targeted site by the blood, where the drug or nutraceutical is absorbed by diffusion [74]. The controlled release systems are usually devised to enable superior control over the drug exposure in certain time, to assist the passage of drug from several physiological barriers, to protect premature elimination of drugs before reaching the target site while minimizing its exposure to other delicate regions of the body [75] Controlled release mechanisms majorly work on the principle of dissolution, diffusion, and ion exchange [76, 77].

20.3.3.1 Dissolution and Diffusion

Dissolution is attributed to the polymer core erosion while diffusion is linked to the concentration gradient of the core material and its solubility. In dissolution-controlled release systems, drugs are usually coated in slowly dissolving polymers. While in diffusion-controlled release systems, drugs are encapsulated and released through diffusion via inert water-insoluble polymers. Dissolution and diffusion based controlled release are sub categorized into two approaches namely reservoir and matrix type [74, 78].

20.3.3.1.1 Reservoir Type Dissolution and Diffusion

In the reservoir approach, the core material is encapsulated with a thick rate controlling membrane (polymeric layer). With the passage of time, the coating material dissolves and pores are formed in the polymer, hence releasing the drug [74, 79]. The drug release is inversely proportional to the thickness of coating. Reservoir type dissolution and diffusion follows zero order kinetics [76].

20.3.3.1.2 Matrix Type Dissolution and Diffusion

In the matrix system, the drug is homogenized and compressed with a slowly dissolving carrier (polymeric matrix) and released gradually over time. Cellulose, polyethylene glycols, poly methylene acrylates and waxes are the example of slowly dissolving coating materials [74]. The matrix type dissolution and diffusion follow first order kinetics [76].

20.3.3.2 Ion Exchange-Controlled Release

The ionized drug binds to the charged resin, forming complex ion exchange resonates. The basic drug combines with cation exchange resin while acidic drug combines with anion exchange resin. When the poly salt resonates in contact with the internal environment of the body, the bond between drug and resin breaks. In this system, the drug is released in the free form and becomes active. The release of drug is influenced by the ionic environment of the resins. Therefore, ion exchange-controlled release of drug is also known as pH dependent control drug delivery system [80].

20.3.4 Citrus Waste Based Drug Delivery Systems

Allopathic medicines and therapeutic agents pose harmful side effects on the human health [81]. Keeping this in view, medical researchers are paying attention to the development of advanced delivery systems originated from natural sources by valorization of low graded foods and food wastes. Citrus waste is of significant importance in the field of pharmaceuticals and medicine because of its potential to revolutionize therapeutic delivery systems. Table 20.3 represents the compilation of experiments conducted to explore the potential of citrus waste in development of controlled release drug delivery systems.

20.4 Current Research and Developments

The following sections discussed the current research and development in utilizing citrus waste by-products for enhanced bioavailability and therapeutic efficacy of nutraceuticals and drugs. Furthermore, evidence-based scientific explanations have been reported to better comprehend the potential of these by-products for further exploration.

Table 20.3 Utilization of citrus waste in drug delivery systems

Citrus waste	Bioactive ingredient/by product	Drug loading technique	Drug delivery system	Compatible polymer(s)	Targeted disease/cell lines	Inference	References
Citrus reticulata peel	Dietary flavanone; hesperidin rich drug.	Ionic gelation followed by pH based coacervation method	Protein-inorganic nanohybrid carrier	Casein	Ovarian cancer (SKOV-3) and Breast cancer (MDA-MB-231) cell lines	Increased cytotoxicity of citrus peel extracted hesperidin drug against SKOV-3 MDA-MB-231 breast cancer cells. Stimuli-responsive (pH dependent) drug release behavior.	[56]
Citrus junos peel	Flavonoid aglycone; Quercetin	Inotropic gelation	Hydrogels	Pectin	Colon targets	Below 1% cumulative quercetin release after exposure to simulated gastric fluid and intestinal fluid. Quercetin was significantly released from the hydrogel beads (65.37–99.54%) when exposed to simulated colonic fluid for 12 h.	[66]
Citrus reticulata peel	Pancreatic lipase inhibitors; Polymethoxyflavonoids (PMFs)	Thin film hydration-high pressure homogenization	Liposomes	Lipids	Obesity	Significantly stronger anti-lipase effect for encapsulated PMFs compared with free PMFs.	[63]
Citrus grandis or Citrus maxima pomelo seeds	Limonoids; Limonin and nomilin	Polymer-polymer encapsulation	Zein-alginate encapsulated limonin and nomilin nanoparticles	Zein and alginate	Cancer	Nanoparticles exhibited improved release in simulated intestinal fluid with better anti-angiogenic and free radical-scavenging activity.	[72]

Citrus medica peel	Phenolics	Ionotropic gelation	Ca-alginate microbeads	Starch, carrageenan, and pectin	–	Pectin based microbeads showed excellent controlled release under *in-vitro* gastrointestinal conditions with highest encapsulation efficiency (90.18%) and antioxidant activity with an Effective Concentration 50% (EC_{50}) of 108.97 μg/mL.	[65]
Citrus limetta peel extract	Alcoholic and phenolic compounds such as carveol, linalool, menthol	–	silver nanoparticles (AgNPs)	Silver metal	Topical bacterial and fungal infections	Potent anti-biofilm properties and effectively permeabilized cell membranes against *Staphylococcus aureus, Escherichia coli. Micrococcus luteus, Streptococcus mutans, Staphylococcus epidermidis,* Antifungal activity against *candida* species.	[52]
Citrus sinensis aqueous fruit extract	L-ascorbic acid or vitamin-C	Microwave irradiation	Non-toxic copper nanoparticles	Copper metal	Bacterial infections	Higher antibacterial effects of on gram-positive (*S. aureus*) bacteria than the gram-negative (*E. coli*). Non-cytotoxic nature on the L929 mouse fibroblast cell line.	[55]

20.4.1 Enhanced Bioavailability and Therapeutic Efficacy of Drugs and Nutraceuticals by Citrus-Based Carriers

Citrus byproducts, rich in compounds like pectin and limonene, serve as carriers in drug delivery systems, enhancing bioavailability. Pectin/chitosan nanoparticles (NPs) are utilized to encapsulate hydrophobic nutraceuticals like quercetin, while pectin-coated beads enable controlled release particularly in the intestine [82, 83]. Citrus-derived NPs transport drugs like Diclofenac, Ibuprofen, Etodolac, and Paracetamol, enhancing their efficacy, and improve the therapeutic potential of compounds like curcumin and resveratrol in nutraceutical delivery systems [84–86].

Citrus essential oils have anti-inflammatory and antioxidant properties. Limonene, a key terpene in citrus oils, enhances drug delivery through its permeation-promoting properties when incorporated into various nanocarriers, allowing better skin permeation and transdermal delivery. Citrus oil/terpene nano formulations enhance transdermal drug delivery and improve antioxidant activity for advanced skin applications [27]. Similarly, pectin can effectively interact with antibiotics through attraction and bonding. This enables the packing of more antibiotics into the carrier, maximizing their content in the polysaccharide matrix. Several studies suggest that pectin-based complexes offer a sustainable alternative and can overcome challenges related to drug bioavailability. Pectin can be modified to enhance its properties and tailor its release kinetics, making it suitable for controlled drug delivery systems. Incorporating citrus byproducts in drug delivery system (DDS) allows regulated molecular delivery of drug [87]. Non-specific distribution of anticancer drugs and their tendency to exhibit various side effects emphasized the need of nanocarriers to enhance specificity. Employing citrus-derived NPs minimizes toxicity concerns and unwanted side effects exhibited by the drug alone. Moreover, green synthesized NPs are biocompatible and biodegradable [88].

Pectin holds various properties that make it a suitable agent to be harnessed for NPs synthesis. Pectin is water soluble, biodegradable, and contains free carboxyl and hydroxyl groups. Due to presence of free carboxyl and hydroxyl groups, pectin binds well with protein and encapsulates the drug. Moreover, pectin is resistant to proteases present in gastrointestinal tract thereby increasing the targeted delivery to colon. Prolonged release of drug results in reduction of drug dose. Owing to its hydrophilic properties, pectin ensures regulated and prolonged release of drug into the system. Pectin improves drug entrapment efficiency thereby minimizing the risk of drug leakage. Higher pectin concentrations result in enhanced adhesiveness of polysaccharide-protein matrices with drug [89]. Pectin breakdown by pectinase in colon facilitates the use of pectin NPs for targeted molecular delivery of a drug or nutraceutical to colon. Hence, pectin is a good candidate to be incorporated in colon specific drug and nutraceutical delivery [90]. Pectin has many galactose residues and can bind to asialoglycoprotein receptor (ASGPR) present in liver parenchyma cells. This property of pectin can be utilized to deliver liver targeted drugs. Moreover, pectin can serve as a good nanocarrier to possibly mitigate hepatic cancer by regulating target delivery to liver [91].

20.4.2 Case Studies and Scientific Experiments Confirming Preventive Healthcare Via Improved Bioavailability

The treatment of metformin loaded pectin NPs showed no signs of hemolysis suggesting its hemocompatibility if administered *in-vivo*. Moreover, the production of reactive oxygen species (ROS) was reduced in response to administration of metformin-loaded pectin NPs. Decreased oxidative stress (OS) may be attributed to the antioxidant properties of pectin. In addition to enhancing Metformin's bioavailability and regulating its targeted delivery, pectin also reduced OS and improved glucose uptake by red blood cells (RBCs) thereby alleviating diabetes mellitus [15] as mentioned in (Table 20.4). The co-adjuvant effect of pectin and metformin may enhance the prognosis of DM [89]. Chlorogenic acid (CGA) is usually subjected to degradation on its way to colon. Using pectin-derived NPs circumvents this issue, owing to its ability to dodge proteases. Therefore, pectin may serve as a good nanocarrier in providing stability to NPs in acidic environment thereby preventing degradation of NPs and drug leakage [92]. Doxorubicin (DOX) is a hydrophobic drug and encapsulating it in pectin NPs enhances its solubility. DOX loaded in pectin NPs ameliorates its ability to cause oxidative damage in HeLa (cervical cancer) cells [93].

According to a study, zein-Vitexin Rhamnoside-pectin nanoparticles were synthesized through hydrogen bonding and electrostatic interactions which were analyzed by SEM (Scanning Electron Microscopy), FTIR (Fourier transform infrared spectroscopy), and AFM (atomic force microscopy). Vitexin Rhamnoside (VR) encapsulation efficiency reached 67%, indicating successful integration into the zein-pectin matrix. The nanoparticles displayed stability across a pH range of 3–8.5 and resistance to lower ionic strength. In both *in-vitro* digestion and *in-vivo* rat intestinal perfusion experiments, the nanoparticles exhibited significant slow-release properties, leading to enhanced VR absorption in the duodenal segment, as reflected in improved absorption rate and permeability, thereby ultimately enhancing bioavailability [94]. In another study, a more effective form of nutraceutical quercetin was synthesized by chemically modifying it with docosahexaenoic acid which was subsequently combined with pectin using an eco-friendly method. Cell tests confirmed that the pectin-quercetin blend had no harmful effects at normal dietary levels and showed potential in decelerating the movement of certain cancer cells. Moreover, this pectin combination augmented quercetin's effectiveness in the body and protected it during transport [82].

The incorporation of pectin contributed to the pH responsiveness in the case of colon-targeted release of Resveratrol [8]. Polymeric nanoparticles (PNPs) were prepared using pectin through nebulization/ionotropic gelation. These pH responsive PNPs demonstrated remarkable control over RES release, exhibiting only 3% release in acidic conditions over 2 hours and achieving an 85% release at pH 6.8 over 30 hours. Furthermore, the permeability increased from 0.15% in the Caco-2 model to 5.5%, indicating enhanced drug transport through the intestinal barrier, particularly in the presence of mucus showcasing the potential for controlled RES

Table 20.4 Use of citrus bioactive compounds in drug and nutraceutical delivery systems

Citrus bioactive compounds	Drug/Nutraceutical	Study Type	Subjects/Cell lines	Effects	References
Pectin	Metformin	*In-vitro*	RBCs and L6 cell lines	↓ ROS ↑ PI3K ↑ GLUT4	[89]
Pectin	SAMe	*In-vitro and In-vivo*	SGF and SCoF2; Wistar male rats	↑ Colonic absorption	[88]
Pectin	CGA	*In-vitro*	HCT-116 cells	↓ HCT-116 cells survival	[92]
Pectin	Cetuximab + Curcumin	*In-vitro*	Caco-2 cells and HCT-116	↑ HCT-116 and Caco-2 apoptosis	[101]
Pectin	DOX	*In-vitro*	HeLa and HEK93 cells	↓ HeLa cells viability	[93]
Pectin	Curcumin	*In-vitro*	HepG2 cells	↓ HepG2 cell growth ↑ Antitumor activity	[85]
Pectin	Resveratrol	*In-vitro*	HepG2 cells	↑ Antioxidant activity	[86]
Pectin	VR	*In-vitro and In-vivo*	Duodenal cells in rats; SGF and SIF	↑ VR absorption and slow release of VR	[94]
Pectin	Quercetin	*In-vitro*	Caco-2 cells, 3 T3 fibroblasts, HepG2	↑ Bioavailability	[82]
Pectin	Baclofen (Muscle Relaxant)	*In-vitro*	Normal fibroblast cell lines.	Controlled drug release	[83]
Pectin	Resveratrol	*In-vitro*	Caco-2 cell model and mucus-secreting triple co-culture model	↑ Permeability, Sustained release of resveratrol	[95]
Pectin	Ceftizoxime	*In-vitro*	Fibroblast cell lines	↑ Drug release ↑ Microbial activity	[96]
Pectin	Insulin	*In-vitro and In-vivo*	Type 1 diabetic rats; SGF and SIF	↑ Drug release control, Stability, and *in-vivo* efficacy	[97]
Pectin	Tetracycline	*In-vitro*	Bacterial cell strains	↑ Drug binding and loading	[100]

(continued)

Table 20.4 (continued)

Citrus bioactive compounds	Drug/Nutraceutical	Study Type	Subjects/Cell lines	Effects	References
Limonene	Carvedilol	*In-vitro and In-vivo*	JB6 mouse epidermal cells; Franz diffusion cells	↑ Skin permeation and potential chemo preventive activity	[98]
Limonene	DOX	*In-vitro*	Cancer cells	↑ Solubilization ↑ Antioxidant activity	[99]

Abbreviations: *ROS* Reactive Oxygen Species, *P13K* Phosphoinositide 3-kinase, *GLUT4* Glucose Transporter Type 4, *SAMe* S-adenosyl-L-methionine, *SGF* Simulated Gastric Fluid, *SCoF2* Simulated Colonic Fluid, *SIF* Simulated Intestinal Fluid, *CGA* Chlorogenic Acid, *VR* Vitexin Rhamnoside, *DOX* Doxorubicin

delivery to the colon, combining pH responsiveness and enhanced permeability [95]. The citrus byproduct pectin significantly enhances drug bioavailability by acting as a biocompatible nanocarrier for ceftizoxime which is a third-generation cephalosporin antibiotic. The Ceftizoxime-loaded pectin nanoparticles (CPNs) exhibited efficient encapsulation, with an encapsulation efficiency of 69.4%, emphasizing pectin's role in enhancing drug delivery and improving the bioavailability of the drug. Pectin's adjustable surface properties contribute to the controlled drug release mechanism. The CPNs also demonstrated effective antimicrobial activity against various bacteria, underscoring the potential of pectin as a carrier for enhanced drug delivery [96].

Pectin plays a crucial role in improving the functionality of insulin by enhancing its availability and effectiveness. It is pivotal in oral insulin delivery due to its strong gelling, mucoadhesive characteristics in the gastrointestinal tract, specifically in developing the folic acid (FA)-modified pectin nanoparticles (INS/DFAN). A method employing dual crosslinking enhances stability and controls the release of insulin, thus preventing premature release in simulated gut conditions. Additionally, pectin improves the transit of insulin across gut cells, enhancing its overall availability. It was therefore observed that orally administered INS/DFAN significantly reduced blood glucose levels in diabetic rats demonstrating improved insulin bioavailability [97] (Fig. 20.2).

Another citrus byproduct, Limonene has been observed to increase skin permeation of carvedilol, which is evidenced by *in-vitro* assays with JB6 mouse epidermal cells and Franz diffusion cell experiments using rat skin. The combination of carvedilol and limonene showed higher colony inhibition in cell assays, and limonene significantly increased the permeation of carvedilol suggesting its ability to enhance drug bioavailability in topical delivery. Overall, limonene, acting as a carrier, improves skin permeation and bioavailability, potentially enhancing drug efficacy [98]. Limonene, as a carrier, significantly enhances solubilization and absorption within the body. Additionally, it provides antioxidant activity, shielding the drug from degradation and oxidative damage, thus ensuring sustained

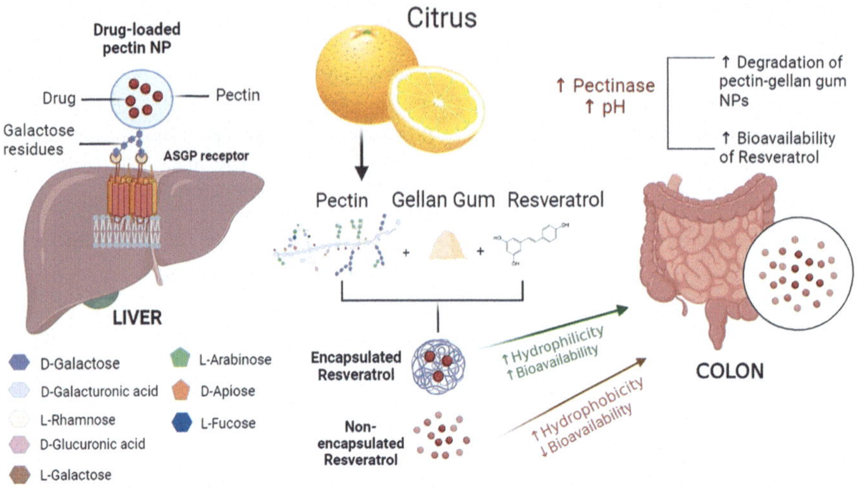

Fig. 20.2 Targeted delivery and enhanced bioavailability of drug loaded pectin nanoparticles in liver and colon

effectiveness. The formation of self-assembled nanoparticles, including nano emulsion and polylactide nanoparticles, further improves drug solubility and stability, resulting in a sustained release profile that prolongs the therapeutic potential of drug [99].

Moreover, the formation of pectin-based complexes with Tetracycline, facilitated by electrostatic attraction and hydrogen bonding, highlights the capacity of pectin to efficiently bind with drugs. This interaction enhances drug loading within the carrier, maximizing the content of Tetracycline in the polysaccharide matrix [100].

20.5 Environmental and Economic Impacts

20.5.1 Environmental Impact of Citrus Byproducts

The waste from citrus processing industries may pose a serious threat to the environment if not managed properly or left unhandled [102]. The improper disposal of citrus waste into water bodies results in the reduction of oxygen availability resulting in deleterious effects on aquatic life. Several characteristics of citrus fruits can elevate the burden of environmental pollution. These include their high organic content, resulting in easy biodegradability and increased susceptibility to fermentation. Hence, ample oxygen would be required to carry out the degradative conversion [103]. High sucrose content present in the citrus waste makes it more susceptible to fungal spoilage. Fungal contamination results in the production of mycotoxins that exacerbate soil quality [104]. Furthermore, high moisture content and low pH of citrus accelerate its spoilage and increase the rate of microbial colonization [17]. Unprocessed citrus waste exhibits detrimental effects on the environment by increasing greenhouse gas (GHG) emissions and impairing soil fertility eventually contributing towards climate change [105]. Due to the perishable nature of citrus waste, soil quality is greatly compromised subsequently exerting negative effects on crops and vegetation [106]. The improper disposal of waste significantly impairs soil quality by depleting nitrogen levels thus impeding the growth and development of plants [14]. Anaerobic decomposition of citrus waste by microorganisms leads to elevated production of methane in landfill sites. Methane gas contributes to global warming by the greenhouse effect. Moreover, low pH of citrus waste leads to pronounced leachate production, that can aggravate soil quality and pose threat to aquatic life [107]. A substantial amount of citrus waste undergoes combustion leading to release of sulfur, carbon and nitrogen oxides thus escalating climatic impacts [103, 108].

Keeping in view of the above observations, valorization of citrus waste can help in mitigating the aforementioned detrimental effects. As elaborated in the former parts of the chapter, citrus waste contains a diverse range of reusable byproducts including dietary fibers, polyphenols, pectin, and other bioactive compounds. This indicates its multifarious nature which can be exploited to produce functional products [109]. Various valorization techniques including controlled anaerobic digestion and fermentation are gaining attention due to their eco-friendly nature [110]. For an instance, enzymatic assisted extraction technique doesn't involve any solvent, thereby reducing environmental burden [102]. The valorization of citrus waste using green extraction techniques such as supercritical fluid extraction (SFE) may minimize risk of environmental concerns and allow mass production of notable biological products.

20.5.2 Economic Feasibility of Citrus Byproducts for Drugs and Nutraceuticals Delivery Systems

The byproducts from citrus industry include numerous bioactive compounds that can be utilized to increase the efficacy of drugs and nutraceutical delivery system. Utilization of these compounds to enhance the bioavailability of drugs and nutraceuticals can significantly reduce the environmental and economic burden by decreasing the overall cost associated with waste management methods [111]. Moreover, employing bioactive compounds from citrus waste to modulate the bioavailability and efficacy of drugs and nutraceuticals can minimize the dependence on synthetic active ingredients [10, 112].

Traditional raw materials used to formulate nanoparticles (NPs) necessitates advanced equipment, expertise, and extensive toxicity analysis procedures to ensure the safety of NPs. Moreover, the conventional methods used to manufacture NPs are energy-demanding involving high temperature & pressure and potentially toxic reducing agents [113, 114]. Owing to the scarcity and instability of conventional raw materials, harnessing abundant and stable citrus byproducts for NPs production is a promising strategy to adopt. Moreover, citrus-derived NPs are considered to be biocompatible and safer, thus minimizing the need of safety tests and alleviating the economic burden [11, 115]. Several bioactive compounds act as reducing agents and enhance the stability of NPs rendering them a sustainable and cost-effective choice for NPs synthesis [111, 116]. Furthermore, flavonoids and other phytochemicals may serve as efficient stabilizing and capping agents, thereby reducing the reliance on external capping or stabilizing agents subsequently minimizing the cost. Production of citrus-derived NPs is quicker than exploiting microorganisms for NPs synthesis [113]. Additionally, utilization of citrus byproducts can promote large-scale production of NPs. A diverse range of metallic NPs of distinct sizes and shapes can be produced in a single-step synthesis procedure [117]. Considering the synergistic effect of citrus byproducts, drug dose could be reduced while enhancing the therapeutic potential of the treatment, ultimately enhancing the economic viability [89]. The synergistic effect of antibiotics and citrus-derived NPs facilitates rapid prognosis that can potentially lower the net treatment costs in long-term clinical settings [118]. Furthermore, use of citrus-derived nanoparticles and encapsulants can foster circular economy and zero waste approach [106]. Conclusively, the use of citrus waste and other agricultural byproducts may potentially enhance the economic sustainability and reduce the climatic impacts associated with the unhandled waste materials.

20.6 Challenges, Future Aspects and Conclusions

Conventional methods such as solid-liquid extraction to obtain bioactive compounds are often unsuitable because of various limitations. These procedures are often time-consuming and associated with reduced yield, elevated cost and high risk of impurity [10, 119]. In this regard, green extraction methods including pulse electric field extraction, ultrasound-assisted extraction and super critical fluid extractions may serve as efficient procedures to circumvent this issue [120]. Owing to high water content in citrus peels, the drying process can be challenging to execute and may also result in releases of volatile organic compounds [121]. Increased quantity of essential oils present in citrus, hinders the efficacy of composting, one of the most applied methods for citrus waste management, by significantly reducing the rate of fermentation [122]. Lack of resources and awareness regarding the potential valorization of citrus waste products into novel compounds are the major challenges in developing countries [9]. Therefore, implementation of cost-effective and eco-friendly waste management practices and increasing awareness about valorization of byproducts to minimize economic and environmental burdens are of paramount importance.

Although there are numerous advantages of adopting green methods for synthesis of nanoparticles, various factors such as incubation time, temperature and pH of the medium can affect the rate of nanoparticle synthesis. Consequently, this may hinder the efficacy of drug and nutraceutical delivery leading to an increased financial burden [113]. Despite these limitations, it is suggested that synthesis of citrus-derived NPs may augment the efficiency of targeted drug and nutraceutical delivery. However, very few studies and research exist to validate this phenomenon. Therefore, there is a growing need to conduct research to test and observe the safety and efficacy of citrus-derived NPs and address any potential economic concerns before incorporating it to global scale.

Using citrus waste to extract bioactive compounds presents several challenges. Standardize the yield and chemical composition of these compounds is difficult due to differences in factors such as extraction technique, type and variety of citrus fruit, and climatic conditions [17, 27]. Further, not much is known about the safety and toxicity of these compounds [20]. Therefore, there is a need for future studies that focus on investigating the potential curative properties of the citrus-derived bioactive compounds, especially to identify their safety limits and interactions with other materials regarding drug delivery systems [17]. While pectin and its nanocomposites exhibit promising therapeutic and drug carrying abilities, still there is limited understanding about the *in-vivo* roles of pectin bioactivity. Meanwhile, further research is being conducted to investigate the role of polymer blending with pectin to enhance its stability and effectiveness as an encapsulant in drug delivery systems. The use of EOs has various limitations such as hydrophobicity, instability, and volatility; however, encapsulation within lipid-based delivery systems is a potential solution to overcome these challenges [25]. Conventional extraction methods necessitate high temperatures and prolonged extraction leading to breakdown of phenolic compounds. In this regard, non-conventional methods like MAE, UAE, and SFE, are better alternatives [40]. Various studies suggest that using a combination of these methods can increase the yield and recovery of bioactive compounds [41].

Since the suggestive source being discussed is the natural citrus byproduct/waste, there is a heightened likelihood of variations and impurities among the extracted byproducts leading to variable results in scientific integrations. Pectin is frequently employed in drugs encapsulations and is also one of the notable citrus byproducts. However, despite being a promising drug delivery technique, challenges persist for pectin-based hydrogels. These limitations include regulatory hurdles, biocompatibility concerns, and intricate drug interactions. Therefore, overcoming these challenges is essential to maximize the potential of pectin-based hydrogels in drug delivery applications [22]. Additionally, research involving suitable animal models is needed to validate the outcomes observed in laboratory settings and substantiate the potential utility of the said nanoparticles in delivering drugs [96]. For topical skin techniques, further testing in animal models is necessary to evaluate the effectiveness of citrus-waste based new formulations against skin cancer induced by UV exposure [98]. Pectin is the most extensively studied citrus bioactive compound in drug and nutraceutical delivery systems. Nonetheless, studies investigating the efficacy and safety of pectin-derived NPs are still limited. Other citrus bioactive compounds like polyphenols are relatively unstable and are susceptible to degradation in response to variations in pH, temperature and exposure to metal ions [123]. Additional investigations are required to validate their effectiveness and delve into the potential roles of citrus bioactive compounds in drug and nutraceutical delivery systems.

References

1. Raţu, R. N., Veleşcu, I. D., Stoica, F., Usturoi, A., Arsenoaia, V. N., Crivei, I. C., et al. (2023). Application of Agri-food by-products in the food industry. *Agriculture, 13*, 1559.
2. Marsh, K., & Bugusu, B. (2007). Food packaging—Roles, materials, and environmental issues. *Journal of Food Science, 72*, R39–R55.
3. Barbulova, A., Colucci, G., & Apone, F. (2015). New trends in cosmetics: By-products of plant origin and their potential use as cosmetic active ingredients. *Cosmetics, 2*, 82–92.
4. Rudzińska, A., Juchaniuk, P., Oberda, J., Wiśniewska, J., Wojdan, W., Szklener, K., et al. (2023). Phytochemicals in cancer treatment and cancer prevention—Review on epidemiological data and clinical trials. *Nutrients, 15*, 1896.
5. Bissim, S. M., Kenmogne, S. B., Tcho, A. T., Lateef, M., Ahmed, A., Happi, E. N., et al. (2019). Bioactive acridone alkaloids and their derivatives from Citrus aurantium (Rutaceae). *Phytochemistry Letters, 29*, 148–153.
6. Rodríguez-García, C., Sánchez-Quesada, C., & Gaforio, J. J. (2019). Dietary flavonoids as cancer chemopreventive agents: An updated review of human studies. *Antioxidants, 8*, 137.
7. Richa, R., Kohli, D., Vishwakarma, D., Mishra, A., Kabdal, B., Kothakota, A., et al. (2023). Citrus fruit: Classification, value addition, nutritional and medicinal values, and relation with pandemic and hidden hunger. *Journal of Agriculture and Food Research, 14*, 100718.
8. Arias, A., Feijoo, G., & Moreira, M. T. (2022). Exploring the potential of antioxidants from fruits and vegetables and strategies for their recovery. *Innovative Food Science & Emerging Technologies, 77*, 102974.
9. Mahato, N., Sharma, K., Sinha, M., & Cho, M. H. (2018). Citrus waste derived nutra−/pharmaceuticals for health benefits: Current trends and future perspectives. *Journal of Functional Foods, 40*, 307–316.
10. Ashraf, H., Iahtisham-Ul-Haq, B. M. S., Nayik, G. A., Ramniwas, S., Damto, T., et al. (2024). Phytochemical and antioxidant profile of citrus peel extracts in relation to different extraction parameters. *International Journal of Food Properties, 27*, 286–299.
11. Saini, R. K., Ranjit, A., Sharma, K., Prasad, P., Shang, X., Gowda, K. G. M., et al. (2022). Bioactive compounds of citrus fruits: A review of composition and health benefits of carotenoids, flavonoids, limonoids, and terpenes. *Antioxidants., 11*, 239.
12. Khan, U. M., Sameen, A., Aadil, R. M., Shahid, M., Sezen, S., Zarrabi, A., et al. (2021). Citrus genus and its waste utilization: A review on health-promoting activities and industrial application. *Evidence-Based Complementary and Alternative Medicine, 2021*, 1–17.
13. Sharma, K., Mahato, N., Cho, M. H., & Lee, Y. R. (2017). Converting citrus wastes into value-added products: Economic and environmently friendly approaches. *Nutrition, 34*, 29–46.
14. Mahato, N., Agarwal, P., Mohapatra, D., Sinha, M., Dhyani, A., Pathak, B., et al. (2021). Biotransformation of citrus waste-II: Bio-sorbent materials for removal of dyes, heavy metals and toxic chemicals from polluted water. *Process, 9*, 1544.
15. Maqbool, Z., Khalid, W., Atiq, H. T., Koraqi, H., Javaid, Z., Alhag, S. K., et al. (2023). Citrus waste as source of bioactive compounds: Extraction and utilization in health and food industry. *Molecules, 28*, 1636.
16. Singh, B., Singh, J. P., Kaur, A., & Yadav, M. P. (2021). Insights into the chemical composition and bioactivities of citrus peel essential oils. *Food Research International, 143*, 110231.
17. Suri, S., Singh, A., & Nema, P. K. (2022). Current applications of citrus fruit processing waste: A scientific outlook. *Applied Food Research, 2*, 100050.

18. Andrade, M. A., Barbosa, C. H., Shah, M. A., Ahmad, N., Vilarinho, F., Khwaldia, K., et al. (2022). Citrus by-products: Valuable source of bioactive compounds for food applications. *Antioxidants, 12*, 38.
19. Kedir, W. M., Deresa, E. M., & Diriba, T. F. (2022). Pharmaceutical and drug delivery applications of pectin and its modified nanocomposites. *Heliyon, 8*, e10654.
20. Andrade, M. A., Barbosa, C. H., Shah, M. A., Ahmad, N., Vilarinho, F., Khwaldia, K., et al. (2023). Citrus by-products: Valuable source of bioactive compounds for food applications. *Antioxidants, 12*, 38.
21. Sultana, N. (2023). Biological properties and biomedical applications of pectin and pectin-based composites: A review. *Molecules, 28*(24), 7974.
22. Han, S. S., Ji, S. M., Park, M. J., Suneetha, M., & Uthappa, U. T. (2022). Pectin based hydrogels for drug delivery applications: A mini review. *Gels, 8*, 834.
23. Odun-Ayo, F., & Reddy, L. (2023). Potential biomedical applications of modified pectin as a delivery system for bioactive substances. *Polysaccharides, 4*, 1–32.
24. Roman-Benn, A., Contador, C. A., Li, M.-W., Lam, H.-M., Ah-Hen, K., Ulloa, P. E., et al. (2023). Pectin: An overview of sources, extraction and applications in food products, biomedical, pharmaceutical and environmental issues. *Food Chemistry Advances, 2*, 100192.
25. Cimino, C., Maurel, O. M., Musumeci, T., Bonaccorso, A., Drago, F., Souto, E. M. B., et al. (2021). Essential oils: Pharmaceutical applications and encapsulation strategies into lipid-based delivery systems. *Pharmaceutics, 13*, 327.
26. Alhasso, B., Ghori, M. U., & Conway, B. R. (2022). Systematic review on the effectiveness of essential and carrier oils as skin penetration enhancers in pharmaceutical formulations. *Scientia Pharmaceutica, 90*, 14.
27. Júlia Scherer, S., & Mateus, P. G. (2023). Citrus essential oils and nanosystems towards skin delivery. In G. Mateus Pereira & S. Júlia Scherer (Eds.), *Citrus research* (p. Ch. 11). IntechOpen.
28. Medeleanu, M. L., Fărcaş, A. C., Coman, C., Leopold, L., Diaconeasa, Z., & Socaci, S. A. (2023). Citrus essential oils – Based nano-emulsions: Functional properties and potential applications. *Food Chemistry X, 20*, 100960.
29. Islam, F., Khan, J., Zehravi, M., Das, R., Haque, M. A., Banu, A., et al. (2024). Synergistic effects of carotenoids: Therapeutic benefits on human health. *Process Biochemistry, 136*, 254–272.
30. Wang, H., Wang, C., Zou, Y., Hu, J., Li, Y., & Cheng, Y. (2020). Natural polyphenols in drug delivery systems: Current status and future challenges. *Giant, 3*, 100022.
31. Multari, S., Carlin, S., Sicari, V., & Martens, S. (2020). Differences in the composition of phenolic compounds, carotenoids, and volatiles between juice and pomace of four citrus fruits from southern Italy. *European Food Research and Technology, 246*, 1991–2005.
32. Wahnou, H., Liagre, B., Sol, V., El Attar, H., Attar, R., Oudghiri, M., et al. (2023). Polyphenol-based nanoparticles: A promising frontier for enhanced colorectal cancer treatment. *Cancers, 15*, 3826.
33. Swapnil, P., Meena, M., Singh, S. K., Dhuldhaj, U. P., & Marwal, A. (2021). Vital roles of carotenoids in plants and humans to deteriorate stress with its structure, biosynthesis, metabolic engineering and functional aspects. *Current Plant Biology, 26*, 100203.
34. El-Sharnouby, G. A., Aleid, S. M., & Al-Otaibi, M. M. (2013). Conversion of processed citrus wastes into nutritional components. *Journal of Food Processing & Technology, 04*(08), 1000259.
35. Saini, R. K., Keum, Y.-S., Daglia, M., & Rengasamy, K. R. (2020). Dietary carotenoids in cancer chemoprevention and chemotherapy: A review of emerging evidence. *Pharmacological Research, 157*, 104830.
36. Dhuique-Mayer, C., Gence, L., Portet, K., Tousch, D., & Poucheret, P. (2020). Preventive action of retinoids in metabolic syndrome/type 2 diabetic rats fed with citrus functional food enriched in β-cryptoxanthin. *Food & Function, 11*, 9263–9271.

37. Metibemu, D. S., & Ogungbe, I. V. (2022). Carotenoids in drug discovery and medicine: Pathways and molecular targets implicated in human diseases. *Molecules, 27*, 6005.
38. Teigiserova, D. A., Tiruta-Barna, L., Ahmadi, A., Hamelin, L., & Thomsen, M. (2021). A step closer to circular bioeconomy for citrus peel waste: A review of yields and technologies for sustainable management of essential oils. *Journal of Environmental Management, 280*, 111832.
39. Brahmi, F., Mokhtari, O., Legssyer, B., Hamdani, I., Asehraou, A., Hasnaoui, I., et al. (2021). Chemical and biological characterization of essential oils extracted from citrus fruits peels. *Materials Today: Proceedings, 45*, 7794–7799.
40. Usman, I., Hussain, M., Imran, A., Afzaal, M., Saeed, F., Javed, M., et al. (2022). Traditional and innovative approaches for the extraction of bioactive compounds. *International Journal of Food Properties, 25*, 1215–1233.
41. Osorio-Tobón, J. F. (2020). Recent advances and comparisons of conventional and alternative extraction techniques of phenolic compounds. *Journal of Food Science and Technology, 57*, 4299–4315.
42. Anticona, M., Blesa, J., Frigola, A., & Esteve, M. J. (2020). High biological value compounds extraction from citrus waste with non-conventional methods. *Food, 9*, 811.
43. Łubek-Nguyen, A., Ziemichód, W., & Olech, M. (2022). Application of enzyme-assisted extraction for the recovery of natural bioactive compounds for nutraceutical and pharmaceutical applications. *Applied Sciences, 12*, 3232.
44. Pagar, K., Chavan, K., Kasav, S., Basnet, P., Rahdar, A., Kataria, N., et al. (2023). Bio-inspired synthesis of CdO nanoparticles using citrus limetta peel extract and their diverse biomedical applications. *Journal of Drug Delivery Science and Technology, 82*, 104373.
45. Chamundeeswari, M., Jeslin, J., & Verma, M. L. (2019). Nanocarriers for drug delivery applications. *Environmental Chemistry Letters, 17*, 849–865.
46. Yetisgin, A. A., Cetinel, S., Zuvin, M., Kosar, A., & Kutlu, O. (2020). Therapeutic nanoparticles and their targeted delivery applications. *Molecules, 25*, 2193.
47. Chandra, P., Yoo, J. J., & Lee, S. J. (2015). Biomaterials in regenerative medicine: Challenges in technology transfer from science to process development. In *Translational regenerative medicine* (pp. 151–167). Elsevier.
48. Yusuf, A., Almotairy, A. R. Z., Henidi, H., Alshehri, O. Y., & Aldughaim, M. S. (2023). Nanoparticles as drug delivery systems: A review of the implication of nanoparticles' physicochemical properties on responses in biological systems. *Polymers, 15*, 1596.
49. Azmy, R. M. (2021). Nanoformulated materials from citrus wastes. In *Waste recycling technologies for nanomaterials manufacturing* (pp. 649–669). Springer.
50. Caballero, S., Li, Y. O., McClements, D. J., & Davidov-Pardo, G. (2022). Encapsulation and delivery of bioactive citrus pomace polyphenols: A review. *Critical Reviews in Food Science and Nutrition, 62*, 8028–8044.
51. Barbhuiya, R. I., Singha, P., Asaithambi, N., & Singh, S. K. (2022). Ultrasound-assisted rapid biological synthesis and characterization of silver nanoparticles using pomelo peel waste. *Food Chemistry, 385*, 132602.
52. Dutta, T., Ghosh, N. N., Das, M., Adhikary, R., Mandal, V., & Chattopadhyay, A. P. (2020). Green synthesis of antibacterial and antifungal silver nanoparticles using Citrus limetta peel extract: Experimental and theoretical studies. *Journal of Environmental Chemical Engineering, 8*, 104019.
53. Sivakavinesan, M., Vanaja, M., Lateef, R., Alhadlaq, H. A., Mohan, R., Annadurai, G., et al. (2022). Citrus limetta Risso peel mediated green synthesis of gold nanoparticles and its antioxidant and catalytic activity. *Journal of King Saud University-Science, 34*, 102235.
54. Baglari, S., Wary, R. R., Kalita, P., & Baruah, M. B. (2024). Green synthesis of CuO nanoparticles using citrus maxima peel aqueous extract. *Materials Today: Proceedings, 103*, 141–146.
55. Jahan, I., Erci, F., & Isildak, I. (2021). Facile microwave-mediated green synthesis of non-toxic copper nanoparticles using Citrus sinensis aqueous fruit extract and their antibacterial potentials. *Journal of Drug Delivery Science and Technology, 61*, 102172.

56. Purushothaman, B. K., & Meera Sheriffa Begum, K. M. (2020). Magnetic casein-CaFe2O4 nanohybrid carrier conjugated with progesterone for enhanced cytotoxicity of citrus peel derived hesperidin drug towards breast and ovarian cancer. *International Journal of Biological Macromolecules, 151*, 293–304.
57. Abdelkawi, A., Slim, A., Zinoune, Z., & Pathak, Y. (2023). Surface modification of metallic nanoparticles for targeting drugs. *Coatings, 13*, 1660.
58. Nazari-Vanani, R., Azarpira, N., & Heli, H. (2018). Development of self-nanoemulsifying drug delivery systems for oil extracts of Citrus aurantium L. blossoms and Rose damascena and evaluation of anticancer properties. *Journal of Drug Delivery Science and Technology, 47*, 330–336.
59. Han, G., Duan, X., Jiang, B., Li, Y., Li, B., Yang, J., et al. (2023). Emulsifying properties, in vitro digestive characteristics, and β-carotene bioaccessibility of mandarin peel pectin emulsions prepared with different carrier oil phases. *International Journal of Biological Macromolecules, 242*, 124961.
60. Muñoz, J., Alfaro-Rodríguez, M.-C., Prieto-Vargas, P., Lobo, C., & Garcia, M. C. (2023). Preparation of Nanoemulgels containing lemon essential oil and pectin: Physical stability and rheological properties. *Applied Sciences, 13*, 12662.
61. Shade, C. W. (2016). Liposomes as advanced delivery systems for nutraceuticals. *Integrative Medicine: A Clinician's Journal, 15*, 33.
62. Prevete, G., Carvalho, L. G., del Carmen, R.-D. M., Verardo, V., Mancini, G., Fiore, A., et al. (2024). Ultrasound assisted extraction and liposome encapsulation of olive leaves and orange peels: How to transform biomass waste into valuable resources with antimicrobial activity. *Ultrasonics Sonochemistry, 102*, 106765.
63. Zheng, G., Wang, K., Chen, B., Liu, M., Yang, W., Ning, J., et al. (2022). The enhanced solubility and anti-lipase activity of citrus peel polymethoxyflavonoids extracts with liposomal encapsulation. *LWT, 161*, 113395.
64. Daraee, H., Etemadi, A., Kouhi, M., Alimirzalu, S., & Akbarzadeh, A. (2016). Application of liposomes in medicine and drug delivery. *Artificial Cells, Nanomedicine, and Biotechnology, 44*, 381–391.
65. Dadwal, V., Joshi, R., & Gupta, M. (2021). Formulation, characterization and in vitro digestion of polysaccharide reinforced Ca-alginate microbeads encapsulating Citrus medica L. phenolics. *LWT, 152*, 112290.
66. Lee, T., & Chang, Y. H. (2020). Structural, physicochemical, and in-vitro release properties of hydrogel beads produced by oligochitosan and de-esterified pectin from yuzu (Citrus junos) peel as a quercetin delivery system for colon target. *Food Hydrocolloids, 108*, 106086.
67. Kulkarni, N. S., Guererro, Y., Gupta, N., Muth, A., & Gupta, V. (2019). Exploring potential of quantum dots as dual modality for cancer therapy and diagnosis. *Journal of Drug Delivery Science and Technology, 49*, 352–364.
68. Choudhury, N., Meghwal, M., & Das, K. (2021). Microencapsulation: An overview on concepts, methods, properties and applications in foods. *Food Frontiers, 2*, 426–442.
69. Arpagaus, C., Collenberg, A., Rütti, D., Assadpour, E., & Jafari, S. M. (2018). Nano spray drying for encapsulation of pharmaceuticals. *International Journal of Pharmaceutics, 546*, 194–214.
70. Ledri, S. A., Milani, J. M., Shahidi, S.-A., & Golkar, A. (2024). Comparative analysis of freeze drying and spray drying methods for encapsulation of chlorophyll with maltodextrin and whey protein isolate. *Food Chemistry: X., 21*, 101156.
71. Banožić, M., Krzywonos, M., Aladić, K., Pińkowska, H., Mucha, I., Złocińska, A., et al. (2023). Physicochemical, structural characterization and evaluation of encapsulated hesperidin from natural sources: Comparison of two encapsulation techniques; spray drying and freeze drying. *Journal of Drug Delivery Science and Technology, 90*, 105098.
72. Bautista, G. F. M., Vidallon, M. L. P., Salamanez, K. C., & Rodriguez, E. B. (2019). Nanodelivery system based on zein-alginate complexes enhances in vitro chemopreventive

activity and bioavailability of pomelo [Citrus maxima (Burm.) Merr.] seed limonoids. *Journal of Drug Delivery Science and Technology, 54*, 101296.
73. Pedroso-Santana, S., & Fleitas-Salazar, N. (2020). Ionotropic gelation method in the synthesis of nanoparticles/microparticles for biomedical purposes. *Polymer International, 69*, 443–447.
74. Adepu, S., & Ramakrishna, S. (2021). Controlled drug delivery systems: Current status and future directions. *Molecules, 26*, 5905.
75. Vargason, A. M., Anselmo, A. C., & Mitragotri, S. (2021). The evolution of commercial drug delivery technologies. *Nature Biomedical Engineering, 5*, 951–967.
76. Kashkooli, F. M., Soltani, M., & Souri, M. (2020). Controlled anti-cancer drug release through advanced nano-drug delivery systems: Static and dynamic targeting strategies. *Journal of Controlled Release, 327*, 316–349.
77. Kamaly, N., Yameen, B., Wu, J., & Farokhzad, O. C. (2016). Degradable controlled-release polymers and polymeric nanoparticles: Mechanisms of controlling drug release. *Chemical Reviews, 116*, 2602–2663.
78. Lee, J. H., & Yeo, Y. (2015). Controlled drug release from pharmaceutical nanocarriers. *Chemical Engineering Science, 125*, 75–84.
79. Fonseca-Santos, B., & Chorilli, M. (2017). An overview of carboxymethyl derivatives of chitosan: Their use as biomaterials and drug delivery systems. *Materials Science and Engineering: C, 77*, 1349–1362.
80. Claus, J., Eickner, T., Grabow, N., Kragl, U., & Oschatz, S. (2020). Ion exchange controlled drug release from polymerized ionic liquids. *Macromolecular Bioscience, 20*, 2000152.
81. Ezike, T. C., Okpala, U. S., Onoja, U. L., Nwike, P. C., Ezeako, E. C., Okpara, J. O., et al. (2023). Advances in drug delivery systems, challenges and future directions. *Heliyon, 9*(6), e17488.
82. Nalini, T., Basha, S. K., Sadiq, A. M., & Kumari, V. S. (2022). Pectin/chitosan nanoparticle beads as potential carriers for quercetin release. *Materials Today Communications, 33*, 104172.
83. Nabipour, H., Batool, S., & Hu, Y. (2023). Pectin-coated baclofen-layered zinc hydroxide nanohybrid as a bio-based nanocomposite carrier for oral delivery. *IEEE Transactions on Nanobioscience, 22*, 63–70.
84. Gupta, R. D., & Raghav, N. (2020). Nano-crystalline cellulose: Preparation, modification and usage as sustained release drug delivery excipient for some non-steroidal anti-inflammatory drugs. *International Journal of Biological Macromolecules, 147*, 921–930.
85. Wu, Y., Liu, H., Li, Z., Huang, D., Nong, L., Ning, Z., et al. (2019). Pectin-decorated selenium nanoparticles as a nanocarrier of curcumin to achieve enhanced physicochemical and biological properties. *IET Nanobiotechnology, 13*, 880–886.
86. Huang, X., Liu, Y., Zou, Y., Liang, X., Peng, Y., McClements, D. J., et al. (2019). Encapsulation of resveratrol in zein/pectin core-shell nanoparticles: Stability, bioaccessibility, and antioxidant capacity after simulated gastrointestinal digestion. *Food Hydrocolloids, 93*, 261–269.
87. Kapoor, D. U., Garg, R., Gaur, M., Pareek, A., Prajapati, B. G., Castro, G. R., et al. (2024). Pectin hydrogels for controlled drug release: Recent developments and future prospects. *Saudi Pharmaceutical Journal, 32*, 102002.
88. Ergin, A. D., Bayindir, Z. S., Ozcelikay, A. T., & Yuksel, N. (2021). A novel delivery system for enhancing bioavailability of S-adenosyl-l-methionine: Pectin nanoparticles-in-microparticles and their in vitro-in vivo evaluation. *Journal of Drug Delivery Science and Technology, 61*, 102096.
89. Chinnaiyan, S. K., Karthikeyan, D., & Gadela, V. R. (2018). Development and characterization of metformin loaded pectin nanoparticles for T2 diabetes mellitus. *Pharmaceutical Nanotechnology, 6*, 253–263.
90. Khotimchenko, M. (2020). Pectin polymers for colon-targeted antitumor drug delivery. *International Journal of Biological Macromolecules, 158*, 1110–1124.

91. Anil, S., & David, G. (2019). Targeting study of HepG2 hepatoma cells in vitro by drug-loaded pectin-based nanoparticles. *bioRxiv*, 628818.
92. Zhu, H., Zhang, L., Kou, F., Zhao, J., Lei, J., & He, J. (2024). Targeted therapeutic effects of oral magnetically driven pectin nanoparticles containing chlorogenic acid on colon cancer. *Particuology, 84*, 53–59.
93. Kaushik, P., Priyadarshini, E., Rawat, K., Rajamani, P., & Bohidar, H. (2020). pH responsive doxorubucin loaded zein nanoparticle crosslinked pectin hydrogel as effective site-specific anticancer substrates. *International Journal of Biological Macromolecules, 152*, 1027–1037.
94. Huang, X., Li, T., & Li, S. (2023). Encapsulation of vitexin-rhamnoside based on zein/pectin nanoparticles improved its stability and bioavailability. *Current Research in Food Science, 6*, 100419.
95. Prezotti, F. G., Boni, F. I., Ferreira, N. N., Silva, D. S., Almeida, A., Vasconcelos, T., et al. (2020). Oral nanoparticles based on gellan gum/pectin for colon-targeted delivery of resveratrol. *Drug Development and Industrial Pharmacy, 46*, 236–245.
96. Kumar, P., Kumar, V., Kumar, R., & Pruncu, C. I. (2020). Fabrication and characterization of ceftizoxime-loaded pectin nanocarriers. *Nanomaterials (Basel), 10*(8), 1452.
97. Zhang, F., Pei, X., Peng, X., Gou, D., Fan, X., Zheng, X., et al. (2022). Dual crosslinking of folic acid-modified pectin nanoparticles for enhanced oral insulin delivery. *Biomaterials Advances, 135*, 212746.
98. Wang, J., Chao, L., Yeung, S., & Huang, Y. (2022). Abstract A022: Limonene as a vehicle in topical delivery of carvedilol for skin cancer chemoprevention. *Cancer Prevention Research, 15*, A022–A022.
99. Assali, M., Jaradat, N., & Maqboul, L. (2022). The formation of self-assembled nanoparticles loaded with doxorubicin and d-limonene for cancer therapy. *ACS Omega, 7*, 42096–42104.
100. Chekunkov, Y. V., Minzanova, S. T., Khabibullina, A. V., Arkhipova, D. M., Mironova, L. G., Voloshina, A. D., et al. (2022). Citrus pectin based complexes for the tetracycline delivery. *Food Hydrocolloids for Health, 2*(2), 100100.
101. Sabra, R., Billa, N., & Roberts, C. J. (2019). Cetuximab-conjugated chitosan-pectinate (modified) composite nanoparticles for targeting colon cancer. *International Journal of Pharmaceutics, 572*, 118775.
102. Satari, B., & Karimi, K. (2018). Citrus processing wastes: Environmental impacts, recent advances, and future perspectives in total valorization. *Resources, Conservation and Recycling, 129*, 153–167.
103. Russo, C., Maugeri, A., Lombardo, G. E., Musumeci, L., Barreca, D., Rapisarda, A., et al. (2021). The second life of citrus fruit waste: A valuable source of bioactive compounds. *Molecules, 26*, 5991.
104. Bashir, M., Hamza, F., & Sale, M. (2020). Postharvest fungal spoilage of some citrus fruits. *Bioengineering and Bioscience, 7*, 10–14.
105. Lamine, M., Hamdi, Z., Zemni, H., Rahali, F. Z., Melki, I., Mliki, A., et al. (2024). From residue to resource: The recovery of high-added values compounds through an integral green valorization of citrus residual biomass. *Sustainable Chemistry and Pharmacy, 37*, 101379.
106. Suri, S., Singh, A., & Nema, P. K. (2021). Recent advances in valorization of citrus fruits processing waste: A way forward towards environmental sustainability. *Food Science and Biotechnology., 30*(13), 1–26.
107. Parvin, F., & Tareq, S. M. (2021). Impact of landfill leachate contamination on surface and groundwater of Bangladesh: A systematic review and possible public health risks assessment. *Applied Water Science, 11*, 100.
108. Zema, D., Calabrò, P., Folino, A., Tamburino, V., Zappia, G., & Zimbone, S. (2018). Valorisation of citrus processing waste: A review. *Waste Management, 80*, 252–273.
109. Di Donna, L., Bartella, L., De Vero, L., Gullo, M., Giuffrè, A. M., Zappia, C., et al. (2020). Vinegar production from Citrus bergamia by-products and preservation of bioactive compounds. *European Food Research and Technology, 246*, 1981–1990.

110. Panwar, D., Panesar, P. S., & Chopra, H. K. (2021). Recent trends on the valorization strategies for the management of citrus by-products. *Food Reviews International, 37*, 91–120.
111. Niluxsshun, M. C. D., Masilamani, K., & Mathiventhan, U. (2021). Green synthesis of silver nanoparticles from the extracts of fruit peel of Citrus tangerina, Citrus sinensis, and Citrus limon for antibacterial activities. *Bioinorganic Chemistry and Applications, 2021*, 6695734.
112. Ashraf, H., Butt, M. S., Nadeem, M., Aadil, R. M., Rusu, A. V., & Trif, M. (2022). Microencapsulated curcumin from Curcuma longa modulates diet-induced hypercholesterolemia in Sprague Dawley rats. *Frontiers in Nutrition, 9*, 1026890.
113. Pal, G., Rai, P., & Pandey, A. (2019). Green synthesis of nanoparticles: A greener approach for a cleaner future. In *Green synthesis, characterization and applications of nanoparticles* (pp. 1–26). Elsevier.
114. Chandrakala, V., Aruna, V., & Angajala, G. (2022). Review on metal nanoparticles as nanocarriers: Current challenges and perspectives in drug delivery systems. *Emergent Materials, 5*, 1593–1615.
115. Ying, S., Guan, Z., Ofoegbu, P. C., Clubb, P., Rico, C., He, F., et al. (2022). Green synthesis of nanoparticles: Current developments and limitations. *Environmental Technology & Innovation, 26*, 102336.
116. Parveen, K., Banse, V., Ledwani, L. (2016). Green synthesis of nanoparticles: Their advantages and disadvantages. In *Proceedings of the AIP conference proceedings.*
117. Nadaroglu, H., Güngör, A. A., & Selvi, İ. (2017). Synthesis of nanoparticles by green synthesis method. *International Journal of Innovative Research and Reviews, 1*, 6–9.
118. Baloch, H., Siddiqua, A., Nawaz, A., Latif, M. S., Zahra, S. Q., Alomar, S. Y., et al. (2023). Synthesis and characterization of Sulfur nanoparticles of Citrus Limon extract embedded in nanohydrogel formulation: In vitro and in vivo studies. *Gels, 9*, 284.
119. Wen, L., Zhang, Z., Sun, D.-W., Sivagnanam, S. P., & Tiwari, B. K. (2020). Combination of emerging technologies for the extraction of bioactive compounds. *Critical Reviews in Food Science and Nutrition, 60*, 1826–1841.
120. Anoopkumar, A., Aneesh, E. M., Sirohi, R., Tarafdar, A., Kuriakose, L. L., Surendhar, A., et al. (2023). Bioactives from citrus food waste: Types, extraction technologies and application. *Journal of Food Science and Technology*, 61, 444–458.
121. Mahato, N., Sharma, K., Sinha, M., Dhyani, A., Pathak, B., Jang, H., et al. (2021). Biotransformation of citrus waste-I: Production of biofuel and valuable compounds by fermentation. *Process, 9*, 220.
122. Ruiz, B., & Flotats, X. (2016). Effect of limonene on batch anaerobic digestion of citrus peel waste. *Biochemical Engineering Journal, 109*, 9–18.
123. Xiao, J. (2022). Recent advances on the stability of dietary polyphenols. *eFood, 3*, e21.

Chapter 21
Valorizations of Citrus Food Waste: Management of Oranges Citrus Peel Waste

Niyonshima Alexandre

21.1 Introduction

The increasing global demand for citrus, particularly oranges, has led to a significant rise in the production and consumption of these fruits. however, this burgeoning market has also resulted in substantial amounts of waste, primarily in the form of citrus peels, which are often discarded during the processing and consumption of oranges. Citrus peel waste is a major contributor to environmental polllution, posing challenges for waste management and necessitating sustainable solutions. Valorization, the process of converting waste into valuable products, offers a promising approach to mitigate the negative impacts of orange peel waste while simultaneously generating economic benefits. by exploring innovative methods for reusing and repurposing citrus peels, the potential for reducing environmental footprints and enhancing resource efficiency is immense. Citrus peels, rich in essential oils, pectin, dietary fiber, and bioactive compounds, can be transformed into a range of valuable products, including natural additives, dietary supplements, and biodegradable materials. this study delves into the management of citrus peel waste, focusing on the valorization of orange by-products through sustainable practices and technologies. It aims to illuminate the potential of valorizing citrus peel waste. The countries that are geographically located in the subtropical, tropical subequatorial, and equatorial climatic belts of the earth play a role in global citrus fruit production and international trace of citrus. In the years 2000 and 2001, the global gross yields were respectively 105,960 million tons and 161,801 million tons. In 2021, The high-ranked countries in citrus production were Brazil, India, China, Mexico, the USA, Spain, Egypt, Indonesia, Iran, and Italy [29]. The common term for citrus includes sweet or sour oranges (C. sinensis, C. aurantium), mandarins (C. reticulata), lemons (C. limon), limes (C. aurantifolia), pomelos (C. maxima) and

N. Alexandre (✉)
Rwanda Agriculture and Animal Resources Development Board (RAB), Kigali, Rwanda
e-mail: alexandre.niyonshima@rab.gov.rw

A. Chauhan et al. (eds.), *Valorization of Citrus Food Waste*,
https://doi.org/10.1007/978-3-031-77999-2_21

grapefruit(C. paradisi). the production of oranges reached 48.8 million tons in the 2023/2024 season [30].

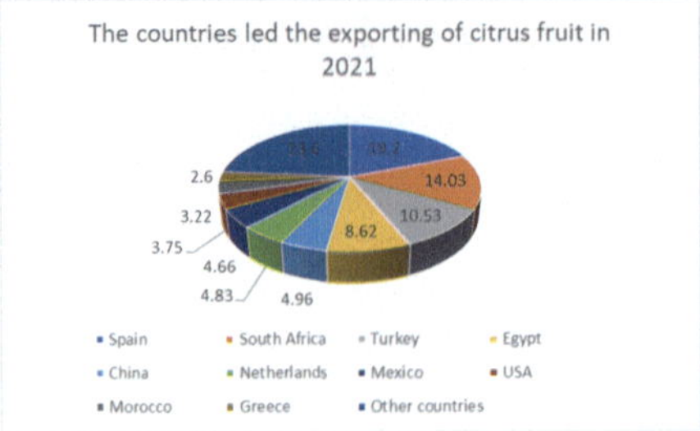

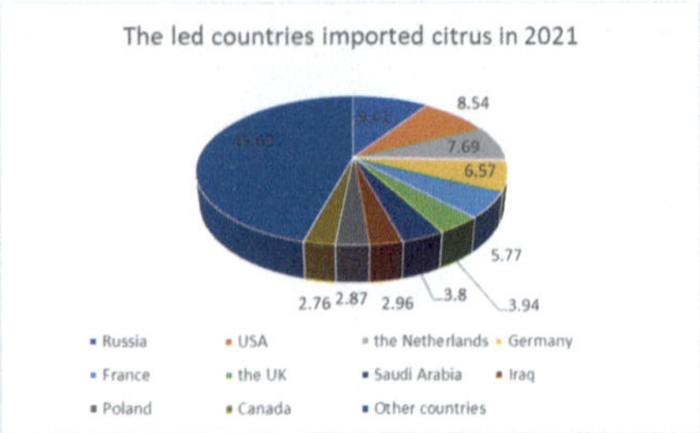

The composition of the orange fruit is determined by the preharvest factors such as the climate, cultivars or rootstocks, the growing condition factors (irrigation, plant nutrition, growth regulators, sunlight, pruning, and pest and diseases), harvest and postharvest management (condition, technology applied too).

In general, the citrus fruit contains N(1–2 g/kg on a wet basis), lipids (oleic, linoleic, linolenic, palmitic, stearic acids, glycerol, and a phytosterol), carbohydrates (Monosaccharide include glucose, fructose and disaccharide such as sucrose), acids (primarily citric and malic, but also tartaric, benzoic, oxalic, and succinic), insoluble carbohydrates (cellulose, pectin), enzymes (pectin-esterase, phosphatase, peroxidase), flavonoids (hesperidin, na ringin), bitter principles (limonin, isolimonin), peel oil (D-limonene), volatile constituents (alcohols, aldehydes, ketones, esters, hydrocarbons, acids), pigments (carotenes, xanthophylls), vitamins (ascorbic acid, Vitamin B complex, carotenoids), and minerals (primarily calcium and potassium). The nutrient content of by-product feedstuff of citrus is influenced by factors that include the source of the fruit and the type of processing to be exposed to [26] (Table 21.1).

In the processed citrus industry, 40–50% of the fruit is consumed while the remaining solid residue composite of the peel (flavedo and albedo), seed, and membrane residues (pulp or juice sac residue, rag or membranes and core) are waste [26, 43]. The dry matter of citrus waste is composed of 60–65% peel, 30–35% segment pulp, and 0–10% seeds [31]. Consequently, the food loss from postharvest management and fruit processing industries presents a threat and foists a severe burden on the ecosystem through environmental pollution and health hazards [37]. To prevent the problem, well-structured guidance on peel management should be in place and escort the processors in the management of citrus peels.

Table 21.1 Products and by products of various tissues of citrus fruit

Fruit part	Description	Biochemical composition	Reference
Whole peel or rind (pericarp)	consists of albedo (inner whity spongy peel, mesocarp) and flavedo (outer yellow peel, epicarp). Pectin is abundant in albedo. Animals can be fed the entire peel in combination with molasses, pulp residue (rag), and/or other ingredients. Food supplements and human foods are also produced with it.	Pectin, Methane, D-limonene, bioethanol, pectic enzymes, singe cell protein, essential oil (monoterpenes, sesquiterpenes and their oxygenated derivatives)	Jeong [20], Singh et al. [39]
Pulp (principal edible portion, endocarp)	is used primarily, following mechanical extraction and screening, to provide raw juice for human nourishment. Known by another name, pulp, this material that is filtered out of the raw juice is typically mixed with other residues to create by-products that are fed to animals.	Organic acids (citric, malic and ascorbic acids), sugars (sucrose, glucose and fructose), phenolic compounds (Hesperidin, narirutin and ferulic acid) polyphenols dietary fiber vitamin C, folate vit A or carotenoid Flavonoids	Kelebek et al. [21], Chanson-Rolle et al. [10], Dugo et al. [14], Gattuso et al. [18]
Pulp residue (called rag in the industry)	comprises the pulp fraction that has been screened, which includes seeds, segment walls or membranes, juice vesicles, and cores. Typically, peel residue and pulp residue are combined to create by-product meals. Citrus processors make press liquor, citrus molasses, citrus pulp, and citrus meal, and feed yeast from the lime-treated mass peel and pulp wastes. Food supplements and human foods are also produced with it.	Soluble and insoluble dietary fiber	Sungsinchai et al. [40]
Seed	sometimes separated from the rag to produce seed oils, seed meals, and dried seed-pressed cake.	protein, ether extract and crude fibre	
Waste waters (aqueous effluent emulsions from processing plants)	has the ability to be used in the manufacturing of goods like yeasts and activated sludge. Additionally, human foods and dietary supplements are produced with it.	Water (75–80%)	Calabrò et al. [9]

21.2 Disposal of Waste Orange Peel

The composition of fruit and vegetable waste is total solid ranging from 8 to 18% and the total volatile solid of 87%. Seventy-five percent (75%) of organic matter found in fruit and vegetable waste consists of sugar and hemicellulose and is easily biodegradable due to the high moisture that facilitates biological treatment. However, under anaerobic digestion remains a trend. The phenolic and non-biodegradable compounds (cellulose and lignin consist of 9% and 5% respectively) resist to biological degradation. Therefore, it requires pre-treatment techniques to minimize the organic loading rate [4].

21.3 Exploring the Treatment Methods

Based on the studies that have been conducted, the following treatments for peels of orange have shown some promise toward the reduction of exposure of high risk. The treatments include heat treatment, evaporation, membrane filtering, anaerobic process, anaerobic co-digestion, biodiesel production, ignition or combustion, supercritical, subcritical fluid, extraction, coagulation, and composting;

21.3.1 Thermal Process

The thermal process consists of the exposure of the peels to heat through heat transfer forms such as convention, conductive, and irradiation. The extraction of individual phytochemicals including hesperidin, limonene, and pectin is effective under temperatures of 180 and 200 °C microwave irradiation of 800 W [6]. Waste orange peels treated under hydrostatic pressure (600 MPa: Come up time of 2,5,10 or 20 Min) and mild heat treatment conditions of 55 or 70 °C have the potential to modify the functionality of orange peels with short processing times such as increasing in soluble dietary fiber, significant conversion of insoluble dietary fiber into the soluble dietary fiber, increasing in swelling capacity and oil holding capacity and the HHP improved the adsorption and desorption water retention of WOP in the range of 0.1–0.93 aw [41]. The heat treatment affects phenolic compounds and the antioxidant capacity of Huyou (Citrus Paradise Changshanhuyou) and the High -performance liquid chromatography coupled with a photodiode array detector has shown that the free fraction of phenolic acids increase and ester, glycoside, and ester-bound fractions decrease and the total flavanone glycoside declined. Furthermore, the antioxidant activity of the methanol extract of huyou peel increased. However, long heating at high temperatures (like 120 °C for 90 min or 150 °C for 30 min) may destroy the flavanone glycosides therefore, to enhance the antioxidant capacity of citrus peel, the proper and reasonable heat treatment should be used [45]. Scheme of microwave hydrothermal pyrolysis of WOP.

21.3.1.1 Scheme of Microwave Hydrothermal Pyrolysis of WOP

Xu et al. [45]

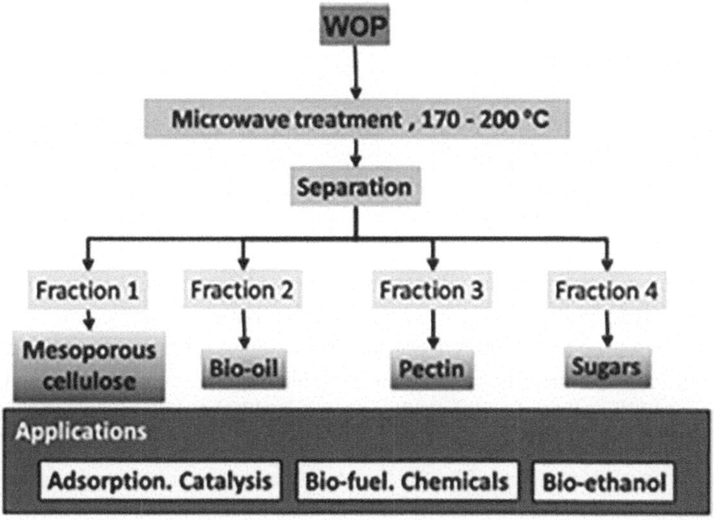

21.3.2 Evaporation

The evaporation or vaporation method is mainly used to treat the wastewater. Through evaporation, the quantity of water or volatile compounds is reduced or completely volatile. The remaining residue such as solid substances or non-volatile substances is left for disposal as hazardous waste.

The evaporation can be classified into three groups depending on the purpose of evaporation:

21.3.2.1 Vaporization

In vaporization, the objective is to produce a vapor from a pure or impure liquid. When a liquid at its boiling point receives additional heat at constant pressure, its molecules gain sufficient energy to overcome the intermolecular forces holding them together in the liquid state. They then escape as individual molecules of vapor until the vaporization process is finished. The vapor itself or its condensate is the necessary primary product. The temperature of a boiling liquid remains constant until all of the liquid has been converted into a gas. The maximum yields of essential oils were found to be 4.40%, 3.47%, and 2.536%, respectively, when steam distillation, water distillation, and solvent extraction were used in the investigations on the extraction processes of essential oils from orange peel. Steam distillation was hence a major method of oil extraction. Notably, the quality of the obtained essential oil depends on the temperature and time factors applied during the extraction [19].

21.3.2.2 Concentration

Focus Making a concentrate from a diluted solution is the goal of concentration. The process of boiling off a solvent—typically water—from the liquid yields a high concentration. The ultimate product that is typically obtained from the evaporation process is concentrate, which can be utilized, for example, to heat another step in the process. But occasionally, as in solvent recovery, the volatile, evaporated component is also a major product. Because carbonated orange peel is biocompatible, carbon-rich, hydrophilic, has low thermal conductivity, and has great sun absorption, it has been used as an efficient photothermal material for interfacial solar steam generation. Excellent heat insulation was provided by the carbonated orange peel's tiny pores, and the upward-concave shape improved light absorption and the ability to float [34].

21.3.2.3 Crystallization

One of the fundamental processes in thermal process technology is crystallization, which aims to create a solid phase. In addition to material separation, cleansing, and formation, inorganic (and occasionally organic) salt recovery is accomplished through crystallization. Depending on the specific needs, the crystalline phase is produced from a solution or melts by chilling crystallization, vacuum cooling, or evaporation. The solid phase is separated via filtration, decanters, or centrifuges. Co-crystallization was used to microencapsulate the orange peel oil extract. It was discovered that sucrose syrups had an encapsulation capability of more than 90% for an oil-to-sugar ratio ranging from 100 to 250 g. The encapsulation effectiveness measured by surface oil ranged from 3350 to 8880 mg oil/kg solids. For every prepared co-crystallize, the bulk density was more than 670 kg/m3 and the moisture content was less than 10 g/kg. After a day of storage at 35 °C, sensory examination revealed that all panelists could detect oxidized flavors in oils without antioxidant addition. After 2 months of storage at room temperature, no evidence of oxidized flavors was seen in the oil that had BHA added before co-crystallization [7, 33].

21.3.3 Membrane Filtering

Filtering technology has the purpose of separating substances based on their size and shape without the application of thermic intervention but by using pressure and special design. Since membrane processes can concentrate desirable by-products like whey proteins, recycled paints and degreased baths in the mechanical engineering industries, retrieved precious metals or aromatic compounds, etc., they have been used for more than 20 years in the food industry's production processes and are steadily gaining ground in de-pollution techniques [5].

Different kinds of membrane separation techniques have been created for particular industrial uses. Reverse osmosis (RO), ultrafiltration (UF), nanofiltration (NF), microfiltration (MF), electrodialysis (ED), gas separation (GS), and pervaporation (PV) are a few of the most often utilized procedures [5, 44]. The organophilic pervaporation was integrated into the process of extracting limonene from orange peels after using polypropylene glycol 240 (PPG). As a result, the selective recovery of limonene free of solvent was obtained [22].

21.3.4 Anaerobic Process and Anaerobic Co-digestion

In the absence of oxygen, the microbial organisms degradate and stabilize the organic materials leading to the formation of biogas including a mixture of carbon dioxide and methane, a renewable energy source, and microbial biomass. The anaerobic process has been used in the management of orange peels for further production of methane [8]. The essential oils are predominated by an antioxidant known as "D-limonene" which is considered a biological activity inhibitor, therefore, optimizing the anaerobic digestion requires the pretreatment method such as ensiling, aeration, thermal and alkaline treatment [9, 27] or co-digestion of orange peel waste that is fresh or ensile with other substrates [17, 28] toward the dilution of the D-limonene load.

21.3.5 Biodiesel Production

Biodiesel production is a complex process and its design depends on the expected quality of biodiesel to fulfill the requirement of ASTM D 6751; in other words, the final product depends on the type of processes, equipment, and the quality of feedstock. The aforementioned parameters affect the transesterification reaction that is mostly used in the production of biodiesel [12].

The transesterification reaction consists of the molar concentration of glycerides to alcohol, catalysts, reaction temperature, reaction time and free fatty acids, and water content of lipids [25]. The other primary methods of biodiesel production are "direct use and blending", "microemulsion" and "thermal cracking (pyrolysis)" [32]. The methyl oleate is a component of biodiesel was successfully obtained by the 96.51 ± 0.4% conversion from oleic acid under the optimized reaction conditions (15:1 MeOH to oleic acid molar ratio, 7 wt.% catalyst loading w.r.t oleic acid, 80 °C reaction temperature, 60 min reaction time) afterward the fine powder of orange peel is partial carbonization-sulfonation treatment were used [23].

21.3.6 Combustion

Waste orange peel (WOP) has been shown in recent studies to be a potentially significant resource that may be converted into high-value goods, the most practical options in terms of yield may include heat generation, bio6methanization, and composting. The study revealed that waste orange peel can effectively valorize into combustion [38]. In addition, the biodiesel obtained from orange peel has the potential to replace pure diesel, and when the proportion of biodiesel to diesel in the volume blend is 10%, combustion and performance are at their best [15].

21.3.7 Supercritical and Subcritical Fluid

Supercritical and subcritical fluid technologies have been used to isolate natural products from the raw material. A fluid is said to as supercritical when it is driven to temperatures and pressures higher than its critical point. The supercritical state of a fluid is thus defined as a state in which the fluid's properties are identical to those of a gas and a liquid, or as a state in which the fluid is compressible, exhibiting characteristics similar to those of a gas despite having a density similar to that of a liquid and, consequently, a similar solvating power. Supercritical fluids have greater transport qualities than liquids because of their low viscosity and relatively high diffusivity. As a result, they may diffuse through solid materials more easily and produce faster extraction yields [5].

Supercritical fluid was first introduced in the mid-1980s. Its extraction efficiency has been scientifically reviewed regarding various matrices including the pollutants from soils, fats from food [1], active plant substances, and polymers' additives. The extraction is based on the conventional liquid extraction by using solvents such as organic solvents, and carbon dioxide, ... This method has been improved like soxhlet. However, the basic concept and the extraction factors were retained [35]. The study on the extraction of orange peel essential oil using supercritical fluid reveals that the optimized extraction time and ideal extraction condition is 317.51 min at 74.85 °C with a ratio of 4 in solvent to sample. The analysis of the extract revealed a significant value for saponification, acid, and free fatty acid are 121.67,4.13,117.54, AND 1.56 mag KOH/g respectively. The chemical composition analysis showed that Limonene, α-pinene, β-myrcene, γ-terpinene, and α-terpineol are 43.96%,15.6%,10.21%,5%, and 4.15% respectively [16].

Subcritical method refers to the treatment of samples with high temperature and high-pressure water. Water is replaced by methanol, ethanol, salts and ionic liquids and the subcritical methods have shown a capacity to extract natural products such as alkaloids, carbohydrates, essential oil, flavonoids, glycosides, lignans, organic acids, polyphenolics, quinones, steroids, and terpenes. The optimal temperature ranges from 130 to 240 °C for extracting the mentioned substances [11]. The subcritical fluid within a temperature of 110–150 °C and water flow rate (10–30 ml/min), extraction of the high maximum of hesperidin and narirutin occur at 150 °C and 10 mL/min [24].

21.3.8 Coagulation

Coagulation and flocculation are a process of neutralizing the charge in colloidal particles and performing polymer bridging. The two types of coagulants include chemical-based ones like Alum and ferric salts and natural coagulants that are derived from plants, animals, and microorganisms. This last category is considered safe and eco-friendly. The four types of mechanisms of natural coagulants are double-layer compression, polymer bridging charge neutralizing, and sweep coagulation [2]. At a pH of 7.5, the powdered orange peels with a dosage of 0.2 g/L showed 97% turbidity removal from dairy wastewater [3]. The coagulation mechanism is influenced by other conditions such as type, dose, basicity of coagulant, pH/alkalinity, presence of anions, and nature of organic water with its fractionation [36]. The basicity of powdered orange peel should be cared for during the pretreatment or management of the orange peel.

Saxena et al. [36]

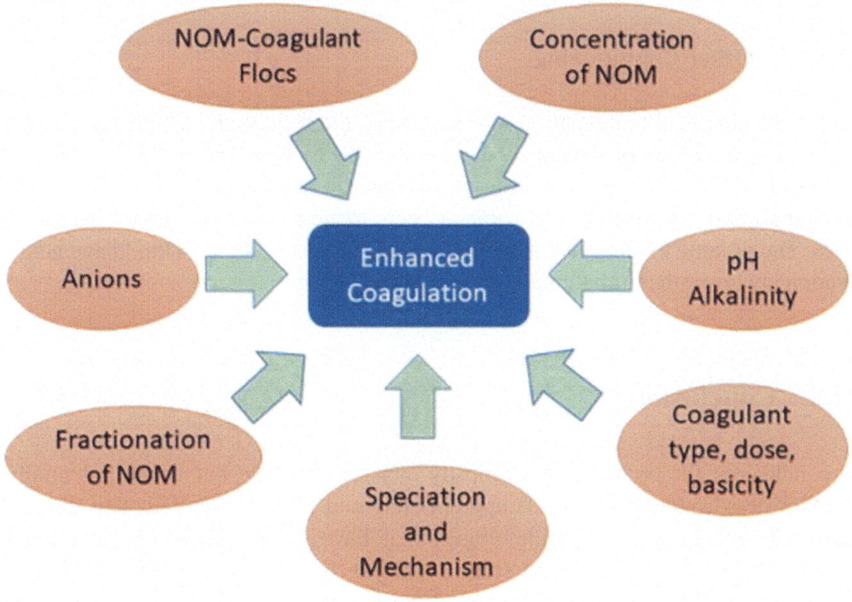

21.3.9 *Composting*

21.3.9.1 Microbial Degradation

The study on the inoculation of citrus peel composting was conducted and the physicochemical properties and the bacterial community structure of citrus peel composting inoculated were studied. C/N, organic matter, moisture, pectin, and cellulose content decreased along with the composing process; however, the pH, soluble protein, and total nutrients showed an opposite trend. The inoculation increases and diversifies the bacterial community and reaches its maximum in 21 days and the bacillus, Sphingobacterium, and Saccharomonospora became dominant genera in the inoculum [42].

21.3.9.2 Biological Degradation

The degradation of organic matter of orange peel through the vermicomposting process was succeeded after 13 weeks and the degradation condition was determined by monitoring of physicochemical parameters such as temperature, pH, humidity, organic pattern, total organic carbon, total nitrogen, and the carbon/nitrogen ratio. To balance the substrate's nutrients, vegetable waste, and eggshells considered as load material were mixed for earthworm (Eisenia fetida) activity. As a result, the total nitrogen had been increased [13].

21.3.9.3 Other Conditioning

Due to the chemical composition of orange peel, the formation of high-value products like methane can be developed and pursued the treatment toward the extraction of D-limonene. Under thermophilic conditions, the methane production rate and biodegradability are higher than orange peels treated under mesophilic temperatures. The organic loading rate is in the range of 1.20–3.67 kg COD/m3d and acidification occurs at the high Organic loading rate [27].

21.4 Conclusion

In conclusion, the orange peel is mostly considered waste but it can potentially benefit to environment when it is appropriately and suitable handled. This chapter reviewed the management technologies that have been scientifically explored. However, there is still much to study regarding of management of orange peel including the efficiency in combination or interference of two or more management methods toward the valorization of orange peel.

References

1. Akanda, M. J. H., Sarker, M. Z. I., Ferdosh, S., Manap, M. Y. A., Ab Rahman, N. N. N., & Ab Kadir, M. O. (2012). Applications of supercritical fluid extraction (SFE) of palm oil and oil from natural sources. *Molecules, 17*(2), 1764–1794.
2. Amran, A. H., Zaidi, N. S., Muda, K., & Loan, L. W. (2018). Effectiveness of natural coagulant in coagulation process: A review. *International Journal of Engineering & Technology, 7*(3), 34–37.
3. Anju, S. (2016). Exploring the use of orange peel and neem leaf powder as alternative coagulant in treatment of dairy wastewater. *International Journal of Scientific & Engineering Research, 7*(4), 238–244.
4. Arvanitoyannis, I. S. (2010). *Waste management for the food industries*. Academic.
5. Arvanitoyannis, I. S., & Varzakas, T. H. (2008). Fruit/fruit juice waste management: Treatment methods and potential uses of treated waste. In *Waste management for the food industries* (Vol. 2, pp. 569–628). Academic Press.
6. Balu, A. M., Budarin, V., Shuttleworth, P. S., Pfaltzgraff, L. A., Waldron, K., Luque, R., & Clark, J. H. (2012). Valorisation of orange peel residues: Waste to biochemicals and nanoporous materials. *ChemSusChem, 5*(9), 1694.
7. Beristain, C. I., Vazquez, A., Garcia, H. S., & Vernon-Carter, E. J. (1996). Encapsulation of orange peel oil by co-crystallization. *LWT-Food Science and Technology, 29*(7), 645–647.
8. Calabrò, P. S., & Panzera, M. F. (2018). Anaerobic digestion of ensiled orange peel waste: Preliminary batch results. *Thermal Science and Engineering Progress, 6*, 355–360.
9. Calabrò, P. S., Paone, E., & Komilis, D. (2018). Strategies for the sustainable management of orange peel waste through anaerobic digestion. *Journal of Environmental Management, 212*, 462–468.
10. Chanson-Rolle, A., Braesco, V., Chupin, J., & Bouillot, L. (2016). Nutritional composition of orange juice: A comparative study between French commercial and home-made juices. *Food and Nutrition Sciences, 7*(04), 252.
11. Cheng, Y., Xue, F., Yu, S., Du, S., & Yang, Y. (2021). Subcritical water extraction of natural products. *Molecules, 26*(13), 4004.
12. Dahiya, A. (2020, January). Biofuel conversion pathways service learning projects and case studies. In *Bioenergy* (pp. 671–690). Academic Press.
13. De Medina-Salas, L., Giraldi-Díaz, M. R., Castillo-González, E., & Morales-Mendoza, L. E. (2020). Valorization of orange peel waste using precomposting and vermicomposting processes. *Sustainability, 12*(18), 7626.
14. Dugo, P., Herrero, M., Giuffrida, D., Ragonese, C., Dugo, G., & Mondello, L. (2008). Analysis of native carotenoid composition in orange juice using C30 columns in tandem. *Journal of Separation Science, 31*(12), 2151–2160.
15. Fasogbon, S. K., Adebayo, P. B., & Ogunsola, A. D. (2021). Combustion and performance analysis of compression ignition engine fuelled with varying orange peel oil-base biodiesel and diesel. *International Journal of Renewable Energy Technology, 12*(2), 130–143.
16. Felicia, W. X. L., Rovina, K., Aqilah, N. M. N., & Jaziri, A. A. (2024). Optimisation of supercritical fluid extraction of orange (Citrus sinenis L.) peel essential oil and its physicochemical properties. *Current Research in Green and Sustainable Chemistry, 8*, 100410.
17. Forgács, G., Pourbafrani, M., Niklasson, C., Taherzadeh, M. J., & Hováth, I. S. (2012). Methane production from citrus wastes: Process development and cost estimation. *Journal of Chemical Technology & Biotechnology, 87*(2), 250–255.
18. Gattuso, G., Barreca, D., Gargiulli, C., Leuzzi, U., & Caristi, C. (2007). Flavonoid composition of citrus juices. *Molecules, 12*(8), 1641–1673.
19. Giwa, S. O., Muhammad, M., & Giwa, A. (2018). Utilizing orange peels for essential oil production. *Journal of Engineering and Applied Sciences, 13*(1), 17–27.
20. Jeong, G. T. (2023). Statistical approach for the optimization of levulinic acid production from orange peel as agricultural waste. *Biofuels, Bioproducts and Biorefining, 17*(5), 1303–1314.

21. Kelebek, H., Selli, S., Canbas, A., & Cabaroglu, T. (2009). HPLC determination of organic acids, sugars, phenolic compositions and antioxidant capacity of orange juice and orange wine made from a Turkish cv. Kozan. *Microchemical Journal, 91*(2), 187–192.
22. Kulkarni, P. S., Brazinha, C., Afonso, C. A., & Crespo, J. G. (2010). Selective extraction of natural products with benign solvents and recovery by organophilic pervaporation: Fractionation of D-limonene from orange peels. *Green Chemistry, 12*(11), 1990–1994.
23. Kumawat, M. K., & Rokhum, S. L. (2022). Biodiesel production from oleic acid using biomass-derived sulfonated orange peel catalyst. *Frontiers in Catalysis, 2*, 914670.
24. Lachos-Perez, D., Baseggio, A. M., Mayanga-Torres, P. C., Junior, M. R. M., Rostagno, M. A., Martínez, J., & Forster-Carneiro, T. (2018). Subcritical water extraction of flavanones from defatted orange peel. *The Journal of Supercritical Fluids, 138*, 7–16.
25. Ma, F., & Hanna, M. A. (1999). Biodiesel production: A review. *Bioresource Technology, 70*(1), 1–15.
26. Mamma, D., & Christakopoulos, P. (2008). Citrus peels: An excellent raw material for the bioconversion into value-added products. *WORLD, 94793*(59041.4), 19224–19229.
27. Martín, M. A., Siles, J. A., Chica, A. F., & Martín, A. (2010). Biomethanization of orange peel waste. *Bioresource Technology, 101*(23), 8993–8999.
28. Martín-González, L., Font, X., & Vicent, T. (2013). Alkalinity ratios to identify process imbalances in anaerobic digesters treating source-sorted organic fraction of municipal wastes. *Biochemical Engineering Journal, 76*, 1–5.
29. Mukhametzyanov, R. R., Brusenko, S. V., Khezhev, A. M., Kelemetov, E. M., & Kirillova, S. S. (2024). Changing the global production and trade of citrus fruits. In *Sustainable development of the agrarian economy based on digital technologies and smart innovations* (pp. 19–24). Springer Nature Switzerland.
30. Ningi, T., Radebe, N., & Nkunjana, T. (2024). Global perspective.
31. Pascual, J. M., & Carmona, J. F. (1980). Composition of citrus pulp. *Animal Feed Science and Technology, 5*(1), 1–10.
32. Rajalingam, A., Jani, S. P., Kumar, A. S., & Khan, M. A. (2016). Production methods of biodiesel. *Journal of Chemical and Pharmaceutical Research, 8*(3), 170–173.
33. Rodríguez, J., Martín, M. J., Ruiz, M. A., & Clares, B. (2016). Current encapsulation strategies for bioactive oils: From alimentary to pharmaceutical perspectives. *Food Research International, 83*, 41–59.
34. Roy, A., Tariq, M. Z., La, M., Choi, D., & Park, S. J. (2024). 3D carbonized orange peel: A self-floating solar absorber for efficient water evaporation. *Desalination, 573*, 117191.
35. Sairam, P., Ghosh, S., Jena, S., Rao, K. N. V., & Banji, D. (2012). Supercritical fluid extraction (SFE)-an overview. *Asian Journal of Research in Pharmaceutical Science, 2*(3), 112–120.
36. Saxena, K., Brighu, U., & Choudhary, A. (2018). Parameters affecting enhanced coagulation: A review. *Environmental Technology Reviews, 7*(1), 156–176.
37. Sharma, P., Vishvakarma, R., Gautam, K., Vimal, A., Gaur, V. K., Farooqui, A., et al. (2022). Valorization of citrus peel waste for the sustainable production of value-added products. *Bioresource Technology, 351*, 127064.
38. Siles, J. A., Vargas, F., Gutiérrez, M. C., Chica, A. F., & Martín, M. A. (2016). Integral valorisation of waste orange peel using combustion, biomethanisation and co-composting technologies. *Bioresource Technology, 211*, 173–182.
39. Singh, B., Singh, J. P., Kaur, A., & Yadav, M. P. (2021). Insights into the chemical composition and bioactivities of citrus peel essential oils. *Food Research International, 143*, 110231.
40. Sungsinchai, S., Niamnuy, C., Jaree, A., & Devahastin, S. (2017). Influences of pretreatment and drying methods on composition, micro/molecular structures and some health-related functional characteristics of dietary fibre powder from orange pulp residues. *International Journal of Food Science & Technology, 52*(10), 2217–2229.
41. Tejada-Ortigoza, V., Garcia-Amezquita, L. E., Serna-Saldívar, S. O., Martín-Belloso, O., & Welti-Chanes, J. (2018). High hydrostatic pressure and mild heat treatments for the modifica-

tion of orange peel dietary fiber: Effects on hygroscopic properties and functionality. *Food and Bioprocess Technology, 11*, 110–121.
42. Wang, J., Liu, Z., Xia, J., & Chen, Y. (2019). Effect of microbial inoculation on physico-chemical properties and bacterial community structure of citrus peel composting. *Bioresource Technology, 291*, 121843.
43. Wilkins, H., Merrilees, B., & Herington, C. (2007). Towards an understanding of total service quality in hotels. *International Journal of Hospitality Management, 26*(4), 840–853.
44. Xiang, H., Min, X., Tang, C. J., Sillanpää, M., & Zhao, F. (2022). Recent advances in membrane filtration for heavy metal removal from wastewater: A mini review. *Journal of Water Process Engineering, 49*, 103023.
45. Xu, G., Ye, X., Chen, J., & Liu, D. (2007). Effect of heat treatment on the phenolic compounds and antioxidant capacity of citrus peel extract. *Journal of Agricultural and Food Chemistry, 55*(2), 330–335.

Chapter 22
Challenges and Opportunities in Citrus Waste Valorization

Ravinder Kumar, Ahmad Ali, and Sumit Singh Sheoran

22.1 Introduction

Citrus fruits, falls under a broad category of fruits including oranges, lemons, limes, grapefruits, and mandarins. These fruits can be found in various parts of the world. The citrus industry generates significant amounts of waste, including peels, seeds, pulp, and other by-products, which pose environmental and economic challenges [1]. However, these waste materials also present substantial opportunities for valorization through various technological and biotechnological processes. This chapter explores the challenges and opportunities in citrus waste (CW) valorization, highlighting recent advancements and potential applications.

In recent decades, the global increase in natural resource consumption has shifted the focus of researchers and stakeholders toward sustainable development. This concept aims to mitigate environmental, economic, and societal costs. Adhering to various rules applied by countries it is really beneficial and important to recover the value added byproducts form the citrus wastes generated worldwide and make positive progress to develop a circular economy. Majority agricultural based industries are converting their old age straight streamlines to more circular economy paradigm to reduce raw material and co-product waste. Agriculture industry waste has significant amount of recoverable by-products ranging from human nutrition, animal feed to energy sources. This promotes sustainability and enhances the efforts towards creating a circular bioeconomy [2].

R. Kumar (✉)
Food Science Program, Division of Food, Nutrition & Exercise Sciences, College of Agriculture, Food & Natural Resources, University of Missouri, Columbia, MO, USA
e-mail: rkz8m@missouri.edu

A. Ali · S. S. Sheoran
Chemical and Biomedical Engineering, College of Engineering, University of Missouri, Columbia, MO, USA

A. Chauhan et al. (eds.), *Valorization of Citrus Food Waste*,
https://doi.org/10.1007/978-3-031-77999-2_22

Citrus fruits are one of the largest grown fruits worldwide. They are grown in around 11.4 million hectares of land globally, thus it is a major contributor to agricultural production. According to Statista, a total of 165.9 million metric tons of citrus fruit was produced in 2022 worldwide, with Asia being the leading producer followed by South America. A large portion of this produce goes to industries for recovery of essential oils and extraction of juices [3]. While production two main types of by products are obtained other than the liquid juice i.e., the citrus waste water and semisolid fraction of pulp.

Numerous reviews have addressed solutions for valorizing the semisolid fraction of citrus waste. Many of them reported that typically the solid waste is disposed or reused, which contains bioactive compounds with various health benefits. It is thus, valuable for industries such as nutrition and pharmaceuticals that can extract essential oils from it, food industries that can extract dietary fibers and minerals from them, agricultural industries can source fertilizers and similarly other industries can use it for their benefit [2]. Moreover, the high organic content of this waste can pose environmental challenges due to its potential for putrescence and while processing this waste many pollutants could be released into the environment such as phenolic acids, carbohydrates, fibers, etc., which may lead to soil degradation and water eutrophication due to their rapid degradation and fermentability [4].

The chemical properties of these wastes can vary significantly depending on the type of method used for their production. Currently, citrus waste water management follows a process, where citrus wastes are treated before beign released to the environment. The management of citrus wastes poses environmental and economic challenges due to acidic nature with a pH of less than 4.5, presence of essential oils and a great amount of organic matter, which can cause greater problems when not properly disposed of. Complying with stringent environmental regulations requires novel intensive treatments, which are costly. There are some easy treatment methods but they cannot be deployed in some countries due to varying rules and their socio-economic constraints [5].

Maximum material on citrus wastewater management focuses on optimizing treatments to minimize environmental impacts. However, world level policies are focusing towards reducing pollution as well as increasing reuse of this citrus waste and make it more circular process. According to the new rule in European union (2020/741) the emphasis is on municipal wastewater and industrial wastewater and its reuse potential mainly for those working in the food sector [3]. A smart move to reuse this wastewater not only recovers valuable byproducts from the wastewater but also saves a lot of fresh water from being used, and these wastewaters after reuse and treatment can be diverted to agricultural fields making it more efficient for circular economy and reducing the stress on fresh water sources.

Several sustainable solutions for citrus wastewater treatment and valorization have been reported. Innovative technologies have shown that citrus wastewaters can be used to extract by-products or treated and reused. Citrus wastewaters have been used to recover polyhydroxyalkanoates (PHA) and agricultural purposes [6]. But the pros and cons of this are not very clear to the scientific community. Published material lacks an important part of integrated approach for sustainable management and emphasizes more on valorization. However, if the integrated management of this citrus wastewater is done it can help recover some value added by-products and can help minimizing waste. With all the advancement in use of citrus wastewater more focus is based on the reuse potential of these wastewaters [7].

From above discussion it is clear that we need an integrated approach to valorize citrus waste to recover as many by products as possible and move forward to the concept of a circular economy. This chapter focuses on providing information of what are the challenges faced in valorization of citrus waste and variety of by products which we can recover from citrus wastes.

22.2 Challenges in Citrus Waste Valorization

Despite the potential benefits, several challenges impede the effective valorization of citrus waste (Fig. 22.1):

22.2.1 High Moisture Content

Citrus waste has a high moisture content, making it prone to microbial spoilage and difficult to store and transport. This necessitates immediate processing or the development of efficient drying methods. Other factors such as transport costs depend on

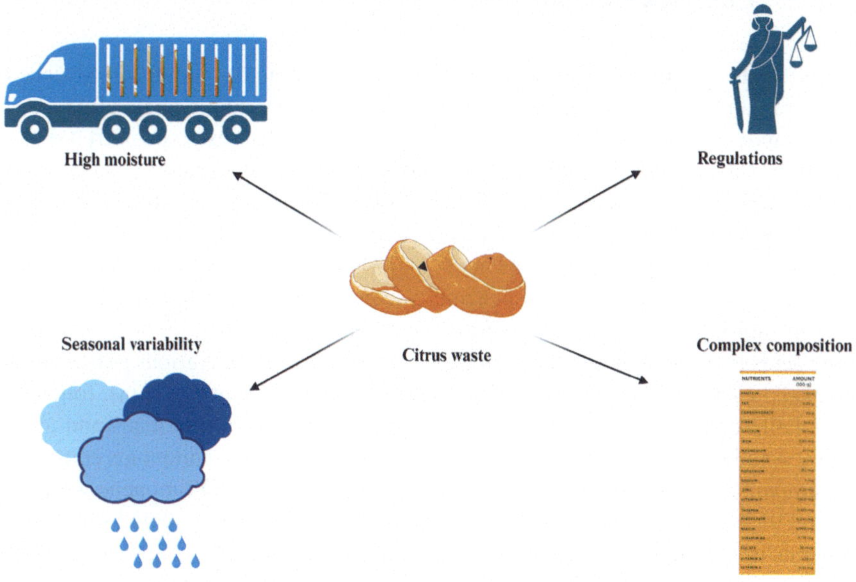

Fig. 22.1 Challenges faces during processing of citrus waste

several factors, including the amount of CWs, size of the trucks, condition of the roads and how apart are the two locations. Larger trucks, while more cost-effective per unit, need good highways which are not typical for the area where these production sites are located. Additionally, the waste is wet that increases its bulk density, and time required for loading and unloading which results in increased costs. Transport costs decrease with drying this waste to lower moisture levels, enabling longer-distance transport and expanding the utilization market, which is one of the reasons most citrus juices are transported as concentrates and then reformulated at the packaging facilities [2, 8].

22.2.2 Seasonal Variability

The seasonal nature of citrus production leads to fluctuations in waste availability, complicating the establishment of continuous valorization processes. Pertaining to the limitation of production at certain times of the year it is difficult to manage the huge load of citrus waste in small period of time and that also varies every year with changing climatic conditions. This is why citrus waste needs dedicated storage areas for effective valorization [9]. This storage helps preserve citrus waste, with available techniques CWs must also be transported from the generator to processing facility.

22.2.3 Complex Composition

Citrus waste comprises a complex mixture of compounds, requiring advanced extraction and separation technologies to isolate valuable components effectively. The composition of CWs is changes with varying time of harvest, type of fruit or the variety of fruit, the stage at which it was plucked, and some variability is also seen with changing climatic conditions. Nutritionally, CWs mainly cellulose and pectin as carbohydrates, multiple types of sugars such as fructose, glucose, galactose and sucrose, multiple types of naturals acids majorly citric and malic acids and other acids in minor amount includes benzoic, succinic, oxalic and tartaric. CWs also contain multiple oils which range from glycerol, stearic acids, phytosterols, palmitic acids to highly popular oleic, linoleic and linolenic acids. Various polyphenols are also present in CWs which are flavones, flavanones and anthocyanins, followed by volatile compounds such as alcohols, esters, hydrocarbons, ketones and aldehydes. Another important class of compounds found in CWs is limonoids constituting limonin and isolimonin. Essential oils in CWs are mainly D-limonene which is around 95% of all the essential oils and the rest is some other essential oils. Most minerals in CWs are calcium, potassium and nitrogen, further in nitrogen containing compounds we can find ammonia nitrogen and nitrates. The bright color in the citrus flesh comes from carotenoids including carotene, xanthophylls, lutein and other pigments. Lastly, CWs also contain enzymes such as peroxidase, phosphatase and pectinesterase, and some vitamins, mainly vitamin B complex [9–12].

22.2.4 Regulatory and Safety Concerns

Different countries have varying sets of regulations that needs to be followed if one is processing agro industrial waste which makes it difficult to use similar strategies in multiple countries. Ensuring the safety and quality of the products is a critical challenge thus it is difficult to use those byproducts for pharmaceuticals, cosmetics and human consumption. This is mainly due to the fact that CWs contain some essential oils, organic matter and other harmful substances (surfactants). When using these wastes there is a release of waste water into fresh water streams or in other areas which might act as breeding grounds for disease causing organism or might affect the aquatic organisms. Regulations imposed by the EU are directive 91/271/EEC that controls the collection, treatment and release of industrial and urban waste water keeping in mind the protection of environment and requires to pre treated before sending to wastewater treatment plants or the environment and these rules are separate for every EU member state [13]. However, another EU regulation numbered 2020/741 was established to release the wastewater for agricultural purposes before making sure it is safe for the environment but it does not cover the industrial waste water [14].

22.3 Opportunities in Citrus Waste Valorization

22.3.1 Nutraceuticals and Functional Foods

In addition to flavonoid and polyphenols, bioactive chemicals from citrus waste possess antioxidant, anti-inflammatory, and anti-bacterial activities, which could be utilized as nutraceuticals and functional foods to meet human demand for health benefits, as well as heightened desire for a natural product. Citrus by-products, such as peel and seeds, with flavonoids as the main component can be employed in making functional food, not only because they contain various dietary components including flavonoids and non-digestible carbohydrates (fiber) but also the variety of polysaccharides within the citrus seeds, such as galacturonic acid, glucuronic acid and other undegraded sugars. It is also well known that polysaccharides contain a high portion of dietary nutrients and additives. Therefore, it may explain why traditional folk medicine of various cuisines still specifies the citrus seeds to be consumed traditionally.

On the other hand, citrus seeds also contained high oil contents, namely essential fatty acids with high nutritional values. These could also be utilized as a source of

Table 22.1 Advantages and challenges in different processes for citrus waste valorization

Products	Process Involved	Relevance/Advantages	Challenges	References
Animal Feed	Drying and pelletizing of peel residues	Nutritional supplement for livestock; reduces feed costs	Presence of anti-nutritional factors	[31]
Biodegradable Films	Extraction and processing of pectin or other polysaccharides	Supports reduction of plastic waste; aligns with eco-friendly initiatives	Mechanical strength and durability of films	[32]
Bioethanol Production	Fermentation of citrus peels	Renewable energy source; reduces dependence on fossil fuels	Inhibition by essential oils; storage and handling of waste	[33]
Biogas Production (Methane)	Anaerobic digestion	Renewable energy source; reduces greenhouse gas emissions	Requires controlled conditions; handling of digestate	[34]
Composting and Agronomic Use (Fertilizer)	Composting, direct application to soil	Enhances soil fertility; supports sustainable agriculture	Requires large-scale operations; potential for contamination	[34, 35]
Essential Oil (limonene)	Cold pressing or distillation	High demand in food, cosmetics, and pharmaceuticals; antimicrobial properties	Seasonal availability; extraction efficiency	[36]
Natural Dyes	Solvent extraction, filtration	Used in food and textile industries; replaces synthetic dyes	Stability and color intensity of natural dyes	[37]
Pectin	Acid extraction, enzymatic hydrolysis	Used as a gelling agent in food industry; high commercial value	Acidic extraction methods can cause equipment corrosion	[38]
Flavonoids and Polyphenols	Solvent extraction, supercritical fluid extraction	Antioxidant properties; used in nutraceuticals and functional foods	Extraction yield; solvent recovery issues	[39]

healthy vegetable oil. In addition, citrus seeds could be processed into flour, notably from dehulled sweet orange (C. sinensis) seeds rich in lipids, carbohydrates, crude fiber, protein and calcium, potassium, sodium, iron and other minerals. These gave high energy values with suitable properties for diets with restricted sodium. Moreover, grapefruit seed extract (GSE) was incorporated into eco-farming as a natural pesticide which controls viruses, bacteria and fungi. It could be utilized in electrostatic deposition coatings for food preservation. The antioxidant-rich oils from lemon and orange seeds were also utilized as natural preservatives with health-promoting benefits. Table 22.1 shows different uses for citrus wastes and also shows what challenges might be face while processing.

22.3.2 Essential Oils and Aromas

Citrus peels are also acknowledged for their high oil content, which is used as raw materials in the flavors, fragrance and pharmaceutical industries. New advanced extraction techniques, including supercritical fluid extraction and cold press, will address the issue of low yield and quality of the citrus essential oils. Essential oils are aromatic fluids composed of volatile substances, primarily found in the cuticles and oil sacs of citrus peels. These oils are rich in terpenic hydrocarbons, such as monoterpenes (primarily limonene) and sesquiterpenes, as well as their oxygenated compounds such as aldehydes, alcohols, ketones, and esters. Limonene, which constitutes approximately 90–98% of essential oils derived using citrus by-products, is notable for its high antimicrobial, biological, and antioxidant properties. The FDA has recognized these oils as safe (GRAS), making them popular as flavoring agents in food and medicine. Advanced extraction methods are also enhancing the extraction of these oils. By incorporating modern methods such as supercritical fluid, ultrasound-assisted and microwave-assisted extraction, as well as quality can be improved. The overall yield and composition of essential oils can vary with extraction methods; for instance, microwave-assisted extraction from C. sinensis peels achieved 0.48% oil at 130 °C in 60 minutes, while microwave-assisted hydrodistillation yielded about 1.8%. Supercritical extraction, particularly with ethanol as a modifier [1], has shown high efficiency, producing yields up to 23.04% from C. junos seeds. Modern methods focus on solvent-free extractions to reduce solvent use and enhance yields. During essential oil recovery, a small amount of oil mixes with distilled water, forming hydrosol, which contains polar oil vapors. Hydrosol, known for high acidity pH 3.5 to pH 6.5, dilute composition, and captivating scent, is used in food, cosmetics, pharmaceuticals, and aromatherapy due to its odor and flavor properties [15]. It consists mainly of oxygenated compounds such as linalool, α-terpineol, and geraniol. Hydrosols also exhibit antibacterial, antifungal, and antioxidant activities. For example, orange blossom hydrosol shows antimicrobial effects against E. coli and Listeria monocytogens. Recent studies on hydrosol from C. aurantium flowers indicate potential antioxidant and antimicrobial applications in food, cosmetic industries and pharmaceuticals. However, further research is needed on in-vivo activity, toxicity and underlying mechanisms.

22.3.3 Pectin Extraction

Pectin, a natural polysaccharide present in citrus peels, is extensively consumed as gelling agent in food industries. To enhance pectin yield and its functional properties, techniques such as microwave, and enzymatic-assisted extraction are used. Pectin is a large polysaccharide (heteropolysaccharide) present in higher plants especially in cell walls. It consists of D-galacturonic connected by α-1,4 glycosidic bonds.

Partial esterification of these polymer undergoes with methanol to produce methyl esters. In commercial applications, pectin is primarily isolated from citrus and apple by-products. However, citrus by-products possess greater pectin content (20–30%) as compared to apple by-products(10–15%, dry weight) [16]. Pectin in citrus by-products present mainly in two forms: insoluble protopectins and a white to light brown variety. It has various applications in the food industry, where it serves as a thickening, emulsifying (used in biodegradable films), and stabilizing agent. It is also used in the production of jams, jellies, bakery fillings, confectionery, and biodegradable films due to its properties. Acid treatment (such as HNO3, HCI or H2SO4, and citric acid (organic)) used to extract citrus by-products at industrials to enhance extraction of pectin and it assisted by controlling temperature at approximately 100 °C. This process facilitates the extraction of insoluble pectin and improves yields. However, the use of acids can lead to corrosion of equipment, environmental pollution, and adverse changes in the physicochemical properties of the pectin [17].

22.3.4 Biofuels and Biochemicals

Citrus waste can be converted into bioethanol and biogas through fermentation and anaerobic digestion process, respectively. Additionally, citrus waste can serve as a feedstock for producing biochemicals, including organic acids and enzymes. Citrus waste contains bulk high value compounds i.e. hemi-cellulose, pectin, cellulose, proteins, soluble sugars and essential oils (EO). A biorefinery can utilize this by-product of agriculture efficiently. Through utilizing advanced conversion technologies, optimizing the use of biomass advanced reducing waste [18]. This approach supports bioeconomy and mitigates environmental impacts from the agro-industry as well [2].

22.3.5 Energy Production

Raw materials rich in carbohydrate and low in lignin are promising for producing second generation gaseous or liquid fuels, and citrus meets these criteria. Alon with that high calorific value (~17 MJ kg^{-1}, dried), CW can be utilized to generate energy for by direct combustion, converting the biomass into heat and power [2]. This process involves burning the agricultural waste to generate thermal energy, which can then be used to produce electricity or provide heating. This method not only helps in managing CW but also contributes to renewable energy generation. CW has also been identified as a suitable substrate for hydrogen production through various fermentation processes. Pre-treatments are usually necessary to reduce inhibiting effects of EOs on biochemical processes [19].

22.3.6 Fuel from Dried CW

CW can be used for electrical energy production or as a fuel for power and heat generation, mainly to minimize the storage and transport costs especially if power stations are located near the biomass production sites. Due to high water levels of CW, pyrolysis and incineration as thermal treatment techniques are not applicable. Furthermore, under 10% moisture content, thermal treatments can be more efficient conducted the evaluation of CW pyrolysis. Citrus waste combustion produces high NO_2 concentrations which is a major disadvantage [12].

22.3.7 Production of Ethanol

The conversion of CW into ethanol has been widely studied. Conversion to ethanol has been extensively studied. Hemicellulosic polysaccharides, cellulose and pectin can be break down into simple sugar and then fermented to produce alcohol [20]. The process involves pretreatment, hydrolysis, and alcoholic fermentation. Pretreatments are necessary to efficiently release fermentable sugars and mitigate EO toxicity. EO content is a major limiting factor in ethanol production, as it inhibits fermentation. EO reduction to 0.08–0.12% is recommended to avoid inhibition. The inhibitory EO concentration varies depending on the microorganism strain [21]. The process of fermentation and enzymatic hydrolysis can be carried out in combined or separate procedures. Combining cellulase and pectinase is more effective for hydrolyzing organic waste.

22.3.8 Production of Biogas

Biogas, which is primarily generated through the anaerobic digestion of organic materials, with CW demonstrating significant methane potential. CWs methane yield depends on various factors, including EO content, which inhibits anaerobic digestion. It can be done by using either thermophilic or mesophilic conditions. Choosing between mesophilic and thermophilic conditions depends on specific case factors. Mesophilic digestion offers stability, while thermophilic digestion increases degradation of organic matter and methane generation rate. Co-digestion and other treatments can help mitigate EO's inhibitory effects, making CW a viable substrate for biogas production [22].

22.3.9 Animal Feed and Agricultural Applications

Citrus waste can be used as animal feed due to its nutritional content. Furthermore, it can be applied as a soil conditioner and fertilizer, enhancing soil health and crop productivity. CWs exhibit substantial intra-annual variation caused by the nature of seasonal citrus production. This variability necessitates the storage of CW to enable its temporal transfer for almost all available valorization options. Storage not only conserves CW but also allows for its effective utilization over time. Additionally, CW needs to be transported to the customer (cattle farms) from the producer [23]. The expense of transport is influenced by several factors, including the quantity of CW, road conditions, distance, and the type of transport used. Larger trucks can reduce costs per unit but they need high quality roads, which might not always be accessible in production regions. Other factors affecting transport costs include the bulk density of wet CW and the time required for loading and administrative tasks.

22.4 Direct Use

22.4.1 Agronomic Utilization

Composting or land spreading methods can be utilized in agriculture for CW. However, due to bioactive compounds present in CW and antimicrobial properties of essential oils (EOs), it can harm soil microorganisms. Studies have shown that EO in CW is toxic to yeasts and inhibits the growth of beneficial bacteria, yeasts, and molds [24–26]. Non-stabilized organic matter application can deplete soil oxygen and cause odors, requiring careful timing before seeding and avoiding incorporation into the soil to prevent groundwater pollution. Proper protocols, composting, or processing bioactive compounds are necessary for safe agronomic use of CW.

22.4.2 Organic Soil Conditioner

Direct land spreading of CW can enhance soil organic matter, air permeability, improving fertility, porosity, water retention, and nutrient consumption. CW releases essential nutrients like nitrogen, phosphorus, and potassium when applied to soil, though additional mineral fertilizers are usually needed. The feasibility of using CW as a soil conditioner influenced by various factors such as transport distance, mechanization level, and available application area. In the past, up to 30–40% of CW was used as soil conditioner in Italy, demonstrating good results in wheat production after soil application without additional fertilizers [27, 28].

22.4.3 Compost Production

CW can be composted to generate high-quality compost, which can fully or partially replacement for non-renewable resources such as peat. The composting process involves a rapid increment of temperature, a short latency phase, and gradual cooling, transforming organic matter into stable compounds. Proper composting eliminates pathogens and weed seeds. Citrus compost has high C/N ratio, greater ligno-cellulosic content, and specific pH levels, making it suitable for agricultural use. Adjustments to the water content, pH and C/N ratio are necessary to optimize composting. Co-composting with municipal solid waste is also a viable option, reducing odor emissions and improving stabilization [12].

22.4.4 *Animal Food*

In the Mediterranean basin, particularly Southern Italy, up to 70–80% of CW is used as animal feed, partially replacing conventional feed like cereals [29]. The use of wet CW is limited to livestock near processing plants due to high transport costs and water content. Advanced farms can exploit CW's nutritional capacity, making transport more economical over short distances. Wet CW can be directly laid on pasture land for animals to consume. Ensiled CW, mixed with other feeds, improves palatability and nutritional value for livestock. Dried CW is more palatable and nutritionally rich, used in various forms for animal feed across different regions [30].

22.4.5 *Dried Feed*

Dried form of CW is preferred for animal feed due to its higher palatability and nutritional value. Fuel-drying CW is not always economical, but solar drying is a viable option in semi-arid regions. Dried CW can be stored and used as supplementary feed, enhancing animal nutrition without affecting palatability or production [30].

22.5 Conclusion

Citrus waste valorization presents both challenges and opportunities. While addressing the technical, regulatory, and logistical hurdles is essential, the potential benefits in terms of environmental sustainability, economic returns, and resource efficiency are substantial. By leveraging technological advancements and adopting a holistic approach, the citrus industry can transform waste into valuable resources, contributing to a sustainable and circular economy.

References

1. Panwar, D., Panesar, P. S., & Chopra, H. K. (2021). Recent trends on the valorization strategies for the management of citrus by-products. *Food Reviews International, 37*(1), 91–120.
2. Satari, B., & Karimi, K. (2018). Citrus processing wastes: Environmental impacts, recent advances, and future perspectives in total valorization. *Resources, Conservation and Recycling, 129*, 153–167.
3. Lucia, C., Laudicina, V. A., Badalucco, L., Galati, A., Palazzolo, E., Torregrossa, M., Viviani, G., & Corsino, S. F. (2022). Challenges and opportunities for citrus wastewater management and valorisation: A review. *Journal of Environmental Management, 321*, 115924.
4. Zayed, A., Badawy, M. T., & Farag, M. A. (2021). Valorization and extraction optimization of Citrus seeds for food and functional food applications. *Food Chemistry, 355*, 129609.
5. Shaw, D., Tripathi, A. D., Paul, V., Agarwal, A., Mishra, P. K., & Kumar, M. (2023). Valorization of essential oils from citrus peel powder using hydro-distillation. *Sustainable Chemistry and Pharmacy, 32*, 101036.
6. Novelli, L. D. D., Sayavedra, S. M., & Rene, E. R. (2021). Polyhydroxyalkanoate (PHA) production via resource recovery from industrial waste streams: A review of techniques and perspectives. *Bioresource Technology, 331*, 124985.
7. Yadav, V., Sarker, A., Yadav, A., Miftah, A. O., Bilal, M., & Iqbal, H. M. (2022). Integrated biorefinery approach to valorize citrus waste: A sustainable solution for resource recovery and environmental management. *Chemosphere, 293*, 133459.
8. Chavan, P., Singh, A. K., & Kaur, G. (2018). Recent progress in the utilization of industrial waste and by-products of citrus fruits: A review. *Journal of Food Process Engineering, 41*(8), e12895.
9. Sharma, P., Vishvakarma, R., Gautam, K., Vimal, A., Gaur, V. K., Farooqui, A., Varjani, S., & Younis, K. (2022). Valorization of citrus peel waste for the sustainable production of value-added products. *Bioresource Technology, 351*, 127064.
10. Salazar Díaz, D. A. (2020). *Citrus waste composition and bioactive compounds: A review*.
11. Sharma, K., Mahato, N., Cho, M. H., & Lee, Y. R. (2017). Converting citrus wastes into value-added products: Economic and environmently friendly approaches. *Nutrition, 34*, 29–46.
12. Zema, D., Calabrò, P., Folino, A., Tamburino, V., Zappia, G., & Zimbone, S. (2018). Valorisation of citrus processing waste: A review. *Waste Management, 80*, 252–273.
13. Preisner, M., Smol, M., & Szołdrowska, D. (2021). Trends, insights and effects of the urban wastewater treatment directive (91/271/EEC) implementation in the light of the polish coastal zone eutrophication. *Environmental Management, 67*(2), 342–354.
14. Berbel, J., Mesa-Pérez, E., & Simón, P. (2023). Challenges for circular economy under the EU 2020/741 wastewater reuse regulation. *Global Challenges, 7*(7), 2200232.
15. Başer, K. H. C., & Demirci, F. (2000). Essential oils. In *Kirk-Othmer encyclopedia of chemical technology* (pp. 1–37). Wiley.
16. Wongkaew, M., Chaimongkol, P., Leksawasdi, N., Jantanasakulwong, K., Rachtanapun, P., Seesuriyachan, P., Phimolsiripol, Y., Chaiyaso, T., Ruksiriwanich, W., & Jantrawut, P. (2021). Mango peel pectin: Recovery, functionality and sustainable uses. *Polymers, 13*(22), 3898.
17. Kumar, S., Konwar, J., Purkayastha, M. D., Kalita, S., Mukherjee, A., & Dutta, J. (2023). Current progress in valorization of food processing waste and by-products for pectin extraction. *International Journal of Biological Macromolecules, 239*, 124332.
18. López, J. Á. S., Li, Q., & Thompson, I. P. (2010). Biorefinery of waste orange peel. *Critical Reviews in Biotechnology, 30*(1), 63–69.
19. Van Dyk, J., Gama, R., Morrison, D., Swart, S., & Pletschke, B. (2013). Food processing waste: Problems, current management and prospects for utilisation of the lignocellulose component through enzyme synergistic degradation. *Renewable and Sustainable Energy Reviews, 26*, 521–531.
20. Roukas, T., & Kotzekidou, P. (2022). From food industry wastes to second generation bioethanol: A review. *Reviews in Environmental Science and Bio/Technology, 21*(1), 299–329.

21. Karaman, K., & Sagdic, O. (2019). Zygosaccharomyces bailii and Z. rouxii induced ethanol formation in apple juice supplemented with different natural preservatives: A response surface methodology approach. *Journal of Microbiological Methods, 163*, 105659.
22. El-Fadel, M., Saikaly, P., & Ghanimeh, S. (2013). Startup and stability of thermophilic anaerobic digestion of OFMSW. *Critical Reviews in Environmental Science and Technology, 43*(24), 2685–2721.
23. Basinas, P., Rusín, J., & Chamrádová, K. (2021). Assessment of high-solid mesophilic and thermophilic anaerobic digestion of mechanically-separated municipal solid waste. *Environmental Research, 192*, 110202.
24. Ahebwa, A., Ahebwa, A., & Kanjanamaneesathian, M. (2020). *Vapour-phase synergistic antifungal activity of selected essential oils against Aspergillus sp., Curvularia sp. and Fusarium sp*. Silpakorn University.
25. Allameh, A., Ziglari, T., & Rasooli, I. (2011). Phytoinhibition of growth and aflatoxin biosynthesis in toxigenic fungi. In *Aflatoxins-Detection, measurement and control* (pp. 283–316). IntechOpen.
26. Stojiljkovic, J., Trajchev, M., Petrovska, M., & Nakov, D. (2019). *The influence of different concentrations of plant essential oils on growth and reproduction of Salmonella enteritidis*. Faculty of Agriculture of Goce Delcev University.
27. Tóthné Bogdányi, F., Boziné Pullai, K., Doshi, P., Erdős, E., Gilián, L. D., Lajos, K., Leonetti, P., Nagy, P. I., Pantaleo, V., & Petrikovszki, R. (2021). Composted municipal green waste infused with biocontrol agents to control plant parasitic nematodes—A review. *Microorganisms, 9*(10), 2130.
28. Algieri, A., Andiloro, S., Tamburino, V., & Zema, D. A. (2019). The potential of agricultural residues for energy production in Calabria (Southern Italy). *Renewable and Sustainable Energy Reviews, 104*, 1–14.
29. de Carvalho, J. C., Molina-Aulestia, D. T., Martinez-Burgos, W. J., Karp, S. G., Manzoki, M. C., Medeiros, A. B. P., Rodrigues, C., Scapini, T., de Souza Vandenberghe, L. P., & Vieira, S. (2022). Agro-industrial wastewaters for algal biomass production, bio-based products, and biofuels in a circular bioeconomy. *Fermentation, 8*(12), 728.
30. Smith, T., Jayasuriya, N., Mlambo, V., Lekule, F., Thomas, D., Owen, E., Pearson, A., Titterton, M. (2005). *Feeds and feeding to improve productivity and survival.*
31. Tamburino, V., Zema, D. A., & Zimbone, S. M. (2007, September). Orange peel utilizations in Southern Italy. In *Proceedings of the CIGR section VI international symposium on food and agricultural products: Processing and innovations, Naples, Italy* (pp. 24–26).
32. Espitia, P. J. P., Du, W. X., de Jesús Avena-Bustillos, R., Soares, N. D. F. F., & McHugh, T. H. (2014). Edible films from pectin: Physical-mechanical and antimicrobial properties-A review. *Food Hydrocolloids, 35*, 287–296.
33. Choi, I. S., Lee, Y. G., Khanal, S. K., Park, B. J., & Bae, H. J. (2015). A low-energy, cost-effective approach to fruit and citrus peel waste processing for bioethanol production. *Applied Energy, 140*, 65–74.
34. Panuccio, M. R., Attinà, E., Basile, C., Mallamaci, C., & Muscolo, A. (2016). Use of recalcitrant agriculture wastes to produce biogas and feasible biofertilizer. *Waste and Biomass Valorization, 7*, 267–280.
35. Guerrero, C. C., de Brito, J. C., Lapa, N., & Oliveira, J. S. (1995). Re-use of industrial orange wastes as organic fertilizers. *Bioresource Technology, 53*(1), 43–51.
36. Ciriminna, R., Forest, B., Meneguzzo, F., Pagliaro, M., & Hamann, M. T. (2020). Technical and economic feasibility of a stable yellow natural colorant production from waste lemon peel. *Applied Sciences, 10*(19), 6812.
37. Espina, L., Somolinos, M., Lorán, S., Conchello, P., García, D., & Pagán, R. (2011). Chemical composition of commercial citrus fruit essential oils and evaluation of their antimicrobial activity acting alone or in combined processes. *Food Control, 22*(6), 896–902.
38. Grohmann, K., & Baldwin, E. A. (1992). Hydrolysis of orange peel with pectinase and cellulase enzymes. *Biotechnology Letters, 14*, 1169–1174.
39. Curto, R. L., Tripodo, M. M., Leuzzi, U., Giuffre, D., & Vaccarino, C. (1992). Flavonoids recovery and SCP production from orange peel. *Bioresource Technology, 42*(2), 83–87.

Index

A. Chauhan et al. (eds.), *Valorization of Citrus Food Waste*,
https://doi.org/10.1007/978-3-031-77999-2

The manufacturer's authorised representative in the EU is Springer Nature Customer Service Centre GmbH, Europaplatz 3, 69115 Heidelberg, Germany. If you have any concerns regarding our products, please contact ProductSafety@springernature.com

Printed and bound by CPI Group (UK) Ltd, Croydon, CR0 4YY
01/10/2026
02225966-0017